T0306090

Steganography in Digital Media

Steganography, the art of hiding of information in apparently innocuous objects or images, is a field with a rich heritage, and an area of rapid current development. This clear, self-contained guide shows you how to understand the building blocks of covert communication in digital media files and how to apply the techniques in practice, including those of steganalysis, the detection of steganography. Assuming only a basic knowledge in calculus and statistics, the book blends the various strands of steganography, including information theory, coding, signal estimation and detection, and statistical signal processing. Experiments on real media files demonstrate the performance of the techniques in real life, and most techniques are supplied with pseudo-code, making it easy to implement the algorithms. The book is ideal for students taking courses on steganography and information hiding, and is also a useful reference for engineers and practitioners working in media security and information assurance.

Jessica Fridrich is Professor of Electrical and Computer Engineering at Binghamton University, State University of New York (SUNY), where she has worked since receiving her Ph.D. from that institution in 1995. Since then, her research on data embedding and steganalysis has led to more than 85 papers and 7 US patents. She also received the SUNY Chancellor's Award for Excellence in Research in 2007 and the Award for Outstanding Inventor in 2002. Her main research interests are in steganography and steganalysis of digital media, digital watermarking, and digital image forensics.

Steganography in Digital Media

Principles, Algorithms, and Applications

JESSICA FRIDRICH

Binghamton University, State University of New York (SUNY)

CAMBRIDGE
UNIVERSITY PRESS

University Printing House, Cambridge CB2 8BS, United Kingdom

Cambridge University Press is part of the University of Cambridge.

It furthers the University's mission by disseminating knowledge in the pursuit of education, learning and research at the highest international levels of excellence.

www.cambridge.org
Information on this title: www.cambridge.org/9780521190190

© Cambridge University Press 2010

First published 2010

A catalogue record for this publication is available from the British Library

ISBN 978-0-521-19019-0 Hardback

To Nicole and Kathy

Time will bring to light whatever is hidden; it will cover up and conceal what is now shining in splendor.

Quintus Horatius Flaccus (65–8 BC)

Contents

Preface

Steganography is another term for covert communication. It works by hiding messages in inconspicuous objects that are then sent to the intended recipient. The most important requirement of any steganographic system is that it should be impossible for an eavesdropper to distinguish between ordinary objects and objects that contain secret data.

Steganography in its modern form is relatively young. Until the early 1990s, this unusual mode of secret communication was used only by spies. At that time, it was hardly a research discipline because the methods were a mere collection of clever tricks with little or no theoretical basis that would allow steganography to evolve in the manner we see today. With the subsequent spontaneous transition of communication from analog to digital, this ancient field experienced an explosive rejuvenation. Hiding messages in electronic documents for the purpose of covert communication seemed easy enough to those with some background in computer programming. Soon, steganographic applications appeared on the Internet, giving the masses the ability to hide files in digital images, audio, or text. At the same time, steganography caught the attention of researchers and quickly developed into a rigorous discipline. With it, steganography came to the forefront of discussions at professional meetings, such as the Electronic Imaging meetings annually organized by the SPIE in San Jose, the IEEE International Conference on Image Processing (ICIP), and the ACM Multimedia and Security Workshop. In 1996, the first Information Hiding Workshop took place in Cambridge and this series of workshops has since become the premium annual meeting place to present the latest advancements in theory and applications of data hiding.

Steganography shares many common features with the related but fundamentally quite different field of digital watermarking. In late 1990s, digital watermarking dominated the research in data hiding due to its numerous lucrative applications, such as digital rights management, secure media distribution, and authentication. As watermarking matured, the interest in steganography and steganalysis gradually intensified, especially after concerns had been raised that steganography might be used by criminals.

Even though this is not the first book dealing with the subject of steganography [22, 47, 51, 123, 142, 211, 239, 250], as far as the author is aware this is the first self-contained text with in-depth exposition of both steganography and steganalysis for digital media files. Even though this field is still developing at a fast pace and many fundamental questions remain unresolved, the foundations have been laid and basic principles established. This book was written to provide the reader with the basic philosophy and building blocks from which many practical steganographic and steganalytic schemes are constructed. The selection of the material presented in this book represents the author's view of the field and is by no means an exhaustive survey of steganography in general. The selected examples from the literature were included to illustrate the basic concepts and provide the reader with specific technical solutions. Thus, any omissions in the references should not be interpreted as indications regarding the quality of the omitted work.

This book was written as a primary text for a graduate or senior undergraduate course on steganography. It can also serve as a supporting text for virtually any course dealing with aspects of media security, privacy, and secure communication. The research problems presented here may be used as motivational examples or projects to illustrate concepts taught in signal detection and estimation, image processing, and communication. The author hopes that the book will also be useful to researchers and engineers actively working in multimedia security and assist those who wish to enter this beautiful and rapidly evolving multidisciplinary field in their search for open and relevant research topics.

The text naturally evolved from lecture notes for a graduate course on steganography that the author has taught at Binghamton University, New York for several years. This pedigree influenced the presentation style of this book as well as its layout and content. The author tried to make the material as self-contained as possible within reasonable limits. Steganography is built upon the pillars of information theory, estimation and detection theory, coding theory, and machine learning. The book contains five appendices that cover all topics in these areas that the reader needs to become familiar with to obtain a firm grasp of the material. The prerequisites for this book are truly minimalistic and consist of college-level calculus and probability and statistics.

Each chapter starts with simple reasoning aimed to provoke the reader to think on his/her own and thus better see the need for the content that follows. The introduction of every chapter and section is written in a narrative style aimed to provide the big picture before presenting detailed technical arguments. The overall structure of the book and numerous cross-references help those who wish to read just selected chapters. To aid the reader in implementing the techniques, most algorithms described in this book are accompanied with a pseudo-code. Furthermore, practitioners will likely appreciate experiments on real media files that demonstrate the performance of the techniques in real life. The lessons learned serve as motivation for subsequent sections and chapters. In order to make the book accessible to a wide spectrum of readers, most technical arguments are presented in their simplest core form rather than the most general fashion, while referring the interested reader to literature for more details. Each chapter is closed with a brief summary that highlights the most important facts. Readers

Cover type	Count
Audio	445
Disk space	416
Images	1689
Network	39
Other files	81
Text	255
Video	86

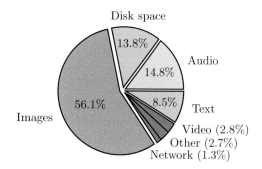

Number of steganographic software applications that can hide data in electronic media as of June 2008. Adapted from [122] and reprinted with permission of John Wiley & Sons, Inc.

can test their newly acquired knowledge on carefully chosen exercises placed at the end of the chapters. More involved exercises are supplied with hints or even a brief sketch of the solution. Instructors are encouraged to choose selected exercises as homework assignments.

All concepts and methods presented in this book are illustrated on the example of digital images. There are several valid reasons for this choice. First and foremost, digital images are by far the most common type of media for which steganographic applications are currently available. Furthermore, many basic principles and methodologies can be readily extended from images to other digital media, such as video and audio. It is also considerably easier to explain the perceptual impact of modifying an image rather than an audio clip simply because images can be printed on paper. Lastly, when compared with other digital objects, the field of image steganography and steganalysis is by far the most advanced today, with numerous techniques available for most typical image formats.

The first chapter contains a brief historical narrative that starts with the rather amusing ancient methods, continues with more advanced ideas for data hiding in written documents as well as techniques used by spies during times of war, and ends with modern steganography in digital files. By introducing three fictitious characters, prisoners Alice and Bob and warden Eve, we informally describe secure steganographic communication as the famous prisoners' problem in which Alice and Bob try to secretly communicate without arousing the suspicion of Eve, who is eagerly eavesdropping. These three characters will be used in the book to make the language more accessible and a little less formal when explaining technical aspects of data-hiding methods. The chapter is closed with a section that highlights the differences between digital watermarking and steganography.

Knowing how visual data is represented in a computer is a necessary prerequisite to understand the technical material in this book. Chapter 2 first explains basic color models used for representing color in a computer. Then, we describe the structure of the most common raster, palette, and transform image formats, including the JPEG. The description of each format is supplied with instructions on how to work with such images in Matlab to give the reader the ability to conveniently implement most of the methods described in this book.

Since the majority of digital images are obtained using a digital camera, camcorder, or scanner, Chapter 3 deals with the process of digital image acquisition through an imaging sensor. Throughout the chapter, emphasis is given to those aspects of this process that are relevant to steganography. This includes the processing pipeline inside typical digital cameras and sources of noise and imperfections. Noise is especially relevant to steganography because the seemingly useless stochastic components of digital images could conceivably convey secret messages.

In Chapter 4, we delve deeper into the subject of steganography. Three basic principles for constructing steganographic methods are introduced: steganography by cover selection, cover synthesis, and cover modification. Even though the focus of this book is on data-hiding methods that embed secret messages by slightly modifying the original (cover) image, all three principles can be used to build steganographic methods in practice. This chapter also introduces basic terminology and key building blocks that form the steganographic channel – the source of cover objects, source of secret messages and secret keys, the data-hiding and data-extraction algorithms, and the physical channel itself. The physical properties of the channel are determined by the actions of the warden Eve, who can position herself to be a passive observant or someone who is actively involved with the flow of data through the channel. Discussions throughout the chapter pave the way towards the information-theoretic definition of steganographic security given in Chapter 6.

The content of Chapter 5 was chosen to motivate the reader to ask basic questions about what it means to undetectably embed secret data in an image and to illustrate various (and sometimes unexpected) difficulties one might run into when attempting to realize some intuitive hiding methods. The chapter contains examples of some early naive steganographic methods for the raster, palette, and JPEG formats, most of which use some version of the least-significant-bit (LSB) embedding method. The presentation of each method continues with critical analysis of how the steganographic method can be broken and why. The author hopes that this early exposure of specific embedding methods will make the reader better understand the need for a rather precise technical approach in the remaining chapters.

Chapter 6 introduces the central concept, which is a formal information-theoretic definition of security in steganography based on the Kullback–Leibler divergence between the distributions of cover and stego objects. This definition puts steganography on a firm mathematical ground that allows methodological development by studying security with respect to a cover model. The concept of security is further explained by showing connections between security and detection theory and by providing examples of undetectable steganographic schemes built using the principles outlined in Chapter 4. We also introduce the concept of a distortion-limited embedder (when Alice is limited in how much she can modify the cover image) and show that some well-known watermarking methods, such as spread-spectrum watermarking and quantization index modulation, can be used to construct secure steganographic schemes. Finally, the reader is presented with an alternative complexity-theoretic definition of steganographic security even though this direction is not further pursued in this book.

Using the definition of security as a guiding philosophy, Chapter 7 introduces several design principles and intuitive strategies for building practical steganographic schemes for digital media files: (1) model-preserving steganography using statistical restoration and model-based steganography, (2) steganography by mimicking natural phenomena or processing, (3) steganalysis-aware steganography, and (4) minimal-impact steganography. The first three approaches are illustrated by describing in detail specific examples of steganographic algorithms from the literature (OutGuess, Model-Based Steganography for JPEG images, stochastic modulation, and the F5 algorithm). Minimal embedding impact steganography is discussed in Chapters 8 and 9.

Chapter 8 is devoted to matrix embedding, which is a general method for increasing security of steganographic schemes by minimizing the number of embedding changes needed to embed the secret message. The reader is first motivated by what appears a simple clever trick, which is later generalized and then reinterpreted within the language of coding theory. The introductory sections naturally lead to the highlight of this chapter – the matrix embedding theorem, which is essentially a recipe for how to turn a linear code into a steganographic embedding method using the principle of syndrome coding. Ample space is devoted to various bounds that impose fundamental limits on the performance one can achieve using matrix embedding.

The second chapter that relates to minimal-impact steganography is Chapter 9. It introduces the important topic of communication with a non-shared selection channel as well as several practical methods for communication using such channels (wet paper codes). A non-shared selection channel refers to the situation when Alice embeds her message into a selected subset of the image but does not (or cannot) share her selection with Bob. This chapter also discusses several diverse problems in steganography that lead to non-shared selection channels and can be elegantly solved using wet paper codes: adaptive steganography, perturbed quantization steganography, a new class of improved matrix embedding methods, public-key steganography, the no-shrinkage F5 algorithm, and the MMx algorithm.

While the first part of this book deals solely with design and development of steganographic methods, the next three chapters are devoted to steganalysis, which is understood as an inherent part of steganography. After all, steganography is advanced through analysis.

In Chapter 10, steganalysis is introduced as the task of discovering the presence of secret data. The discussion in this chapter is directed towards explaining general principles common to many steganalysis techniques. The focus is on statistical attacks in which the warden reaches her decision by inspecting statistical properties of pixels. This approach to steganalysis provides connections with the abstract problem of signal detection and hypothesis testing, which in turn allows importing standard signal-detection tools and terminology, such as the receiver operating characteristic. The chapter continues with separate sections on targeted and blind steganalysis. The author lists several general strategies that one can follow to construct targeted attacks and highlights the important class of quantitative attacks, which can estimate the number of embedding changes. The section on blind steganalysis contains a list of general principles for constructing steganalysis features as well as description of several diverse applications of blind steganalyzers, including construction of targeted attacks, steganography design, multi-class steganalysis, and benchmarking. The chapter is closed with discussion of forensic steganalysis and system attacks on steganography in which the attacker relies on protocol weaknesses of a specific implementation rather than on statistical artifacts computed from the pixel values.

Chapter 11 contains examples of targeted steganalysis attacks and their experimental verifications. Experiments on real images are used to explain various issues when constructing a practical steganography detector and to give the reader a sense of how sensitive the attacks are. The chapter starts with the Sample Pairs Analysis, which is a targeted quantitative attack on LSB embedding in the spatial domain. The derivation of the method is presented in a way that makes the algorithm appear as a rather natural approach that logically follows from the strategies outlined in Chapter 10. Next, the approach is generalized by formulating it within the structural steganalysis framework. This enables several important generalizations that further improve the method's accuracy. The third attack, the Pairs Analysis, is a quantitative attack on steganographic methods that embed messages into LSBs of palette images, such as EzStego. The concept of calibration is used to construct a quantitative attack on the F5 embedding algorithm. The chapter is closed with description of targeted attacks on ±1 embedding in the spatial domain based on the histogram characteristic function.

Chapter 12 is devoted to the topic of blind attacks, which is an approach to steganalysis based on modeling images using features and classifying cover and stego features using machine-learning tools. Starting with the JPEG domain, the features are introduced in a natural manner as statistical descriptors of DCT coefficients by modeling them using several different statistical models. The JPEG domain is also used as an example to demonstrate two options for constructing blind steganalyzers: (1) the cover-versus-all-stego approach in which a binary classifier is trained to recognize cover images and a mixture of stego images produced by a multitude of steganographic algorithms, and (2) a one-class steganalyzer trained only on cover images that classifies all images incompatible with covers as stego. The advantages and disadvantages of both approaches are discussed with reference to practical experiments. Blind steganalysis in the spatial domain is illustrated on the example of a steganalyzer whose features are computed from image noise residuals. This steganalyzer is also used to demonstrate how much statistical detectability in practice depends on the source of cover images.

Chapter 13 discusses the most fundamental problem of steganography, which is the issue of computing the largest payload that can be securely embedded in an image. Two very different concepts are introduced – the steganographic capacity and secure payload. Steganographic capacity is the largest rate at which perfectly secure communication is possible. It is not a property of one specific steganographic scheme but rather a maximum taken over all perfectly secure schemes. In contrast, secure payload is defined as the number of bits that can be communicated at a given security level using a specific imperfect steganographic scheme. The secure payload grows only with the square root of the number of pixels in the image. This so-called square-root law is experimentally demonstrated on a specific steganographic scheme that embeds bits in the JPEG domain. The secure payload is more relevant to practitioners because all practical steganographic schemes that hide messages in real digital media are not likely to be perfectly secure and thus fall under the squre-root law.

To make this text self-contained, five appendices accompany the book. Their style and content are fully compatible with the rest of the book in the sense that the student does not need any more prerequisites than a basic knowledge of calculus and statistics. The author anticipates that students not familiar with certain topics will find it convenient to browse through the appendices and either refresh their knowledge or learn about certain topics in an elementary fashion accessible to a wide audience.

Appendix A contains the basics of descriptive statistics, including statistical moments, the moment-generating function, robust measures of central tendency and spread, asymptotic laws, and description of some key statistical distributions, such as the Bernoulli, binomial, Gaussian, multivariate Gaussian, generalized Gaussian, and generalized Cauchy distributions, Student's t-distribution, and the chi-square distribution.

As some of the chapters rely on basic knowledge of information theory, Appendix B covers selected key concepts of entropy, conditional entropy, joint entropy, mutual information, lossless compression, and KL divergence and some of its key properties, such as its relationship to hypothesis testing and Fisher information.

The theory of linear codes over finite fields is the subject of Appendix C. The reader is introduced to the basic concepts of a generator and parity-check matrix, covering radius, average distance to code, sphere-covering bound, orthogonality, dual code, systematic form of a code, cosets, and coset leaders.

Appendix D contains elements of signal detection and estimation. The author explains the Neyman–Pearson and Bayesian approach to hypothesis testing, the concepts of a receiver-operating-characteristic (ROC) curve, the deflection coefficient, and the connection between hypothesis testing and Fisher information. The appendix continues with composite hypothesis testing, the chi-square test, and the locally most powerful detector. The topics of estimation theory covered in the appendix include the Cramer–Rao lower bound, least-square estimation, maximum-likelihood and maximum a posteriori estimation, and the Wiener filter. The appendix is closed with the Cauchy–Schwartz inequality in Hilbert spaces with inner product, which is needed for proofs of some of the propositions in this book.

Readers not familiar with support vector machines (SVMs) will find Appendix E especially useful. It starts with the formulation of a binary classification problem and introduces linear support vector machines as a classification tool. Linear SVMs are then progressively generalized to non-separable problems and then put into kernelized form as typically used in practice. The weighted form of SVMs is described as well because it is useful to achieve a trade-off between false alarms and missed detections and for drawing an ROC curve. The appendix also explains practical issues with data preprocessing and training SVMs that one needs to be aware of when using SVMs in applications, such as in blind steganalysis.

Because the focus of this book is strictly on steganography in digital signals, methods for covert communication in other objects are not covered. Instead, the author refers the reader to other publications. In particular, linguistic steganography and data-hiding aspects of some cryptographic applications are covered in [238, 239]. The topic of covert channels in natural language is also covered in [18, 25, 41, 161, 182, 227]. A comprehensive bibliography of all articles published on covert communication in linguistic structures, including watermarking applications, is maintained by Bergmair at `http://semantilog.ucam.org/biblingsteg/`. Topics dealing with steganography in Internet protocols are studied in [106, 162, 163, 165, 177, 216]. Covert timing channels and their security are covered in [26, 34, 100, 101]. The intriguing topic of steganography in Voice over IP applications, such as Skype, appears in [6, 7, 58, 147, 150, 169, 251]. Steganographic file systems [4, 170] are useful tools to thwart "rubber-hose attacks" on cryptosystems when a person is coerced to reveal encryption keys after encrypted files have been found on a computer system. A steganographic file system allows the user to plausibly deny that encrypted files reside on the disk. In-depth analysis of current steganographic software and the topics of data hiding in elements of operating systems are provided in [142]. Finally, the topics of audio steganography and steganalysis appeared in [9, 24, 118, 149, 187, 202].

Acknowledgments

I would like to acknowledge the role of several individuals who helped me commit to writing this book. First of all and foremost, I am indebted to Richard Simard for encouraging me to enter the field of steganography and for supporting research on steganography. This book would not have materialized without the constant encouragement of George Klir and Monika Fridrich. Finally, the privilege of co-authoring a book with Ingemar Cox [51] provided me with energy and motivation I would not have been able to find otherwise.

Furthermore, I am happy to acknowledge the help of my PhD students for their kind assistance that made the process of preparing the manuscript in TeX a rather pleasant experience instead of the nightmare that would for sure have followed if I had been left alone with a TeX compiler. In particular, I am immensely thankful to TeX guru Tomáš Filler for his truly significant help with formatting the manuscript, preparing the figures, and proof-reading the text, to Tomáš Pevný for contributing material for the appendix on support vector machines, and to Jan Kodovský for help with combing the citations and proof-reading. I would also like to thank Ellen Tilden and my students from the ECE 562 course on Fundamentals of Steganography, Tony Nocito, Dae Kim, Zhao Liu, Zhengqing Chen, and Ran Ren, for help with sanitizing this text to make it as free of typos as possible.

Discussions with my colleagues, Andrew D. Ker, Miroslav Goljan, Andreas Westfeld, Rainer Böhme, Pierre Moulin, Neil F. Johnson, Scott Craver, Patrick Bas, Teddy Furon, and Xiaolong Li were very useful and helped me clarify some key technical issues. The encouragement I received from Mauro Barni, Deepa Kundur, Slava Voloshynovskiy, Jana Dittmann, Gaurav Sharma, and Chet Hosmer also helped with shaping the final content of the manuscript. Special thanks are due to George Normandin and Jim Moronski for their feedback and many useful discussions about imaging sensors and to Josef Sofka for providing a picture of a CCD sensor. A special acknowledgement goes to Binghamton University Art Director David Skyrca for the beautiful cover design.

Finally, I would like to thank Nicole and Kathy Fridrich for their patience and for helping me to get into the mood of sharing.

1 Introduction

A woman named Alice sends the following e-mail to her friend Bob, with whom she shares an interest in astronomy:

My friend Bob,
   until yesterday I was using binoculars for stargazing. Today, I decided to try my new telescope. The galaxies in Leo and Ursa Major were unbelievable! Next, I plan to check out some nebulas and then prepare to take a few snapshots of the new comet. Although I am satisfied with the telescope, I think I need to purchase light pollution filters to block the xenon lights from a nearby highway to improve the quality of my pictures.
   Cheers,
   Alice.

At first glance, this letter appears to be a conversation between two avid amateur astronomers. Alice seems to be excited about her new telescope and eagerly shares her experience with Bob. In reality, however, Alice is a spy and Bob is her superior awaiting critical news from his secret agent. To avoid drawing unwanted attention, they decided not to use cryptography to communicate in secrecy. Instead, they agreed on another form of secret communication – steganography.

   Upon receiving the letter from Alice, Bob suspects that Alice might be using steganography and decides to follow a prearranged protocol. Bob starts by listing the first letters of all words from Alice's letter and obtains the following sequence:

$$mfbuyiwubfstidttmnttgilaumwuniptcosnatpttafsotncaiaswttitintplpftbt$$
$$xlfanhtitqompca.$$

Then, he writes down the decimal expansion of π

$$\pi = 3.141592653589793\ldots$$

and reads the message from the extracted sequence of letters by putting down the third letter in the sequence, then the next first letter, the next fourth letter, etc. The resulting message is

$$buubdlupnpsspx.$$

Finally, Bob replaces each letter with the letter that precedes it in the alphabet and deciphers the secret message

$$attack\ tomorrow.$$

Let us take a look at the tasks that Alice needs to carry out to communicate secretly with Bob. She first encrypts her message by substituting each letter with the one that follows it in the English alphabet (e.g., a is substituted with b, b with c, ..., and z with a). Note that this simple substitution cipher could be replaced by a more secure encryption algorithm if desired. Then, Alice needs to write an almost arbitrary but meaningful(!) letter while making sure that the words whose location is determined by the digits of π start with the letters of the encrypted message. Of course, instead of the decimal expansion of π, Alice and Bob could have agreed on a different integer sequence, such as one generated from a pseudo-random number generator seeded with a shared key. The shared information that determines the location of the message letters is called the steganographic key or stego key. Without knowing this key, it is not only difficult to read the message but also difficult for an eavesdropper to prove that the text contains a secret message.

Note that the hidden message is unrelated to the content of the letter, which only serves as a decoy or "cover" to hide the very fact that a secret message is being sent. In fact, this is the defining property of steganography:

Steganography can be informally defined as the practice of undetectably communicating a message in a cover object.

We now elaborate on the above motivational example a little more. If Alice planned to send a very long message, the above steganographic method would not be very practical. Instead of hiding the message in the body of the e-mail using her creative writing skills, Alice could hide her message by slightly modifying pixels in a digital picture, such as an image of a galaxy taken through her telescope, and attach the modified image to her e-mail. Of course, that would require a different hiding procedure shared with Bob. A simple method to hide a binary message would be to encode the message bits into the colors of individual pixels in the image so that even values represent a binary 0 and odd values a binary 1. Alice could achieve this by modifying each color by at most one. Here, Alice relies on the fact that such small modifications will likely be imperceptible. This method of covert communication allows Alice to send as many bits as there are pixels in the image without the need to painstakingly form a plausible-looking cover letter. She could even program a computer to insert the message, which could be an arbitrary electronic file, into the image for her.

Digital images acquired using a digital camera or scanner provide a friendly environment to the steganographer because they contain a slight amount of noise that helps mask the modifications that need to be carried out to embed a secret message. Moreover, attaching an image to an e-mail message is commonly done and thus should not be suspicious.

This book deals with steganography of signals represented in digital form, such as digital images, audio, or video. Although the book focuses solely on images, many principles and methods can be adopted to the other multimedia objects.

1.1   Steganography throughout history

The word *steganography* is a composite of the Greek words *steganos*, which means "covered," and *graphia*, which means "writing." In other words, steganography is the art of concealed communication where the very existence of a message is secret. The term steganography was used for the first time by Johannes Trithemius (1462–1516) in his trilogy *Polygraphia* and in *Steganographia* (see Figure 1.1). While the first two volumes described ancient methods for encoding messages (cryptography), the third volume (1499) appeared to deal with occult powers, black magic, and methods for communication with spirits. The volume was published in Frankfurt in 1606 and in 1609 the Catholic Church put it on the list of "*libri prohibiti*" (forbidden books). Soon, scholars began suspecting that the book was a code and attempted to decipher the mystery. Efforts to decode the book's secret message came to a successful end in 1996 and 1998 when two researchers independently [65, 201] revealed the hidden messages encoded in numbers through several look-up tables included in the book [145]. The messages turned out to be quite mundane. The first one was the Latin equivalent of "The quick brown fox jumps over the lazy dog," which is a sentence that contains every letter of the alphabet. The second message was: "The bearer of this letter is a rogue and a thief. Guard yourself against him. He wants to do something to you." Finally, the third was the start of the 21st Psalm.

The first written evidence about steganography being used to send messages is due to Herodotus [109], who tells of a slave sent by his master, Histiæus, to the Ionian city of Miletus with a secret message tattooed on his scalp. After the tattooing of the message, the slave grew his hair back in order to conceal the message. He then traveled to Miletus and, upon arriving, shaved his head to reveal the message to the city's regent, Aristagoras. The message encouraged Aristagoras to start a revolt against the Persian king.

Herodotus also documented the story of Demeratus, who used steganography to alert Sparta about the planned invasion of Greece by the Persian Great King Xerxes. To conceal his message, Demeratus scraped the wax off the surface of a wooden writing tablet, scratched the message into the wood, and then coated the tablet with a fresh layer of wax to make it appear to be a regular blank writing tablet that could be safely carried to Sparta without arousing suspicion.

Aeneas the Tactician [226] is credited with inventing many ingenious steganographic techniques, such as hiding messages in women's earrings or using pigeons to deliver secret messages. Additionally, he described some simple methods for hiding messages in text by modifying the height of letter strokes or by marking letters in a text using small holes.

Hiding messages in text is called linguistic steganography or acrostics. Acrostics was a very popular ancient steganographic method. To embed a unique "signature" in their work, some poets encoded secret messages as initial letters of sentences or successive tercets in a poem. One of the best-known examples

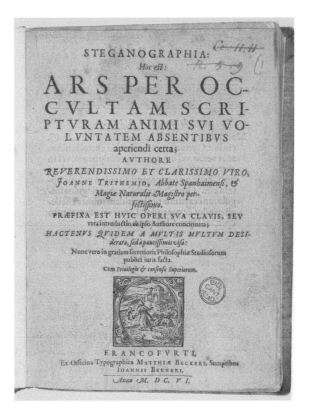

Figure 1.1 The title page of *Steganographia* by Johannes Trithemius, the inventor of the word "steganography." Reproduced by kind permission of the Syndics of Cambridge University Library.

is *Amorosa visione* by Giovanni Boccaccio [247]. Boccaccio encoded three sonnets (more than 1500 letters) into the initial letters of the first verse of each tercet from other poems. The linguistic steganographic scheme described at the beginning of this chapter is an example of Cardan's Grille, which was originally conceived in China and reinvented by Cardan (1501–1576). The letters of the secret message form a random pattern that can be accessed simply by placing a mask over the text. The mask plays the role of a secret stego key that has to be shared between the communicating parties.

Francis Bacon [15] described a precursor of modern steganographic schemes. Bacon realized that by using italic or normal fonts, one could encode binary representation of letters in his works. Five letters of the cover object could hold five bits and thus one letter of the alphabet. The inconsistency of sixteenth-century typography made this method relatively inconspicuous.

A modern version of this steganographic principle was described by Brassil [29]. He described a method for data hiding in text documents by slightly shifting the lines of text up or down by 1/300 of an inch. It turns out that such subtle changes are not visually perceptible, yet they are robust enough to survive photocopying.

This way, the message could be extracted even from printed or photocopied documents.

In 1857, Brewster [31] proposed a very ingenious technique that was actually used in several wars in the nineteenth and twentieth centuries. The idea is to shrink the message so much that it starts resembling specks of dirt but can still be read under high magnification. The technological obstacles to use of this idea in practice were overcome by the French photographer Dragon, who developed technology for shrinking text to microscopic dimensions. Such small objects could be easily hidden in nostrils, ears, or under fingernails [224]. In World War I, the Germans used such "microdots" hidden in corners of postcards slit open with a knife and resealed with starch. The modern twentieth-century microdots could hold up to one page of text and even contain photographs. The Allies discovered the usage of microdots in 1941. A modern version of the concept of the microdot was recently proposed for hiding information in DNA for the purpose of tagging important genetic material [45, 212]. Microdots in the form of dust were also recently proposed to identify car parts [1].

Perhaps the best-known form of steganography is writing with invisible ink. The first invisible inks were organic liquids, such as milk, urine, vinegar, diluted honey, or sugar solution. Messages written with such ink were invisible once the paper had dried. To make them perceptible, the letter was simply heated up above a candle. Later, more sophisticated versions were invented by replacing the message-extraction algorithm with safer alternatives, such as using ultraviolet light.

In 1966, an inventive and impromptu steganographic method enabled a prisoner of war, Commander Jeremiah Denton, to secretly communicate one word when he was forced by his Vietnamese captors to give an interview on TV. Knowing that he could not say anything critical of his captors, as he spoke, he blinked his eyes in Morse code, spelling out T-O-R-T-U-R-E.

Steganography became the subject of a dispute during the match between Viktor Korchnoi and Anatoly Karpov for the World Championship in chess in 1978 [117]. During one of the games, Karpov's assistants handed him a tray with yogurt. This was technically against the rules, which prohibited contact between the player and his team during play. The head of Korchnoi's delegation, Petra Leeuwerik, immediately protested, arguing that Karpov's team could be passing him secret messages. For example, a violet yogurt could mean that Karpov should offer a draw, while a sliced mango could inform the player that he should decline a draw. The time of serving the food could also be used to send additional messages (steganography in timing channels). This protest, which was a consequence of the extreme paranoia that dominated chess matches during the Cold War, was taken quite seriously. The officials limited Karpov to consumption of only one type of yogurt (violet) at a fixed time during the game. Using the terminology of this book, we can interpret this protective measure as an act of an active warden to prevent usage of steganography.

Figure 1.2 Symbols on an American patchwork quilt on display at the National Cryptologic Museum near Washington, D.C.

In the 1990s, the story of a "quilt code" allegedly used in the Underground Railroad surfaced in the media. The Underground Railroad appeared spontaneously as a clandestine network of secret pathways and safe houses that helped black slaves in the USA escape from slavery during the first part of the nineteenth century. According to the story told by a South Carolina woman named Ozella Williams [230], people sympathetic to the cause displayed quilts on their fences to non-verbally inform the escapees about the direction of their journey or which action they should take next. The messages were supposedly hidden in the geometrical patterns commonly found in American patchwork quilts (see Figure 1.2). Since it was common to air quilts on fences, the master or mistress would not be suspicious about the quilts being on display.

The recent explosion of interest in steganography is due to a rather sudden and widespread use of digital media as well as the rapid expansion of the Internet (Figure 1.3 shows the annual count of research articles on the subject of steganography published by the IEEE). It is now a common practice to share pictures, video, and sound with our friends and family. Such objects provide a very favorable environment for concealing secret messages for one good reason: typical digital media files consist of a large number of individual samples (e.g., pixels) that can be imperceptibly modified to encode a secret message. And there is no need to develop technical expertise for those who wish to use steganography because the hiding process itself can be carried out by a computer program that anyone can download from the Internet for free. As of writing this book in late 2008, one can

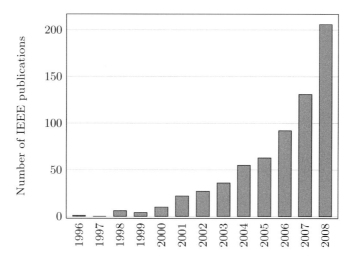

Figure 1.3 The growth of the field is witnessed by the number of articles annually published by IEEE that contain the keywords "steganography" or "steganalysis."

select from several hundreds of steganographic products available on the Internet. Figure 1.4 shows the number of newly released applications or new versions of existing programs capable of hiding data in digital media and text. Some software applications that focus on security, privacy, and anonymity offer the possibility to hide encrypted messages in pictures and music as an additional layer of protection. Examples of such programs are Steganos (`http://www.steganos.com/`) and Stealthencrypt (`http://www.stealthencrypt.com`). An updated list of selected currently available steganographic programs for various platforms can be obtained from `http://www.stegoarchive.com/`.

In the next section, the reader is informally introduced to some key concepts and principles on which modern steganography is built. The author also feels that it is important at this point to explain the differences between steganography and other related privacy and security applications, such as cryptography and digital watermarking. No attempt is made at this point to be rigorous. The goal is to entice the reader and gently introduce some of the challenges elaborated upon in this book.

1.2   Modern steganography

Because electronic communication is very susceptible to eavesdropping and malicious interventions, the issues of security and privacy are more relevant today than ever. Traditional solutions are based on cryptography [207], which is a mature, well-developed field with rigorous mathematical foundations. The cryptographic approach to privacy is to make the exchanged information unreadable to those who do not have the right decryption key. When an encrypted message

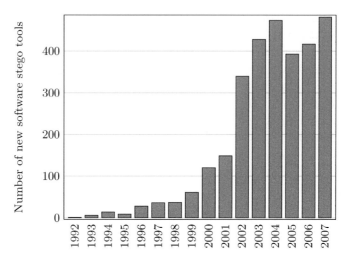

Figure 1.4 The number of newly released steganographic software applications or new versions per year. Adapted from [122] and reprinted with permission of John Wiley & Sons, Inc.

is intercepted, even though the content of the message is protected, the fact that the subjects are communicating secretly is obvious. In some situations, it may be important to avoid drawing attention and instead embed sensitive data in other objects so that the fact that secret information is being sent is not obvious in the first place. This is the approach taken by steganography.

Every steganographic system discussed in this book consists of two basic components – the embedding and extraction algorithms. The embedding algorithm accepts three inputs – the secret message to be communicated, the secret shared key that controls the embedding and extraction algorithms, and the *cover object*, which will be modified to convey the message. The output of the embedding algorithm is called the *stego object*. When the stego object is presented as an input to the message-extraction algorithm, it produces the secret message.

Steganography offers a feasible alternative to encryption in oppressive regimes where using cryptography might attract unwanted attention or in countries where the use of cryptography is legally prohibited. An interesting documented use of steganography was presented at the 4th International Workshop on Information Hiding [209]. Two subjects developed a steganographic scheme of their own to hide messages in uncompressed digital images and then used it successfully for several years when one of them was residing in a hostile country that explicitly prohibited use of encryption. The reason for their paranoia was a story told by their friend who already resided in the area, who had tried to send an encrypted e-mail only to have it returned to him by the local Internet service provider with the message appended, "Please, don't send encrypted emails – we can't read them."

In the early 1980s, Simmons [214] described intriguing political implications of the possibility to send data through a covert communication channel. According to the disarmament treaty SALT, the USA and Soviet Union mutually agreed to equip their nuclear facilities with sensors that would inform the other country about the number of missiles but not some other information, such as their location. All communications were required to be protected using standard digital signatures to prevent unauthorized modification of the sensors' readings. However, both sides quickly became concerned about the possibility to hide additional information through so-called subliminal channels that existed in most digital signature schemes at that time. This triggered research into developing digital signatures free of subliminal channels [57].

1.2.1 The prisoners' problem

The most important property of a steganographic system is undetectability, which means that it should be impossible for an eavesdropper to tell whether Alice and Bob are engaging in regular communication or are using steganography. Simmons provided a popular formulation of the steganography problem through his famous prisoners' problem [214]. Alice and Bob are imprisoned in separate cells and want to hatch an escape plan. They are allowed to communicate but their communication is monitored by warden Eve. If Eve finds out that the prisoners are secretly exchanging messages, she will cut the communication channel and throw them into solitary confinement. The prisoners resort to steganography as a means to exchange the details of their escape. Note that in the prisoners' problem, all that Eve needs to achieve is to detect the presence of secret messages rather than know their content. In other words, when Eve discovers that Alice and Bob communicate secretly, the steganographic system is considered broken. This is in contrast to encryption, where a successful attack means that the attacker gains access to the decrypted content or partially recovers the encryption key.

In the prisoners' problem, it is usually assumed that Eve has a complete knowledge of the steganographic algorithm that Alice and Bob might use, with the exception of the secret stego key, which Alice and Bob agreed upon before imprisonment. The requirement that the steganographic algorithm be known to Eve is Kerckhoffs' principle imported from cryptography. This seemingly strong and paranoid principle states that the security of the communication should not lie in the secrecy of the system but only in the secret key. The principle stems from many years of experience that taught us that through espionage the encryption (steganographic) algorithm or device may fall into the hands of the enemy and, if this happens, the security of the secret channel should not be compromised.

1.2.2   Steganalysis is the warden's job

Steganography is a privacy tool and as such it naturally provokes the human mind to attack it. The effort concerned with developing methods for detecting the presence of secret messages and eventually extracting them is called *steganalysis*. Positioning herself into the role of a *passive warden*, Eve passively monitors the communication between Alice and Bob. She is not only allowed to visually inspect the exchanged text or images, but also can apply some statistical tests to find out whether the distribution of colors in the image follows the expected statistics of natural images.

This field started developing more rapidly after the terrorist attacks of September 11, 2001, when speculations spread through the Internet that terrorists might use steganography for planning attacks [146, 48]. The only publicly documented use of a rather primitive form of steganography for planning terrorist activities was described by *The New York Times* in an article from November 11, 2006. Dhiren Barot, an Al Qaeda operative, filmed reconnaissance video between Broadway and South Street and concealed it before distribution by splicing it into a copy of the Bruce Willis movie *Die Hard: With a Vengeance*. In a different criminal case in 2000, the commercial steganographic tool S-Tools had been used for distribution of child porn.[1] The suspect was successfully prosecuted using steganalysis methods published in [119].

As we already know, steganography is considered broken even when the mere *presence* of the secret message is detected. This is because the primary goal of steganography is to conceal the communication itself. Often, identifying the subjects who are communicating using steganography can be of vital importance despite the fact that the content of the secret message may still be unknown. When the warden discovers the use of steganography, she may choose to cut the communication channel, if such an act is in her power, as in the prisoners' problem, or she may exercise other options. Eve may assume the role of an *active warden* and slightly modify the communicated objects to prevent the prisoners from using steganography. For example, if the prisoners are embedding messages in images, the warden may process the images by slightly resizing them, cropping, and recompressing in hope of preventing the recipient from reading any secret messages. She can also rephrase the content of a letter using synonyms or by changing the order of words in a sentence. During World War I, US Post Office censors used to rephrase telegrams to prevent people from sending hidden messages. In one case, the censor replaced the text "father is dead" with "father is deceased," which prompted the recipient to reply with "is father dead or deceased?" A more recent example of an active warden was given by Gina Fisk and her coworkers from Los Alamos National Laboratory [72], who described an active warden system integrated with a firewall designed to eliminate covert channels in network protocols.

---

[1] Personal communication by Neil F. Johnson, 2007.

The actions of an active warden will likely inform Alice and Bob that they are under surveillance. Instead of actively blocking the covert communication channel, Eve may decide not to intervene at all and instead try to extract the messages to learn about the prisoners' escape plans. Effort directed towards extracting the secret message belongs to the field of *forensic steganalysis*. If Eve is successful and gains access to the stego (encryption) key, a host of other options opens up for her. She can now be devious and impersonate the prisoners to trick them to reveal more secrets. Such a warden is called malicious.

1.2.3   Steganographic security

The defining property of steganography is the requirement that Eve should not be able to decide whether or not a given object conveys a secret message. Formalizing this requirement mathematically, however, is far from easy.

According to Kerckhoffs' principle, the warden has a complete knowledge of the steganographic algorithm, which means that she also has all details about the source of cover objects used by both prisoners. At an abstract level, the properties of the cover source could be described by a statistical distribution on the space of all cover objects that the prisoners can possibly exchange. For example, our amateur astronomer Alice, who by the way really dislikes winter, is much more likely to send a picture of the Moon or Saturn than a snowy landscape. This could be formulated by stating that the probability distribution of cover images used by Alice will have higher values on images with astronomical themes and much lower values on images with winter scenery. Given the fact that Eve knows this distribution, she can run statistical tests to see whether the images sent by Alice are compliant with the expected distribution. Interpreting the results of her test, she can decide that, at a certain confidence level, Alice does or does not communicate using steganography.

Because digital images are quite complex high-dimensional objects consisting of millions of pixels, it is not feasible to obtain even a rough approximation to this hypothetical distribution. Modern steganalysis works with simplified models consisting of a set of statistical quantities derived from images, such as sample histograms of pixel values or various types of higher-order statistics computed from adjacent pairs of pixels. Eve calculates these quantities and compares them with their expected values estimated from cover images that would be sent during legitimate use of the channel. Statistically significant deviations from the expected values are then interpreted as evidence that the image has been modified by a steganographic algorithm.

While this quantitative view of steganographic security permits precise mathematical formulation and rigorous study, it is only an approximation of the concept of undetectability. For example, statistical quantities do not well describe the semantic meaning of the communicated images. Imagine the situation when Alice writes her message on paper, takes a picture of it, and attaches the image to her e-mail. Because this image was never modified, its statistical properties should be compliant with those of other images produced by her camera. Thus, any automatic steganalysis system only inspecting statistical properties of images would label the image as compliant with the legitimate use of the channel. A human warden will, of course, have full access to the communicated message.

It is intuitively clear that Alice and Bob can increase their sense of security and decrease the chance of being caught by Eve if they communicate only very short messages. This would, however, make their communication less efficient and quite likely impractical. Therefore, Alice and Bob need to know how large a message they can hide in a given object without introducing artifacts that would trigger Eve's detector. The size of the critical message is called the *steganographic capacity*. The research in steganography focuses on design of algorithms that permit sending messages that are as long as possible without making the stego objects statistically distinguishable from cover objects.

1.2.4   Steganography and watermarking

At this point, we wish to stress one important difference between steganography and a related data-hiding field called watermarking [16, 51, 186]. Even though watermarking and steganography share some fundamental similarities in that they both secretly hide information, they address very different applications. In steganography, the cover image is a mere decoy and has no relationship to the secret message. In contrast, a watermark usually carries supplemental information about the cover image or some other data related to the cover, such as labels identifying the sender or the receiver. For example, when purchasing an MP3 song over the Internet, information about the seller and/or buyer can be inserted into the song in the form of an inaudible but robust watermark. The watermark may be used later to trace illegally distributed copies of the song. Watermarks can also convey information about the song itself (e.g., its robust hash or digest) so that the song's integrity can later be verified by comparing the song with the watermark payload.

The second and perhaps even more important difference between steganography and watermarking is the issue of the existence of a secret message in an image. While in steganography it is of utmost importance to make sure the image does not exhibit any traces of hidden data, the presence of a watermark is often advertised to deter illegal activity, such as unauthorized copying or redistribution. Additionally, steganography is a mode of communication and as such needs to allow sending large amounts of data. On the contrary, even a very short digital watermark can be quite useful. For example, the presence of a watermark (a one-bit payload) may testify about the image's ownership. These very different requirements imposed on these two applications make their design and analysis quite different.

Summary

- Steganography is the practice of communicating a secret message by hiding it in a cover object.
- Steganography is usually described as the prisoners' problem in which two prisoners, Alice and Bob, want to hatch an escape plan but their communication is monitored by the warden (Eve), who will cut the communication once she suspects covert exchange of data.
- The most important property of steganography is statistical undetectability, which means that it should be impossible for Eve to prove the existence of a secret message in a cover. Statistically undetectable steganographic schemes are called secure.
- A warden who merely observes the traffic between Alice and Bob is called passive. An active or malicious warden tampers with the communication in order to prevent the prisoners from using steganography or to trick them into revealing their communication.
- Digital watermarking is a data-hiding application that is related to steganography but is fundamentally quite different. While in steganography the secret message has usually no relationship to the cover object, which plays the role of a mere decoy, watermarks typically supply additional information about the cover. Moreover, and most importantly, watermarks do not have to be embedded undetectably.

2   Digital image formats

Digital images are commonly represented in four basic formats – raster, palette, transform, and vector. Each representation has its advantages and is suitable for certain types of visual information. Likewise, when Alice and Bob design their steganographic method, they need to consider the unique properties of each individual format. This chapter explains how visual data is represented and stored in several common image formats, including raster and palette formats, and the most popular format in use today, the JPEG. The material included in this chapter was chosen for its relevance to applications in steganography and is thus necessarily somewhat limited. The topics covered here form the minimal knowledge base the reader needs to become familiar with. Those with sufficient background may skip this chapter entirely and return to it later on an as-needed basis. An excellent and detailed exposition of the theory of color models and their properties can be found in [74]. A comprehensive description of image formats appears in [32].

In Section 2.1, the reader is first introduced to the basic concept of color as perceived by humans and then learns how to represent color quantitatively using several different color models. Section 2.2 provides details of the processing needed to represent a natural image in the raster (BMP, TIFF) and palette formats (GIF, PNG). Section 2.3 is devoted to the popular transform-domain format JPEG, which is the most common representation of natural images today. For all three formats, the reader is instructed how to work with such images in Matlab.

2.1   Color representation

Visible light is a superposition of electromagnetic waves with wavelengths spanning the interval between approximately 380 nm and 750 nm. Each color can be associated with the spectral density function $P(\lambda)$, which describes the amount of energy present at wavelength λ. Thus, even though one could say that there are infinitely (or even uncountably) many different colors, each color corresponding to a different density function $P(\lambda)$, the human eyes are capable of distinguishing only a relatively small subset of all possible colors. There are three different receptors in the eye retina called cones, with peak sensitivity to red, green, and blue light. The cones that register the blue light have the smallest sensitivity to light intensity, while the cones that respond to green light have the highest sensitivity. Electrical signals produced by the cones are fed to the brain, allowing us to perceive color. This is the tristimulus theory of color perception.

This theory leads to the so-called additive color model. According to this model, any color is obtained as a linear combination of three basic colors (or color channels) – red, green, and blue. Denoting the amount of each color as R, G, and B, where each number is from the interval $[0, 1]$ (zero intensity to full intensity), each color can be represented as a three-dimensional vector in the RGB color cube $(R, G, B) \in [0, 1]^3$. Hardware systems that emit light are usually modeled as additive. For example, old computer monitors with the Cathode-Ray Tube (CRT) screens create colors by combining three RGB phosphores on the screen. Liquid-Crystal Display (LCD) panels combine the light from three adjacent pixels. Full intensity of all three colors is perceived as white, while low intensity in all is perceived as dark or black.

The subtractive color model is used for hardware devices that create colors by absorption of certain wavelengths rather than emission of light. A good example of a subtractive color device is a printer. The standard basic colors for subtractive systems are, by convention, cyan, magenta, and yellow, leading to color representation using the vector CMY. These three colors are obtained by removing from white the colors red, green, and blue, respectively. The CMY system is augmented with a fourth color, black (abbreviated as K) to improve the printing contrast and save on color toners.

The following relationship holds between the additive RGB and subtractive CMY systems:

$$C = 1 - R, \tag{2.1}$$
$$M = 1 - G, \tag{2.2}$$
$$Y = 1 - B. \tag{2.3}$$

Although the additive RGB color model describes the colors perceivable by humans quite well, it is redundant because the three signals are highly correlated among themselves and is thus not the most economical for transmission. A very popular color system is the YUV model originally developed for transmission of color TV signals. The requirement of backward compatibility with old black-and-white TVs led the designers to form the color TV signal as luminance augmented with chrominance signals. The reader is forewarned that from now on the letter Y will always stand for luminance and not yellow as in the CMY(K) color system.

The luminance Y is defined as a weighted linear combination of the RGB channels with weights determined by the sensitivity of the human eye to the three RGB colors,

$$Y = 0.299R + 0.587G + 0.114B. \tag{2.4}$$

The chrominance components are the differences

$$U = R - Y,$$ (2.5)
$$V = B - Y$$ (2.6)

conveying the color information. The transformation between the RGB and YUV systems is linear,

$$\begin{pmatrix} Y \\ U \\ V \end{pmatrix} = \begin{pmatrix} 0.299 & 0.587 & 0.114 \\ 0.701 & -0.587 & -0.114 \\ -0.299 & -0.587 & 0.886 \end{pmatrix} \begin{pmatrix} R \\ G \\ B \end{pmatrix},$$ (2.7)

$$\begin{pmatrix} R \\ G \\ B \end{pmatrix} = \begin{pmatrix} 1 & 1 & 0 \\ 1 & -0.509 & -0.194 \\ 1 & 0 & 1 \end{pmatrix} \begin{pmatrix} Y \\ U \\ V \end{pmatrix}.$$ (2.8)

Note that if the RGB colors are represented using 8-bit integers in the range $\{0, \ldots, 255\}$, the luminance Y shares the same range, while U and V fall into the range $\{-179, \ldots, 179\}$. To adjust all three components to the same range representable using 8 bits, the chrominance components are further linearly transformed to C_r and C_b, obtaining thus the YC_rC_b color model

$$\begin{pmatrix} Y \\ C_r \\ C_b \end{pmatrix} = \begin{pmatrix} 0 \\ 128 \\ 128 \end{pmatrix} + \begin{pmatrix} 0.299 & 0.587 & 0.114 \\ 0.5 & -0.419 & -0.081 \\ -0.169 & -0.331 & 0.5 \end{pmatrix} \begin{pmatrix} R \\ G \\ B \end{pmatrix}.$$ (2.9)

Because human eyes are much less sensitive to changes in chrominance than in luminance, the chrominance signals are often represented with fewer bits without introducing visible distortion into the image. This fact is utilized in the JPEG compression format and it is also used for TV signals, where a smaller bandwidth is allocated to the chrominance signals and a wider bandwidth is used for luminance. Digital image formats that use the YC_rC_b model include IIF, TIFF, JFIF, JPEG, and MPEG (motion JPEG).

2.1.1 Color sampling

Even though some image formats allow arbitrarily accurate representation of the color intensity values (IIF, PostScript), most formats represent the intensities in a quantized form using a fixed number of n_c bits, which allows capturing 2^{n_c} different shades. The most appropriate color sampling is heavily dependent on the application. At one extreme, most fax machines work with only two colors (black and white), while high-resolution satellite imagery or photo-realistic synthetic images may use up to 16 bits per color channel (48 bits for the color). Examples of typical color sampling values for selected applications are listed in Table 2.1.

In this book, we will mostly deal with raster images in BMP (Bitmap), TIFF (Tagged Image File Format), and PNG (Portable Network Graphics) formats, palette (indexed) images in GIF (Graphics Interchange Format) or PNG, and

Table 2.1. Typical color bit depth for various applications. For $n_c \geq 8$, the sampling of color images is in bits per color channel.

n_c	Colors	Application
1	2	Fax, black-and-white drawings
2	4	
4	16	Line drawings, charts, cartoons
8	256	Grayscale images, true-color natural images
12	4096	Grayscale medical images, digital sensor output, scans
14	16384	Digital sensor output, scans, film-quality digital images
16	65536	Photo-realistic synthetic images, scans, satellite imagery

Table 2.2. Color sampling for selected popular image formats.

	Raster (color)	Raster (grayscale)	Palette
BMP	24	8	1, 4, 8
TIFF	24, 30, 36, 42, 48	4–8	1–8
PNG	24, 48	1, 2, 4, 8, 16	1, 2, 4, 8
GIF	–	–	1–8
JPEG	24	8, 12	–

the JPEG format. Table 2.2 shows the color bit depth allowed by each format. For palette formats, the bit depth represents the range of palette indices.

2.2   Spatial-domain formats

The majority of readers would probably agree that the most intuitive way to represent natural images in a computer is to sample the colors on a sufficiently dense rectangular grid. This approach also nicely plays into how digital images are usually acquired through an imaging sensor (Chapter 3). Images stored in such spatial-domain formats form very large files that allow the steganographer to hide relatively large messages. In this section, we describe the details of the raster and palette representations.

2.2.1   Raster formats

In a raster format, the image data is typically stored in a row-by-row manner with one or more bytes (or bits) per pixel depending on the format and the number of bits allocated per pixel. While three bytes are necessary for each pixel of a 24-bit true-color image, grayscales typically require 8 bits (one byte) per pixel. Monochrome (black-and-white) images need only one bit per pixel. The most common formats that allow raster image representation are BMP, TIFF, and PNG. Their color sampling is shown in Table 2.2. These formats may use lossless compression [206] to provide a smaller file size. For example, BMP may use runlength compression (optionally), PNG uses DEFLATE, while TIFF allows multiple different compression schemes.

In this book, a grayscale image in raster format will be represented using an $M \times N$ matrix of integers from the range $\{0, \ldots, 2^{n_c} - 1\}$, where typically $n_c = 8$. A true-color BMP image will be represented with three such matrices. The Matlab command[1] for importing a BMP image is X = imread('my_image.bmp'). Alternatively, saving a uint8 matrix of integers X to a BMP image is obtained using the command imwrite(X, 'my_stego_image.bmp', 'bmp'). The reader is urged to check the Matlab help facility for the list of formats supported by his/her version of Matlab.

2.2.2   Palette formats

Palette formats are typically used for images with low color depth, such as computer-generated graphics, line drawings, and cartoons. For this type of visual information, the format is lossless (there is no loss of fidelity). The image consists of the header, image palette, and image data. The palette can have up to 256 colors stored as 8-bit RGB triples. The image data is a rectangular $M \times N$ array of 8-bit pointers to the palette.

When converting an image with more than 256 colors to a palette image, two separate procedures are employed: creation of the color palette and converting each pixel color to a palette color. The palette could be a fixed array of colors independent of the image content or it could be derived from the image using color-quantization algorithms. The latter option always gives much more visually pleasing results. Note that color quantization is a lossy process.

There exist many different algorithms for color quantization [111]. The popularity algorithm calculates the histogram and takes the most frequently occurring 256 colors as the palette. The median-cut algorithm recursively fits a box around the colors (in the RGB color cube $\{0, \ldots, 255\}^3$), splitting it along its longest dimension at the median in that dimension. The recursive process ends when $2^8 = 256$ boxes are obtained. The color palette is then formed by the center of gravity of colors from each box. This algorithm produces better results than the popularity algorithm. It is possible to split the box according to other criteria, such as minimizing the spread in each box, etc.

Once the image palette has been obtained, the color of every pixel needs to be mapped onto the newly created palette. Again, a number of approaches exist, ranging from simple truncation to the nearest neighbor to stochastic dithering or dithering along a space-filling curve. In this book, we describe only the simplest dithering methods to explain the main concepts.

---

[1] Matlab Image Processing Toolbox is required.

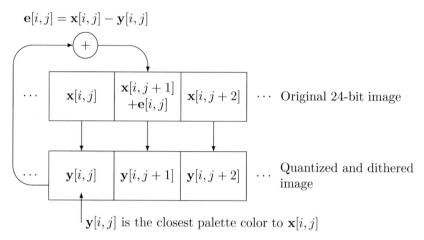

$$\mathbf{e}[i,j] = \mathbf{x}[i,j] - \mathbf{y}[i,j]$$

$\mathbf{y}[i,j]$ is the closest palette color to $\mathbf{x}[i,j]$

Figure 2.1 Simple dithering mechanism.

Let us denote the original pixel colors as $\mathbf{x}[i]$, $i = 1, \ldots, M \times N$, assuming here that the sequence $\mathbf{x}[i]$ has been obtained by scanning the image in some continuous manner (for example, by rows). Let us further denote the palette colors in the RGB model as $\mathbf{c}[k] = (\mathbf{r}[k], \mathbf{g}[k], \mathbf{b}[k])$, $k = 0, \ldots, 255$, $\mathbf{r}[k], \mathbf{g}[k], \mathbf{b}[k] \in \{0, \ldots, 255\}$. During dithering, depicted in Figure 2.1, the pixel values are processed one-by-one using the following formulas ($\mathbf{y}[i]$ denotes the dithered image with colors from the palette):

$$\mathbf{y}[i] = \mathbf{c}[k], \text{ where } \mathbf{c}[k] \text{ is the closest color to } \mathbf{x}[i], \tag{2.10}$$

$$\mathbf{e}[i] = \mathbf{x}[i] - \mathbf{y}[i] \text{ is the color approximation error at pixel } i, \tag{2.11}$$

$$\mathbf{x}[i+1] = \mathbf{x}[i] + \mathbf{e}[i] \text{ is the next pixel value corrected by}$$

$$\text{the approximation error made at } \mathbf{x}[i]. \tag{2.12}$$

In this algorithm, the pixel colors are truncated to the closest palette color, and, at the same time, the next pixel to be visited is modified by the truncation error at the current pixel. If a pixel is truncated, say, to a color that is less red, a small amount of red is added to the next pixel to locally preserve the overall color balance in the image.

This process can be further refined by spreading the truncation error among more pixels. We just need to make sure that the pixels are spatially close to the current pixel and that no pixels are modified that have already been visited. Weights can be assigned as fixed numbers or random variables with sum equal to 1 across all pixels receiving a portion of the truncation error.

One of the most popular dithering algorithms is Floyd–Steinberg dithering. To explain this algorithm, we now represent the pixels in the image as a two-dimensional array and assume that the dithering algorithm starts in the upper

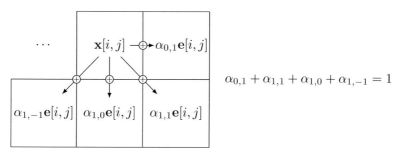

Figure 2.2 Floyd–Steinberg dithering.

left corner with pixel indices $i = 0$, $j = 0$ (follow Figure 2.2):

$$\mathbf{y}[i,j] = \mathbf{c}[k] \text{ where } \mathbf{c}[k] \text{ is the closest color to } \mathbf{x}[i,j], \tag{2.13}$$

$$\mathbf{e}[i,j] = \mathbf{x}[i,j] - \mathbf{y}[i,j], \tag{2.14}$$

$$\mathbf{x}[i,j+1] = \mathbf{x}[i,j+1] + \alpha_{0,1}\mathbf{e}[i,j], \tag{2.15}$$

$$\mathbf{x}[i+1,j+1] = \mathbf{x}[i+1,j+1] + \alpha_{1,1}\mathbf{e}[i,j], \tag{2.16}$$

$$\mathbf{x}[i+1,j] = \mathbf{x}[i+1,j] + \alpha_{1,0}\mathbf{e}[i,j], \tag{2.17}$$

$$\mathbf{x}[i+1,j-1] = \mathbf{x}[i+1,j-1] + \alpha_{1,-1}\mathbf{e}[i,j]. \tag{2.18}$$

Typical values of the coefficients are $\alpha_{0,1} = \frac{7}{16}$, $\alpha_{1,1} = \frac{1}{16}$, $\alpha_{1,0} = \frac{5}{16}$, and $\alpha_{1,-1} = \frac{3}{16}$. The dithering process basically arranges for a trade-off between limited color resolution and spatial resolution. Because human eyes have the ability to integrate colors in a small patch when looking from a distance, dithering allows us to perceive new shades of color not present in the palette. The dithering process introduces characteristic structures or patterns (noisiness) that may become visible in areas of small color gradient. The stochastic error spread helps by breaking any regular dithering patterns that may otherwise arise, thereby creating a more visually pleasing image. As an example, in Figure 2.3 we show a magnified portion of a true-color image after storing it as a GIF image with 256 colors in the palette obtained using the median-cut algorithm (the color image is displayed in Plate 1). The colors were dithered using Floyd–Steinberg dithering.

The Image Processing Toolbox of Matlab offers a number of routines that make working with GIF images in Matlab very easy. A GIF image can be read using the command [Ind, Map] = imread('my_image.gif'). The variable Map is an $n \times 3$ double array of palette colors consisting of $n \leq 256$ colors. Each row of Map is one palette color in the RGB format scaled so that $R, G, B \in [0, 1]$. The variable Ind is the array of indices to the palette of the same dimensions as the image. Modified versions of both arrays can be used to write the modified image to disk as a GIF file using the command imwrite(Ind, Map, 'my_stego_image.gif', 'gif').

Figure 2.3 Magnified portion of a true-color image (left) and the same portion after storing the image as GIF (right). A color version of this figure appears in Plate 1.

2.3   Transform-domain formats (JPEG)

People perceive natural images as a collection of segments filled with texture rather then as matrices of pixels. In particular, tests on human subjects showed that our visual system is fairly insensitive to small changes in color or high-spatial-frequency noise. Thus, it is highly inefficient to store natural images as rectangular matrices of colors. Engineers working in data compression have long realized this fact and proposed several much more efficient image formats that work by transforming the image into a different domain where it can be represented in an easily compressible "sparse" form. Such formats are typically lossy, meaning that the format conversion introduces some perceptual loss that is imperceptible under regular viewing conditions. Substantial savings in storage space justify the slight loss of fidelity. The two most commonly used transforms today are the Discrete Cosine Transform (DCT) and the Discrete Wavelet Transform (DWT). The DCT is at the heart of the JPEG format, while the DWT is used in JPEG2000 [228]. In this section, we introduce only the JPEG format as JPEG2000 steganography is currently not well developed.

JPEG stands for the Joint Photographic Experts Group that finalized the standard in 1992. In this section, we review basic properties of the format relevant to steganography. A detailed description of the format can be found in [185].

JPEG compression consists of the following five basic steps.

1. **Color transformation.** The color is transformed from the RGB model to the $YC_{\mathrm{r}}C_{\mathrm{b}}$ model (2.9). Although this step is not necessary (JPEG can work directly with the RGB representation) it is typically used because it enables higher compression ratios at the same fidelity.
2. **Division into blocks and subsampling.** The luminance signal Y is divided into 8×8 blocks. The chrominance signals C_{r} and C_{b} may be subsampled before dividing into blocks (more details are provided below).

3. **DCT transform.** The YC_rC_b signals from each block are transformed from the spatial domain to the frequency domain with the DCT. The DCT can be thought of as a change of basis representing 8×8 matrices.

4. **Quantization.** The resulting transform coefficients are quantized by dividing them by an integer value (quantization step) and rounded to the nearest integer. The luminance and chrominance signals may use different quantization tables. Larger values of the quantization steps produce a higher compression ratio but introduce more perceptual distortion.

5. **Encoding and lossless compression.** The quantized DCT coefficients are arranged in a zig-zag order, encoded using bits, and then losslessly compressed using Huffman or arithmetic coding. The resulting bit stream is prepended with a header and stored with the extension '.jpg' or '.jpeg.' For applications in steganography, it will not be necessary to understand the details of this last step.

To view a JPEG image, we first need to obtain the spatial-domain representation from the JPEG file, which is achieved essentially by reversing the five steps above. The JPEG bit stream is first parsed, then decompressed, and then the two-dimensional array of quantized DCT coefficients is formed. The coefficients in each block are then multiplied by the quantization steps and the inverse DCT is applied to produce the raw pixel values. Finally, the values are rounded to integers from a certain dynamic range (usually the set $\{0, \ldots, 255\}$). While lossless compression and the DCT are reversible processes, the quantization in Step 4 is irreversible and, in general, the decompressed image will not be identical to the original image before compression.

2.3.1   Color subsampling and padding

Because human eyes are less sensitive to changes in color than in luminance, the chrominance signals, C_r, C_b, are typically downsampled before applying the DCT to achieve a higher compression ratio. This is executed by initially dividing the image into macroblocks of 16×16 pixels. Each macroblock produces four 8×8 luminance blocks and 1, 2, or 4 blocks for each chrominance, depending on how the chrominance in the macroblock is subsampled. If it is subsampled by a factor of 2 in each direction, each macroblock will have only one 8×8 chrominance C_r block and one chrominance C_b block (and, of course, four luminance blocks). This is usually written in an abbreviated form $4 : 1 : 1$. If neither chrominance signal is subsampled, we have the $4 : 4 : 4$ representation. Both C_r and C_b can be subsampled only along one direction, which would lead to two chrominance 8×8 blocks in every macroblock, $4 : 2 : 2$. Other possibilities are allowed by the format, such as subsampling C_r only in one direction and C_b in both directions $(4 : 2 : 1)$.

If the image dimensions, $M \times N$, are not multiples of 8, the image is padded to the nearest larger multiples, $8 \lceil M/8 \rceil \times 8 \lceil N/8 \rceil$. During decompression, the padded parts are not displayed. Also, before applying the DCT, all pixel values are shifted by subtracting 128 from them.

2.3.2   Discrete cosine transform

We now provide a mathematical description of selected steps in JPEG compression. We start with the DCT and explain its properties. For an 8×8 block of luminance (or chrominance) values $\mathbf{B}[i,j]$, $i,j = 0, \ldots, 7$, the 8×8 block of DCT coefficients $\mathbf{d}[k,l]$, $k,l = 0, \ldots, 7$ is computed as a linear combination of luminance values,

$$\mathbf{d}[k,l] = \sum_{i,j=0}^{7} \mathbf{f}[i,j;k,l]\mathbf{B}[i,j] \tag{2.19}$$

$$= \sum_{i,j=0}^{7} \frac{\mathbf{w}[k]\mathbf{w}[l]}{4} \cos \frac{\pi}{16}k(2i+1) \cos \frac{\pi}{16}l(2j+1)\mathbf{B}[i,j], \tag{2.20}$$

where $\mathbf{f}[i,j;k,l] = (\mathbf{w}[k]\mathbf{w}[l]/4) \cos \frac{\pi}{16}k(2i+1) \cos \frac{\pi}{16}l(2j+1)$ and $\mathbf{w}[0] = 1/\sqrt{2}$, $\mathbf{w}[k > 0] = 1$. The coefficient $\mathbf{d}[0,0]$ is called the DC coefficient (or the DC term), while the remaining coefficients with $k + l > 0$ are called AC coefficients.

The DCT is invertible and the inverse transform (IDCT) is

$$\mathbf{B}[i,j] = \sum_{k,l=0}^{7} \frac{\mathbf{w}[k]\mathbf{w}[l]}{4} \cos \frac{\pi}{16}k(2i+1) \cos \frac{\pi}{16}l(2j+1)\mathbf{d}[k,l]. \tag{2.21}$$

The fact that the transform involves real numbers rather than integers increases its complexity and memory requirements, which could be an issue for mobile electronic devices, such as digital cameras or cell phones. Fortunately, the JPEG format allows various implementations of the transform that work only with integers and are thus much faster and more easily implemented in hardware. The fact that there exist various implementations of the transform means that one image could be compressed to several slightly different JPEG files. In fact, this difference may not be that small for some images and may influence the statistical distribution of DCT coefficients [95].

The DCT can be interpreted as a change of basis in the vector space of all 8×8 matrices, where the sum of matrices and multiplication by a scalar are defined in the usual elementwise manner and the dot product between matrices $\mathbf{X}$ and $\mathbf{Y}$ is $\mathbf{X} \cdot \mathbf{Y} = \sum_{i,j=0}^{7} \mathbf{X}[i,j]\mathbf{Y}[i,j]$. For a fixed pair (k,l), we call the 8×8 matrix $\mathbf{f}[i,j;k,l]$ the (k,l)th basis pattern. All 64 such patterns, depicted in Figure 2.4, form an orthonormal system because

$$\sum_{i,j=0}^{7} \mathbf{f}[i,j;k,l]\mathbf{f}[i,j;k',l'] = \delta(k - k')\delta(l - l') \text{ for all } k, k', l, l' \in \{0, \ldots, 7\},$$

$$\tag{2.22}$$

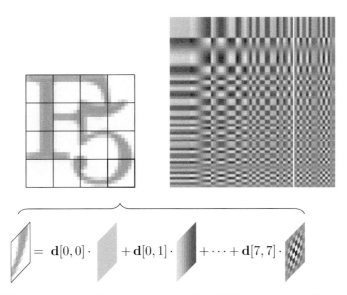

Figure 2.4 All 64 orthonormal basis patterns used in JPEG compression. Below, an example of an expansion of a pattern into a linear combination of basis patterns. Image provided courtesy of Andreas Westfeld.

where δ is the Kronecker delta,

$$\delta(x) = \begin{cases} 1 & \text{when } x = 0 \\ 0 & \text{when } x \neq 0. \end{cases} \tag{2.23}$$

Equation (2.20) can then be understood as decomposition of the pixel block $\mathbf{B}$ into the basis. The DCT coefficients $\mathbf{d}[k, l]$ are coefficients in a linear combination of the patterns that produces the pixel block $\mathbf{B}$. Figure 2.4 demonstrates this fact graphically.

We also wish to remark that the two-dimensional DCT can be built from the one-dimensional DCT by taking tensor products of one-dimensional basis vectors. For a fixed pair (k, l), $\mathbf{f}[i, j; k, l] = \mathbf{u}[i] \otimes \mathbf{v}[j]$, where $\mathbf{u}[i] = (\mathbf{w}[k]/2) \cos \frac{\pi}{16} k(2i + 1)$ and $\mathbf{v}[j] = (\mathbf{w}[l]/2) \cos \frac{\pi}{16} l(2j + 1)$ and $\mathbf{a} \otimes \mathbf{b}[i, j] = \mathbf{a}[i]\mathbf{b}[j]$. In fact, most two- and higher-dimensional transforms are constructed in this manner.

2.3.3 Quantization

The purpose of quantization is to enable representation of DCT coefficients using fewer bits, which necessarily results in loss of information. During quantization, the DCT coefficients $\mathbf{d}[k, l]$ are divided by quantization steps from the quantization matrix $\mathbf{Q}[k, l]$ and rounded to integers

$$\mathbf{D}[k, l] = \text{round}\left(\frac{\mathbf{d}[k, l]}{\mathbf{Q}[k, l]}\right), \quad k, l \in \{0, \dots, 7\}, \tag{2.24}$$

 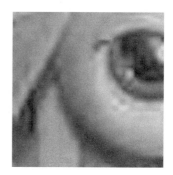

Figure 2.5 A magnified portion of a true-color image in BMP format (left) and the same portion after compressing with JPEG quality factor $q_f = 20$ (right). A color version of this figure appears in Plate 2.

where we denoted the operation of rounding x to the closest integer as round(x). The larger the quantization step, the fewer bits can be allocated to each DCT coefficient and the larger the loss and with it the perceptual distortion. JPEG introduces very characteristic compression artifacts that manifest themselves as "blockiness" or spatial discontinuities at the boundary of the 8×8 blocks (see Figure 2.5 or Plate 2 for the color version). The blockiness becomes more pronounced with coarser quantization. At fine quantization (low compression ratio or high quality factor), the blockiness artifacts are not perceptible even when inspected under magnification.

The JPEG standard recommends a set of quantization matrices indexed by a quality factor $q_f \in \{1, 2, \ldots, 100\}$. These matrices became known as the "standard" quantization matrices. Denoting the 8×8 matrix of ones with boldface $\mathbf{1}$, the standard quantization matrices are obtained using the following formula:

$$\mathbf{Q}_{q_f} = \begin{cases} \max\left\{\mathbf{1}, \text{round}\left(2\mathbf{Q}_{50}\left(1 - q_f/100\right)\right)\right\}, & q_f > 50 \\ \min\left\{255 \cdot \mathbf{1}, \text{round}\left(\mathbf{Q}_{50}50/q_f\right)\right\}, & q_f \leq 50, \end{cases} \quad (2.25)$$

where the 50% quality standard JPEG quantization matrix (for the luminance component Y) is

$$\mathbf{Q}_{50}^{(\text{lum})} = \begin{pmatrix} 16 & 11 & 10 & 16 & 24 & 40 & 51 & 61 \\ 12 & 12 & 14 & 19 & 26 & 58 & 60 & 55 \\ 14 & 13 & 16 & 24 & 40 & 57 & 69 & 56 \\ 14 & 17 & 22 & 29 & 51 & 87 & 80 & 62 \\ 18 & 22 & 37 & 56 & 68 & 109 & 103 & 77 \\ 24 & 35 & 55 & 64 & 81 & 104 & 113 & 92 \\ 49 & 64 & 78 & 87 & 103 & 121 & 120 & 101 \\ 72 & 92 & 95 & 98 & 112 & 100 & 103 & 99 \end{pmatrix}. \quad (2.26)$$

The chrominance quantization matrices are obtained using the same mechanism with the 50% quality chrominance quantization matrix,

$$
\mathbf{Q}_{50}^{(\mathrm{chr})} =
\begin{pmatrix}
17 & 18 & 24 & 47 & 99 & 99 & 99 & 99 \\
18 & 21 & 26 & 66 & 99 & 99 & 99 & 99 \\
24 & 26 & 56 & 99 & 99 & 99 & 99 & 99 \\
47 & 66 & 99 & 99 & 99 & 99 & 99 & 99 \\
99 & 99 & 99 & 99 & 99 & 99 & 99 & 99 \\
99 & 99 & 99 & 99 & 99 & 99 & 99 & 99 \\
99 & 99 & 99 & 99 & 99 & 99 & 99 & 99 \\
99 & 99 & 99 & 99 & 99 & 99 & 99 & 99
\end{pmatrix} .
\tag{2.27}
$$

The JPEG format allows arbitrary (non-standard) quantization matrices that can be stored in the header of the JPEG file. Many digital cameras use such non-standard matrices. An example of a non-standard luminance quantization matrix from a Kodak DC 290 camera is

$$
\begin{pmatrix}
5 & 5 & 5 & 5 & 5 & 6 & 6 & 8 \\
5 & 5 & 5 & 5 & 5 & 6 & 7 & 8 \\
5 & 5 & 5 & 5 & 6 & 7 & 8 & 9 \\
5 & 5 & 5 & 6 & 7 & 8 & 9 & 10 \\
5 & 5 & 6 & 7 & 8 & 9 & 11 & 12 \\
6 & 6 & 7 & 8 & 9 & 11 & 13 & 14 \\
6 & 7 & 8 & 9 & 11 & 13 & 15 & 16 \\
8 & 8 & 9 & 10 & 12 & 14 & 16 & 19
\end{pmatrix} .
\tag{2.28}
$$

We will denote the (k,l)th DCT coefficient in the bth block as $\mathbf{D}[k,l,b]$, $b \in \{1, \ldots, N_B\}$, where N_B is the number of all 8×8 blocks in the image. Note that for a color image, there will be three such three-dimensional arrays, one for luminance and two for the chrominance signals. The pair $(k,l) \in \{0, \ldots, 7\} \times \{0, \ldots, 7\}$ is called the spatial frequency (or mode) of the DCT coefficient.

Before the DCT coefficients are encoded using bits and losslessly compressed, the blocks are ordered from the upper left corner to the bottom right corner and the individual coefficients from each block are arranged by scanning the block using a zig-zag scan that starts at the spatial frequency $(0,0)$ and proceeds towards $(7,7)$.

2.3.4   Decompression

The decompression works in the opposite order. After reading the quantized DCT blocks from the JPEG file, each block of quantized DCT coefficients $\mathbf{D}$ is multiplied by the quantization matrix $\mathbf{Q}$, $\tilde{\mathbf{d}}[k,l] = \mathbf{Q}[k,l]\mathbf{D}[k,l]$, $k,l \in \{0, \ldots, 7\}$, and the inverse DCT is applied to the 8×8 matrix $\tilde{\mathbf{d}}$. The values are finally rounded to integers and truncated to a finite dynamic range (usually $\{0, \ldots, 255\}$). The block of decompressed pixel values $\tilde{\mathbf{B}}$ is thus

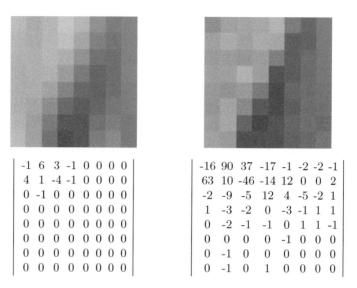

$$
\begin{array}{cccccccc}
\text{-1} & 6 & 3 & \text{-1} & 0 & 0 & 0 & 0 \\
4 & 1 & \text{-4} & \text{-1} & 0 & 0 & 0 & 0 \\
0 & \text{-1} & 0 & 0 & 0 & 0 & 0 & 0 \\
0 & 0 & 0 & 0 & 0 & 0 & 0 & 0 \\
0 & 0 & 0 & 0 & 0 & 0 & 0 & 0 \\
0 & 0 & 0 & 0 & 0 & 0 & 0 & 0 \\
0 & 0 & 0 & 0 & 0 & 0 & 0 & 0 \\
0 & 0 & 0 & 0 & 0 & 0 & 0 & 0
\end{array}
\qquad
\begin{array}{cccccccc}
\text{-16} & 90 & 37 & \text{-17} & \text{-1} & \text{-2} & \text{-2} & \text{-1} \\
63 & 10 & \text{-46} & \text{-14} & 12 & 0 & 0 & 2 \\
\text{-2} & \text{-9} & \text{-5} & 12 & 4 & \text{-5} & \text{-2} & 1 \\
1 & \text{-3} & \text{-2} & 0 & \text{-3} & \text{-1} & 1 & 1 \\
0 & \text{-2} & \text{-1} & \text{-1} & 0 & 1 & 1 & \text{-1} \\
0 & 0 & 0 & 0 & \text{-1} & 0 & 0 & 0 \\
0 & \text{-1} & 0 & 0 & 0 & 0 & 0 & 0 \\
0 & \text{-1} & 0 & 1 & 0 & 0 & 0 & 0
\end{array}
$$

Figure 2.6 An 8×8 luminance block of pixels and its quantized DCT coefficients for JPEG quality factor $q_f = 20$ (left) and $q_f = 90$ (right).

$$\tilde{\mathbf{B}} = \text{trunc}\left(\text{round}\left(\text{IDCT}(\tilde{\mathbf{d}})\right)\right), \tag{2.29}$$

where $\text{IDCT}(\cdot)$ is the inverse DCT (2.21) and $\text{trunc}(x)$ is the operation of truncating integers to a finite dynamic range ($\text{trunc}(x) = x$ for $x \in [0, 255]$, $\text{trunc}(x) = 0$ for $x < 0$, and $\text{trunc}(x) = 255$ for $x > 255$). Due to quantization, rounding, and truncation, $\tilde{\mathbf{B}}$ will in general differ from the original block $\mathbf{B}$.

2.3.5   Typical DCT block

The quantized DCT coefficients in a JPEG file are represented using integers in the range $[-1023, 1024]$. If we were to compress an image resembling white noise, the 8×8 matrix of quantized DCT coefficients in each block, b, $\mathbf{D}[k, l, b]$, $0 \le k, l \le 7$, would be filled with many non-zero integers because the spectrum of white noise is flat (all frequencies contribute the same amount of energy). Because natural images consist of objects rather than random textures, most of their energy is concentrated in low spatial frequencies. Consequently, the non-zero elements in each 8×8 block of DCT coefficients will be concentrated in the upper left corner with $(k, l) = (0, 0)$ (the low-spatial-frequency corner). To illustrate this claim, in Figure 2.6 we show an 8×8 luminance block and the block of quantized DCT coefficients for the same block.

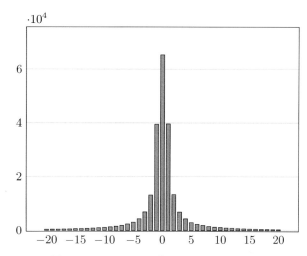

Figure 2.7 Histogram of luminance DCT coefficients for the image shown in Figure 5.1 stored as 95% quality JPEG.

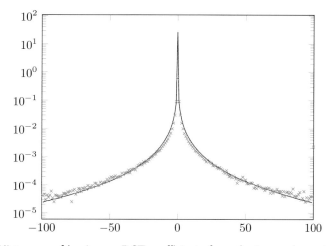

Figure 2.8 Histogram of luminance DCT coefficients from the image shown in Figure 5.1 compressed as 95% quality JPEG and the generalized Gaussian fit.

2.3.6   Modeling DCT coefficients

DCT coefficients of natural images follow a distribution with a spike around zero (see Figure 2.7), which is often modeled using the generalized Gaussian or Cauchy distribution (see Appendix A for description of these distributions).

The method of moments in Section A.8 is a simple approach that can be used to determine the parameters of the generalized Gaussian fit from data. In Figure 2.8, we show the histogram of DCT coefficients from the image shown in Figure 5.1 and the generalized Gaussian fit obtained using the method of moments.

2.3.7   Working with JPEG images in Matlab

Many steganographic methods for JPEG images work by directly manipulating quantized DCT coefficients extracted from the JPEG file and saving the modified array again as JPEG. Steganalysis methods for JPEG images also need access to the quantized DCT coefficients rather than the JPEG image decompressed to the spatial domain. However, most image-editing programs, including Matlab, do not provide direct access to the coefficients. The `imread` command in Matlab returns the spatial-domain representation of the JPEG file (the decompressed file) rather than its DCT coefficients.

The Independent JPEG group (`http://www.ijg.org`) provides access to its libjpeg library with C routines for parsing a JPEG file, extracting header data, DCT coefficients, and quantization tables, and writing data back to a JPEG file. Most readers of this book are probably familiar with programming in Matlab and would thus prefer to carry out the above-mentioned tasks within Matlab itself. This is, indeed, possible using the Matlab JPEG Toolbox written by Phil Sallee (`http://www.philsallee.com`). It essentially gives the user the ability to use the libjpeg library through Matlab functions. Among others, two routines included in the toolbox are `jpeg_read` and `jpeg_write`. They are Matlab MEX wrappers for the libjpeg library. The toolbox includes precompiled MEX binaries for Windows. For other operating systems, the user can download the libjpeg library and compile the MEX routines.

Below, we give an example of how to use the Matlab JPEG Toolbox to extract the array of luminance and chrominance DCT coefficients and the quantization matrices from a JPEG file, and how to write the data back to a JPEG file.

```
im=jpeg_read('color_image.jpg');
Lum=im.coef_arrays{im.comp_info(1).component_id};
ChromCr=im.coef_arrays{im.comp_info(2).component_id};
ChromCb=im.coef_arrays{im.comp_info(3).component_id};
Lum_quant_table=im.quant_tables{im.comp_info(1).quant_tbl_no};
Chrom_quant_table=im.quant_tables{im.comp_info(2).quant_tbl_no};
...
jpeg_write(im_stego,'my_stego_image.jpg');
```

The array of luminance DCT coefficients `Lum` is obtained by replacing every 8×8 pixel block with the corresponding block of quantized DCT coefficients.

Summary

- Human eyes are sensitive to electromagnetic radiation in the range of approximately 380 nm to 750 nm and each color is uniquely captured by the spectral power within this range.

- According to the tristimulus theory of human perception, each color that humans can perceive can be obtained as a superposition of three basic colors – red, green, and blue (RGB).
- There exist other color models, such as CMYK (cyan, magenta, yellow, black), YUV, and YC_rC_b (luminance and two chrominance signals), that are suitable for various applications.
- There are four main types of image formats: raster, palette (indexed), transform, and vector formats.
- Raster formats represent a digital image as a rectangular array of integers sampled using a fixed number of bits. Typical raster image formats are BMP, TIFF, and PNG. Images in raster formats are often large despite the fact that the formats use lossless compression to decrease the amount of data that needs to be stored.
- Palette images consist of two parts – a color palette and an array of indices to the palette. Typical palette formats are GIF and PNG. Palette images are convenient for representing charts, computer art, and other images with low color depth.
- Images in transform formats are represented through transform coefficients quantized using a fixed number of bits rather than using pixels directly. The most popular transform format is JPEG, which uses the discrete cosine transform (DCT). The coefficient quantization is controlled through a quantization table, which in turn can be controlled using a quality factor. Transform formats enable more efficient storage of visual data through lossy compression.
- Vector formats, such as WMF, EPS, and PS, can represent objects in the image using parametric description.
- Many formats enable several different image representations. For example, BMP and PNG can be either raster or palette, while EPS and TIFF are very general formats that allow multiple image representations.
- The raster and transform formats are most suitable for steganography.
- DCT coefficients in a JPEG file follow a symmetrical distribution with a spike around zero that is well modeled using generalized Gaussian or Cauchy distributions.

Exercises

2.1 [**Palette ordering**] Write a Matlab routine that orders a palette (represented as an $n \times 3$ array) by luminance. Use the conversion formula (2.4) for the ordering.

2.2 [**Draw palette**] Write a routine that displays the palette colors of a GIF image as 16×16 little squares arranged by rows (from left to right, top to bottom) into a square pattern. Inspect Figure 5.5 or Plate 6 for an example of the output.

2.3 [JPEG quantization table] Take a picture of a highly textured image with your digital camera set to take images in the highest-quality JPEG format. A closeup of grass or foliage would be good. Load the JPEG image in Matlab using the routine `jpeg_read` from the Matlab JPEG Toolbox. Extract the luminance quantization matrix. Is the matrix a standard matrix? Repeat the experiment by taking another JPEG image at the same JPEG quality, but this time choose a smoother content, such as a picture of blue sky, sea, or a wall painted with one color. Compare the quantization tables. Since most cameras use different quantization matrices for different images depending on their content, the two matrices will probably be different.

2.4 [JPEG histogram] Using the JPEG Matlab Toolbox, write a routine that displays a histogram of DCT coefficients from a chosen DCT mode (k, l). Use your routine and analyze a picture of a regular scene. Plot the histograms of DCT modes $(1, 2)$, $(1, 4)$, and $(1, 6)$. Notice that the histograms become progressively "spikier." This is because modes with higher spatial frequencies are more often quantized to zero.

2.5 [Generalized Gaussian Fit] Write a routine that fits the generalized Gaussian distribution to a histogram of DCT coefficients using the method of moments (Section A.8). For debugging purposes, apply it to artificially generated Gaussian data to see whether you are getting the expected result. Then apply it to the three histograms obtained in the previous project. The shape parameter should decrease with increasing spatial frequency of the DCT mode, quantifying thus your observation that the histogram becomes spikier.

3 Digital image acquisition

This book focuses on steganographic methods that embed messages in digital images by slightly modifying them. In this chapter, we explain the process by which digital images are created. This knowledge will help us design more secure steganography methods as well as build more sensitive detection schemes (steganalysis).

Fundamentally, there exist two mechanisms through which digital images can be created. They can be synthesized on a computer or acquired through a sensor. Computer-generated images, such as charts, line drawings, diagrams, and other simple graphics generated using drawing tools, could, in principle, be made to hold a small amount of secret data by the selection of colors, object types (line type, fonts), their positions or dimensions, etc. Realistic-looking computer graphics generated from three-dimensional models (or measurements) using specialized methods, such as ray-tracing or radiosity, are typically not very friendly for steganography as they are generated by deterministic algorithms using well-defined rules. In this book, we will mostly deal with images acquired with cameras or scanners because they are far more ubiquitous than computer-generated images and provide a friendlier environment for steganography. As with any categorization, the boundary between the two image types (real versus computer-generated) is blurry. For example, it is not immediately clear how one should classify a digital-camera image processed in Photoshop to make it look like Claude Monet's style of painting or a collage of computer-generated and real images.

This chapter is devoted to digital imaging sensors, which form the heart of most common digital image-acquisition devices today, such as scanners, digital cameras, and digital videocameras. We emphasize that from the point of view of the steganographers (Alice and Bob), we are interested in any imperfections, noise sources, or variations that enter the image-acquisition process because these uncertainties could be utilized for steganography. The steganalyst, Eve, on the other hand, would like to know what patterns, periodicities, or dependences exist among pixels of a digital image because the steganographer may disrupt these structures by embedding, allowing Eve to construct a steganography detector. The reader should read this chapter with both views in mind.

Section 3.1 explains the physical process of registering light by an imaging sensor. After registering the light, the signal created at each pixel needs to be transferred from the sensor for further processing. This topic is described in Section 3.2. The processing of the acquired signal in the camera is the subject of Sections 3.3 and 3.4 that explain how sensors register color through a color filter array and how they process the signal for it to be viewable on a computer monitor. Finally, in Section 3.5 we describe a topic of special interest to steganographers – imperfections and stochastic processes involved in image acquisition.

3.1 CCD and CMOS sensors

The imaging sensor is a silicon semiconductor device that forms an image by capturing photons, converting them into electrons, transferring them, and eventually converting to voltage, which is turned into digital output through quantization in an A/D converter. There exist two competing imaging sensor technologies: the Charge-Coupled Device (CCD) and the Complementary Metal–Oxide–Semiconductor (CMOS). Both were invented in the late 1960s and early 1970s from research originally focused on solid-state memory devices. The CCD, which was proposed and tested at Bell Labs by Boyle and Smith [2, 28], was originally preferred due to its superior image quality when compared with CMOS images. Today, these two technologies coexist because the CMOS technology offers lower cost, faster readout, and greater flexibility that allows on-chip processing. The two technologies use the same process for generating electrons from photons but differ in how the electrons are transferred.

The reason why imaging sensors are made of silicon is its responsiveness to light in the visible spectrum (380 nm to 750 nm). The physics behind the sensors is the photoelectric effect. When photons collide with silicon, electron–hole pairs are created. In theory, one photon of visible light would release exactly one electron. In practice, sensors are not 100% efficient and thus one photon will release less than one electron. And even when an electron is released, it may not be captured and processed by the sensor. Thus, image sensors always have less than 100% quantum efficiency. One important factor that influences quantum efficiency is the quality and purity of the silicon wafer used for manufacturing the sensor. Homogeneous crystal lattices aligned in the same direction will allow the silicon to conduct electrons more efficiently.

Image sensors typically consist of a rectangular array of pixels also called photosites. Each pixel has a photosensitive area (photodetector) that receives photons and converts them to electrons. The electrons accumulate in a potential or electric-charge well (pixel well). The charge depends on the light intensity and the exposure time (also called integration time). The photodetectors are usually equipped with a miniature lens (microlens) whose purpose is to increase the photodetector's sensitivity to light coming from different incident angles.

The number of pixels on the sensor determines the resolution of the image the sensor produces. Smaller pixel size enables higher resolution but may lead to higher noise levels in the resulting image. This is mainly due to decreased

Figure 3.1 Kodak KAI 1100 CM-E CCD sensor with 11 megapixels.

well capacity and lower quantum efficiency and due to the fact that the various electronic components are packed closer to each other and thus influence each other more. Inhomogeneities of the silicon also become more influential with decreased pixel size. Sensors with larger pixels (10 microns or larger) produce images with a higher signal-to-noise ratio (SNR) but are more expensive.

3.2   Charge transfer and readout

To capture an image, both CCD and CMOS sensors perform a sequence of individual tasks. They first absorb photons, generate a charge using the photoelectric phenomenon, then collect the charge, transfer it, and finally convert the charge to a voltage. The CCD and CMOS sensors differ in how they transfer the charge and convert it to voltage.

CCDs transfer charges between pixel wells by shifting them from one row of pixels of the array to the next row, from top to bottom (see Figure 3.2). The transfer happens in a parallel fashion through a vertical shift-register architecture. The charge transfer is "coupled" (hence the term charge-coupled device) in the sense that as one row of charge is moved down, the next row of charge (which is coupled to it) shifts into the vacated pixels. The last row is a horizontal shift register that serially transfers the charge out of the sensor for further processing. Each charge is converted to voltage, amplified, and then sent to an A/D converter, which converts it into a bit string. Even though the charge-transfer process itself is not completely lossless, the Charge-Transfer Efficiency (CTE) in today's CCD sensors is very high (0.99999 or higher) and thus influences the resulting image quality in a negligible manner at least as far as steganographic applications are concerned. The charge conversion and amplification, however, do introduce noise into the signal in the form of readout and amplifier noise. Both noise signals can be well modeled as a sequence of iid Gaussian variables.

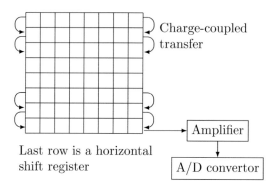

Figure 3.2 Charge transfer in a CCD.

The CCD array can be either a one-dimensional strip of photodetectors or a two-dimensional array. To obtain an image using a linear CCD, it needs to be mechanically moved across the image (as in a flat-bed scanner). Spy satellites use the same mode of acquisition known as Time Delay and Integration (TDI) imaging. The sensor movement introduces additional sources of imperfections and variations that could potentially be used for steganography. Array CCDs use pixels that are usually rectangular. One exception is Fuji's Super CCD that uses a "honeycomb" array of octagonal pixels that maximizes the use of silicon in the sensor.

The charge transfer and readout in a CMOS sensor is very different from a CCD. In a CMOS sensor, it is possible to read out the charge directly at each individual pixel. This offers direct rather than sequential access to image data, which gives CMOS sensors greater flexibility. The individual pixel readout (similar in principle to random-access memory) was made possible using technology called an Active Pixel Sensor (APS). An APS has a readout amplifier transistor at each pixel. Besides signal amplification at each pixel, the sensor can also perform other functions, such as adjustment of gain under low-light conditions, white balance, noise reduction (as already mentioned above), or even A/D conversion. The random access offers the possibility to read out only a targeted area of the sensor (windowing readout) or subsample the image at acquisition. CMOS sensors can easily accommodate processing right on the imaging chip (which CCDs cannot due to their process limitations). The down side of having all this extra on-chip circuitry is an increased level of noise produced by electronic components, such as transistor and diode leakage, cross-talk, and charge injection.

3.3   Color filter array

Because a photodetector registers all incident photons in the visible spectrum, it registers all colors and thus the sensor produces a grayscale image. To produce a

Figure 3.3 The Bayer color filter array.

color image, each photodetector has a filter layer bonded to the silicon that allows only light of a certain color to pass through, absorbing all other wavelengths. The filters are assigned to pixels in a two-dimensional pattern called the Color Filter Array (CFA). Most sensors use the Bayer pattern developed by Kodak in the 1970s. It is obtained by tiling the sensor periodically using 2×2 squares as depicted in Figure 3.3. The Bayer pattern has twice as many green pixels as red or blue, reflecting the fact that human eyes are more sensitive to green than red or blue.

To form a complete digital image, the other two missing colors at each pixel must be obtained by interpolation (also called demosaicking) from neighboring pixels. A very simple (but not particularly good) color-interpolation algorithm computes the missing colors as in Table 3.1. There exist numerous very sophisticated content-adaptive color-interpolation algorithms that perform much better than this simple algorithm.

Color-interpolation algorithms may introduce artifacts into the resulting image, such as aliasing (moiré patterns) or misaligned colors in the neighborhood of edges. These artifacts have a very characteristic structure and thus cannot be used for steganography.

A very important consequence of color interpolation for steganography is that no matter how sophisticated the demosaicking algorithm is, it is a type of filtering, and it will inevitably introduce *dependences among neighboring pixels*. After all, the red color at a green pixel is some function of the neighboring colors, etc. Small modifications of the colors due to data embedding may disrupt these dependences and thus become statistically detectable.

We note that not all cameras use CFAs. Some high-end digital videocameras use a prism that splits the light into three beams and sends them to three separate sensors, each sensor registering one color at every pixel. This approach is not usually taken in compact cameras because it makes the camera more bulky and expensive. There also exist special sensors that can register all three colors at every pixel, capitalizing on the fact that red, green, and blue light penetrates to different depths of the silicon layer. By reading out the charge from each layer separately, rather than from the whole photodetector as in conventional sensors, all three colors are obtained at every pixel. This design is incorporated in the Foveon X3 sensor, for example, in the Sigma SD9 camera.

To capture color, scanners typically use trilinear CCDs consisting of three adjacent linear CCDs, each equipped with a different color filter.

Table 3.1. A simple example of a color-interpolation algorithm for the Bayer pattern.

	Color-interpolation algorithm
At red pixel	$G = (G_N + G_E + G_S + G_W)/4$
	$B = (B_{NW} + B_{NE} + B_{SE} + B_{SW})/4$
At green pixel	$R = (R_E + R_W)/2$
	$B = (B_N + B_S)/2$
At blue pixel	$R = (R_{NW} + R_{NE} + R_{SE} + R_{SW})/4$
	$G = (G_N + G_E + G_S + G_W)/4$

3.4   In-camera processing

The signal registered at each photodetector due to incoming photons goes through a complicated chain of processing before the actual digital image in some viewable image format is written to the camera memory device. We already know that the photons create a charge, which is subsequently converted to voltage, which is amplified and converted to a digital form in an A/D converter. Some cameras are able to export this raw signal onto the memory card to give the user more control over the final processing stage, which is done off-line on a computer, usually using manufacturer-supplied software. For example, Canon and Nikon use the CRW and NEF raw formats, respectively. The raw output is always a grayscale image, usually sampled at higher bit rates, such as 10–14 bits per pixel, or higher for professional cameras, such as those used in astronomy. One can view this raw sensor output as a digital equivalent of a negative. In fact, Adobe has been trying to standardize the format for the raw sensor output by including a Photoshop plug-in that can handle Adobe's DNG (Digital NeGative) format.

Most consumer-end digital cameras do not output the raw sensor signal and instead perform a host of various processing operations to obtain a visually pleasing image that is stored in some common format, such as TIFF or JPEG. Even though the type of processing may greatly vary among cameras, most cameras apply white balance, demosaicking, color correction, gamma correction, denoising, and filtering (e.g., sharpening). The white balance is a multiplicative adjustment designed to correct for differences in the spectrum of the ambient light,

$$R \leftarrow g_R R, \qquad (3.1)$$

$$G \leftarrow g_G G, \qquad (3.2)$$

$$B \leftarrow g_B B, \qquad (3.3)$$

where g_R, g_G, g_B are the gains for each color channel.

After demosaicking the signal using color-interpolation algorithms designed for the CFA, the signal is further processed using so-called color correction, whose purpose is to adjust the amounts of red, green, and blue so that the image is correctly displayed on a computer monitor. Color correction is a linear transformation

$$
\begin{pmatrix} R \\ G \\ B \end{pmatrix} \leftarrow \begin{pmatrix} c_{11} & c_{12} & c_{13} \\ c_{21} & c_{22} & c_{23} \\ c_{31} & c_{32} & c_{33} \end{pmatrix} \begin{pmatrix} R \\ G \\ B \end{pmatrix}. \tag{3.4}
$$

Because the CCD response is linear in the sense that the charge is proportional to the light intensity[1] (number of photons registered), it is incompatible with the human visual system, which has a logarithmic response to light. Thus, the signal must be corrected in a non-linear fashion, usually using gamma correction,

$$
R \leftarrow R^{\gamma}, \tag{3.5}
$$
$$
G \leftarrow G^{\gamma}, \tag{3.6}
$$
$$
B \leftarrow B^{\gamma}, \tag{3.7}
$$

where typically $\gamma = 2.2$.

These basic processing steps are usually supplied with other actions, such as denoising, defects removal, sharpening, etc. Finally, the signal is quantized to either a true-color 24-bit image in TIFF or JPEG format. For applications in steganography, one should realize that denoising, filtering, and JPEG compression introduce additional local dependences whose presence is quite fundamental for steganography and steganalysis. If the data-embedding algorithm does not preserve the nature of these dependences, the embedding changes will become statistically detectable.

3.5   Noise

There are numerous noise sources that influence the resulting image produced by the sensor. Some are truly random processes, such as the shot noise caused by quantum properties of light, while other sources are systematic in the sense that they would be the same if we were to take the same image twice. It should be clear that while systematic imperfections cannot be used for steganography, random components can and are thus very fundamental for our considerations. If we knew the random noise component exactly, in principle we could subtract it from the image and replace it with an artificially created noise signal with the same statistical properties that would carry a secret message (Chapter 7). We now review each noise source individually, pointing out their role in steganography.

Dark current is the image one would obtain when taking a picture in complete darkness. The main factors contributing to dark current are impurities in the silicon wafer or imperfections in the silicon crystal lattice. Heat also leads to

---

[1] Some sensors use Anti-Blooming Gates (ABGs) when the charge exceeds 50% of well capacity to prevent blooming. This causes the photodetector response to become non-linear at high intensity.

Figure 3.4 A magnified portion of a dark frame.

dark noise (with the noise energy doubling with an increase in temperature of 6–8°C). The thermal noise can be suppressed by cooling the sensor, which is typically done in astronomy. The number of thermal electrons is also proportional to the exposure time. Some consumer cameras take a dark frame with a closed shutter when the camera is powered up and subtract it from every image the camera takes. One interesting example worth mentioning here is the KAI 2020 chip developed by Kodak. This chip calculates the dark current (sensor output when not exposed to light) and subtracts it from the illuminated image. This method is frequently used in CMOS sensors to suppress noise as well as other artifacts.

Figure 3.4 shows an example of dark current on the raw sensor output obtained using a 60-second exposure with the SBIG STL-1301E camera equipped with a 1280×1024 CCD sensor cooled to -15°C.

Photo-Response Non-Uniformity (PRNU). Due to imperfections in the manufacturing process and the silicon wafer, the dimensions as well as the quantum efficiency of each pixel may slightly vary. Additional imperfections may be introduced by anomalies in the CFA and microlenses. Therefore, even when taking a picture of an absolutely uniformly illuminated scene (e.g., the blue sky), the sensor output will be slightly non-uniform even if we eliminated all other sources of noise. This non-uniformity may have components of low spatial frequencies, such as a gradient or darkening at image corners, circular or irregular blobs due to dirty optics or dust on the protective glass of the sensor (Figure 3.5), and a stochastic component resembling white noise due to anomalies in the CFA and varying quantum efficiency among pixels. The PRNU is a systematic artifact in the sense that two images of exactly the same scene would exhibit approximately the same PRNU artifacts. (This is why it is sometimes called the fixed pattern noise.) Thus, the PRNU does not increase the amount of information we can embed in an image because it is systematic and not random.

Figure 3.5 Dust particles on the sensor protective glass show up as fuzzy dark spots (circled).

Figure 3.6 Magnified portion of the stochastic component of PRNU in the red channel for a Canon G2 camera. For display purposes, the numerical values were scaled to the range $[0, 255]$ and rounded to integers to form a viewable grayscale image.

It is worth mentioning that the PRNU can be used as a sensor fingerprint for matching an image to the camera that took it [43] in the same way as bullet scratches can be used to match a bullet to the gun barrel that fired it. Figure 3.6 shows a magnified portion of the PRNU in the red channel from a Canon G2 camera. To isolate the PRNU and suppress random noise sources, the pattern was obtained by averaging the noise residuals of 300 images acquired in the TIFF format. The noise residual for each image was obtained by taking the difference between the image and its denoised version after applying a denoising filter.

Shot noise is the result of the quantum nature of light and makes the number of electrons released at each photodetector essentially a random variable due to the random variations of photon arrivals. Shot noise is a fundamental limitation that cannot be circumvented. The presence of random components during image acquisition has some fundamental consequences for steganography that we elaborate upon in Section 6.2.1.

The number of photons captured by the photodetector during the exposure of Δt seconds is a random variable ξ that follows the Poisson distribution (law of rare events)

$$\Pr\{\xi = k\} = \frac{e^{-\lambda \Delta t}(\lambda \Delta t)^k}{k!} = \mathbf{p}[k] \tag{3.8}$$

with mean value and variance (see Exercise 3.2)

$$E[\xi] = \sum_{k \geq 0} \mathbf{p}[k]k = \lambda \Delta t, \tag{3.9}$$

$$\mathrm{Var}[\xi] = \sum_{k \geq 0} \mathbf{p}[k]k^2 - (\lambda \Delta t)^2 = \lambda \Delta t, \tag{3.10}$$

where $\lambda > 0$ is the expected number of photons captured in a unit time interval. With increased number of photons, $\lambda \Delta t$, the relative (percentual) variations of ξ decrease because the ratio

$$\frac{E[\xi]}{\sqrt{\mathrm{Var}[\xi]}} = \frac{1}{\sqrt{\lambda \Delta t}} \to 0. \tag{3.11}$$

This means that the shot noise decreases with increased pixel size and with longer exposure times. Also, we note that for large $\lambda \Delta t$ the Poisson distribution is well approximated with a Gaussian distribution $N(\lambda \Delta t, \lambda \Delta t)$.

Charge-transfer efficiency. The transfer of charge to the output amplifier in a CCD sensor is not a completely lossless phenomenon. This results in an additional source of variations in the final collected charge. The charge-transfer efficiency in the latest CCD designs is very close to 1 (e.g., 0.99999), which means that it is entirely reasonable to simply neglect this effect for applications in steganography and assume that the charge-transfer efficiency is 1.

Amplifier noise. The charge collected at each pixel is amplified using an on-chip amplifier. This can be done in the last row of photodetectors for a CCD sensor or at each pixel in a CMOS sensor. The amplifier noise is well modeled as a Gaussian random variable with zero mean. It dominates the shot noise at low light conditions.

Quantization noise. The amplified signal on the sensor output is further transformed through a complicated chain of processing, such as demosaicking, gamma correction, low-pass filtering to prevent aliasing during subsequent resampling, etc. Finally, the image data can be converted to one of the image formats, such as TIFF or JPEG, which introduces quantization errors.

The most important fact we need to realize for steganography is that the in-camera processing will introduce local dependences among neighboring pixels. This is what makes steganography in digital images very challenging because the exact nature of these dependences is quite complex and different for each camera.

Consequently, it is not clear how to perform embedding so that the embedding modifications stay compatible with the processing (and are thus undetectable).

Defective pixels. Some pixels or their associated circuitry may be faulty and always generate the same signal. Pixels that constantly output the highest signal values are called hot, while pixels that always output a zero signal are called dead pixels. Defective pixels can be quite visible in images under a close inspection (see Figure 3.7). Some cameras may attempt to identify such pixels and eliminate their effect from the final image (the color at a defective pixel is replaced with an interpolated value). It is also possible to upgrade the firmware with the information about defective pixels and thus eliminate the defects from images at acquisition.

Figure 3.7 Hot pixel in an image (top) and its closeup (bottom). Because the hot pixel had a red filter in front of it, the hot pixel appears red. Note the spread of the red color due to demosaicking and other in-camera processing. For a color version of this figure, see Plate 3.

Blooming, cross-talk. There are other types of imperfections whose nature is not stochastic (noise-like) and are thus not interesting for steganography. We include them here for completeness. When a photodetector is illuminated with sufficient intensity, the charge generated at that site may overflow from the potential well and spread into the neighboring sites (see Figure 3.8). This digital equivalent of overexposure is called blooming. Cross-talk pertains to the situation when a photon striking a photodetector at an angle passes through the CFA and eventually hits a neighboring photodetector. This undesirable interference can be suppressed by optically shielding the photodetectors.

Figure 3.8 Blooming artifacts due to overexposure. Also, notice the green artifact in the lower right corner caused by multiple reflections of light in camera optics. A color version of this figure is displayed in Plate 4.

Summary

- The photoelectric effect in silicon is the main physical principle on which imaging sensors are based.
- There exist two competing sensor technologies – CCD and CMOS.
- Each imaging sensor consists of millions of individual photosensitive detectors (photodiodes, photodetectors, or pixels).
- The light creates a charge at each pixel that is transferred, converted to voltage, amplified, and quantized. CCD and CMOS differ mainly in how the charge is transferred. In a CCD, the charge is transferred out of the sensor in a sequential manner, while CMOS sensors are capable of transferring the charge in a parallel fashion.
- The quantized signal generated by the sensor is further processed through a complex chain of processing that involves white-balance (gain) adjustment, demosaicking, color correction, denoising, filtering, gamma correction, and finally conversion to some common image format (JPEG, TIFF).
- The processing introduces local dependences into the image.
- The image-acquisition process is influenced by many sources of imprecision and noise due to physical properties of light (shot noise), slight differences in pixel dimensions and silicon inhomogeneities (pixel-to-pixel non-uniformity), optics (vignetting, chromatic aberration), charge transfer and readout (read-

out noise, amplifier noise, reset noise), quantization noise, pixel defects (hot and dead pixels), and defects due to charge overflow (blooming).

- Some sources of imprecision and noise are random in nature (e.g., shot noise, readout noise), while others are systematic components that repeat from image to image.
- The presence of truly random noise components in images acquired using imaging sensors is quite fundamental and has direct implications for steganography (Section 6.2.1).

Exercises

3.1 [Law of rare events] Assume that photons arrive sequentially in time in an independent fashion with a constant average rate of arrival. Let λ be the probability that one photon arrives in a unit time interval. Show that the probability that k events occur in a unit time interval is

$$\Pr\{k \text{ events}\} = \frac{e^{-\lambda}\lambda^k}{k!}. \tag{3.12}$$

Hint: Divide the unit time interval into n subintervals of length $1/n$. Assuming that the probability that two events occur in one subinterval is negligible, the probability that exactly k photons arrive can be obtained from the binomial distribution $Bi(n, \lambda/n)$

$$\Pr\{k \text{ events}\} = \binom{n}{k} \left(\frac{\lambda}{n}\right)^k \left(1 - \frac{\lambda}{n}\right)^{n-k}. \tag{3.13}$$

The result is then obtained by taking the limit $n \to \infty$ for a fixed k.

3.2 [Poisson random variable] Show that the mean and variance of a Poisson random variable ξ

$$\Pr\{\xi = k\} = \frac{e^{-\lambda\Delta t}(\lambda\Delta t)^k}{k!} = \mathbf{p}[k] \tag{3.14}$$

are

$$E[\xi] = \sum_{k \geq 0} \mathbf{p}[k]k = \lambda\Delta t, \tag{3.15}$$

$$\text{Var}[\xi] = \sum_{k \geq 0} \mathbf{p}[k]k^2 - (\lambda\Delta t)^2 = \lambda\Delta t. \tag{3.16}$$

3.3 [Analysis of sensor imperfections] Take your digital camera and adjust its settings to the highest resolution and highest JPEG quality. Then take $N \geq 10$ images of blue sky, $\mathbf{I}[i, j; k]$, $i = 1, \ldots, m$, $j = 1, \ldots, n$, $k = 1, \ldots, N$, where indices i and j determine the pixel at the position (i, j) in the kth image. You might want to zoom in but not to the range of a digital zoom if your camera has this capability. Make sure that the images do not contain stray objects, such as airplanes, birds, or stars/planets. Compute the sample variance of all pixels across all N images

$$\hat{\sigma}[i,j] = \frac{1}{N-1} \sum_{k=1}^{N} \left(\mathbf{I}[i,j;k] - \bar{\mathbf{I}}[i,j] \right)^2 , \tag{3.17}$$

where

$$\bar{\mathbf{I}}[i,j] = \frac{1}{N} \sum_{k=1}^{N} \mathbf{I}[i,j;k] \tag{3.18}$$

is the sample mean at pixel (i,j). Plot the histogram of the sample variance. The variations at each pixel are due to combined random noise sources, such as the shot noise or readout noise.

Extract the noise residual $\mathbf{W}$ from all images using the Wiener filter

$$\mathbf{W}[\cdot,\cdot;k] = \mathbf{I}[\cdot,\cdot;k] - W(\mathbf{I}[\cdot,\cdot;k]). \tag{3.19}$$

In Matlab, the Wiener filter W is accessed using the routine `wiener2.m`. Then, average all N noise residuals

$$\mathbf{K}[i,j] = \frac{1}{N} \sum_{k=1}^{N} \mathbf{W}[i,j;k]. \tag{3.20}$$

The averaging suppresses random noise components in $\mathbf{K}$, while the systematic components become more pronounced (pattern noise, PRNU, defective pixels, dark current). First, find the location of outlier values in $\mathbf{K}$. They will likely correspond to defective pixels. Use the routine `mat2gray.m` to convert $\mathbf{K}$ to the range $[0,1]$ and view the result using `imshow.m` (after casting to `uint8`). Zoom in so that you can discern individual pixels and pan over the pattern. Do you see any regular patterns? Is the pattern $\mathbf{K}$ random or does it exhibit local stochastic structures? Do you see any artifacts around the border of $\mathbf{K}$? Focus on the neighborhood of outlier pixels and describe their appearance.

4 Steganographic channel

The main goal of steganography is to communicate secret messages without making it apparent that a secret is being communicated. This can be achieved by hiding messages in ordinary-looking objects, which are then sent in an overt manner through some communication channel. In this chapter, we look at the individual elements that define steganographic communication.

Before Alice and Bob can start communicating secretly, they must agree on some basic communication protocol they will follow in the future. In particular, they need to select the type of cover objects they will use for sending secrets. Second, they need to design the message-hiding and message-extraction algorithms. For increased security, the prisoners should make both algorithms dependent on a secret key so that no one else besides them will be able to read their messages. Besides the type of covers and the inner workings of the steganographic algorithm, Eve's ability to detect that the prisoners are communicating secretly will also depend on the size of the messages that Alice and Bob will communicate. Finally, the prisoners will send their messages through a channel that is under the control of the warden, who may or may not interfere with the communication.

We recognize the following five basic elements of every steganographic channel[1] (see Figure 4.1):

- Source of covers,
- Data-embedding and -extraction algorithms,
- Source of stego keys driving the embedding/extraction algorithms,
- Source of messages,
- Channel used to exchange data between Alice and Bob.

Because in this book the covers are digital images, the cover-source attributes include the image format, origin, resolution, typical content type, etc. The cover-source properties are determined by objects that Alice and Bob would be exchanging if they were not secretly communicating. Modern steganography typically makes some fundamental assumption about the cover source that enables formal analysis. For example, considering the cover source as a random variable allows analysis of steganography using information theory (this topic is elabo-

---

[1] In this book, we sometimes also use the shorter term stegosystem.

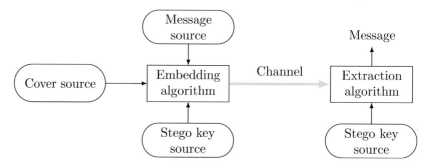

Figure 4.1 Elements of the steganographic channel.

rated upon in Section 6.1). Alternatively, the cover source can be interpreted as an oracle that can be queried. This leads to the complexity-theoretic study of steganography explained in Section 6.4.

The embedding algorithm is a procedure through which the sender determines an image that communicates the required secret message. The procedure may depend on a secret shared between Alice and Bob called the stego key. This key is needed to correctly extract a secret message from the stego image. For example, Alice can embed her secret bit stream as the least significant bits of pixels chosen along a non-intersecting pseudo-randomly generated path through the image (determined by the stego key).

The protocol that Alice and Bob use to select the stego keys is usually modeled with a random variable on the space of all keys. For example, a reasonable strategy is to select the stego key randomly (with uniform distribution) from the set of all possible stego keys.

The message source has a major influence on the security of the steganographic channel. Imagine two extreme situations. On the one hand, Alice and Bob always communicate only a short message, say 16 bits, in every stego image. On the other hand, Alice and Bob have a need to communicate large messages and frequently embed as many bits into the image as the embedding algorithm allows. Intuitively, the prisoners are at much greater risk in the latter case. The distribution of messages can be modeled using a random variable on the space of all possible messages.

The actual communication channel used to send the images is assumed to be monitored by a warden (Eve). Eve can assume three different roles. Positioning herself into the role of a passive observer, she merely inspects the traffic and does not interfere with the communication itself. This is called the passive-warden scenario. Alternatively, Eve may suspect that Alice and Bob might use steganography and she can preventively attempt to disrupt the steganographic channel by intentionally distorting the images exchanged by Alice and Bob. For example, she may compress the image using JPEG, resize or crop the image, apply a slight amount of filtering, etc. Unless Alice and Bob use steganography that is resistant (robust) to such processing, the steganographic channel would be broken by Eve's actions. This is called the active-warden scenario. Finally, Eve can be even more devious and she may try to guess the steganographic method that Alice and Bob use and attempt to impersonate Alice or Bob or otherwise intervene to confuse the communicating parties. There is a difference between this so-called malicious warden and the active warden. The actions of an active warden are aimed at making steganography impossible for Alice and Bob, while a malicious warden does not necessarily intend to entirely disrupt the stego channel but rather use it to her advantage to determine whether or not steganography is taking place. In this book, we focus mainly on the passive-warden scenario in which the communication channel is assumed error-free.[2] This is the case of communication via standard Internet protocols, where error-correction and authentication tools guarantee error-free data transmission.

The discussion above underlines the need to view steganographic communication in a wider context. When designing a steganographic scheme, the prisoners have a choice in selecting the basic elements to form their communication scheme. As will be seen in the next chapter, it is not very difficult to write a computer program that hides a large amount of data in an image. On the other hand, writing a program that does so without introducing any detectable artifacts is quite hard.

> *The problem of steganography can thus be formulated as finding embedding and extraction algorithms for a given cover source that enable communication of reasonably large messages without introducing any embedding artifacts that could be detected by the warden. In other words, the goal is to embed secret messages undetectably.*

The concept of undetectability will be made rather precise in Chapter 6 that deals with a formal definition of steganographic security. Until then, we will assume some intuitive meaning of these concepts.

The embedding and extraction algorithms are obviously the most important parts of any stegosystem. Steganographic algorithms can utilize three different fundamental architectures that determine the internal mechanism of the embedding and extraction algorithms. Alice can "perform embedding" by choosing a cover image that already has the desired message hidden inside. This is called steganography by cover selection and is the subject of Section 4.1. Alternatively, Alice can construct an object that conveys her message. This strategy is called steganography by cover synthesis and is elaborated upon in Section 4.2. The third option, which is the most practical for communicating large amounts of data, is steganography by cover modification, explained in Section 4.3. This is the mainstream approach to steganography that is also the central focus of this book.

---

[2] The active- and malicious-warden scenarios are discussed in [14, 53, 66, 72, 116, 178] and the references therein.

4.1   Steganography by cover selection

In steganography by cover selection, Alice has available a fixed database of images from which she selects one that communicates the desired message. For example, one bit of information could be sent by choosing a picture in a landscape or portrait orientation. Alternatively, the presence of an animal in the picture could have hidden meaning, such as "attack tomorrow," etc. The embedding algorithm can work simply by randomly drawing images from the database till an image is found that communicates the desired message. The stego key here is essentially the set of rules that tell Alice and Bob how to interpret the images.

An important case of steganography by cover selection involves message-digest (hash) functions. Alice selects an image from the database and applies to it a message-digest function (this function may depend on a key and must be shared with Bob). If the digest matches the desired message bit stream, the image is forwarded to Bob, otherwise Alice selects a different image till she obtains a match. The expected number of tries needed to obtain a match depends exponentially on the length of the digest and thus quickly becomes impractically large. Upon receiving an image, Bob simply extracts the digest to read the message. The advantage of this approach is that the cover is always "100% natural" because it is a real image that was not modified in any way. An obvious disadvantage is an impractically low payload.

Under close inspection, one can realize one potential problem with steganography by cover selection that prevents us from proclaiming it "truly undetectable." To make the issue more apparent, we use a very simple digest function formed by the least significant bits of the first three pixels in the image,

$$h(\mathbf{x}) = \{\mathbf{x}[1] \bmod 2, \mathbf{x}[2] \bmod 2, \mathbf{x}[3] \bmod 2\}. \tag{4.1}$$

Note that the digest consists of three bits. The problem may arise when Alice decides to use this technique to communicate not just once but repetitively. If Alice is equally likely to send each triple of bits out of all eight possible triples of bits (which is a reasonable assumption if she is sending parts of an encrypted document, for example), the stego images sent by her will equally likely produce any of the 8-bit triples as the digest. How do we know, however, that in natural images the distribution of LSBs of the first three pixels in the upper left corner follows this distribution? Most likely, it does not because these pixels are likely to belong to a piece of sky and thus their values are far from being independent. Note that this problem arises only because we consider multiple uses of the steganographic channel and allow Eve to inspect all transmissions from Alice rather than considering one image at a time. This observation is, in fact, quite fundamental and will lead us to a formal definition of steganographic security using information theory in Section 6.1.

The reader might suspect that had Alice used a better digest, such as the first three bits of a cryptographic hash MD5 or SHA applied to the whole image rather than just three pixels, intuitively, the above issue would be largely eliminated. We revisit this simple thought experiment in Section 6.4.1 dealing with the complexity-theoretic definition of steganography.

4.2   Steganography by cover synthesis

In steganography by cover synthesis, Alice creates the cover so that it conveys the desired message. There were some speculations in the press that Bin Laden's videos may contain hidden messages communicated, for example, by his clothes, the position of his rifle, or the choice of words in his speech [203]. This would be an example of steganography by cover synthesis.

Another quite interesting example that involves text as the cover rather than digital images is encoding messages so that they resemble spam. Here, the steganographer uses the fact that spam often contains incoherent or unusual wording, which can be used to encode a message. The program SpamMimic (www.spammimic.com) uses mimic functions [238, 239] to hide messages in artificially created spam-looking stego text.

Steganography by cover synthesis could be combined with steganography by cover selection to alleviate the exponential complexity of embedding by hashing [188]. Let us assume that we can take a large number of images of exactly the same scene with the same digital camera. For example, we could fix the camera on a tripod and take multiple images under constant light conditions. Let us assume that the images are 8-bit grayscale with $\mathbf{x}_j[i]$ standing for the intensity of the ith pixel in the jth image, $i = 1, \ldots, n$, $j = 1, \ldots, K$. We now make the key observation that the light intensity values at a fixed pixel, i, when viewed across all images, j, will slightly vary due to random noise sources present in images, such as shot noise, readout noise, and noise due to on-chip electronics (see Figure 4.2 and the discussions in Section 3.5).

Alice will use a cryptographic hash function modified to return 4 bits when applied to 16 pixels. For example, she could use the last 4 bits from the MD5 hash [207]. Here, the values 4 and 16 are chosen rather arbitrarily and other values can certainly be used. In order to embed her message, Alice divides every image into disjoint blocks of 4×4 pixels and assembles a new image in a block-by-block fashion so that each 4×4 block conveys 4 message bits. To embed the first 4 bits in the first 4×4 block of pixels, Alice searches through the hashes $h(\mathbf{x}_j[1], \ldots, \mathbf{x}_j[16])$, $j \in \{1, 2, \ldots, K\}$ till she finds a match between the hash of the first 16 pixels and the message, which will happen for image number j_1. Then, she moves to the next block and finds $j_2 \in \{1, \ldots, K\}$ so that $h(\mathbf{x}_j[17], \ldots, \mathbf{x}_j[32])$ matches the next 4 message bits, etc. The final stego image $\mathbf{y}$ will be a mosaic assembled from blocks from different images $\mathbf{y} = (\mathbf{x}_{j_1}[1], \ldots, \mathbf{x}_{j_1}[16], \mathbf{x}_{j_2}[17], \ldots, \mathbf{x}_{j_2}[32], \mathbf{x}_{j_3}[33], \ldots)$. The probability of finding a match in one particular block among all K images is $1 - (1 - 1/16)^K$. The probability of being able to embed the whole message, which consists of

Figure 4.2 The same 16×16 pixel block from four (uncompressed) TIFF images of blue sky taken with the same camera within a short time interval. First note that the blocks are not uniform even though the scene is perfectly uniform. Second, the blocks appear different in every picture due to random noise sources.

$n/16$ groups of 4 bits is thus $(1 - (1 - 1/16)^K)^{n/16}$. For increasing number of images, K, this probability can be made arbitrarily close to 1. For example, for a square $n = 512 \times 512$ image and only $K = 400$ images, this probability is $0.9999998993\ldots$.

This method is more a theoretical construct rather than a practical steganographic technique. This is because it is in fact difficult to obtain multiple samplings of one signal (identical photographs of one scene accurate to within a fraction of a pixel width). Small variations in the exposure time due to the mechanics of the shutter and other mechanical vibrations will necessarily limit the alignment between different pictures.

The scheme makes an implicit assumption that pixels at the block boundaries are independent, or, more accurately, that the random components of the images are independent. As we already know from Chapter 3, various types of in-camera processing, such as color interpolation or filtering, create dependences among neighboring pixels and thus among their noise components. Consequently, Eve may start constructing an attack by inspecting the dependences among pixels spanning the block boundaries and compare them with dependences among neighboring pixels from the interior of the blocks.

A qualitatively different realization of steganography by cover synthesis is called data masking. The idea is that all steganalysis tools are in the end automated and work by extracting a set of numerical features from the image that is later analyzed for compatibility with features typically obtained from cover images (see Chapter 10). Thus, all that Alice needs to achieve to evade detection is to make the stego image look like a cover in the feature space. The stego "image" does not have to look like a natural image because, as long as its features are compatible with features of natural images, it should pass through the detector. A practical data-masking method was constructed in [200] by applying a time-varying inverse Wiener filter shaping every 1024 message bits to match a reference audio frame.

4.3   Steganography by cover modification

In this book, we deal mainly with steganography by cover modification because it is by far the most studied steganography paradigm today. Postponing specific examples of steganographic schemes to Chapters 5–7, we now introduce basic definitions and concepts.

Alice starts with a cover image and makes modifications to it in order to embed secret data. Alice and Bob work with the set of all possible covers and the sets of keys and messages that may, in the most general case, depend on each cover:

$$\mathcal{C} \ldots \text{ set of cover objects } \mathbf{x} \in \mathcal{C}, \tag{4.2}$$

$$\mathcal{K}(\mathbf{x}) \ldots \text{ set of all stego keys for } \mathbf{x}, \tag{4.3}$$

$$\mathcal{M}(\mathbf{x}) \ldots \text{ set of all messages that can be communicated in } \mathbf{x}. \tag{4.4}$$

Following the diagram displayed in Figure 4.3, a steganographic scheme is a pair of embedding and extraction functions Emb and Ext,

$$\text{Emb}: \quad \mathcal{C} \times \mathcal{K} \times \mathcal{M} \to \mathcal{C}, \tag{4.5}$$

$$\text{Ext}: \quad \mathcal{C} \times \mathcal{K} \to \mathcal{M}, \tag{4.6}$$

such that for all $\mathbf{x} \in \mathcal{C}$, and all $\mathbf{k} \in \mathcal{K}(\mathbf{x})$, $\mathbf{m} \in \mathcal{M}(\mathbf{x})$,

$$\text{Ext}\,(\text{Emb}(\mathbf{x}, \mathbf{k}, \mathbf{m}), \mathbf{k}) = \mathbf{m}. \tag{4.7}$$

In other words, Alice can take any cover $\mathbf{x} \in \mathcal{C}$ and embed in it any message $\mathbf{m} \in \mathcal{M}(\mathbf{x})$ using any key $\mathbf{k} \in \mathcal{K}(\mathbf{x})$, obtaining the stego image $\mathbf{y} = \text{Emb}(\mathbf{x}, \mathbf{k}, \mathbf{m})$. The number of messages that can be communicated in a specific cover $\mathbf{x}$ depends on the steganographic scheme and it may also depend on the cover itself. For example, if $\mathcal{C}$ is the set of all 512×512 grayscale images and Alice embeds one message bit per pixel, then $\mathcal{M} = \{0, 1\}^{512 \times 512}$ and $|\mathcal{M}(\mathbf{x})| = 2^{512 \times 512}$ for all $\mathbf{x} \in \mathcal{C}$. On the other hand, if $\mathcal{C}$ is the set of all 512×512 grayscale JPEG images with quality factor $q_\mathrm{f} = 75$ and Alice embeds one bit per each non-zero quantized DCT coefficient, the number of messages that can be embedded in a specific cover depends on the cover itself because the number of non-zero DCT coefficients in a JPEG file depends on the image content.

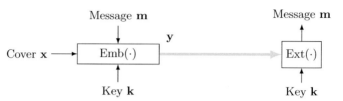

Figure 4.3 Steganography by cover modification (passive-warden case).

Thus, we define the embedding capacity (payload) of cover $\mathbf{x}$ in bits as

$$\log_2 |\mathcal{M}(\mathbf{x})|, \tag{4.8}$$

and the relative embedding capacity is

$$\frac{\log_2 |\mathcal{M}(\mathbf{x})|}{n}, \tag{4.9}$$

where n is the number of elements in $\mathbf{x}$, such as the number of pixels or non-zero DCT coefficients. For raster image formats, relative capacity is often expressed in bpp (bits per pixel). For JPEG images, we use the unit bpnc (bits per non-zero DCT coefficient). By taking expectations in (4.8) and (4.9) with respect to some distribution of covers $\mathbf{x}$, we can speak of the average embedding and relative embedding capacity.

Perhaps the most fundamental concept in steganography is the steganographic capacity defined as the maximal number of bits that can be embedded without introducing statistically detectable artifacts. For now, we stay with this rather vague definition, leaving the precise definition of this advanced concept to Chapter 13. Typically, the steganographic capacity is much smaller than the embedding capacity.

Embedding algorithms of many steganographic schemes require a representation of cover and stego images using bits or, more generally, symbols from some alphabet $\mathcal{A}$ using a symbol-assignment function π,

$$\pi : \mathcal{X} \to \mathcal{A}, \tag{4.10}$$

where $\mathcal{X}$ is the range of individual cover elements, such as pixels or DCT coefficients. One frequently used bit-assignment (parity) function is the least significant bit

$$\mathrm{LSB}(x) = x \bmod 2. \tag{4.11}$$

Throughout this book, the reader will learn about other symbol-assignment functions.

If the embedding algorithm is designed to avoid making embedding changes to certain areas of the cover image, we speak of adaptive steganography. For example, we may wish to skip flat areas in the image and concentrate embedding changes to textured regions. The subset of the image where embedding changes are allowed is called the selection channel. Another example of a selection channel is when the message bits are embedded along a pseudo-random path through the image generated from the stego key. In general, it is in the interest of both Alice and Bob not to reveal any information or as little as possible about the selection channel as this knowledge can help an attacker. If the selection channel is available to Alice but not to Bob, it is called a non-shared selection channel.

Steganography by cover modification introduces distortion into the cover. The distortion is typically measured with a mapping $d(\mathbf{x}, \mathbf{y})$, $d : \mathcal{C} \times \mathcal{C} \to [0, \infty)$. One commonly used family of distortion measures is parametrized by $\gamma \geq 1$,

$$d_\gamma(\mathbf{x}, \mathbf{y}) = \sum_{i=1}^{n} |\mathbf{x}[i] - \mathbf{y}[i]|^\gamma. \tag{4.12}$$

For $\gamma = 1$,

$$d_1(\mathbf{x}, \mathbf{y}) = \sum_{i=1}^{n} |\mathbf{x}[i] - \mathbf{y}[i]| \tag{4.13}$$

is the L_1 norm between the vectors $\mathbf{x}$ and $\mathbf{y}$, while

$$d_2(\mathbf{x}, \mathbf{y}) = \sum_{i=1}^{n} |\mathbf{x}[i] - \mathbf{y}[i]|^2 \tag{4.14}$$

is the energy of embedding changes.

The measure

$$\vartheta(\mathbf{x}, \mathbf{y}) = \sum_{i=1}^{n} 1 - \delta(\mathbf{x}[i] - \mathbf{y}[i]) \tag{4.15}$$

is the number of embedding changes, where δ is the Kronecker delta (2.23). Note that if the amplitude of embedding changes is $|\mathbf{x}[i] - \mathbf{y}[i]| = 1$, d_γ and ϑ coincide for all γ.

The distortion measures above are absolute in the sense that they measure the total distortion. Often, it is useful to express distortion per cover element (e.g., per pixel), in which case we speak of relative distortion

$$\frac{d(\mathbf{x}, \mathbf{y})}{n}. \tag{4.16}$$

The quantity

$$\beta = \frac{\vartheta(\mathbf{x}, \mathbf{y})}{n} \tag{4.17}$$

is called the change rate and will typically be denoted β.

Two popular relative measures are the Mean-Square Error (MSE)

$$\text{MSE} = \frac{d_2(\mathbf{x}, \mathbf{y})}{n} = \frac{1}{n} \sum_{i=1}^{n} |\mathbf{x}[i] - \mathbf{y}[i]|^2, \tag{4.18}$$

and the Peak Signal-to-Noise Ratio (PSNR)

$$\text{PSNR} = 10 \log_{10} \frac{x_{\max}^2}{\text{MSE}}, \tag{4.19}$$

where x_{max} is the maximum value that $\mathbf{x}[i]$ can attain. For example, for 8-bit grayscale images, $x_{\mathrm{max}} = 255$.

The average embedding distortion is the expected value $E\left[d(\mathbf{x}, \mathbf{y})\right]$ taken over all $\mathbf{x} \in \mathcal{C}, \mathbf{k} \in \mathcal{K}, \mathbf{m} \in \mathcal{M}$ selected according to some fixed probability distributions from their corresponding sets.

A very important characteristic of a stegosystem that has a major influence on its security is the embedding efficiency. We define it here rather informally as the average number of bits embedded per average unit distortion

$$e = \frac{E_{\mathbf{x}}[\log_2 |\mathcal{M}(\mathbf{x})|]}{E_{\mathbf{x},\mathbf{m}}\left[d(\mathbf{x}, \mathbf{y})\right]}. \tag{4.20}$$

The complexity of computing the embedding efficiency depends on the steganographic scheme, on its inner mechanism, and on the cover source. For the simplest embedding schemes, such as LSB embedding or ± 1 embedding in the spatial domain (see Sections 5.1 and 7.3.1), the embedding efficiency can be easily evaluated analytically. On the other hand, for some steganographic methods the maximal payload as well as the distortion may be a rather complex function of the cover content or even the stego key, which makes computing the embedding efficiency analytically virtually impossible due to our imprecise knowledge of the cover source. In such cases, the embedding efficiency is determined experimentally. Finally, we note that the embedding efficiency depends on the distortion measure d.

In the next chapter, we give examples of a few simple steganographic schemes that embed messages by slightly modifying the pixels, pointers, or DCT coefficients in the image file. This will illustrate the concepts defined above as well as provide motivation for the reader.

Summary

- A steganographic channel consists of the source of covers, the message source, embedding and extraction algorithms, the source of stego keys, and the communication channel.
- If the channel is error-free, we speak of a passive warden. An active warden intentionally distorts the communication with the hope of preventing usage of steganography. A malicious warden tries to trick the communicating parties, e.g., by impersonation.
- There exist three types of embedding algorithms: steganography by cover selection, cover synthesis, and cover modification.
- The embedding capacity for a given cover is the maximal number of bits that can be embedded in it.
- The relative embedding capacity is the ratio between the embedding capacity and the number of elements in the cover where a message can be embedded.

- The steganographic capacity is the maximal payload that can be embedded in the cover without introducing detectable modifications.
- The symbol-assignment function is used to convert individual cover elements to bits or alphabet symbols.
- The embedding distortion is a measure of the strength of embedding changes.
- PSNR and MSE are relative distortion measures (distortion per cover element).
- The embedding efficiency is the average number of bits embedded per unit distortion.

Exercises

4.1 [Distortion due to noise adding] Consider a steganographic scheme for which the impact of embedding is equivalent to adding iid realizations of a random variable η with pdf $f(x)$ to individual cover elements and rounding the result to the nearest integer,

$$\mathbf{y}[i] = \text{round}(\mathbf{x}[i] + \boldsymbol{\eta}[i]), \tag{4.21}$$

where $\text{round}(x)$ is x rounded to its nearest integer. Neglecting the boundary issue that $\mathbf{y}[i]$ may get out of the dynamic range of allowed values, show that the expected value of the embedding distortion is

$$E\left[d_1(\mathbf{x}, \mathbf{y})\right] = \sum_{k=-\infty}^{\infty} k\mathbf{p}[k], \tag{4.22}$$

$$E\left[d_2(\mathbf{x}, \mathbf{y})\right] = \sum_{k=-\infty}^{\infty} k^2\mathbf{p}[k], \tag{4.23}$$

where $\mathbf{p}[k] = \int_{k-\frac{1}{2}}^{k+\frac{1}{2}} f(x)\mathrm{d}x$.

4.2 [Exponentially decaying distortion] Show that when η in Exercise 4.1 is exponential with pdf $f(x) = \lambda e^{-\lambda x}$ for $x > 0$ and $f(x) = 0$ for $x \leq 0$, $\lambda > 0$,

$$E\left[d_1(\mathbf{x}, \mathbf{y})\right] = \frac{2e^{\lambda}}{(e^{\lambda} - 1)^2} \sinh \frac{\lambda}{2}, \tag{4.24}$$

$$E\left[d_2(\mathbf{x}, \mathbf{y})\right] = \frac{2e^{\lambda}(e^{\lambda} + 1)}{(e^{\lambda} - 1)^3} \sinh \frac{\lambda}{2}, \tag{4.25}$$

where $\sinh x = (e^x - e^{-x})/2$ is the hyperbolic sine function.

4.3 [Rule for adding PSNR] Let $\mathbf{x}[i]$ be an n-dimensional vector of real numbers and let $\boldsymbol{\eta}_1[i] \sim N(0, \sigma_1^2)$ and $\boldsymbol{\eta}_2[i] \sim N(0, \sigma_2^2)$ be two iid Gaussian sequences with zero mean. Show that for $n \to \infty$, the PSNR between $\mathbf{x}$ and $\mathbf{y} = \mathbf{x} + \boldsymbol{\eta}_1 + \boldsymbol{\eta}_2$ satisfies

$$10^{-\frac{\text{PSNR}}{10}} = 10^{-\frac{\text{PSNR}_1}{10}} + 10^{-\frac{\text{PSNR}_2}{10}}, \tag{4.26}$$

where PSNR_1 is between $\mathbf{x}$ and $\mathbf{x} + \boldsymbol{\eta}_1$ and PSNR_2 is between $\mathbf{x} + \boldsymbol{\eta}_1$ and $\mathbf{x} + \boldsymbol{\eta}_1 + \boldsymbol{\eta}_2$. Use the fact that the variance of the sum of two independent Gaussian variables is equal to the sum of their variances. Also, notice that with $n \to \infty$, $\text{MSE} = \hat{\sigma}^2 \to \sigma^2$, where $\hat{\sigma}^2$ is the sample variance.

5 Naive steganography

The first steganographic techniques for digital media were constructed in the mid 1990s using intuition and heuristics rather than from specific fundamental principles. The designers focused on making the embedding imperceptible rather than undetectable. This objective was undoubtedly caused by the lack of steganalytic methods that used statistical properties of images. Consequently, virtually all early naive data-hiding schemes were successfully attacked later. With the advancement of steganalytic techniques, steganographic methods became more sophisticated, which in turn initiated another wave of research in steganalysis, etc. This characteristic spiral development can be expressed through the following quotation:

Steganography is advanced through analysis.

In this chapter, we describe some very simple data-hiding methods to illustrate the concepts and definitions introduced in Chapter 4 and especially Section 4.3. At the same time, we point out problems with these simple schemes to emphasize the need for a more exact fundamental approach to steganography and steganalysis.

In Section 5.1, we start with the simplest and most common steganographic algorithm – Least-Significant-Bit (LSB) embedding. The fact that LSB embedding is not a very secure method is demonstrated in Section 5.1.1, where we present the histogram attack. Section 5.1.2 describes a different attack on LSB embedding in JPEG images that can not only detect the presence of a secret message but also estimate its size.

Some of the first steganographic methods were designed for palette images, which is the topic of Section 5.2. We discuss six different ideas for hiding information in palette images and point out their weaknesses as well as other problematic issues pertaining to their design. The palette format here serves as a useful educational platform to stimulate the reader to think about certain fundamental problems that are common to steganography in general and not necessarily specific to any image format.

Algorithm 5.1 Embedding message $\mathbf{m} \in \{0,1\}^m$ in cover image $\mathbf{x} \in \mathcal{X}^n$.

```
// Initialize a PRNG using stego key
// Input: message m ∈ {0,1}ᵐ, cover image x ∈ 𝒳ⁿ
Path = Perm(n);
// Perm(n) is a pseudo-random permutation of {1, 2, ..., n}
y = x;
m = min(m, n);
// If message longer than available capacity, truncate it
for i = 1 to m {
    y[Path[i]] = x[Path[i]] + m[i] − x[Path[i]] mod 2;
}
// y is the stego image with m embedded message bits
```

5.1   LSB embedding

Arguably, LSB embedding is the simplest steganographic algorithm. It can be applied to any collection of numerical data represented in digital form. Let us assume that $\mathbf{x}[i] \in \mathcal{X} = \{0, \ldots, 2^{n_c} - 1\}$ is a sequence of integers. For example, $\mathbf{x}[i]$ could be the light intensity at the ith pixel in an 8-bit grayscale image ($n_c = 8$), an index to a palette in a GIF file ($n_c = 8$), or a quantized DCT coefficient in a JPEG file ($n_c = 11$). Depending on the image format and the bit depth chosen for representing the individual values, each $\mathbf{x}[i]$ can be represented using n_c bits $\mathbf{b}[i, 1], \ldots, \mathbf{b}[i, n_c]$,

$$\mathbf{x}[i] = \sum_{k=1}^{n_c} \mathbf{b}[i, k] 2^{n_c - k}. \tag{5.1}$$

Thus, one can think of the sequence $(\mathbf{b}[i, 1], \ldots, \mathbf{b}[i, n_c])$ as the binary representation of $\mathbf{x}[i]$ in big-endian form (the most significant bit $\mathbf{b}[i, 1]$ is first). The LSB is the last bit $\mathbf{b}[i, n_c]$.

LSB embedding, as its name suggests, works by replacing the LSBs of $\mathbf{x}[i]$ with the message bits $\mathbf{m}[i]$, obtaining in the process the stego image $\mathbf{y}[i]$. Algorithm 5.1 shows a pseudo-code for embedding a bit stream in an image along a pseudo-random path generated from a secret key shared between Alice and Bob.

Note that in a color image the number of elements in the cover, n, is three times larger than for a grayscale image. Thus, the pseudo-random path is chosen across all pixels and color channels. The message embedded using Algorithm 5.1 can be extracted with the pseudo-code in Algorithm 5.2.

The amplitude of changes in LSB embedding is 1, $\max_i |\mathbf{x}[i] - \mathbf{y}[i]| = 1$, which is the smallest possible change for any embedding operation. Under typical viewing conditions, the embedding changes in an 8-bit grayscale or true-color image are not visually perceptible. Moreover, because natural images contain a small amount of noise due to various noise sources present during image acquisition

---

Algorithm 5.2 Extracting message **m** from stego image **y**.

```
// Initialize a PRNG using stego key
// Input: stego image y ∈ X^n
Path = Perm(n);
// Perm(n) is a pseudo-random permutation of {1, 2, . . . , n}
for i = 1 to m {
    m[i] = y[Path[i]] mod 2;
}
// m is the extracted secret message
```

Table 5.1. Relative counts of bits, neighboring bit pairs, and triples from an LSB plane of the image shown in Figure 5.1.

	Frequency of occurrence
Bits	0.49942, 0.50058
Pairs	0.24958, 0.24984, 0.24984, 0.25073
Triples	0.1246, 0.1250, 0.1247, 0.1251, 0.1250, 0.1249, 0.1251, 0.1256

(see Chapter 3), the LSB plane of raw, never-compressed natural images already looks random. Figure 5.1 (and color Plate 5) shows the original cover image and its LSB plane $\mathbf{b}[i, n_c]$ for the red channel.

Table 5.1 contains the frequencies of occurrence of single bits in $\mathbf{b}[i, n_c]$, $i = 1, \ldots, n$, pairs of neighboring bits ($\mathbf{b}[i, n_c], \mathbf{b}[i + 1, n_c]$), and triples of neighboring bits ($\mathbf{b}[i, n_c], \mathbf{b}[i + 1, n_c], \mathbf{b}[i + 2, n_c]$). The data are consistent with the claim that the LSB plane is random. Even though this is not a proof of randomness,[1] the argument is convincing enough to make us intuitively believe that any attempts to detect the act of randomly flipping a subset of bits from the LSB plane are doomed to fail. This seemingly intuitive claim is far from truth because LSB embedding in images can be very reliably detected (see Chapter 11 on targeted steganalysis). For now, we provide only a small hint.

Even if the LSB plane of covers was truly random, it may still be possible to detect embedding changes due to flipping LSBs if, for example, the second LSB plane $\mathbf{b}[i, n_c - 1]$ and the LSB plane were somehow dependent! In the most extreme case of dependence, if $\mathbf{b}[i, n_c - 1] = \mathbf{b}[i, n_c]$ for each i, detecting LSB changes would be trivial. All we would have to do is to compare the LSB plane with the second LSB plane.

LSB embedding belongs to the class of steganographic algorithms that embed each message bit at one cover element. In other words, each bit is located at a

---

[1] Pearson's chi-square test from Section D.4 could replace intuition by verifying that the distributions are uniform at a certain confidence level.

Figure 5.1 A true-color 800×548 image and the LSB plane of its red channel. Plate 5 shows the color version of the image.

certain element. The embedding proceeds by visiting individual cover elements and applying the embedding operation (flipping the LSB) if necessary, to match the LSB with the message bit. Not all steganographic schemes must follow this simple embedding paradigm. Methods that use syndrome coding utilize multiple cover elements to embed each bit in the sense that the extraction algorithm needs to see more than one element to extract one bit (see Chapter 8).

The embedding operation of flipping the LSB can be written mathematically in many different ways:

$$\mathrm{LSBflip}(x) = \begin{cases} x+1 & \text{when } x \text{ even} \\ x-1 & \text{when } x \text{ odd} \end{cases} \tag{5.2}$$

$$= x + 1 - 2(x \bmod 2) \tag{5.3}$$

$$= x + (-1)^x. \tag{5.4}$$

LSBflip is an idempotent operation satisfying $\mathrm{LSBflip}(\mathrm{LSBflip}(x)) = x$. This means that repetitive LSB embedding will partially cancel itself out and thus there is a limit on the maximal expected distortion that can be introduced by repetitive LSB embedding. Many other embedding operations do not have this property. Partial cancellation of embedding changes can be used to attack schemes that use LSB embedding [87]. Exercise 5.6 quantifies the effect of partial cancellation.

LSB embedding also induces $2^{n_c - 1}$ disjoint LSB pairs on the set of all possible element values $\{0, 1, \ldots, 2^{n_c} - 1\}$,

$$\{0, 1\}, \{2, 3\}, \ldots, \{2^{n_c} - 2, 2^{n_c} - 1\}. \tag{5.5}$$

Note that if $\mathbf{x}[i]$ is in LSB pair $\{2k, 2k + 1\}$, it must stay there after embedding because the pair elements differ only in their LSBs ($2k \leftrightarrow 2k + 1$). This simple observation is the starting point of many powerful attacks on LSB embedding (Chapter 11).

For any steganographic method, it is often valuable to mathematically express the impact of embedding on the image histogram. Many steganographic techniques introduce characteristic artifacts into the histogram and these artifacts can be used to detect the presence of secret messages (construct attacks). Let $\mathbf{h}[j]$, $j = 0, \ldots, 2^{n_c} - 1$, be the histogram of elements from the cover image

$$\mathbf{h}[j] = \sum_{i=1}^{n} \delta(\mathbf{x}[i] - j), \tag{5.6}$$

where δ is the Kronecker delta (2.23). We will assume that Alice is embedding a stream of m random bits. The assumption of randomness is reasonable because Alice naturally wants to minimize the impact of embedding and thus compresses the message and probably also encrypts to further improve the security of communication. We denote by $\alpha = m/n$ the relative payload Alice communicates. Assuming she embeds the bits along a pseudo-random path through the image, the probability that a pixel is not changed is equal to the probability that it is not selected for embedding, $1 - \alpha$, plus the probability that it is selected, α, multiplied by the probability that no change will be necessary, which happens with probability $\frac{1}{2}$ because we are embedding a random bit stream. Thus, for any j,

$$\Pr\{\mathbf{y}[i] = j | \mathbf{x}[i] = j\} = 1 - \alpha + \frac{\alpha}{2} = 1 - \frac{\alpha}{2}, \tag{5.7}$$

$$\Pr\{\mathbf{y}[i] \neq j | \mathbf{x}[i] = j\} = \frac{\alpha}{2}. \tag{5.8}$$

Because during LSB embedding the pixel values within one LSB pair $\{2k, 2k + 1\}$, $k = 0, \ldots, 2^{n_c - 1} - 1$, are changed into each other but never to any other value, the sum $\mathbf{h}_\alpha[2k] + \mathbf{h}_\alpha[2k + 1]$ stays unchanged for any α and thus forms an invariant under LSB embedding. Here, we denoted the histogram of the stego image as $\mathbf{h}_\alpha$. Thus, the expected value of $\mathbf{h}_\alpha[2k]$ is equal to the number of pixels with values $2k$ that stay unchanged plus the number of pixels with values $2k + 1$

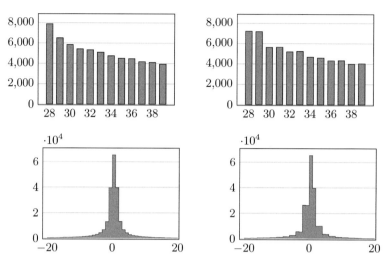

Figure 5.2 Effect of LSB embedding on histogram. Top: Magnified portion of the histogram of the image shown in Figure 5.1 after converting it to an 8-bit grayscale before and after LSB embedding. (Shown are histogram values for grayscales between 28 and 39.) Bottom: Histogram of quantized DCT coefficients of the same image before and after embedding using Jsteg (see Section 5.1.2). Left figures correspond to cover images, right figures to stego.

that were flipped to $2k$:

$$E\left[\mathbf{h}_\alpha[2k]\right] = \left(1 - \frac{\alpha}{2}\right)\mathbf{h}[2k] + \frac{\alpha}{2}\mathbf{h}[2k+1], \tag{5.9}$$

$$E\left[\mathbf{h}_\alpha[2k+1]\right] = \frac{\alpha}{2}\mathbf{h}[2k] + \left(1 - \frac{\alpha}{2}\right)\mathbf{h}[2k+1]. \tag{5.10}$$

Note that if Alice fully embeds her cover image with n bits ($\alpha = 1$), we have

$$E\left[\mathbf{h}_1[2k]\right] = E\left[\mathbf{h}_1[2k+1]\right] = \frac{\mathbf{h}[2k] + \mathbf{h}[2k+1]}{2}, \ k = 0, \ldots, 2^{n_c-1} - 1. \tag{5.11}$$

We say that LSB embedding has a tendency to even out the histogram within each bin. This leads to a characteristic staircase artifact in the histogram of the stego image (Figure 5.2), which can be used as an identifying feature for images fully embedded with LSB embedding. This observation is quantified in the so-called histogram attack [246], which we now describe.

5.1.1   Histogram attack

In a fully embedded stego image ($\alpha = 1$), we expect

$$\mathbf{h}_\alpha[2k] \approx \overline{\mathbf{h}}[2k] = \frac{\mathbf{h}_\alpha[2k] + \mathbf{h}_\alpha[2k+1]}{2}, \ k = 0, \ldots, 2^{n_c-1} - 1. \tag{5.12}$$

Formally, the histogram attack amounts to the following composite hypothesis-testing problem:

$$H_0: \quad \mathbf{h}_\alpha \sim \overline{\mathbf{h}}, \tag{5.13}$$

$$H_1: \quad \mathbf{h}_\alpha \not\sim \overline{\mathbf{h}}, \tag{5.14}$$

which we approach using Pearson's chi-square test [221] (also, see Appendix D). This test determines whether the even grayscale values in the stego image follow the known distribution $\overline{\mathbf{h}}[2k]$, $k = 0, \ldots, 2^{n_c-1} - 1$. The chi-square test first computes the test statistic S,

$$S = \sum_{k=0}^{d-1} \frac{(\mathbf{h}_\alpha[2k] - \overline{\mathbf{h}}[2k])^2}{\overline{\mathbf{h}}[2k]}, \tag{5.15}$$

where $d = 2^{n_c-1}$. Under the null hypothesis, the even grayscale values follow the probability mass function, $\overline{\mathbf{h}}[2k]$, and the test statistic (5.15) approximately follows the chi-square distribution, $S \sim \chi^2_{d-1}$, with $d - 1$ degrees of freedom (see Section A.9 on the chi-square distribution). That is as long as all d bins, $\mathbf{h}[2k]$, $k = 0, \ldots, 2^{n_c-1} - 1$, are sufficiently populated. Any unpopulated bins must be merged so that $\overline{\mathbf{h}}[2k] > 4$ for all k, to make S approximately chi-square distributed.

One can intuitively see that if the even grayscales follow the expected distribution, the value of S will be small, indicating the fact that the stego image is fully embedded with LSB embedding. Large values of S mean that the match is poor and notify us that the image under inspection is not fully embedded. Thus, we can construct a detector of images fully embedded using LSB embedding by setting a threshold γ on S and decide "cover" when $S > \gamma$ and "stego" otherwise. The probability of failing to detect a fully embedded stego image (probability of missed detection) is the conditional probability that $S > \gamma$ for a stego image,

$$P_{\text{MD}}(\gamma) = \Pr\{S > \gamma | \mathbf{x} \text{ is stego}\}. \tag{5.16}$$

We set the threshold γ so that the probability of a miss is at most P_{MD}. Denoting the probability density function of χ^2_{d-1} as $f_{\chi^2_{d-1}}(x)$, the threshold is determined from (5.16),

$$P_{\text{MD}}(\gamma) = \int_\gamma^\infty f_{\chi^2_{d-1}}(x)\mathrm{d}x = \int_\gamma^\infty \frac{e^{-\frac{x}{2}} x^{\frac{d-1}{2}-1}}{2^{\frac{d-1}{2}} \Gamma\left(\frac{d-1}{2}\right)} \mathrm{d}x. \tag{5.17}$$

The value $P_{\text{MD}}(\gamma)$ is called the p-value and it measures the statistical significance of γ. It is the probability that a chi-square-distributed random variable with $d - 1$ degrees of freedom would attain a value larger than or equal to γ.

The histogram attack can identify images fully embedded with random messages ($\alpha = 1$) and it can also be used to detect messages with $\alpha < 1$ if the order of embedding is known (e.g., sequential). In this case, the $m = \alpha n$ message bits are

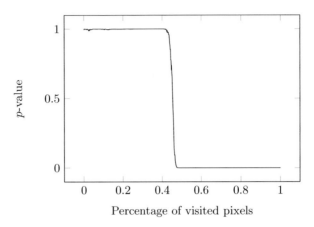

Figure 5.3 The p-value for the histogram attack on a sequentially embedded 8-bit grayscale image with relative message length $\alpha = 0.4$.

embedded along a known path in the image represented using the vector of indices $\mathbf{Path}[i]$, $i = 1, \ldots, n$. Evaluating the ith p-value $\mathbf{p}_\mathrm{v}[i]$ from the histogram of $\{\mathbf{x}[\mathbf{Path}[1]], \ldots, \mathbf{x}[\mathbf{Path}[i]]\}$ from the stego image, after a short transient phase, $\mathbf{p}_\mathrm{v}[i]$ will reach a value close to 1. It will suddenly fall to zero when we arrive at the end of the message (at approximately $i \approx \alpha n$) and it will stay at zero until we exhaust all pixels. This is because the test statistic S will cease to follow the chi-square distribution. Figure 5.3 shows $\mathbf{p}_\mathrm{v}[i]$ for a sequentially embedded message with $\alpha = 0.4$ (the cover image is shown in Figure 5.1). Thus, for sequential embedding this test not only determines with very high probability that a random message has been embedded but also estimates the message length.

If the embedding path is not known, the histogram attack is ineffective unless the majority of pixels have been used for embedding. Attempts to generalize this attack to randomly spread messages include [199, 242]. The most accurate steganalysis methods for LSB embedding are the detectors discussed in Chapter 11.

5.1.2   Quantitative attack on Jsteg

LSB embedding can be applied to any collection of numerical data. If the cover elements follow a distribution about which we have some a priori knowledge, we can use it to construct an attack in the following manner [255]. Let us describe our a priori knowledge about the cover image using some function of the cover histogram $\mathbf{h}$,

$$F(\mathbf{h}) = 0. \qquad (5.18)$$

From equations (5.9) and (5.10), $\mathbf{h}$ can be expressed using $E\left[\mathbf{h}_\alpha\right]$ by solving the system of two linear equations for $\mathbf{h}[2k]$ and $\mathbf{h}[2k+1]$,

$$\mathbf{h}[2k] = aE[\mathbf{h}_\alpha[2k]] - bE[\mathbf{h}_\alpha[2k+1]], \tag{5.19}$$

$$\mathbf{h}[2k+1] = -bE[\mathbf{h}_\alpha[2k]] + aE[\mathbf{h}_\alpha[2k+1]], \tag{5.20}$$

with $a = (1 - \alpha/2)/(1 - \alpha)$ and $b = (\alpha/2)/(1 - \alpha)$. Using the approximation $\mathbf{h}_\alpha \approx E[\mathbf{h}_\alpha]$, we can substitute (5.19) and (5.20) into (5.18) and thus obtain an equation for the unknown relative message length α. This equation will contain only the histogram of the stego image, $\mathbf{h}_\alpha$, which is known.

Note that this attack provides an estimate of the unknown message length independently of whether or not the message placement is known. Steganalysis methods that estimate the message length, or, more accurately, the number of embedding changes, are called *quantitative*. We illustrate this general method by attacking the steganographic algorithm Jsteg (`http://zooid.org/~paul/crypto/jsteg/`), which embeds data in JPEG images.

Jsteg uses the LSB embedding principle applied to quantized DCT coefficients with the exception that the LSB pair $\{0,1\}$ is skipped because allowing embedding into 0s would lead to quite disturbing artifacts. From Chapter 2, we know that the histogram of quantized DCT coefficients in a JPEG file is approximately symmetrical. This a priori knowledge can be expressed as

$$\mathbf{h}[j] - \mathbf{h}[-j] = 0, \ j = 1, 2, \ldots . \tag{5.21}$$

Moreover, the histogram is monotonically increasing for $j < 0$ and decreasing for $j > 0$. Because the LSB pairs are $\ldots, \{-4, -3\}, \{-2, -1\}, \{0, 1\}, \{2, 3\}, \ldots$ and because LSB embedding evens out the differences in counts in each LSB pair, $\mathbf{h}[2k]$ decrease and $\mathbf{h}[2k+1]$ increase with embedding for $k > 0$ and the effect is the opposite for $k < 0$ ($\mathbf{h}[2k]$ increase and $\mathbf{h}[2k+1]$ decrease). Thus, we use the following function of the histogram to attack Jsteg:

$$F(\mathbf{h}) = \sum_{k>0} \mathbf{h}[2k] + \sum_{k<0} \mathbf{h}[2k+1] - \sum_{k\geq 0} \mathbf{h}[2k+1] - \sum_{k<0} \mathbf{h}[2k]. \tag{5.22}$$

Note that for the cover image $F(\mathbf{h}) = 0$ as required. Also, with embedding, F increases due to the impact of LSB embedding on positive and negative even and odd DCT values. By substituting (5.19) and (5.20) into (5.22) and using the approximation $\mathbf{h}_\alpha \approx E\left[\mathbf{h}_\alpha\right]$, we obtain[2]

$$\sum_{k>0} (a\mathbf{h}_\alpha[2k] - b\mathbf{h}_\alpha[2k+1]) + \sum_{k<0} (-b\mathbf{h}_\alpha[2k] + a\mathbf{h}_\alpha[2k+1])$$

$$- \sum_{k>0} (-b\mathbf{h}_\alpha[2k] + a\mathbf{h}_\alpha[2k+1]) - \sum_{k<0} (a\mathbf{h}_\alpha[2k] - b\mathbf{h}_\alpha[2k+1]) = \mathbf{h}_\alpha[1]. \tag{5.23}$$

---

[2] Note that because the DCT coefficients equal to 1 are skipped, $\mathbf{h}[1] = \mathbf{h}_\alpha[1]$.

Rearranging the terms, an equation for the unknown relative message length α is obtained,

$$(a+b)\sum_{k>0}(\mathbf{h}_\alpha[2k]-\mathbf{h}_\alpha[2k+1])+(a+b)\sum_{k<0}(\mathbf{h}_\alpha[2k+1]-\mathbf{h}_\alpha[2k])=\mathbf{h}_\alpha[1],$$
$$(5.24)$$

where $a+b=1/(1-\alpha)$ (recall that $\mathbf{h}_\alpha$ is known as it is the histogram of the stego image). Solving for α, we finally get for its estimate

$$\hat{\alpha}=1-\frac{\sum_{k\neq0}\Delta\mathbf{h}_\alpha[k]}{\mathbf{h}_\alpha[1]},\qquad(5.25)$$

where

$$\Delta\mathbf{h}_\alpha[k]=\mathbf{h}_\alpha[2k]-\mathbf{h}_\alpha[2k+1]\quad\text{for }k>0,\qquad(5.26)$$
$$\Delta\mathbf{h}_\alpha[k]=\mathbf{h}_\alpha[2k+1]-\mathbf{h}_\alpha[2k]\quad\text{for }k<0.\qquad(5.27)$$

This estimate can be used to formulate the following hypothesis-testing problem:

$$\mathrm{H}_0:\quad\hat{\alpha}=0,\qquad(5.28)$$
$$\mathrm{H}_1:\quad\hat{\alpha}>0,\qquad(5.29)$$

where H_0 and H_1 correspond to the hypotheses that $\mathbf{h}_\alpha$ is a histogram of a cover and a stego image, respectively.

Figure 5.4 shows the histograms of the detector response $\hat{\alpha}$ for $\alpha=0,0.1$, $0.2,0.3$, and 0.5 on a database of 954 images, 315, 320, and 319 of which came from Canon G2, Canon S40, and Kodak DC 290 digital cameras with dimensions 2272×1704, for both Canon images, and 1792×1200 for images taken by Kodak DC 290. All images were acquired in the raw format, then off-line converted to grayscale and saved as 75% quality JPEGs before the experiment. The estimated message length $\hat{\alpha}$ appears to be unbiased and is overall a good detector capable of distinguishing cover images from images embedded with Jsteg. Note that the distribution of the estimator error is far from Gaussian and exhibits quite thick tails. The reader is referred to Section 10.3.2, which describes statistical properties of error of quantitative steganalyzers.

5.2   Steganography in palette images

Palette images, such as GIF or the indexed form of PNG, represent the image data using pointers to a palette of colors stored in the header. It is possible to hide messages both in the palette and in the image data.

5.2.1   Embedding in palette

Reordering the image palette and reindexing the image data correspondingly is a modification that does not change the visual appearance of the image. Thus, it

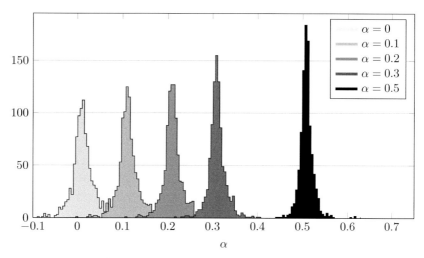

Figure 5.4 Histogram of estimated message length $\hat{\alpha}$ for cover images and stego images embedded with Jsteg and relative payload $0.1, 0.2, 0.3, 0.5$. The data was computed from 954 grayscale JPEG images with quality factor $q_f = 75$.

is possible to hide short messages as permutations of the palette. This method is implemented in the steganographic program Gifshuffle (`http://www.darkside.com.au/gifshuffle/`). Gifshuffle uses the palette order to hide up to $\log_2 256! \approx 1684$ bits ≈ 210 bytes in the palette by permuting its entries.

While this steganographic method does not change the appearance of the image, its security is low because many image-editing programs order the palette according to luminance, frequency of occurrence, or some other scalar factor. A randomly ordered palette will thus immediately raise suspicion. Also, displaying the image and resaving it may erase the information because the palette may be reordered. Another disadvantage of Gifshuffle is that its capacity is quite small and independent of the image size.

5.2.2 Embedding by preprocessing palette

For most palette images, it is not possible to apply LSB embedding directly to the image colors because new colors that are not in the palette would be obtained and, once the total number of unique colors exceeds 256, the image could no longer be stored as a GIF. One simple solution to this problem is to preprocess the palette before embedding by decreasing the color depth to 128, 64, or 32 colors. This way, when the LSBs of one, two, or three color channels, respectively, are perturbed, the total number of newly created colors will be at most 256. Thus, it will be possible to embed one, two, or three bits per pixel without introducing artifacts that are too disturbing. This method was implemented in the earlier versions of S-Tools (`ftp://ftp.ntua.gr/pub/crypt/mirrors/idea.sec.dsi.unimi.it/code/s-tools4.zip`).

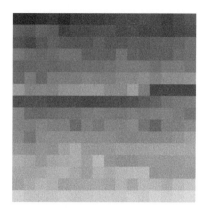

Figure 5.5 Palette colors of image shown in Figure 5.1 saved as GIF. A color version of this figure is shown in Plate 6.

Although this method provides high capacity, its steganographic security is low because the palette of the stego image will have very unusual structure [121] that is unlikely to occur naturally during color quantization (see Section 2.2.2). It will contain suspicious groups of 2, 4, or 8 close colors depending on the technique. It is thus relatively easy to identify stego images simply by analyzing the palette. What is even worse is that the detection will be equally reliable even for very short messages.

5.2.3   Parity embedding in sorted palette

Many problems associated with applying LSB-like embedding to palette images can be somewhat alleviated by presorting the palette colors so that neighboring colors are close and applying simple LSB embedding to the pointers. For example, the EzStego program (http://www.fqa.com/stego_com/) orders the palette by luminance. However, since luminance is a linear combination of three basic colors, occasionally colors with similar luminance values may be relatively far from each other (e.g., the RGB colors $[6, 98, 233]$ and $[233, 6, 98]$ have the same luminance but represent two visually very different colors). When this happens, visible suspicious artifacts result in the stego image.

Figures 5.5 and 5.6 (see Plates 6 and 7 for color versions of both figures) show the original palette colors and palette colors sorted by luminance arranged in a row-by-row fashion into a square 16×16 array. Notice that the luminance-sorted palette contains many very different neighboring colors. Not surprisingly, LSB embedding into indices to the sorted palette creates quite disturbing artifacts (see Figure 5.7(b) or Plate 8 for its color version).

There were attempts to modify the EzStego algorithm to eliminate this problem by ordering the palette using more sophisticated algorithms that would minimize the differences between neighboring palette colors in the palette. The task

Figure 5.6 Palette colors sorted by luminance (by rows). A color version of this figure is shown in Plate 7.

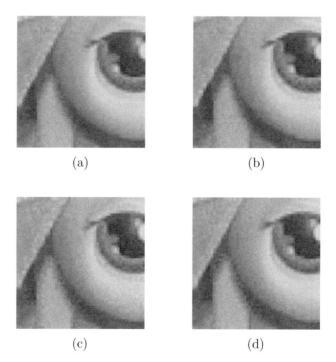

(a)   (b)

(c)   (d)

Figure 5.7 A small magnified portion of the image shown in Figure 5.1 saved as 256-color GIF (a), the same portion after embedding a maximum-length message using EzStego (b), optimal-parity embedding (c), and embedding while dithering (d). A color version of this figure is shown in Plate 8.

of finding the optimal ordering leads to the traveling-salesman problem, which is known to be NP-complete.

After a little thought, it should be clear that we do not really need to order the palette to obtain the minimal embedding distortion. In fact, all that is needed is to assign parities (bits 0 and 1) to each palette color so that the closest color to each palette color has the opposite parity. If this could be achieved, we can simply embed messages as color parities of indices instead of their LSBs by swapping a color for its closest neighbor from the palette. This is the idea behind optimal-parity embedding [80], explained next.

5.2.4   Optimal-parity embedding

Let the image palette contain N_p colors represented as RGB triples, $\mathbf{c}[j] = (\mathbf{r}[j], \mathbf{g}[j], \mathbf{b}[j])$, with parities $\mathbf{P}[j] \in \{0, 1\}$, $j = 1, \ldots, N_p$. We define the isolation, $\mathbf{s}[j]$, for each color $\mathbf{c}[j]$ as the distance from $\mathbf{c}[j]$ to its closest neighbor from the palette, $\mathbf{s}[j] = \min_{j' \neq j} d(\mathbf{c}[j], \mathbf{c}[j'])$. For example, one can use the Euclidean distance in the RGB cube as the measure of distance between two colors,

$$d_{\mathrm{RGB}}(\mathbf{c}[j], \mathbf{c}[j']) = \sqrt{\|\mathbf{c}[j] - \mathbf{c}[j']\|^2} \tag{5.30}$$

$$= \sqrt{(\mathbf{r}[j] - \mathbf{r}[j'])^2 + (\mathbf{g}[j] - \mathbf{g}[j'])^2 + (\mathbf{b}[j] - \mathbf{b}[j'])^2}. \tag{5.31}$$

We define parity assignment in the following manner.

1. Calculate the matrix of distances between all pairs of colors $\mathbf{d}_{\mathrm{RGB}}[k, l] = d_{\mathrm{RGB}}(\mathbf{c}[k], \mathbf{c}[l])$. Set $\mathcal{P} = \{\emptyset\}$.
2. Order all distances $\mathbf{d}_{\mathrm{RGB}}[k, l]$, $k > l$, to a non-decreasing sequence $\mathcal{D} = \mathbf{d}_{\mathrm{RGB}}[k_1, l_1] \leq \mathbf{d}_{\mathrm{RGB}}[k_2, l_2] \leq \ldots$. For a unique order, resolve ties, for example, using alphabetical order.
3. Iteratively repeat Step 4 until $\mathcal{P}$ contains all N_p palette colors.
4. Choose the next distance $\mathbf{d}_{\mathrm{RGB}}[k, l]$ in $\mathcal{D}$, such that either $\mathbf{c}[k] \notin \mathcal{P}$ or $\mathbf{c}[l] \notin \mathcal{P}$. If no such $\mathbf{d}_{\mathrm{RGB}}[k, l]$ can be found, this means that $\mathcal{P}$ already contains all N_p colors and we are done.
   a. If both $\mathbf{c}[k] \notin \mathcal{P}$ and $\mathbf{c}[l] \notin \mathcal{P}$, assign two opposite parities to both $\mathbf{c}[k]$ and $\mathbf{c}[l]$ and update $\mathcal{P} = \mathcal{P} \cup \{\mathbf{c}[k]\} \cup \{\mathbf{c}[l]\}$.
   b. If $\mathbf{c}[k] \notin \mathcal{P}$ and $\mathbf{c}[l] \in \mathcal{P}$, set $\mathbf{P}[k] = 1 - \mathbf{P}[l]$ and update $\mathcal{P} = \mathcal{P} \cup \{\mathbf{c}[k]\}$.
   c. If $\mathbf{c}[k] \in \mathcal{P}$ and $\mathbf{c}[l] \notin \mathcal{P}$, set $\mathbf{P}[l] = 1 - \mathbf{P}[k]$ and update $\mathcal{P} = \mathcal{P} \cup \{\mathbf{c}[l]\}$.

Note that once a parity of a color has been defined, it cannot be changed later by the algorithm. It is also clear that at the end all colors will have assigned parities. We now show that for any color $\mathbf{c}[j]$, $\mathbf{s}[j] = d_{\mathrm{RGB}}(\mathbf{c}[j], \mathbf{c}[j'])$ for some palette color $\mathbf{c}[j']$ with the opposite parity $\mathbf{P}[j'] \neq \mathbf{P}[j]$. Let k_a, l_a be the index pair of the first occurrence of index j in the sequence $k_1, l_1, k_2, l_2, \ldots$. Without loss of generality, let us assume that $k_a = j$. Because it is the first occurrence of j, $\mathbf{P}[j]$ has not been assigned and thus $\mathbf{P}[j] = 1 - \mathbf{P}[l_a]$. Also, there is no color in the palette that is strictly closer to $\mathbf{c}[j]$ than $\mathbf{c}[l_a]$ because, if there were such a color, we would have encountered color $\mathbf{c}[j]$ earlier in the process because $\mathcal{D}$ is non-decreasing and thus k_a would not be the first occurrence of index j. This means that the parity assignment constructed using the above algorithm guarantees that for any palette color there is another color in the palette that is closest and has the opposite parity. Thus, we can construct a steganographic algorithm that embeds one message bit at each palette index (each pixel) as the color parity and this algorithm induces the smallest possible distortion. This is because if the message bit does not match the color parity, we can swap the color for another palette color that is the closest to it.

Note that there may be more than one parity assignment with the above optimal property. The algorithm above, however, is deterministic and will always produce one specific assignment. Also note that the optimal parity assignment is only a function of the palette and not of the frequency with which the colors appear in the image. This means that the recipient can construct the same assignment from the stego image as the sender and thus read the message.

Assuming that color $\mathbf{c}[j]$ occurs in the cover image with frequency[3] $\mathbf{p}[j]$, $\sum_{j=1}^{N_\mathrm{p}} \mathbf{p}[j] = 1$, if the message-carrying pixels are selected pseudo-randomly, the expected embedding distortion per visited pixel is

$$\frac{E\left[d_2(\mathbf{x}, \mathbf{y})\right]}{m} = \frac{1}{m} \sum_{i=1}^{n} E\left[d_{\mathrm{RGB}}^2(\mathbf{x}[i], \mathbf{y}[i])\right] = \frac{1}{2} \sum_{j=1}^{N_\mathrm{p}} \mathbf{p}[j]\mathbf{s}^2[j], \qquad (5.32)$$

because for a message of length m, there will be on average $m\mathbf{p}[j]$ pixels of color $\mathbf{c}[j]$ containing message bits and one half of them will have to be modified to $\mathbf{c}[j']$, inducing distortion $\mathbf{s}^2[j]$.

5.2.5   Adaptive methods

Quite often, palette images contain large areas of uniform color, where no embedding should take place to avoid introducing easily detectable artifacts. Thus, it seems that steganography for palette images would benefit from methods that use adaptive selection channels and limit the embedding changes to textured areas while avoiding simple structures, such as segments of uniform color. Adaptive selection channels determined by the content of the cover image, however, create a potential problem with message recovery because the recipient does not have the cover image. We illustrate this issue on a simple adaptive method for palette images.

The image is divided into disjoint blocks, $\mathbf{B}$, formed, for example, by 3×3 blocks completely covering the image. Using the optimal parity assignment from the previous section, we define block parity as the eXclusive OR (XOR) of parities of all pixels in the block, $\mathbf{P}(\mathbf{B}) = \bigoplus_{\mathbf{x}[i]\in\mathbf{B}} \mathbf{P}(\mathbf{x}[i])$. We also define a measure of texture, which is a map, $t(\mathbf{B}) \to [0, \infty)$, that assigns a scalar value to each block. For example, we could define $t(\mathbf{B})$ as the number of unique colors in $\mathbf{B}$. If the texture measure exceeds a certain threshold, $t(\mathbf{B}) > \gamma$, we check whether the block parity matches the message bit. If it does not, one of the colors is changed to adjust its parity and thus the parity of the whole block. For example, we can

---

[3] One can also say that $\mathbf{p}$ is the normalized color histogram or sample pmf of colors.

Algorithm 5.3 Optimal parity assignment, $\mathbf{P}$, for a palette consisting of N_{p} colors $\mathbf{c}[j] = (\mathbf{r}[j], \mathbf{g}[j], \mathbf{b}[j])$, $j = 1, \ldots, N_{\mathrm{p}}$.

```
// Input: image palette c[j] = (r[j], g[j], b[j]),  j = 1,...,Np
for k = 1 to Np {
    for l = 1 to Np {
```
$$\mathbf{d}_{\mathrm{RGB}}[k,l] = \sqrt{(\mathbf{r}[k] - \mathbf{r}[l])^2 + (\mathbf{g}[k] - \mathbf{g}[l])^2 + (\mathbf{b}[k] - \mathbf{b}[l])^2}.$$
```
    }
}
for k = 1 to Np  dRGB[k,k] = Inf;
```
$\mathbf{inc} = (0, \ldots, 0);$ `//` vector of zeros of length N_{p}
`while` $\sum_{k=1}^{N_{\mathrm{p}}} \mathbf{inc}[k] < N_{\mathrm{p}}$ `{`
  $[k_{\mathtt{min}}, l_{\mathtt{min}}] = \arg\min_{k,l} \mathbf{d}_{\mathrm{RGB}}[k,l];$
  $c = \mathbf{inc}[k_{\mathtt{min}}] + \mathbf{inc}[l_{\mathtt{min}}];$
  `if` $(c = 1)$ `{`
    `if` $(\mathbf{inc}[k_{\mathtt{min}}] = 0)$ `{`
      $\mathbf{inc}[k_{\mathtt{min}}] = 1; \mathbf{P}[k_{\mathtt{min}}] = 1 - \mathbf{P}[l_{\mathtt{min}}];$
    `}` `else` `{`
      $\mathbf{inc}[l_{\mathtt{min}}] = 1; \mathbf{P}[l_{\mathtt{min}}] = 1 - \mathbf{P}[k_{\mathtt{min}}];$
    `}`
  `}`
  `if` $(c = 0)$ `{`
    $\mathbf{inc}[k_{\mathtt{min}}] = 1; \mathbf{P}[k_{\mathtt{min}}] = 0;$
    $\mathbf{inc}[l_{\mathtt{min}}] = 1; \mathbf{P}[l_{\mathtt{min}}] = 1;$
  `}`
  $\mathbf{d}_{\mathrm{RGB}}[k_{\mathtt{min}}, l_{\mathtt{min}}] = \mathtt{Inf};$
`}`
`//` $\mathbf{P}$ is the optimal parity assignment

---

choose the color with the smallest isolation to minimize the overall embedding distortion. The recipient simply reads the message by following the same steps, extracting message bits from the parity of all blocks whose texture measure is above the threshold.

The problem with this scheme is that the act of embedding may change the texture measure to fall below the threshold and thus the recipient will not read the bit embedded in that block. This problem is common to many steganographic methods that use adaptive selection rules. For the scheme above, the solution is simple. After embedding each bit, we need to check whether the block texture is still above the threshold. If it is, the embedding can continue. If the texture falls below the threshold, we need to embed the same bit in the next block because the block where we just embedded will be skipped by the recipient. This modification decreases the embedding efficiency because sometimes changes are made that do not embed any bits. However, the decrease in embedding efficiency is usually very small because the chances that the block's texture will fall below the threshold after embedding are small.

The problem above is a specific example of a selection channel that is not completely shared between the sender and the recipient. Chapter 9 is devoted to the problem of communicating with non-shared selection channels, where a general solution is presented using so-called wet paper codes.

5.2.6   Embedding while dithering

We use the palette format to demonstrate one more important type of steganographic schemes that embed messages while applying an information-reducing operation to the cover image, such as quantization. The embedding minimizes at the same time the combined distortion due to quantization and embedding.

Let us assume that the cover image is a true-color 24-bit image, for example in the BMP format, and we wish to save it as a GIF. As explained in Chapter 2, this conversion involves color quantization, to "round" the pixel colors to palette colors, and dithering, which spreads the quantization error. The embedding scheme starts with computing the optimal parity assignment for the palette. To embed m bits, m pixels are pseudo-randomly selected to carry the message bits. Finally, color quantization and dithering are performed as usual by scanning the image by rows with one exception. At each message-carrying pixel, its color is quantized to the closest palette color with the parity that matches the message bit. The combined quantization and embedding are thus diffused to neighboring pixels. This way, both the quantization error and the error due to message embedding will be diffused through the whole image. An example of an image embedded at 1 bpp with this method is shown in Figure 5.7(d) (see Plate 8 for its color version).

The concept of embedding a message while the cover image is being processed is a very powerful one and can be greatly generalized. When applied to other image-processing operations, however, a more advanced embedding mechanism is required. This topic is further elaborated upon in Chapter 9 on non-shared selection channels.

Finally, we would like to state that palette images with a small number of colors are not very suitable for steganography. It is difficult (if not impossible) to design secure schemes with reasonable capacity. Moreover, palette images are typically used for storing computer art and line drawings, where embedding changes are usually easily detectable due to the semantic meaning of the objects in the image. Thus, the emphasis in this book will be given especially to the ubiquitous JPEG format and raster formats. We included steganography of GIF images in this chapter because the GIF format allowed us to demonstrate many important embedding principles and issues that will be revisited throughout this book.

Summary

- The simplest steganographic method is Least-Significant-Bit (LSB) embedding.
- The effect of LSB embedding on an image histogram can be quantified. The embedding evens out the populations of both values from the same LSB pair $\{2k, 2k+1\}$.
- Histogram attack is an attack on LSB embedding when the message placement is known.
- Jsteg is a steganographic algorithm for JPEG images that uses LSB embedding. It can be attacked by utilizing a priori knowledge about the cover-image histogram, such as its symmetry.
- Messages can be embedded in palette images either in the palette or in the indices (image data). Hiding in palette provides limited capacity independent of the image size.
- Embedding in indices to palette offers larger capacity but often creates easily discernible artifacts.
- Optimal-parity embedding is an assignment of bits to palette colors that can be used to minimize the embedding distortion.
- Embedding while dithering is an embedding method for palette images that minimizes the total distortion due to color quantization and embedding.
- A possible way to improve security is to confine the embedding changes to more textured or noisy areas of the cover (adaptive steganography).

Exercises

5.1 [Embedding in two LSBs (LSB2)] Consider the following steganographic scheme that embeds bits into the least two LSBs of pixels

$$\mathcal{C} = \{0, \ldots, 255\}^n, \tag{5.33}$$
$$\mathcal{M} = \{0, 1\}^{2n}, \tag{5.34}$$
$$\mathcal{K} = \{\emptyset\}. \tag{5.35}$$

Emb : Two bits are embedded sequentially at each pixel by replacing two least significant bits with two message bits. For example, if $\mathbf{x}[i] = 14 = (00001110)_2$ and we want to embed bit pairs 00, 01, 10, 11, $\mathbf{x}[i]$ is changed to 12, 13, 14, 15, respectively. If $\mathbf{x}[i] = 32 = (00100000)_2$, we embed the same bit pairs by changing $\mathbf{x}[i]$ to 32, 33, 34, and 35, etc. Ext : Two bits are extracted sequentially as the least two LSBs from the pixels to form the message.

Calculate the embedding efficiency for both d_1 and d_2 distortion under the assumption that the message is a random bit stream and the pixel values are uniformly distributed in $\{0, \ldots, 255\}$.

Hint: Write the expected distortion for $\mathbf{x}[i] = 4k, 4k+1, 4k+2, 4k+3$, and then use the assumption that the pixel values are uniformly distributed to obtain the average distortion per pixel.

5.2 [Alternative embedding in 2 LSBs (LSB2–)] Consider the following steganographic scheme with the same $\mathcal{C}, \mathcal{M}, \mathcal{K}$ as in Exercise 5.1 with embedding function Emb : Two message bits are sequentially embedded at each pixel by always modifying the pixel value to the closest value with the required least two LSBs. For example, if $\mathbf{x}[i] = 14 = (00001110)_2$ and we want to embed bit pairs 00, 01, 10, 11, $\mathbf{x}[i]$ is changed to 12, 13, 14, 15, respectively. If $\mathbf{x}[i] = 32 = (00100000)_2$, we embed the same bit pairs by changing $\mathbf{x}[i]$ to 32, 33, 34, and 31. Note that the last change from 32 to 31 will lead to modification of all six LSBs! The extraction function is the same as in the previous exercise. Calculate the embedding efficiency for both d_1 and d_2 distortion under the assumption that the message is a random bit stream. For simplicity, ignore the boundary issues. Compare the embedding efficiency of LSB2 and LSB2– for each distortion measure. Prove that the LSB2– method has higher embedding efficiency than the LSB2 method for distortion measure $d_\gamma(\mathbf{x}, \mathbf{y}) = \sum_{i=1}^{n} |\mathbf{x}[i] - \mathbf{y}[i]|^\gamma$ for any $\gamma > 0$.

5.3 [LSB embedding of biased bit stream] Assume that the cover image is an 8-bit grayscale image. Suppose that the secret message is a random biased bit stream, i.e., the bits are iid realizations of a Bernoulli random variable $\nu \sim B(p_0)$ with probability mass function

$$\Pr\{\nu = 0\} = p_0, \tag{5.36}$$
$$\Pr\{\nu = 1\} = p_1, \tag{5.37}$$

where $p_0 + p_1 = 1$. Let $\mathbf{h}_\alpha$ be the image histogram after LSB embedding a secret message of relative length α in the cover image. Show that

$$E[\mathbf{h}_\alpha[2k]] = (1 - \alpha p_1)\mathbf{h}[2k] + \alpha p_0 \mathbf{h}[2k+1], \tag{5.38}$$
$$E[\mathbf{h}_\alpha[2k+1]] = \alpha p_1 \mathbf{h}[2k] + (1 - \alpha p_0)\mathbf{h}[2k+1]. \tag{5.39}$$

5.4 [Histogram attack for biased message] Assume that an 8-bit grayscale stego image is fully embedded ($\alpha = 1$) with biased bit stream from the previous exercise. First show

$$\frac{E[\mathbf{h}_\alpha[2k]]}{E[\mathbf{h}_\alpha[2k+1]]} = \frac{p_0}{p_1} \quad \text{for all } k = 0, \ldots, 127, \tag{5.40}$$

$$E[\mathbf{h}_\alpha[2k]] = p_0(\mathbf{h}_\alpha[2k] + \mathbf{h}_\alpha[2k+1]), \tag{5.41}$$
$$E[\mathbf{h}_\alpha[2k+1]] = (1 - p_0)(\mathbf{h}_\alpha[2k] + \mathbf{h}_\alpha[2k+1]). \tag{5.42}$$

The distribution now depends on the unknown message bias p_0. Estimate p_0

$$\hat{p}_0 = \arg\max_{\lambda \in \mathbb{R}} \sum_{k=0}^{d-1} \left(\mathbf{h}_\alpha[2k] - 2\lambda\overline{\mathbf{h}}[2k]\right)^2 + \left(\mathbf{h}_\alpha[2k+1] - 2(1-\lambda)\overline{\mathbf{h}}[2k]\right)^2, \tag{5.43}$$

where $\overline{\mathbf{h}}[2k]$ is defined in (5.12), and compute the chi-square test statistic

$$S = \sum_{k=0}^{d-1} \frac{\left(\mathbf{h}_\alpha[2k] - 2\hat{p}_0\overline{\mathbf{h}}[2k]\right)^2}{2\hat{p}_0\overline{\mathbf{h}}[2k]}, \tag{5.44}$$

which now follows the chi-square distribution with $d - 2 = 126$ degrees of freedom because we had to estimate the unknown parameter p_0 from the data.

5.5 [Power of parity] Assume that the cover image has a biased distribution of LSBs: the fraction r of pixels has LSBs equal to 1 and the fraction $1 - r$ has LSBs equal to 0, $0 < r < 1$. Let $\mathbf{x}[i]$ be the LSBs of pixels ordered along a pseudo-random path. Consider the following sequence of bits $\mathbf{b}[i]$ obtained as the XOR of LSBs of disjoint groups of m consecutive pixels $\mathbf{b}[1] = \mathbf{x}[1] \oplus \mathbf{x}[2] \oplus \cdots \oplus \mathbf{x}[m]$, $\mathbf{b}[2] = \mathbf{x}[m + 1] \oplus \mathbf{x}[m + 2] \oplus \cdots \oplus \mathbf{x}[2m], \ldots$. Show that the bit stream $\mathbf{b}[i]$ becomes unbiased exponentially fast with m by proving that

$$|\Pr\{\mathbf{b}[i] = 0\} - \Pr\{\mathbf{b}[i] = 0\}| = |1 - 2r|^m. \tag{5.45}$$

Hint: For m even, $\Pr\{\mathbf{b}[i] = 0\} = \Pr\{\mathbf{x}[1] + \mathbf{x}[2] + \cdots + \mathbf{x}[m]$ is even$\}$ and express the probabilities using r and $1 - r$.

5.6 [Repetitive LSB embedding] Let $\mathbf{x}$ be an 8-bit grayscale cover image and $\mathbf{y}$ be the stego image after LSB embedding in $\mathbf{x}$ a random unbiased message of relative length α_1. Continue by embedding in $\mathbf{y}$ using LSB embedding another random unbiased message of relative length α_2, obtaining the doubly-embedded stego image $\mathbf{z}$. Assume that the paths for both embeddings are pseudo-randomly chosen and independent of each other. The stego image $\mathbf{z}$ will appear to have been embedded with a random unbiased message of relative length $L(\alpha_1, \alpha_2)$. Show that

$$L(\alpha_1, \alpha_2) = \alpha_1 + \alpha_2 - \alpha_1\alpha_2. \tag{5.46}$$

Furthermore, consider the case when you embed messages of relative length α repetitively into the same image k times. Show that the relative length $L_k(\alpha)$ of the message that appears to have been embedded after k repetitive embeddings is

$$L_k(\alpha) = 1 - (1 - \alpha)^k. \tag{5.47}$$

Note that $L_k(\alpha) \to 1$ as $k \to \infty$ exponentially fast, for any α positive.

5.7 [Upper bound on message length] Let $\mathbf{h}_\alpha$ be the histogram of an 8-bit grayscale image embedded using LSB embedding with a random unbiased message of relative length α. Prove the following upper bound:

$$\alpha \leq 2\frac{\min\{\mathbf{h}_\alpha[2k], \mathbf{h}_\alpha[2k+1]\}}{\mathbf{h}_\alpha[2k] + \mathbf{h}_\alpha[2k+1]} \quad \text{for all } k = 0, \ldots, 127. \tag{5.48}$$

Hint: $\mathbf{h}_\alpha[2k] \approx (1 - \alpha/2)\,\mathbf{h}[2k] + (\alpha/2)\mathbf{h}[2k+1] \geq (\alpha/2)\mathbf{h}[2k] + (\alpha/2)\mathbf{h}[2k+1]$ because $1 - \alpha/2 \geq \alpha/2$ for $0 \leq \alpha \leq 1$.

5.8 **[LSB embedding as noise adding]** Explain why the impact of LSB embedding in cover $\mathbf{x}$ cannot be written as adding to $\mathbf{x}$ an iid noise, $\mathbf{y} = \mathbf{x} + \boldsymbol{\xi}$. **Hint:** Are $\mathbf{x}$ and $\boldsymbol{\xi}$ independent?

5.9 **[View bit planes]** Write a routine in Matlab that displays a selected bit plane (a two-color black-and-white image) of an image represented using a `uint8` array. For a color image, display the bit planes of red, green, and blue channels separately as three two-color images. You may wish to use green–white combination for the bit plane of the green channel, red–white combination for the red channel, etc.

6 Steganographic security

In the previous chapter, we saw a few examples of simple steganographic schemes and successful attacks on them. We learned that the steganographic scheme called LSB embedding leaves a characteristic imprint on the image histogram that does not occur in natural images. This observation lead to an algorithm (a detector) that could decide whether or not an image contains a secret message. The existence of such a detector means that LSB embedding is not secure. We expect that for a truly secure steganography it should be impossible to construct a detector that could distinguish between cover and stego images. Even though this statement appears reasonable at first sight, it is vague and allows subjective interpretations. For example, it is not clear what is meant by "could distinguish between cover and stego images." We cannot construct a detector that will always be 100% correct because it is hardly possible to detect the effects of flipping one LSB, at least not reliably in every cover. Just how reliable must a detector be to pronounce a steganographic method insecure?

Even though there are no simple practical solutions to the questions raised in the previous paragraph, they can in principle be studied within the framework of information theory. Imagine that Alice and Bob are engaging in a legitimate communication and do not use steganography. Let us suppose that they exchange grayscale 512×512 images in raster format that were never compressed. If we observed their communication for a sufficiently long time, the images would sample out a probability distribution P_c in the space of all covers $\mathcal{C} = \mathcal{X}^{512 \times 512}$, $\mathcal{X} = \{0, \ldots, 255\}$. This distribution captures legitimate communication between Alice and Bob. On the other hand, if Alice and Bob embed secrets in images, again over a long time period, the images will appear to follow a different distribution P_s over $\mathcal{C}$. Intuitively, Alice wants to design the stego method while making sure that P_s is as close to P_c as possible to prevent Eve from discovering the fact that she communicates secretly with Bob. There exists a fundamental relationship between the "distance" between P_c and P_s and Eve's ability to detect images with steganographic content. This distance can be taken as a measure of steganographic security and it will impose constraints on the reliability of the best detector Eve can ever build.

At this point, we ignore the fundamental question of whether it is feasible to assume that the distributions P_c and P_s can be estimated in practice or even whether they are appropriate descriptions of the cover-image source. We simply assume that Eve knows the steganographic channel (Kerckhoffs' principle) and thus knows both P_c and P_s. This rather strong assumption is justified because in real life the prisoners can never be sure how much Eve knows (she may be a government agency with significant resources, for example) and thus it is prudent to grant her omnipotence. Under this assumption, in Section 6.1 we define steganographic security as the KL divergence between the distributions of cover and stego images, P_c and P_s. The importance of this information-theoretic quantity will become apparent later when we show how the KL divergence imposes fundamental limits on Eve's detector and how it can be used for comparing steganographic schemes. In Section 6.2, we discuss several specific examples of perfectly secure stegosystems and point out an interesting relationship between perfect security and perfect compression. Section 6.3 investigates secure stegosystems under the condition that the embedding distortion introduced by Alice is limited (the case of a so-called distortion-limited embedder). In the same section, we also show how certain algorithms originally proposed for robust watermarking can be modified into secure stegosystems.

Even though the information-theoretic definition of security is well developed and widely accepted in the steganographic community, there exist important alternative approaches, which we also mention in this paragraph. Inspired by the concept of security of public-key cryptosystems, in Section 6.4 we explain a complexity-theoretic approach to steganographic security, which makes an important connection between security in steganography and properties of some common cryptographic primitives, such as one-way functions. This research direction arose due to critique of the information-theoretic definition of security, which ignores the important issue of computational complexity and Eve's ability to actually implement an attack.

6.1   Information-theoretic definition

We now give a mathematically precise form to the thoughts presented above [35]. Recall from Chapter 4 that every steganographic scheme uses a pair of embedding and extraction mappings $\text{Emb} : C \times K \times M \to C$, $\text{Ext} : C \times K \to M$ defined on the sets of all possible covers C, stego keys K, and the set of messages M. The mappings are assumed to satisfy $\text{Ext}(\text{Emb}(\mathbf{x}, \mathbf{k}, \mathbf{m}), \mathbf{k}) = \mathbf{m}$ for any message $\mathbf{m} \in M$, cover $\mathbf{x} \in C$, and key $\mathbf{k} \in K$. In other words, we will work under the simplifying assumption that the set of keys and messages is the same for every cover $\mathbf{x}$. The object obtained as a result of embedding is called the stego object $\mathbf{y} = \text{Emb}(\mathbf{x}, \mathbf{k}, \mathbf{m})$.

Assuming the cover is drawn from C with probability distribution P_c and the stego key, as well as the message, are drawn according to distributions P_k, P_m over their corresponding spaces K, M, the distribution of stego objects will be denoted as P_s. Note that, even though for steganography by cover selection or cover synthesis the stego object $\mathbf{y}$ does not have to be obtained as a modification of some $\mathbf{x} \in \mathcal{C}$, the steganographic method will still generate some distribution P_s over $\mathbf{y} \in \mathcal{C}$.

6.1.1 KL divergence as a measure of security

We intuitively expect that if P_c is "close" to P_s, then Eve must make erroneous decisions increasingly more often. Given an object $\mathbf{x}$, Eve must decide between two hypotheses: H_0, which represents the hypothesis that $\mathbf{x}$ does not contain a hidden message, and H_1, which stands for the hypothesis that $\mathbf{x}$ does contain a hidden message. Under hypothesis H_0, the observation $\mathbf{x}$ is drawn from the distribution P_c, $\mathbf{x} \sim P_c$. Conversely, under H_1, $\mathbf{x} \sim P_s$.

The two distributions can be compared using their Kullback–Leibler divergence, also called KL distance or relative entropy (see Appendix B and [50]),

$$D_{\mathrm{KL}}(P_c||P_s) = \sum_{\mathbf{x} \in \mathcal{C}} P_c(\mathbf{x}) \log \frac{P_c(\mathbf{x})}{P_s(\mathbf{x})}, \tag{6.1}$$

which is a fundamental concept from information theory measuring how different the two distributions are. Here, the log can be to the base 2, in which case the KL divergence is measured in bits, or it could be the natural logarithm and the unit is a "nat."

When $D_{\mathrm{KL}}(P_c||P_s) = 0$, we call Alice's stegosystem perfectly secure (undetectable) because in this case the distribution of the stego objects P_s created by Alice is identical to the distribution P_c assumed by Eve. Thus, it is impossible for Eve to distinguish between covers and stego objects. If $D_{\mathrm{KL}}(P_c \parallel P_s) \leq \epsilon$, then we say the steganographic system is ϵ-secure.

To better understand what is meant by ϵ-security, consider that Eve has a stego detector, which is a mapping $F : \mathcal{C} \rightarrow \{0, 1\}$. The response of Eve's detector is binary – it answers either 0 for cover or 1 for stego. The detector can make two types of error. The first type of error is a false alarm (false positive), which occurs when Eve decides that a hidden message is present when in fact it is absent. The second type of error is a missed detection (false negative) which occurs when Eve decides that a hidden message is absent, when in fact it is present. Let P_{FA} and P_{MD} denote the probabilities of false alarm and missed detection, respectively.

Assuming the detector is fed only covers distributed according to P_c, Eve will decide 0 or 1 with probabilities $p_c(0) = 1 - P_{\mathrm{FA}}$ and $p_c(1) = P_{\mathrm{FA}}$, respectively. On stego objects distributed according to P_s, Eve's detector assigns 0 and 1 with probabilities $p_s(0) = P_{\mathrm{MD}}$ and $p_s(1) = 1 - P_{\mathrm{MD}}$, respectively. The KL divergence between the two distributions of Eve's detector

$$p_c = (p_c(0), p_c(1)) = (1 - p_c(1), p_c(1)), \tag{6.2}$$

$$p_s = (p_s(0), p_s(1)) = (p_s(0), 1 - p_s(0)) \tag{6.3}$$

is given by (6.1) with $\mathcal{C} = \{0, 1\}$, $p_c(1) = P_{FA}$, and $p_s(0) = P_{MD}$,

$$d_{KL}(P_{FA}, P_{MD}) \triangleq D_{KL}(p_c \| p_s) \tag{6.4}$$

$$= (1 - P_{FA}) \log \frac{1 - P_{FA}}{P_{MD}} + P_{FA} \log \frac{P_{FA}}{1 - P_{MD}}. \tag{6.5}$$

Because Eve's detector is a type of processing and processing cannot increase the KL divergence (Proposition B.6 in Appendix B), for an ϵ-secure stego system we must have

$$d_{KL}(P_{FA}, P_{MD}) \leq D_{KL}(P_c \| P_s) \leq \epsilon. \tag{6.6}$$

This inequality imposes a fundamental limit on the performance of any detector Eve can build. Requiring that the probability of false alarm for Eve's detector cannot be larger than some fixed value P_{FA}, $0 < P_{FA} < 1$, the smallest possible probability of missed detection she can achieve is

$$P_{MD}(P_{FA}) = \arg \min_{P_{MD} \in [0,1]} \{P_{MD} | d_{KL}(P_{FA}, P_{MD}) \leq \epsilon\}, \tag{6.7}$$

where the minimum is taken over all detectors whose probability of false alarm is P_{FA}. Figure 6.1 shows the probability of detection of a stego object

$$P_D(P_{FA}) = 1 - P_{MD}(P_{FA}) \tag{6.8}$$

as a function of P_{FA} for various values of ϵ. The curves are called Receiver-Operating-Characteristic (ROC) curves (see Appendix D for the definition). The region under each curve is the range within which the performance of Eve's detector must fall for a given value of ϵ. Note that Eve cannot minimize both errors at the same time as there appears to be a trade-off between the two types of error. In particular, if Eve is not allowed to falsely accuse Alice or Bob of using steganography, $P_{FA} = 0$, Eve's detector will fail to detect ϵ-secure steganographic communication with probability at least

$$P_{MD} \geq e^{-\epsilon}. \tag{6.9}$$

This can be easily seen by setting $P_{FA} = 0$ in (6.6). Thus, the smaller ϵ is, or the closer the two distributions, P_c and P_s, are, the greater the likelihood that a covert communication will not be detected. This motivates choosing the KL divergence as a measure of steganographic security.

It is also instructive to inspect what kind of detector Eve obtains for secure stegosystems with $\epsilon = 0$. Because the KL divergence is always non-negative, from (6.6) we have that $d_{KL}(P_{FA}, P_{MD}) = 0$. As shown in Appendix B, this can happen only if distributions p_c and p_s are the same, or $P_{MD} = 1 - P_{FA}$. A detector whose false alarms and missed detections satisfy this relationship amounts to a detector that is just randomly guessing. To see this, imagine the following family of detectors parametrized by $p \in [0, 1]$:

$$F_p(\mathbf{x}) = \begin{cases} 1 & \text{with probability } p \\ 0 & \text{with probability } 1 - p. \end{cases} \tag{6.10}$$

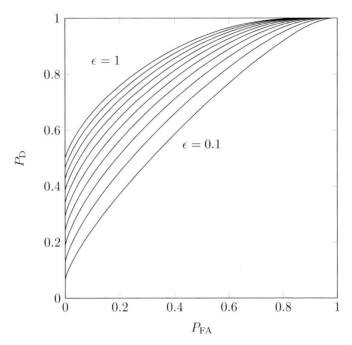

Figure 6.1 Probability of detection P_D of a stego image as a function of false alarms, P_FA. The ROC curves correspond to $\epsilon = 0.1, 0.2, \ldots, 1$ with $\epsilon = 1$ corresponding to the top curve and $\epsilon = 0.1$ to the bottom curve.

When a cover image is sent to this detector, F_p flips a biased coin and decides "stego" with probability p (the false-alarm probability is $P_\mathrm{FA} = p$). On the other hand, presenting the detector with a stego object, it is detected as cover with probability $1 - p$ (the missed detection rate is $P_\mathrm{MD} = 1 - p$). Thus, this randomly guessing detector satisfies $P_\mathrm{MD} = 1 - P_\mathrm{FA}$.

6.1.2   KL divergence for benchmarking

It is of great interest to Alice and Bob to minimize the chance that their secret communication will be detected by Eve. The prisoners would thus prefer to use the most secure stegosystem currently available. For this, they need some means of comparing security of stegosystems.

In this chapter, we defined the security of a stegosystem as the KL divergence between the distributions of cover and stego objects. It is thus tempting to use it for comparing (benchmarking) steganographic algorithms in the following manner. Given two stegosystems, $S^{(1)}$ and $S^{(2)}$, that share the same set of covers, we would say that $S^{(1)}$ is more secure than $S^{(2)}$ whenever $D_\mathrm{KL}(P_\mathrm{c}||P_\mathrm{s}^{(1)}) \leq D_\mathrm{KL}(P_\mathrm{c}||P_\mathrm{s}^{(2)})$, where $P_\mathrm{s}^{(1)}$ and $P_\mathrm{s}^{(2)}$ are the distributions of their stego objects. This approach to benchmarking would be justified if a larger KL divergence implied the existence of a better steganography detector. This is asymptotically correct due to the Chernoff–Stein lemma (Appendix B) in the limit when n, the number of objects exchanged by Alice and Bob, approaches infinity. The lemma says that if Eve imposes a bound on the probability of false alarms, P_{FA}, the best detector she can build will misclassify n observed stego objects as covers with probability P_{MD} satisfying

$$\lim_{n\to\infty} \frac{1}{n} \log P_{\mathrm{MD}}(P_{\mathrm{FA}}) = -D_{\mathrm{KL}}(P_{\mathrm{c}}||P_{\mathrm{s}}). \tag{6.11}$$

Alternatively, one could also say that the probability of a miss decays exponentially with n, $P_{\mathrm{MD}}(P_{\mathrm{FA}}) \approx e^{-nD_{\mathrm{KL}}(P_{\mathrm{c}}||P_{\mathrm{s}})}$.

We stress that this result holds only in the limit of a large number of observations, n. For a finite number of observations, larger KL divergence does not necessarily imply the existence of a better detector. Exercise 6.3 shows an example of two families of distributions, g_N and h_N on $\mathcal{C} = \{1, \ldots, N\}$, for which $D_{\mathrm{KL}}(g_N||h_N) = N$ and for which any detector based on one observation will always have $P_{\mathrm{D}} - P_{\mathrm{FA}} \leq \delta_N$, where $\delta_N \to 0$. In other words, despite the fact that the KL divergence between distributions g_N and h_N grows to infinity with N, our ability to decide between them *on the basis of a single observation* diminishes to random guessing.

Benchmarking steganographic systems using their KL divergence as hinted above is, however, still not without problems. Ignoring for now the issue of numerically evaluating the KL divergence for a real stegosystem, the KL divergence is a function of the relative payload (or, more accurately, the distribution of the change rate β). Thus, it is conceivable that two stegosystems compare differently for different payloads.

On the other hand, if the prisoners wish to stay undetected, with time they must start embedding smaller and smaller payloads, otherwise Eve would detect the covert communication with certainty. To see this, imagine that the prisoners communicate using change rate bounded from below, $\beta \geq \beta_0 > 0$. Then, the KL divergence between cover and stego objects would be bounded from below by $D_{\mathrm{KL}}(P_{\mathrm{c}}||P_{\beta_0})$.[1] Invoking the Chernoff–Stein lemma again, Eve's detector would thus achieve an arbitrarily small probability of missed detection, P_{MD}, at any bound on false alarms P_{FA}. In other words, the prisoners would be caught with probability approaching 1.

Thus, it makes sense to define the steganography benchmark by properties of Eve's best detector in the limit of $\beta \to 0$. Let us assume for simplicity that covers are represented by n iid realizations $\mathbf{x}[i]$, $i = 1, \ldots, n$, of a scalar random variable x with pmf $P_{\mathrm{c}} \equiv P_0$. If the embedding modifications are also independent of each other, the stego object is a sequence of iid realizations that follow distribution P_β, where β is the change rate. In the simplest case, Eve knows the payload and

---

[1] P_{β_0} stands for the distribution of stego objects modified with change rate β_0.

solves the following simple binary hypothesis-testing problem:

$$H_0 : \beta = 0, \tag{6.12}$$

$$H_1 : \beta > 0 \text{ known.} \tag{6.13}$$

The optimal detector for this problem is the likelihood-ratio test (see Appendix D, equation (D.10))

$$L_\beta(\mathbf{x}) = \sum_{i=1}^{n} \log \frac{P_\beta(\mathbf{x}[i])}{P_0(\mathbf{x}[i])}. \tag{6.14}$$

It is shown in Section D.2 that for small β and large n, $L_\beta(\mathbf{x})/n$ approaches the Gaussian distribution

$$\frac{1}{n} L_\beta(\mathbf{x}) \sim \begin{cases} N\left(-\frac{1}{2}\beta^2 I(0), \frac{1}{n}\beta^2 I(0)\right) & \text{under } H_0 \\ N\left(\frac{1}{2}\beta^2 I(0), \frac{1}{n}\beta^2 I(0)\right) & \text{under } H_1, \end{cases} \tag{6.15}$$

where $I(\beta)$ is the Fisher information for one observation (see Section D.2 for more details about Fisher information),

$$I(\beta) = \sum_x \frac{1}{P_\beta(x)} \left(\frac{\partial P_\beta(x)}{\partial \beta}\right)^2. \tag{6.16}$$

The connection between Fisher information and KL divergence is seen through the following equation, which holds up to second order in β (see Proposition B.7):

$$D_{\mathrm{KL}}(P_0||P_\beta) = D_{\mathrm{KL}}(P_\beta||P_0) = \frac{\beta^2}{2} I(0). \tag{6.17}$$

Because of the Gaussian character of the detection statistic, the performance of Eve's optimal steganography detector is completely described using the deflection coefficient (see Section D.1.2)

$$d^2 = \frac{\left(\frac{1}{2}\beta^2 I(0) + \frac{1}{2}\beta^2 I(0)\right)^2}{\beta^2 I(0)/n} = n\beta^2 I(0). \tag{6.18}$$

We note that a larger deflection coefficient means a more accurate detector and thus a less secure stegosystem. Therefore, we can conclude that for small β stegosystems with larger Fisher information $I(0)$ are more detectable than those with smaller $I(0)$, which makes this quantity useful for benchmarking. Moreover, through the Cramer–Rao lower bound (Section D.6), $I(0)$ imposes a lower bound on the variance of any unbiased estimator of the change rate $\beta \approx 0$ from an n-element stego object,

$$\mathrm{Var}[\hat\beta] \geq \frac{1}{nI(0)}. \tag{6.19}$$

In Exercises 6.4 and 6.5, the reader is guided to derive the Fisher information, $I(0)$, for LSB embedding and ± 1 embedding in iid cover sources.

In this section, we worked under the assumption that the cover was an iid sequence. Fisher information can also be used for benchmarking stegosystems with covers exhibiting dependences in the form of a Markov chain [69, 70]. For such stegosystems, similar limiting behavior for small change rates can be established for the KL divergence between the Markov chain of covers and the hidden Markov chain of stego objects.[2]

Fisher information could potentially be used for benchmarking stegosystems with real-life cover sources, such as digital images. Here, there is little hope, however, that we could compute it analytically due to the difficulty of modeling images. A plausible practical option is to model images using numerical features, such as features used in blind steganalysis (see Chapter 12), and compute the Fisher information experimentally from a database of cover and stego images embedded with varying change rate. This is currently an active area of research and the interested reader is referred to [136, 191] and the references therein.

6.2   Perfectly secure steganography

In the previous section, we introduced the concept of a perfectly secure stegosystem by demanding that such systems preserve the distribution of covers. At this point, it is not clear whether non-trivial secure systems indeed exist. Thus, in this section we describe several examples of secure and ϵ-secure stegosystems. Other examples can be found in Section 6.3. We also point out the relationship between perfect security and perfect compression.

Consider the following simple one-time-pad steganographic system with $\mathcal{C} = \mathcal{X}^n$, $\mathcal{X} = \{0,1\}$, and P_c the uniform distribution on $\mathcal{X}^n$ [35]. Given a secret message $\mathbf{m} \in \mathcal{X}^n$, Alice selects $\mathbf{k} \in \mathcal{K} = \mathcal{X}^n$ at random and synthesizes the stego object as XOR (eXclusive OR) $\mathbf{y} = \mathbf{k} \oplus \mathbf{m}$. The message is extracted by Bob as $\mathbf{m} = \mathbf{k} \oplus \mathbf{y}$. Obviously, Alice and Bob need to preagree on the set of secret keys and their selection prior to starting the communication. The stego objects will again follow a uniform distribution on $\mathcal{C}$. This system is, however, not very useful because if random bit strings were commonly exchanged there would be no need for steganography as secrecy could be simply achieved using cryptographic means.

For general cover sources, an ϵ-secure stegosystem can be obtained using the principle of cover selection in the following manner [35]. We first split the set of all covers, $\mathcal{C}$, into two subsets of approximately the same probability, $\mathcal{C} = \mathcal{C}_0 \cup \mathcal{C}_1$, $P_c(\mathcal{C}_0) \approx P_c(\mathcal{C}_1)$. For example, two such sets can be found as

$$\mathcal{C}_0 = \arg\min_{\mathcal{C}' \subset \mathcal{C}} \left| P_c(\mathcal{C}') - P_c(\mathcal{C} - \mathcal{C}') \right|, \ \mathcal{C}_1 = \mathcal{C} - \mathcal{C}_0. \tag{6.20}$$

---

[2] Steganographic embedding into a Markov chain produces stego objects that are no longer Markov and instead form a hidden Markov chain (see, e.g., [213]).

Let $P_{c,0}(\mathbf{x})$ be the distribution P_c constrained on $\mathcal{C}_0$,

$$P_{c,0}(\mathbf{x}) = \begin{cases} P_c(\mathbf{x})/P_c(\mathcal{C}_0) & \mathbf{x} \in \mathcal{C}_0 \\ 0 & \text{otherwise.} \end{cases} \tag{6.21}$$

Similarly, let $P_{c,1}$ be P_c constrained on $\mathcal{C}_1$. If Alice wants to send bit m, she simply generates a cover using $P_{c,m}$ and sends it to Bob. Bob shares with Alice the breakup of $\mathcal{C}$ into $\mathcal{C}_0 \cup \mathcal{C}_1$ and extracts the bit by noting in which subset the stego object lies. Unless $P_c(\mathcal{C}_0) = P_c(\mathcal{C}_1)$, this stegosystem will not be perfectly secure but only ϵ-secure. We now compute the KL divergence between P_c and the stego distribution P_s.

Let $\delta = P_c(\mathcal{C}_0) - P_c(\mathcal{C}_1)$. Then, because $P_c(\mathcal{C}_0) + P_c(\mathcal{C}_1) = 1$, we have $P_c(\mathcal{C}_0) = (1+\delta)/2$ and $P_c(\mathcal{C}_1) = (1-\delta)/2$. Thus, the probability that $\mathbf{x} \in \mathcal{C}_0$ is sent as a stego object is $P_s(\mathbf{x}) = \frac{1}{2}P_{c,0}(\mathbf{x}) = \frac{1}{2}P_c(\mathbf{x})/P_c(\mathcal{C}_0) = P_c(\mathbf{x})/(1+\delta)$. Similarly, $P_s(\mathbf{x}) = P_c(\mathbf{x})/(1-\delta)$ for $\mathbf{x} \in \mathcal{C}_1$. Thus, we can write for the KL divergence

$$D_{\mathrm{KL}}(P_c||P_s) = \sum_{\mathbf{x} \in \mathcal{C}_0} P_c(\mathbf{x})\log(1+\delta) + \sum_{\mathbf{x} \in \mathcal{C}_1} P_c(\mathbf{x})\log(1-\delta) \tag{6.22}$$

$$= P_c(\mathcal{C}_0)\log(1+\delta) + P_c(\mathcal{C}_1)\log(1-\delta) \tag{6.23}$$

$$= \frac{1+\delta}{2}\log(1+\delta) + \frac{1-\delta}{2}\log(1-\delta) \tag{6.24}$$

$$\leq \frac{1+\delta}{2}\delta - \frac{1-\delta}{2}\delta = \delta^2. \tag{6.25}$$

Here, we used the log inequality $\log(1+x) \leq x$ for all real x. Thus, the stegosystem is δ^2-secure.

This construction can be generalized by splitting $\mathcal{C}$ into more than two subsets, which would allow Alice to communicate $\log_2 k$ bits if k subsets were used. A variant of this method in which $\mathcal{C}_0$ and $\mathcal{C}_1$ form two interleaved lattices can be used to design ϵ-secure stegosystems with distortion-limited embedder as investigated in Section 6.3.

6.2.1   Perfect security and compression

If Alice could somehow arrange that her stego objects follow the cover distribution P_c, her stego scheme would be perfectly secure. In this case, Alice would be on average sending $H(P_c)$ bits in every cover, where $H(P_c) = -\sum_{\mathbf{x} \in \mathcal{C}} P_c(\mathbf{x})\log P_c(\mathbf{x})$ is the entropy of the cover source. Alice could theoretically construct such a steganographic method using the principle of embedding by cover synthesis [3] (Chapter 4). The reader is encouraged to read Section B.3 to better appreciate the following arguments.

Let us assume that Alice has a perfect prefix-free compression scheme. Here, by perfect we understand that the covers can be on average encoded using $H(P_c)$ bits. Alice feeds her encrypted (and thus "random") message bits into the decompressor one-by-one and, as soon as she obtains a codeword, she simply sends to

Bob the object from $\mathcal{C}$ encoded by that codeword. Then, she continues with the remainder of the message bits till the complete message has been sent. Bob reads the secret message bits by compressing Alice's objects using the same compression scheme and concatenating the bit strings. Because Alice uses a perfect prefix-free compressor, she will be sending objects that exactly follow the required distribution P_c and thus $P_s = P_c$, which means that the steganographic method is perfectly secure. Note that the average steganographic capacity over all covers is the entropy $H(P_c)$.

This stegosystem is rather academic because the complexity and dimensionality of covers formed by digital media objects, such as images, will prevent us from determining even a rough approximation to the distribution P_c. It is, however, possible to realize this idea within a sufficiently simple model of covers (see Section 7.1.2 on model-based steganography). The scheme also tells us something quite fundamental. In Chapter 3, we learned that the process of image acquisition using imaging sensors is influenced by multiple sources of imperfections and noise. Some of the noise sources, such as the shot noise, are truly *random* phenomena caused by the quantum properties of light.[3] Thus, in the absence of other imperfections, even if we took multiple images of exactly the same scene with identical camera settings, we would obtain slightly different images. The images could be described as a superposition of the true scene $\mathbf{S}[i]$ and a two-dimensional field of some (not necessarily independent or identically distributed) random variables $\boldsymbol{\eta}[i]$,

$$\mathbf{x}[i] = \mathbf{S}[i] + \boldsymbol{\eta}[i], \; i = 1, \ldots, n, \tag{6.26}$$

where n is the number of pixels in the image. The variables $\boldsymbol{\eta}[i]$ will be locally dependent and also dependent on $\mathbf{S}$ due to demosaicking, in-camera processing, and artifacts, such as blooming, and possibly due to JPEG compression. Because the dependence is only local, $\boldsymbol{\eta}[i]$ and $\boldsymbol{\eta}[j]$ will be independent as long as the distance between the pixels i and j is larger than some fixed threshold determined by the physical phenomena inside the sensor and the character of the dependences. If we model the dependences as a Markov Random Field (MRF), it is known that the entropy of a MRF increases linearly with the number of random variables [50]. Thus, with increasing number of pixels, n, the entropy of $\boldsymbol{\eta}$ will be proportional to the number of pixels,

$$H(\boldsymbol{\eta}) = O(n). \tag{6.27}$$

We can thus conclude that fundamentally the steganographic capacity of images obtained using an imaging sensor increases *linearly* with the number of pixels. The reader is encouraged to compare this finding with the results from Chapter 13, which is dedicated to the issue of steganographic capacity.

---

[3] Experiments convincingly violating Bell's inequalities [8] showed that quantum mechanics is free of "hidden" parameters and is thus indeterministic.

This thought experiment also tells us that with every digital image, $\mathbf{x}$, acquired by a sensor, there is a large cluster of natural images that slightly differ from $\mathbf{x}$ in their noise components. And the size of this cluster increases exponentially with the number of pixels in the image. Essentially all steganographic schemes try, in one way or another, to reach into this image cluster by following certain design principles and elements as discussed in Chapters 7–9.

6.2.2   Perfect security with respect to model

Constructing secure steganographic schemes for digital-media objects, such as digital images, is not an easy task. The problem is that very little is typically known about the distribution P_c, which describes the cover source. In the classical prisoners' scenario, if Alice hides messages in images taken with her own digital camera, the only legitimate images she can send are those taken from her prison cell with her camera, potentially processed in the computer (e.g., cropped, compressed, filtered, etc.). In this case, the cover source is rather narrow but still too complex to model it analytically. Due to the very high dimensionality of images, it is equally impossible to even obtain an accurate sampled version of P_c.

To make this rather complex problem more manageable, one usually adopts a model for the cover source and proves statistical undetectability *within the model.* (This would force Eve to use a better model to detect the hidden data.) A common simplification is to model the statistics of *individual pixels* (or DCT coefficients) in the image rather than the statistics of images as a whole. For example, images represented in the transform domain using wavelet, Fourier, or DCT coefficients can be reasonably well modeled as a sequence of iid random variables because the transform coefficients are largely decorrelated.[4] Thus, an often-made assumption is to model the cover image $\mathbf{x}$ as an iid sequence $\mathbf{x}[i] \sim f(x; \boldsymbol{\theta})$, where f is the probability mass function that depends on a vector parameter $\boldsymbol{\theta}$. For example, the distribution of wavelet coefficients is often modeled using the generalized Gaussian variable or the generalized Cauchy variable (Section A.8). If the value of $\boldsymbol{\theta}$ is the same across all cover images, we speak of a homogeneous cover source. A more realistic cover-source model that allows dependences among neighboring pixels is obtained by modeling $\mathbf{x}[i]$ as a Markov chain [70, 71, 225]. In this case, $\boldsymbol{\theta}$ would be the transition-probability matrix. An even more realistic cover model is the heterogeneous [22] model, in which $\boldsymbol{\theta}$ depends on the image. Imposing a probabilistic model on $\boldsymbol{\theta}$ leads to a mixture model for the entire cover source. One of the few theoretical steganalysis studies that use this mixture model is [148].

The power of the models described in the previous paragraph is that one can obtain their sample approximations from a single image (for homogeneous

---

[4] For block transforms, such as the DCT in JPEG compression, one can use a more general model and consider the image as 64 parallel channels, each modeled as an iid sequence with a different distribution.

models) or a large image database to also estimate the distribution of $\boldsymbol{\theta}$ for heterogeneous models. Adopting this approach, Alice can then guarantee that her steganography will be undetectable within the model if her embedding preserves the essential statistics. For an iid source, she needs to preserve the histogram, while for the Markov model, she needs to preserve the transition-probability matrix and the Markov property. This approach to steganography enables exact mathematical proofs of security as well as progressive, methodological improvement. There is some hope that with advances in statistical modeling of images, this approach will lead to secure stegosystems.

So far, however, all steganographic schemes for digital images that follow this paradigm (see the methods in Chapter 7) have been broken. This is because all that Eve needs to do to attack a stegosystem that is provably secure within a given model is to use a better model of covers that is not preserved by embedding. Often, it is sufficient to merely identify a *single statistical quantity* that is predictably perturbed by embedding, which is usually not very hard given the complexity of typical digital media. We discuss these issues from the point of view of steganalysis in Chapters 10–12.

6.3   Secure stegosystems with limited embedding distortion

Note that the information-theoretic definition of steganographic security described in Section 6.1 is not concerned with *fidelity*. The embedding distortion can be arbitrary as long as the source of stego objects is statistically indistinguishable from the cover source. Indeed, the perfectly secure systems from Section 6.2 are based on the paradigms of steganography by cover synthesis and cover selection rather than by cover modification and thus lack the concept of embedding distortion. However, in steganography based on cover modification, which is the paradigm for most practical steganographic schemes, Alice starts with a specific cover object and then modifies it to embed the secret message. Here, care is usually being taken to keep the embedding distortion low to prevent introducing atypical statistical characteristics into the stego object. It thus makes sense to consider steganographic security when Alice has to comply with a bound on embedding distortion. (In this case, she is called a distortion-limited embedder.) The idea to study the security of such stegosystems appeared for the first time in [179] by drawing an analogy to the methodology employed in robust watermarking.

In this section, we formulate secure steganography with a distortion-limited embedder and we do so for the more general case of an active warden (the passive-warden scenario is a special case of this formulation) [180, 235]. We also give examples of provably secure stegosystems for some simple models of the cover source. Although the constructions in this chapter are more important from the theoretical point of view and so far have not produced secure steganographic

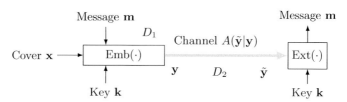

Figure 6.2 A diagram of steganographic communication with an embedder whose distortion is limited by D_1 and noisy channel $A(\cdot|\cdot)$ with distortion limited by D_2.

schemes that would work for real digital media, the results provide quite valuable fundamental insight.

Covers will be represented as sequences of n elements from some alphabet $\mathcal{X}$, $\mathbf{x} \in \mathcal{C} = \mathcal{X}^n$. The steganographic channel, captured with the embedding mapping Emb, extraction mapping Ext, source of covers, messages, and secret keys, is now augmented with two more requirements that bound the expected embedding distortion and the expected channel distortion that the stego object experiences on its way from Alice to Bob,

$$E\left[d(\mathsf{x},\mathbf{y})\right] = \sum_{\mathbf{x}} P_\mathrm{c}(\mathbf{x}) \sum_{\mathbf{k}} P_\mathrm{k}(\mathbf{k}) \sum_{\mathbf{m}} P_\mathrm{m}(\mathbf{m}) d\left(\mathbf{x}, \mathrm{Emb}(\mathbf{x},\mathbf{k},\mathbf{m})\right) \leq D_1, \quad (6.28)$$

$$\sum_{\mathbf{y},\tilde{\mathbf{y}}} P_\mathrm{s}(\mathbf{y}) A(\tilde{\mathbf{y}}|\mathbf{y}) d(\tilde{\mathbf{y}},\mathbf{y}) \leq D_2, \quad (6.29)$$

where $d(\mathbf{x},\mathbf{y})$ is some measure of distortion per cover element, such as the energy $\frac{1}{n} d_2(\mathbf{x},\mathbf{y}) = \frac{1}{n} \sum_{i=1}^{2}(\mathbf{x}[i] - \mathbf{y}[i])^2$. The matrix $A(\tilde{\mathbf{y}}|\mathbf{y})$ captures the probabilistic active warden (noisy channel) and stands for the conditional probability that the noisy stego object, $\tilde{\mathbf{y}}$, is received by Bob when stego object $\mathbf{y}$ was sent by Alice. The scalar values, D_1 and D_2, are the bounds on the per-element embedding and channel distortion. Again, the stegosystem is considered secure if $P_\mathrm{c} = P_\mathrm{s}$. A diagram showing the main elements of the communication channel is displayed in Figure 6.2.

Secure stegosystems with distortion-limited embedder exist and two examples, for the passive-warden case, are given below.[5]

6.3.1   Spread-spectrum steganography

Let us assume that covers are sequences of n iid Gaussian random variables, $\mathbf{x}[i] \sim N(0, \sigma_c^2)$, $i = 1,\ldots,n$. Using a secret key $\mathbf{k}$, Alice seeds a cryptographically secure PRNG with her stego key and generates n iid realizations $\mathbf{w}[i]$, $i = 1,\ldots,n$, of a Gaussian random variable $N(0, D_\mathrm{w})$. Then, Alice embeds one secret message bit $m \in \{0,1\}$ by either adding $\mathbf{w}$ or $-\mathbf{w}$ to the cover,

$$\mathrm{Emb}(\mathbf{x},\mathbf{k},m) = \gamma\mathbf{x} + (2m - 1)\mathbf{w} = \mathbf{y} \qquad (6.30)$$

---

[5] These examples appeared in [235].

for some scalar γ. We now need to determine γ and D_{w} so that the cover model is preserved and the embedding distortion is below D_1.

Because $\mathbf{w}$ is independent of $\mathbf{x}$, the stego object $\mathbf{y}$ is a sequence of iid Gaussian random variables $N(0, \gamma^2 \sigma_{\mathrm{c}}^2 + D_{\mathrm{w}})$. To obtain a perfectly secure stegosystem, we need to preserve the Gaussian model of the cover source. Because $\mathbf{y}$ is a zero-mean Gaussian signal, we must preserve the variance $\gamma^2 \sigma_{\mathrm{c}}^2 + D_{\mathrm{w}} = \sigma_{\mathrm{c}}^2$. The expected value of the embedding distortion is

$$E\left[\frac{1}{n}d_2(\mathbf{x}, \mathbf{y})\right] = \frac{1}{n}E\left[\sum_{i=1}^{n}(\mathbf{x}[i] - \mathbf{y}[i])^2\right] \tag{6.31}$$

$$= \frac{1}{n}E\left[\sum_{i=1}^{n}((1-\gamma)\mathbf{x}[i] - (2m-1)\mathbf{w}[i])^2\right] \tag{6.32}$$

$$= (1-\gamma)^2 \sigma_{\mathrm{c}}^2 + D_{\mathrm{w}}, \tag{6.33}$$

which will be bounded by D_1 when

$$\sigma_{\mathrm{c}}^2 - 2\gamma\sigma_{\mathrm{c}}^2 + \gamma^2\sigma_{\mathrm{c}}^2 + D_{\mathrm{w}} \le D_1, \tag{6.34}$$

or

$$\gamma \le 1 - \frac{D_1}{2\sigma_{\mathrm{c}}^2}. \tag{6.35}$$

Thus, the parameters γ and D_{w} that provide perfect security are

$$\gamma = 1 - \frac{D_1}{2\sigma_{\mathrm{c}}^2}, \tag{6.36}$$

$$D_{\mathrm{w}} = \sigma_{\mathrm{c}}^2(1-\gamma^2) = D_1\left(1 - \frac{D_1}{4\sigma_{\mathrm{c}}^2}\right). \tag{6.37}$$

To finish the description of the stegosystem, we need to describe the extraction mapping Ext. Bob first generates $\mathbf{w}$ using the stego key, computes the correlation $\rho = \frac{1}{n}\sum_{i=1}^{n}\mathbf{y}[i]\mathbf{w}[i]$, and extracts the message bit using the following rule:

$$\rho > 0 \Rightarrow m = 1, \tag{6.38}$$

$$\rho \le 0 \Rightarrow m = 0. \tag{6.39}$$

As shown in Section D.1.2, correlation is the likelihood-ratio test for Bob's simple binary hypothesis test

$$H_0 : m = 0 \tag{6.40}$$

$$H_1 : m = 1 \tag{6.41}$$

given the observed $\mathbf{y}[i]$, $i = 1, \ldots, n$. The test statistic ρ follows the Gaussian distribution with mean $(2m-1)D_1$ and variance proportional to $1/\sqrt{n}$. Thus, with increasing n, the message bit will be correctly extracted with probability approaching 1.

Note that this low-capacity steganographic method is also robust with respect to channel noise. Let us assume that the stego object is subject to noise and

Bob receives $\tilde{\mathbf{y}} = \mathbf{y} + \mathbf{z}$, where each $\mathbf{z}[i] \sim N(0, D_2)$. The distribution of the test statistic ρ has the same mean but a higher variance. Nevertheless, with increasing n, its variance is again proportional to $1/\sqrt{n}$.

Finally, we note that, even though the spread-spectrum embedding method is perfectly secure with a distortion-limited embedder, it has a low embedding capacity far below the theoretical steganographic capacity for such a channel [49, 235] (also, see Chapter 13).

6.3.2   Stochastic quantization index modulation

Quantization Index Modulation (QIM) is a data-embedding method originally proposed for robust watermarking [42]. It can also be used for design of secure steganographic schemes [49, 235]. The central concept is the notion of an N-dimensional lattice Λ, which is a set of discrete points in a Euclidean space $\mathbb{R}^N$, usually spaced in some regular pattern. The lattice defines a quantizer, Q_Λ, that maps any point in the space to the closest point from the lattice, $Q_\Lambda(\mathbf{x}) = \mathbf{y}$, $\mathbf{y} = \arg\min_{\mathbf{x}' \in \Lambda} \|\mathbf{x} - \mathbf{x}'\|$. The set of points that quantize to the origin $(0, 0)$ is called the Voronoi cell and is defined as

$$\mathcal{V} = \{\mathbf{x} \in \mathbb{R}^N \,|\, \|\mathbf{x}\| \le \|\mathbf{x} - \mathbf{x}'\| \text{ for all } \mathbf{x}' \in \Lambda\}, \tag{6.42}$$

where $\|\mathbf{x}\|$ is some norm in $\mathbb{R}^N$. In Figure 6.3, left, the Voronoi cell for lattice $\Lambda_\circ$ is highlighted in gray (in L_1 norm).

The QIM works with a family of interleaved lattices indexed by the message that Alice wants to send to Bob. For each message m, we define a dither vector $\mathbf{d}_m \in \mathbb{R}^N$ and a shifted lattice $\Lambda_m = \Lambda + \mathbf{d}_m$ called the mth coset of Λ. The union of cosets $\cup \Lambda_m = \Lambda_{\mathrm{f}}$ is called the fine lattice and it forms the backbone of the QIM data-embedding method. We will denote the Voronoi cell of the fine lattice as $\mathcal{V}_{\mathrm{f}}$.

In the simplest, classical version of the QIM, the cover is quantized to the lattice determined by the message Alice wants to send. Alice divides the cover into blocks of N elements and quantizes each block $\mathbf{x} \in \mathbb{R}^N$ to Λ_m in order to embed m in $\mathbf{x}$. Formally, the stego block is $\mathbf{y} = Q_{\Lambda_m}(\mathbf{x})$. The lattice Λ and the set of dither vectors are shared with Bob, which enables him to extract the message as the index of the coset to which the stego block belongs. Notice that this embedding method is robust with respect to small amounts of noise. As long as the noisy block $\tilde{\mathbf{y}}$ does not get out of its shifted Voronoi cell $\mathcal{V} = \mathbf{y} + \Lambda_m$ that surrounds $\mathbf{y}$, Bob can quantize the distorted block $\tilde{\mathbf{y}}$ to Λ_{f} before extracting the message. By changing the spacing between the points in Λ, one can obtain a trade-off between the embedding distortion and resistance to channel noise.

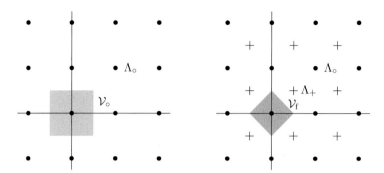

Figure 6.3 Left: Lattice $\Lambda_\circ$ (circles) with its Voronoi cell $\mathcal{V}_\circ$. Right: The fine lattice $\Lambda_\circ \cup \Lambda_+$ and its Voronoi cell $\mathcal{V}_f$. Note that $\Lambda_+ = \Lambda_\circ + (1,1)$.

Example 6.1: [Stochastic QIM with two lattices] Figure 6.3 shows an example of a regular lattice in $\mathbb{R}^2$,

$$\Lambda = \Lambda_\circ = \{(2i, 2j) \mid i, j \in \mathbb{Z}\} \tag{6.43}$$

and two dither vectors $\mathbf{d}_\circ = (0,0)$, $\mathbf{d}_+ = (1,1)$. Note that $\Lambda_+ = \{(2i+1, 2j+1) \mid i, j \in \mathbb{Z}\}$, $\mathcal{V}_\circ = [-1,1] \times [-1,1]$, and $\mathcal{V}_f = \{\mathbf{x} \mid |\mathbf{x}[1] + \mathbf{x}[2]| < 1\}$ if we use the L_1 norm. The message that can be sent in each block consisting of two cover elements is one bit, $m \in \{0, 1\}$.

This simple version of QIM is not suitable for steganography as the stego images confined to the fine lattice would certainly not follow the distribution of covers and could easily be identified by Eve as suspicious. Alice needs to spread the stego objects around the lattice points to preserve the distribution. This is the idea behind the stochastic QIM [235]. For now, we will assume that the covers are $\mathbf{x} \in \mathbb{R}^N$ rather than considering them broken up into disjoint blocks of N elements each.

The Euclidean space is divided into M regions that are translates of each other,

$$\mathcal{R}_m = \Lambda_m + \mathcal{V}_f, \ 1 \le m \le M. \tag{6.44}$$

Note that $\mathcal{R}_m$ is the region of all noisy stego objects that carry the same message, m. Let $\mathbf{p}[m] = P_c(\mathcal{R}_m)$ be the probability of a cover being in $\mathcal{R}_m$. Let us further assume that the probability that message m is sent is exactly $\mathbf{p}[m]$. This could be arranged, for example, by prebiasing the stream of originally uniformly distributed message symbols using an entropy decoder (see more details in Section 7.1.2 on model-based steganography).

We now define the embedding function $\mathrm{Emb}(\mathbf{x}, m)$ for any cover $\mathbf{x} \in \mathbb{R}^N$ and message m. Alice first identifies $\mathbf{z} \in \Lambda_m$ that is closest to $\mathbf{x}$. If $\mathbf{z}$ already belongs to $\mathcal{R}_m$, no embedding change is required and Alice simply sends $\mathbf{y} = \mathbf{x}$. In the opposite case, Alice generates $\mathbf{y}$ randomly from the shifted Voronoi cell $\mathbf{z} + \mathcal{V}_f$ according to the distribution P_c constrained to the cell $\mathbf{z} + \mathcal{V}_f$, $P_c(\cdot)/P_c(\mathbf{z} + \mathcal{V}_f)$. Note that the bound on embedding distortion, D_1, determines how close to each other the lattice points must be.

It should be clear that the stego object $\mathbf{y}$ ends up in $\mathcal{R}_m$ with probability $\mathbf{p}[m]$. Also, within each cell, the probability distribution has the correct shape. The only question is with what probability $\mathbf{y}$ appears in a given cell, or $\Pr\{\mathbf{y} \in \mathbf{z} + \mathcal{V}_f\}$. It can be shown that for a well-behaved P_c, $\Pr\{\mathbf{y} \in \mathbf{z} + \mathcal{V}_f\} \approx P_c(\mathbf{z} + \mathcal{V}_f)$. More accurately, for any $\epsilon > 0$, $|\Pr\{\mathbf{y} \in \mathbf{z} + \mathcal{V}_f\} - P_c(\mathbf{z} + \mathcal{V}_f)| < \epsilon$ for sufficiently small D_1. Rather than proving this for the general case, we provide a qualitative argument for the simple case of the fine lattice shown in Figure 6.3, right. It shall be clear that the argumentation applies to more general cases.

Let us compute the probability of $\mathbf{y} \in \mathbf{z} + \mathcal{V}_f$. In Figure 6.4, $\mathbf{z} + \mathcal{V}_f$ is the diamond. Since $\mathbf{p}[0] = \mathbf{p}[1] = \frac{1}{2}$, $\Pr\{\mathbf{y} \in \mathbf{z} + \mathcal{V}_f\}$ is $\frac{1}{2}$ times the probability that $\mathbf{x}$ falls into the cell directly plus $\frac{1}{2}$ times the probability that we end up in one of the four triangles and move to the cell. Denoting the union of the four triangular regions as $\mathcal{T}$, we have

$$P_s(\mathbf{z} + \mathcal{V}_f) = \Pr\{\mathbf{y} \in \mathbf{z} + \mathcal{V}_f\} = \frac{1}{2}P_c(\mathbf{z} + \mathcal{V}_f) + \frac{1}{2}P_c(\mathcal{T}). \qquad (6.45)$$

If the distribution P_c were locally linear around $\mathbf{z}$, we would have the equality $P_s(\mathbf{z} + \mathcal{V}_f) = \Pr\{\mathbf{y} \in \mathbf{z} + \mathcal{V}_f\} = P_c(\mathbf{z} + \mathcal{V}_f)$. For general but well-behaved P_c,[6] the difference between the probabilities $P_s(\mathbf{z} + \mathcal{V}_f)$ and $P_c(\mathbf{z} + \mathcal{V}_f)$ can be made arbitrarily small for sufficiently small D_1. (Recall that D_1 determines the spacing between the points in the lattice.) In summary, we will have $|P_c(\mathbf{x}) - P_s(\mathbf{x})| < \epsilon$ for all $\mathbf{x}$ for sufficiently small D_1. Assuming $P_c(\mathbf{x}) > \kappa > 0$ whenever $P_c(\mathbf{x}) \neq 0$ (which will be satisfied for any finite cover set $\mathcal{C}$), we can use the result of Exercise 6.2 to claim that the KL divergence $D_{\mathrm{KL}}(P_c\|P_s) < \epsilon/\kappa$.

6.3.3   Further reading

Steganography with distortion-limited embedder in the active-warden scenario has been investigated in [49, 180, 235, 237]. Here, we provide only a brief summary of the achievements while referring the reader to technical papers for more information.

The authors of [180] define the concept of steganographic capacity and study it for two alternative definitions of a distortion-limited embedder and two definitions of a distortion-limited warden. The rather surprising result is that the capacity can be computed in the same manner no matter what combination of the definitions is taken. A portion of this result appears in Chapter 13 dealing with the subject of steganographic capacity. The authors also give a specific construction of secure steganographic schemes for the cover source formed by

---

[6] E.g., for smooth P_c with bounded partial derivatives.

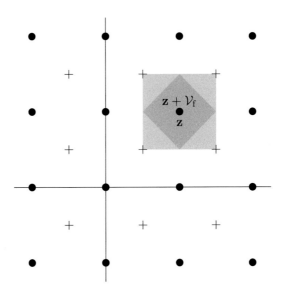

Figure 6.4 Example of a cell $\mathcal{V}_f$ surrounding a point $\mathbf{z}$ from the fine lattice formed by circles and crosses. The four triangular regions shaded light gray are the regions from where the cover $\mathbf{x}$ can be mapped to the cell during embedding.

Bernoulli sequences $B\left(\frac{1}{2}\right)$ (binary sequences with probability of 0 equal to $\frac{1}{2}$) both under an active and under a passive warden. Covers with a multivariate Gaussian distribution and their security and steganographic capacity are studied in [235]. This work also contains a rather fundamental result that block embedding in such Gaussian covers is insecure in the sense that the KL divergence grows linearly with the number of blocks. Perfectly secure and ϵ-secure steganographic methods for Gaussian covers are studied in [49]. The authors compute a lower bound on the capacity of perfectly secure stegosystems and study the increase in capacity for ϵ-secure steganography. They also describe a practical lattice-based construction for high-capacity secure stegosystems. The work is contrasted with results obtained for digital watermarking where the steganographic constraint $P_c = P_s$ is absent. Codes for construction of high-capacity stegosystems are described in [237]. The authors also show how their approach can be generalized from iid sources to Markov sources and sources over continuous alphabets.

6.4   Complexity-theoretic approach

The information-theoretic definition of steganographic security is based on two idealizations that are rather strong. The first one is the assumption that covers can be described by a probability distribution, which is completely known to Eve. This assumption is accepted as the worst-case scenario because one may argue that Eve can observe the channel traffic for sufficiently long to learn this distribution with arbitrary accuracy. Eve may also obtain additional information through espionage. Thus, our paranoia dictates that we should assume that Eve knows all details of the distribution. Realistically, however, this can only be reasonable within a sufficiently simple model of covers in some artificially conceived communication channel rather than in reality, where the covers are very complex objects, such as digital-media files. Additionally, one may attack the very assumption that covers can be described by a random variable at all. We choose not to delve into this rather philosophical issue and instead refer the reader to [22] for an intriguing discussion of this issue from the perspective of epistemology.

Second, the information-theoretic approach by definition ignores complexity issues. It is concerned only with the possibility of constructing an attack rather than its practical realization. It is quite feasible that, even though an attack on a stegosystem can be mounted in principle, its practical realization may require resources so excessive that no warden can implement the attack. For example, if the computational complexity of the attack grows exponentially with the security parameters of the stegosystem (stego key length and number of samples in the cover), one could choose the key length and covers large enough to make sure that any warden with polynomially bound resources will be destined to random guessing. This idea is the basis of definitions of steganographic security based on complexity-theoretic principles [113, 124].

To help explain the point further, we take a look at the related field of cryptography. According to Shannon's groundbreaking work on cryptosystems, the only unconditionally secure cryptosystems are those whose key has the same length as the text to be encrypted. This is what security defined using information theory leads to. However, there exists an important and very useful class of asymmetric cryptosystems (so-called public-key systems [207]) whose security stems from the excessive computational complexity of constructing the attack rather than the fundamental impossibility to construct one. In this section, we explain a similar approach to defining steganographic security that is based on complexity-theoretic considerations.

In 2002, Hopper *et al.* [113] and Katzenbeisser and Petitcolas [124] independently proposed a complexity-theoretic definition of steganographic security. These two proposals share several important novel ideas. First, the requirement of knowing the probability distribution of covers is replaced with a weaker assumption of availability of an oracle O that samples from the set of covers according to their distribution over the channel. Second, the security of a stegosystem is established by means of a probabilistic game between a judge and the warden. The warden is allowed to sample the oracle O and the embedding oracle is implemented as a black box seeded with an unknown stego key. The warden is then essentially asked to distinguish between the outputs of the two oracles. The advantage of the warden is defined as the probability of correct decision minus $\frac{1}{2}$. The stegosystem is secure if the warden's advantage is *negligible* (it falls to zero faster than any power of $1/k$, where k is the security parameter, such as the number of bits in the stego key or the number of elements in the cover). We first describe the approach that appeared in [113].

6.4.1   Steganographic security by Hopper et al.

The communication channel is considered in a rather general framework as a distribution over sequences of timestamped bits[7] and thus extends to steganographic methods that potentially use timing between individual bits for message hiding. The oracle O is allowed to make partial draws from the channel – it can sample the channel distribution conditioned on channel history and provide the next b timestamped bits on the basis of the previous h bits. Such an oracle will be denoted $O_{h,b}$ and is assumed to be available to the prisoners as well as the warden.

A stegosystem is defined as a pair of probabilistic algorithms[8] $\mathrm{PEmb}(h, \mathbf{k}, \mathbf{m})$ and $\mathrm{PExt}(h, \mathbf{k}, \{0, 1\}^\star)$.[9] The embedding algorithm PEmb takes on its input a k-bit stego key $\mathbf{k} \in \{0, 1\}^k$, the secret message $\mathbf{m} \in \{0, 1\}^\star$, a history of h timestamped bits, and the oracle $O_{h,b}$. When applied to lb timestamped cover bits, PEmb returns a sequence of l blocks of b timestamped stego bits. The extraction algorithm takes a key $\mathbf{k}$, a history of h bits, and the sequence of lb stego bits, and returns the hidden message $\mathbf{m}$.

Security of the stegosystem with respect to a passive warden is established by the following game played between a judge and the warden. The warden is given access to oracle $O_{h,b}$ and is allowed to make as many queries as she likes. Then, the judge seeds the embedding oracle $\mathrm{PEmb}(\cdot, \mathbf{k}, \cdot)$ with a fixed stego key and implements it as a black box with two inputs marked with a dot. The first dot is an integer, h, while the second dot can be any bit string $\mathbf{m} \in \{0, 1\}^\star$. The judge then prepares a second oracle, $O(\cdot, \cdot)$, also with two inputs defined[10] as $O(h, \mathbf{m}) = O_{h,|\mathrm{PEmb}(h,\mathbf{k},\mathbf{m})|}$. In other words, $O(h, \mathbf{m})$ samples the next $|\mathrm{PEmb}(h, \mathbf{k}, \mathbf{m})|$ bits from the channel conditioned on the channel history of h bits. The judge then uniformly randomly selects either $\mathrm{PEmb}(\cdot, \mathbf{k}, \cdot)$ or $O(\cdot, \cdot)$ and gives it to the warden. The warden, who does not know whether she was given the first or the second oracle, is allowed to make at most q queries and draw at most l bits in total. She then uses the results of the queries to make a decision regarding whether she is observing an output from the stegosystem, PEmb, or the cover oracle O. Hopper defines advantage against a stegosystem

---

[7] A timestamped bit is a pair $\{\mathbf{m}[i], \mathbf{t}[i]\}$, where $\mathbf{t}[i]$ is the time of sending the ith bit $\mathbf{m}[i]$.

[8] A probabilistic algorithm uses randomness as part of its reasoning, usually realized using a PRNG.

[9] We use the prefix P in the embedding and extraction algorithms to stress the difference that the PEmb and PExt are now probabilistic algorithms rather than mappings as in the previous four sections or in Chapter 4. The symbol $\star$ in PExt means that it can operate on bit strings of any length.

[10] $|\mathbf{s}|$ denotes the length of bit string $\mathbf{s}$.

and ϵ-insecurity on the basis of the warden's ability to correctly identify the oracle using polynomial-complexity calculations.

One of the most intriguing implications of this complexity-theoretic view of steganographic security is the fact that secure stegosystems exist if and only if secure one-way (hash) functions exist under the assumption that the minimal entropy[11] of the cover source is larger than 2. This ties the security in steganography to the security of common cryptographic primitives. The proof of this statement is constructive and the embedding method is essentially a combination of steganography by cover selection and cover synthesis (see Chapter 4) that proceeds by embedding one bit in a block of b cover bits.

Alice uses a key-dependent hash function[12] $h(\mathbf{k}, \cdot)$ that maps b timestamped bits to $\{0, 1\}$. The hash function enters the embedding process in the following manner. Having synthesized (embedded) h timestamped stego bits, Alice generates the next b bits, $\mathbf{y}$, by querying the oracle $O_{h,b}$ until $h(\mathbf{k}, \mathbf{y})$ returns the required message. The number of calls to the oracle is limited by a fixed upper bound and thus there is a small probability that a message bit will not be embedded in the block. This embedding algorithm, which is called rejection sampling, is an example of steganography by cover selection. The extraction algorithm simply applies the same hash function to each block of b timestamped stego bits to recover one message bit from each block. Under some rather technical assumptions about the hash function, which are satisfied if secure one-way functions exist, this embedding algorithm is secure in the sense of Section 6.4.1.

6.4.2   Steganographic security by Katzenbeisser and Petitcolas

A similar complexity-theoretic definition of steganographic security has been given by Katzenbeisser and Petitcolas [124]. Here, we only summarize their approach. The judge first prepares the embedding oracle PEmb$(\cdot, \mathbf{k}, \cdot)$ implemented as a black box with a randomly chosen stego key $\mathbf{k}$. Here, the first argument is an input for the cover, $\mathbf{x}$, and the third argument for the message, $\mathbf{m}$. This oracle returns the stego object containing the message $\mathbf{m}$ embedded in $\mathbf{x}$ using key $\mathbf{k}$. The judge then gives the warden both the embedding oracle and the cover oracle O. The warden can query O an arbitrary (but finite) number of times, obtaining a sequence of covers that follow the distribution of covers. She can also query the embedding oracle with any messages and covers, including pathological covers or messages, obtaining thus a sequence of stego objects. Again, she is not limited in number of queries but performs only polynomial-complexity calculations. The queries to both oracles can be arbitrarily interwoven and based on the previous queries. After the warden has finished her reasoning, the judge queries O twice, obtaining two covers, $\mathbf{x}_1$ and $\mathbf{x}_2$, selects a random message, $\mathbf{m}$, and computes

[11] The minimal entropy, $H_{\min}$, of a source with pmf $\mathbf{p}[i]$ is defined as $H_{\min}(\mathbf{p}) = \min_i \log_2 1/\mathbf{p}[i]$.
[12] In [113], the authors use the term pseudo-random function.

$\text{PEmb}(\mathbf{x}_2, \mathbf{k}, \mathbf{m})$. Then, the judge flips a fair coin and gives either the cover $\mathbf{x}_1$ or the stego object $\text{PEmb}(\mathbf{x}_2, \mathbf{k}, \mathbf{m})$ to the warden. The warden performs a polynomial probabilistic test to decide whether she is observing a cover or stego object. The advantage of the warden is the probability of a correct guess minus $\frac{1}{2}$. The stegosystem is secure for oracle O if the warden's advantage is negligible.

It is possible to construct secure steganographic systems S by reducing an intractable problem P_{int} to the steganographic decision problem for S. The proof of security can then be realized by contradiction in the following manner. Under the assumption that S is not secure in the above sense, there exists a probabilistic game Z between the warden and the judge that allows the warden to detect stego objects with non-negligible probability. If the stegosystem S is constructed in such a way that the existence of Z implies that instances of P_{int} can be solved with non-negligible probability, we obtain a contradiction with the intractability of P_{int}. The interested reader is referred to [124], where the authors use this approach to construct a stegosystem whose insecurity would lead to an attack on the RSA cryptosystem [207].

6.4.3   Further reading

A number of authors expanded the initial work of Hopper on the complexity-theoretic approach to steganography. Most of the results are quite technical, requiring the reader to be closely familiar with advanced concepts used in study of public-key cryptosystems. A complexity-theoretic model for public-key steganography [153, 154, 234] with an active warden was proposed by Backes and Cachin [14] and further studied in [112]. Dedic *et al.* [56] showed that for secure stegosystems the number of queries the sender must carry out is exponential in the relative payload. Furthermore, they provide constructions of stegosystems that nearly match the achievable payloads. The assumption of availability of an oracle that samples conditionally on the channel history was criticized in [166]. The authors study how steganographic security is impacted by imperfect samplers that can only sample with limited history.

Having presented several alternative definitions of security in steganography, we note that the information-theoretic definition is the most widely accepted approach in investigation of the security of multimedia objects, such as digital images. It is possible that the concept of security in secret-key steganography will follow the same path as in cryptography. Practical symmetric cryptographic schemes, such as DES (Data Encryption Standard) or AES (Advanced Encryption Standard), cannot be proved secure but are nevertheless widely used because they offer properties that make them very useful in applications. Their security lies in the fact that nobody has so far been able to produce an attack substantially faster than a brute-force search for the key. Their design is based on principles and elements developed through joint research in cryptography and cryptanalysis. Steganography seems to be taking the same course. Theoretical models and existing attacks give the designers of stegosystems guidance and specific means for construction of the next generation of steganographic schemes. In practice, security is thus often understood in a much less rigorous sense as the inability to practically construct a reliable steganographic detector using existing attacks or their modifications.

Summary

- The information-theoretic definition of steganographic security assumes that the cover source is a random variable with known distribution P_c.
- Feeding covers into a stegosystem according to their distribution P_c, the distribution of stego objects follows distribution P_s.
- A steganographic system is perfectly secure (or ϵ-secure) if the Kullback–Leibler divergence $D_{KL}(P_c||P_s) = 0$ (or $D_{KL}(P_c||P_s) \leq \epsilon$).
- The KL divergence is a measure of how different the two distributions are.
- The existence of perfect compression of the cover source implies the existence of perfectly secure steganographic schemes.
- Describing the covers using a simplified model, we obtain a concept of security with respect to a model. Stegosystems preserving this model are undetectable within the model.
- Spread-spectrum methods and lattice-based methods (quantization index modulation) can be used to construct stegosystems with distortion-limited embedder and distortion-limited active warden.
- Steganographic security defined in the information-theoretic sense is concerned only with the possibility to mount an attack and not its feasibility (computational complexity).
- An alternative definition of security is possible in which access to the distributions of covers is replaced with availability of an oracle that generates covers. Security is then defined as the inability of a polynomially bounded warden to construct a reliable attack.
- The complexity-theoretic security of stegosystems can be proved by reducing the steganalysis problem to a known intractable problem or by using secure one-way functions.

Exercises

6.1 **[KL divergence between two Gaussians]** Let $f_1(x)$ and $f_2(x)$ be the pdfs of two Gaussian random variables $N(\mu_1, \sigma_1^2)$ and $N(\mu_2, \sigma_2^2)$. Prove for their KL divergence

$$D_{KL}(f_1||f_2) = \int f_1(x) \log \frac{f_1(x)}{f_2(x)} \mathrm{d}x = \frac{1}{2} \left(\log \frac{\sigma_2^2}{\sigma_1^2} + \frac{\sigma_1^2}{\sigma_2^2} - 1 + \frac{(\mu_2 - \mu_1)^2}{\sigma_2^2} \right).$$
$$(6.46)$$

6.2 [KL divergence is continuous in max norm] Let g and h be two distributions on $\mathcal{C}$ with $h(\mathbf{x}) > \kappa > 0$ whenever $g(\mathbf{x}) > 0$. Let h_ϵ be a small perturbation of h such that $\max_\mathbf{x} |h(\mathbf{x}) - h_\epsilon(\mathbf{x})| < \epsilon$. Then, for sufficiently small ϵ,

$$\left| D_{\mathrm{KL}}(g||h) - D_{\mathrm{KL}}(g||h_\epsilon) \right| < \frac{\epsilon}{\kappa}. \tag{6.47}$$

Hint: Use the log inequality $\log(1 + x) \le x$, which holds for all $x \in \mathbb{R}$.

6.3 [KL divergence and detectability (counterexample)] Let $\mathcal{C} = \{1, \ldots, N\}$ and define two pmfs on $\mathcal{C}$,

$$g_N[i] = \frac{1}{N} \text{ for } i \in \mathcal{C}, \tag{6.48}$$

$$h_N[i] = \begin{cases} 1/N + \delta_N & \text{for } i = 1 \\ 1/N - \delta_N & \text{for } i = 2 \\ 1/N & \text{for } i > 2. \end{cases} \tag{6.49}$$

First, show that

$$D_{\mathrm{KL}}(g_N||h_N) = -\frac{1}{N} \log(1 - N^2 \delta_N^2). \tag{6.50}$$

Then, use the likelihood-ratio test

$$L(x) = \frac{h_N(x)}{g_N(x)} \tag{6.51}$$

to decide between two hypotheses $\mathrm{H}_0 : x \sim g_N$ and $\mathrm{H}_1 : x \sim h_N$ for one observation x. Show that $P_{\mathrm{FA}} + P_{\mathrm{MD}} \ge 1 - \delta_N$ no matter what threshold is used in the likelihood-ratio test. Finally, choose

$$\delta_N = \frac{1}{N} \sqrt{1 - e^{-N^2}} \tag{6.52}$$

to show that $D_{\mathrm{KL}}(g_N||h_N) = N$.

6.4 [KL divergence for LSB embedding in iid source] Assume that the cover image is a sequence of independent and identically distributed random variables with pmf $\mathbf{p}_0[i]$. After embedding relative payload $\alpha = 2\beta$ using LSB embedding, the stego image is a sequence of iid random variables with pmf

$$\mathbf{p}_\beta[2i] = (1 - \beta)\mathbf{p}_0[2i] + \beta\mathbf{p}_0[2i + 1], \tag{6.53}$$

$$\mathbf{p}_\beta[2i + 1] = \beta\mathbf{p}_0[2i] + (1 - \beta)\mathbf{p}_0[2i + 1]. \tag{6.54}$$

Show that

$$D_{\mathrm{KL}}(\mathbf{p}_0||\mathbf{p}_\beta) = \frac{\beta^2}{2} \sum_{i=0}^{127} (\mathbf{p}_0[2i] - \mathbf{p}_0[2i + 1])^2 \left(\frac{1}{\mathbf{p}_0[2i]} + \frac{1}{\mathbf{p}_0[2i + 1]} \right) + O(\beta^3). \tag{6.55}$$

Hint: Use Proposition B.7.

6.5 **[KL divergence for ±1 embedding in iid source]** Assume that the cover image is a sequence of independent and identically distributed random variables with range $\{0, \ldots, 255\}$ and pmf $\mathbf{p}_0[i]$. After embedding relative payload $\alpha = 2\beta$ using ± 1 embedding (see Section 7.3.1), the stego image becomes a sequence of iid random variables with pmf computed in Exercise 7.1. Show that the KL divergence satisfies

$$
\begin{aligned}
D_{\mathrm{KL}}(\mathbf{p}_0 \| \mathbf{p}_\beta) = &\sum_{i=2}^{253} \frac{(\mathbf{p}_0[i-1] - 2\mathbf{p}_0[i] + \mathbf{p}_0[i+1])^2}{8\mathbf{p}_0[i]} \beta^2 \\
&+ \frac{(\mathbf{p}_0[1] - 2\mathbf{p}_0[0])^2}{8\mathbf{p}_0[0]} \beta^2 + \frac{(2\mathbf{p}_0[0] - 2\mathbf{p}_0[1] + \mathbf{p}_0[2])^2}{8\mathbf{p}_0[1]} \beta^2 \\
&+ \frac{(\mathbf{p}_0[253] - 2\mathbf{p}_0[254] + \mathbf{p}_0[255])^2}{8\mathbf{p}_0[254]} \beta^2 \\
&+ \frac{(\mathbf{p}_0[254] - 2\mathbf{p}_0[255])^2}{8\mathbf{p}_0[255]} \beta^2 + O(\beta^3). \quad (6.56)
\end{aligned}
$$

Hint: Use Proposition B.7.

7 Practical steganographic methods

The definition of steganographic security given in the previous chapter should be a guiding design principle for constructing steganographic schemes. The goal is clear – to preserve the statistical distribution of cover images. Unfortunately, digital images are quite complicated objects that do not allow accurate description using simple statistical models. The biggest problem is their non-stationarity and heterogeneity. While it is possible to obtain simple models of individual small flat segments in the image, more complicated textures often present an insurmountable challenge for modeling because of a lack of data to fit an accurate local model. Moreover, and most importantly, as already hinted in Chapter 3, digital images acquired using sensors exhibit many complicated local dependences that the embedding changes may disturb and leave statistically detectable artifacts. Consequently, the lack of good image models gives space to heuristic methods.

In this chapter, we discuss four major guidelines for construction of practical steganographic schemes:

- Preserve a *model* of the cover source (Section 7.1);
- Make the embedding resemble some natural process (Section 7.2);
- Design the steganography to resist known steganalysis attacks (Section 7.3);
- Minimize the impact of embedding (Section 7.4).

Steganographic schemes from the first class are based on a simplified model of the cover source. The schemes are designed to preserve the model and are thus undetectable *within this model*. The remaining three design principles are heuristic. The goal of the second principle is to masquerade the embedding as some natural process, such as noise superposition during image acquisition. The third principle uses known steganalysis attacks as guidance for the design. Finally, the fourth principle first assigns a cost of making an embedding change at each element of the cover and then embeds the secret message while minimizing the total cost (impact) of embedding. It is also possible and, in fact, advisable, to take into consideration all four principles.

We now describe each design philosophy in more detail and give examples of specific embedding schemes.

7.1   Model-preserving steganography

This principle follows directly from the definition of steganographic security. The designer first chooses a model of cover images and then makes the steganographic scheme preserve this model. This will guarantee that the stego scheme will be undetectable as long as the chosen model completely describes the covers. Arguably, the simplest model is formed by a sequence of independent and identically distributed (iid) random variables. In this model, the cover is completely described by the probability distribution function. This means that for a given cover image, we need to preserve its first-order statistics or histogram. Note that this will lead to undetectable stegosystems for both homogeneous and heterogeneous cover sources (Section 6.2.2).

There exist many approaches that one could take to design a histogram-preserving steganographic scheme [64, 75, 110, 183, 218, 232]. The approach that we explain next is based on the general idea of statistical restoration in which a portion of the image is reserved and not used during embedding so that it can be utilized later to guarantee preservation of the first-order statistics.

7.1.1   Statistical restoration

We choose to illustrate the principle of statistical restoration on the example of the steganographic algorithm OutGuess originally introduced by Provos [198]. OutGuess embeds messages into a JPEG image by slightly modifying the quantized DCT coefficients. It also preserves the histogram of all DCT coefficients. The iid model can be heuristically justified by the argument that DCT coefficients in an individual 8×8 block are largely decorrelated and the fact that inter-block dependences among DCT coefficients are much weaker than dependences among neighboring pixels in the spatial domain.

Steganographic methods based on statistical restoration, such as OutGuess, are two-pass procedures. In the first (embedding) pass, a stego key is used to select a pseudo-random subset, $\mathcal{D}_e$, of all DCT coefficients (both luminance and chrominance coefficients) that will be used for embedding. Similar to the first JPEG steganographic algorithm Jsteg (Section 5.1.2), OutGuess embeds the message bits using simple LSB embedding (Section 5.1) into the coefficients from $\mathcal{D}_e$ while skipping over all coefficients equal to 0 or 1 to avoid introducing disturbing artifacts. In the second pass, corrections are made to the DCT coefficients outside of the set $\mathcal{D}_e$ to match the histogram of the stego image with the cover-image histogram.

Before embedding starts, OutGuess calculates the maximum length of a randomly spread message (the maximal correctable payload) that can be embedded in the cover image during the first pass, while making sure that there will be enough coefficients for the correction phase to adjust the histogram to its original values. The reader is encouraged to verify that the maximal correctable payload is determined by the most imbalanced LSB pair, which is the pair $\{2k, 2k+1\}$ with the largest ratio

$$\frac{\max\{\mathbf{h}[2k], \mathbf{h}[2k+1]\}}{\min\{\mathbf{h}[2k], \mathbf{h}[2k+1]\}}. \tag{7.1}$$

Because the histogram of DCT coefficients in a single-compressed JPEG image has a spike at zero (see the discussions in Chapter 2) and because the LSB pair $\{0, 1\}$ is skipped during embedding, the most imbalanced LSB pair is $\{-2, -1\}$. Because $\mathbf{h}[-2] < \mathbf{h}[-1]$ in typical cover images, after embedding the maximum correctable payload *all* remaining coefficients with value -2 will have to be modified to -1 in the correction phase.

Writing n_{01} for the number of all DCT coefficients not equal to 0 or 1, the maximal correctable payload will be $\alpha_{\max}n_{01}, 0 \leq \alpha_{\max} \leq 1$. Because all $\alpha_{\max}n_{01}$ coefficients are selected pseudo-randomly in the embedding phase, the number of unused DCT coefficients with value -2 after embedding is $(1 - \alpha_{\max})\mathbf{h}[-2]$. In order to restore the number of -1s in the stego image, this value must be larger than or equal to the expected decrease in the number of coefficients with value -1, which is $(\alpha_{\max}/2)\mathbf{h}[-1] - (\alpha_{\max}/2)\mathbf{h}[-2]$. This is because, assuming a random message is embedded, the probability that a message bit will match the LSB of -1 is $\frac{1}{2}$ and thus on average $(\alpha_{\max}/2)\mathbf{h}[-1]$ coefficients with value -1 will be unchanged by embedding and the same number of them will be modified to -2. The second term, $(\alpha_{\max}/2)\mathbf{h}[-2]$ is the expected number of coefficients with value -2 that will be modified to -1 during embedding. Because $\mathbf{h}[-2] < \mathbf{h}[-1]$, the expected drop in the number of coefficients with value -1 is the difference $(\alpha_{\max}/2)\mathbf{h}[-1] - (\alpha_{\max}/2)\mathbf{h}[-2]$. Thus, we obtain the following inequality and eventually an upper bound on the maximum correctable payload $\alpha_{\max}$:

$$(1 - \alpha_{\max})\mathbf{h}[-2] \geq \frac{\alpha_{\max}}{2}\mathbf{h}[-1] - \frac{\alpha_{\max}}{2}\mathbf{h}[-2], \tag{7.2}$$

$$\alpha_{\max} \leq \frac{2\mathbf{h}[-2]}{\mathbf{h}[-1] + \mathbf{h}[-2]}. \tag{7.3}$$

This condition guarantees that at the end of the embedding phase on average there will be enough unused coefficients with magnitude -2 that can be flipped back to -1 to make sure that the occurrences of the LSB pair $\{-2, -1\}$ are preserved after embedding. As this is the most imbalanced LSB pair, the occurrences of virtually all other LSB pairs can be preserved using the same correction step, as well. We note that some very sparsely populated histogram bins in the tails of the DCT histogram may not be restored correctly during the second phase, but, since their numbers are statistically insignificant, the impact on statistical detectability is negligible. The average capacity $\alpha_{\max}$ of OutGuess for typical natural images is around 0.2 bpnc (bits per non-zero DCT coefficient).

Steganographic schemes that embed messages in the spatial domain require more complex models because neighboring pixels are more correlated than DCT coefficients in a JPEG file. These correlations cannot be captured using first-order statistics. Instead, one can use the joint statistics of neighboring pixel pairs [75], statistics of differences between neighboring pixels (see Section 11.1.3 on structural steganalysis), or Markov chains.

7.1.2   Model-based steganography

Steganography based on statistical restoration chooses the sample statistics (e.g., the histogram) as the model to preserve. In contrast, model-based steganography [204] fits a parametric model through the sample data and preserves this data-driven model and does so without the need for a correction step. The cover image is modeled as a random variable that can be divided into two components, $\mathbf{x} = (\mathbf{x}_{\mathrm{inv}}, \mathbf{x}_{\mathrm{emb}})$, where $\mathbf{x}_{\mathrm{inv}}$ is invariant with respect to embedding and $\mathbf{x}_{\mathrm{emb}}$ may be modified during embedding. We denote the range of each random variable as $\mathcal{X}_{\mathrm{inv}}$ and $\mathcal{X}_{\mathrm{emb}}$. For example, we can think of LSB embedding in 8-bit grayscale images where $\mathbf{x}_{\mathrm{inv}} \in \{0,1\}^7 = \mathcal{X}_{\mathrm{inv}}$ are the 7 most significant bits (or the index of the LSB pair) and $\mathbf{x}_{\mathrm{emb}} \in \{0,1\} = \mathcal{X}_{\mathrm{emb}}$, the LSB of $\mathbf{x}$.

The cover model is formed by the conditional probabilities $\Pr\{\mathbf{x}_{\mathrm{emb}}|\mathbf{x}_{\mathrm{inv}}\}$. These probabilities will be needed to extract the message and thus must be known to the recipient. This can be arranged by making $\Pr\{\mathbf{x}_{\mathrm{emb}}|\mathbf{x}_{\mathrm{inv}}\}$ depend only on the invariant component $\mathbf{x}_{\mathrm{inv}}$.

First, the set of all cover elements (pixels or DCT coefficients) is written as a union of disjoint subsets

$$\bigcup_{\mathbf{x}_{\mathrm{inv}} \in \mathcal{X}_{\mathrm{inv}}} \mathcal{C}(\mathbf{x}_{\mathrm{inv}}), \tag{7.4}$$

where $\mathcal{C}(\mathbf{x}_{\mathrm{inv}})$ is the set of cover elements whose invariant part is $\mathbf{x}_{\mathrm{inv}}$. The embedding algorithm embeds a portion of the message in each $\mathcal{C}(\mathbf{x}_{\mathrm{inv}})$ in the following manner. First, the message bits are encoded using symbols from $\mathcal{X}_{\mathrm{emb}}$ and then the symbols are prebiased so that they appear with probabilities $\Pr\{\mathbf{x}_{\mathrm{emb}}|\mathbf{x}_{\mathrm{inv}} = \mathbf{x}_{\mathrm{inv}}\}$. This is achieved by running the message symbols through an entropy *decompressor*, for a compression scheme designed to compress symbols from $\mathcal{X}_{\mathrm{emb}}$ distributed according to the same conditional probabilities.[1] When the decompressor is fed with $\mathbf{x}_{\mathrm{emb}}$ distributed uniformly in $\mathcal{X}_{\mathrm{emb}}$, it will output symbols with probabilities $\Pr\{\mathbf{x}_{\mathrm{emb}}|\mathbf{x}_{\mathrm{inv}} = \mathbf{x}_{\mathrm{inv}}\}$. Thus, when $\mathbf{x}_{\mathrm{emb}}$ of all cover elements from $\mathcal{C}(\mathbf{x}_{\mathrm{inv}})$ are replaced with the transformed symbols, $\mathbf{x}'_{\mathrm{emb}}$, the stego image elements will follow the cover-image model as desired.

The fraction of bits that can be embedded in each element of $\mathcal{C}(\mathbf{x}_{\mathrm{inv}})$ is the entropy of $\Pr\{\mathbf{x}_{\mathrm{emb}}|\mathbf{x}_{\mathrm{inv}} = \mathbf{x}_{\mathrm{inv}}\}$ or

$$H\left(\Pr\{\mathbf{x}_{\mathrm{emb}}|\mathbf{x}_{\mathrm{inv}} = \mathbf{x}_{\mathrm{inv}}\}\right) =$$
$$-\sum_{\mathbf{x}_{\mathrm{emb}}} P_{\mathbf{x}_{\mathrm{emb}}|\mathbf{x}_{\mathrm{inv}}}(\mathbf{x}_{\mathrm{emb}}|\mathbf{x}_{\mathrm{inv}}) \log_2 P_{\mathbf{x}_{\mathrm{emb}}|\mathbf{x}_{\mathrm{inv}}}(\mathbf{x}_{\mathrm{emb}}|\mathbf{x}_{\mathrm{inv}}), \tag{7.5}$$

---

[1] In practice, one can use, for example, arithmetic compression.

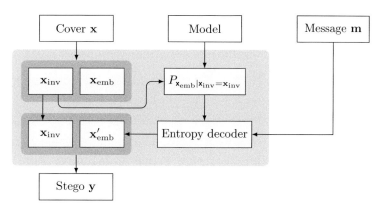

Figure 7.1 Model-based steganography (embedding).

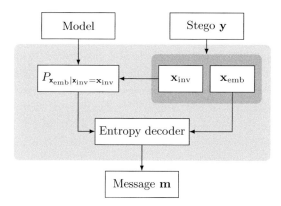

Figure 7.2 Model-based steganography (extraction).

where we denoted for brevity

$$P_{\mathbf{x}_{\mathrm{emb}}|\mathbf{x}_{\mathrm{inv}}}(\mathbf{x}_{\mathrm{emb}}|\mathbf{x}_{\mathrm{inv}}) = \Pr\{\mathbf{x}_{\mathrm{emb}} = \mathbf{x}_{\mathrm{emb}}|\mathbf{x}_{\mathrm{inv}} = \mathbf{x}_{\mathrm{inv}}\}. \tag{7.6}$$

Thus, the total embedding capacity is

$$\sum_{\mathbf{x}_{\mathrm{inv}}} |\mathcal{C}(\mathbf{x}_{\mathrm{inv}})| H\left(\Pr\{\mathbf{x}_{\mathrm{emb}}|\mathbf{x}_{\mathrm{inv}} = \mathbf{x}_{\mathrm{inv}}\}\right), \tag{7.7}$$

where $|\mathcal{C}(\mathbf{x}_{\mathrm{inv}})|$ is the cardinality of $\mathcal{C}(\mathbf{x}_{\mathrm{inv}})$.

To illustrate the model-based approach, we describe a specific realization for steganography in JPEG images [204]. Similar to OutGuess or Jsteg, DCT coefficients equal to 0 or 1 are not used for embedding and the embedding mechanism is LSB flipping. Also, $\mathcal{X}_{\mathrm{inv}} = \{0,1\}^7$ and $\mathcal{X}_{\mathrm{emb}} = \{0,1\}$. The model-based-steganography paradigm is applied to each DCT mode separately. Thus, at the beginning the cover JPEG file is decomposed into 64 subsets corresponding to 64 DCT modes (spatial frequencies). Let $\mathbf{h}[i]$ be the histogram of DCT coefficients for one fixed mode. Because the sums $\mathbf{h}[2i] + \mathbf{h}[2i+1]$ are invariant under LSB embedding for all i, we can use this invariant and fit a parametric model, $h(x)$, through the points $((2i + 2i + 1)/2, (\mathbf{h}[2i] + \mathbf{h}[2i + 1])/2)$. For example, we can model the DCT coefficients using the generalized Cauchy model with pdf (see Appendix A)

$$h(x) = \frac{p-1}{2s} \left(1 + \frac{|x|}{s} \right)^{-p} \tag{7.8}$$

and determine the parameters using maximum-likelihood estimation (see Example D.8). Denoting the 7 most significant bits of an integer a as $\mathrm{MSB}_7(a)$, we define the model using the conditional probabilities

$$\Pr\{\mathbf{x}_{\mathrm{emb}} = 0 | \mathbf{x}_{\mathrm{inv}} = \mathrm{MSB}_7(2i)\} = \frac{h(2i)}{h(2i) + h(2i + 1)}, \tag{7.9}$$

$$\Pr\{\mathbf{x}_{\mathrm{emb}} = 1 | \mathbf{x}_{\mathrm{inv}} = \mathrm{MSB}_7(2i)\} = \frac{h(2i + 1)}{h(2i) + h(2i + 1)}. \tag{7.10}$$

The sender continues with the embedding process by selecting all $\mathbf{h}[2i] + \mathbf{h}[2i + 1]$ DCT coefficients at the chosen DCT mode that are equal to $2i$ or $2i + 1$. Their LSBs are replaced with a segment of the message that was decompressed using an arithmetic decompressor to the length $\mathbf{h}[2i] + \mathbf{h}[2i + 1]$. The decompressor is designed to transform a sequence of uniformly distributed message bits to a biased sequence with 0s and 1s occurring with probabilities (7.9)–(7.10). Because we are replacing a bit sequence with another bit sequence with the same distribution, the model (7.9)–(7.10) will be preserved.

The recipient first constructs the model and computes the probabilities given by (7.9)–(7.10). This can be achieved because $\mathbf{h}[2i] + \mathbf{h}[2i + 1]$ is invariant with respect to embedding changes! Individual message segments are extracted by feeding the LSBs for each DCT mode and each LSB pair $\{2i, 2i + 1\}$ into the arithmetic *compressor* and concatenating them.

This specific example of model-based steganography is designed to preserve the *model* of the histograms of *all* individual DCT modes. This is in contrast with OutGuess that preserves the sample histogram of all DCT coefficients and not necessarily the histograms of the DCT modes. For 80% quality JPEG images, the average embedding capacity of this algorithm is approximately 0.8 bpnc, which is remarkably large considering the scope of the model and four times larger than for OutGuess.

We now calculate the embedding efficiency of this algorithm defined as the average number of bits embedded per unit distortion. In order to simplify the notation, we set

$$p_0 = \Pr\{\mathbf{x}_{\mathrm{emb}} = 0 | \mathbf{x}_{\mathrm{inv}} = \mathrm{MSB}_7(2i)\}. \tag{7.11}$$

The average number of bits embedded in one DCT coefficient equal to $2i$ or $2i + 1$ for one fixed DCT mode is

$$H(p_0) = -p_0 \log_2 p_0 - (1 - p_0) \log_2(1 - p_0). \tag{7.12}$$

An embedding change is performed when the coefficient's LSB does not match the prebiased message bit. Because the probability that the prebiased message bit is 0 is the same as the probability that the LSB of the coefficient is 0, which is p_0, the embedding needs to change the LSB with probability $p_0(1 - p_0) + (1 - p_0)p_0 = 2p_0(1 - p_0)$. Therefore, the embedding efficiency is

$$e(p_0) = \frac{-p_0 \log_2 p_0 - (1 - p_0) \log_2(1 - p_0)}{2p_0(1 - p_0)}. \qquad (7.13)$$

Note that $e(p_0)$ is always greater than or equal to 2 (follow Figure 7.3). Surprisingly, this is higher than the embedding efficiency of simple LSB embedding, which is 2 because in LSB embedding every other LSB is modified, on average.

To summarize, this specific example of model-based steganography preserves the models of histograms of all 64 DCT modes and does so while providing embedding efficiency larger than 2. This is quite an improvement over the naive Jsteg described in Chapter 5.

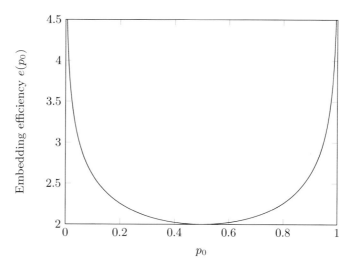

Figure 7.3 Embedding efficiency of Model-Based Steganography for JPEG images.

There exists a more advanced version of this algorithm [205] that attempts to preserve one higher-order statistic called "blockiness" defined as the sum of discontinuities along the boundaries of 8×8 pixel blocks in the spatial domain (see (12.9) in Chapter 12). This is achieved using the idea of statistical restoration by making additional modifications to unused DCT coefficients in an iterative manner to adjust the blockiness to its original value in the cover image.[2] In Chapter 12, the original version of Model-Based Steganography is abbreviated MBS1, while the more advanced version with deblocking is denoted as MBS2.

---

[2] These additional changes, however, make this version of Model-Based Steganography more detectable (see Chapter 12 and [190, 210]).

7.2   Steganography by mimicking natural processing

Model-preserving steganographic schemes are undetectable within the chosen model. However, unless the model comprehensively captures the cover source, all that is needed to construct a steganalysis algorithm is to identify a statistical quantity that is disturbed by the embedding. It turns out that finding such a statistic is usually not very difficult. Often, in steganography by statistical restoration the additional changes in the correction phase only make things worse and in fact make steganalysis easier [190, 210, 245]. Preserving a more complex cover source is not really a practical answer to this problem because schemes based on statistical restoration do not scale well with the complexity of the model. For example, it is not immediately clear how to at the same time preserve the histogram of DCT coefficients and the statistics of DCT coefficient pairs from neighboring 8×8 blocks without sacrificing the embedding capacity. Even though a practical embedding method capable of approximately preserving multiple statistics was described in [144] (the Feature Correction Method), the approach has not led to secure steganographic algorithms so far.

The lack of accurate models justifies heuristic approaches to steganography, such as those that attempt to mask the embedding as a natural process. Presumably, if the effect of embedding were indistinguishable from some natural processing, stego images should stay compatible with the distribution of cover images. In this section, we describe a practical realization of this idea by masking embedding as superposition of noise with given statistical properties.

7.2.1   Stochastic modulation

In Chapter 3, we learned that the process of digital image acquisition is affected by multiple noise sources. Even when taking two images of exactly the same scene under the same conditions and identical camera settings, the images will differ in their noise component due to the presence of random phenomena caused by quantum properties of light (shot noise) and noise present in electronic components of the sensor. This suggests the idea of constructing a steganographic method so that the impact of embedding resembles superposition of sensor noise during image acquisition [77, 82]. The steganalyst would then be required to distinguish whether the image noise component is solely due to image acquisition or contains a component due to message embedding. This is essentially a heuristic plan to produce stego images compatible with the distribution of the cover source.

Our goal is to construct an embedding scheme so that the impact of embedding is equivalent to adding to the cover image a signal obtained by independent realizations of a random variable η with a given probability density function f_η. Because digital images are represented with a finite bit depth, the noise will have to be adequately quantized as well. Assuming for simplicity and without loss of generality that we work with grayscale images, we denote by $\mathbf{r}[i] = \text{round}(\boldsymbol{\eta}[i])$ the realizations of η after rounding them to integers from the set $\{0, \ldots, 255\}$. The rounded noise sequence $\mathbf{r}[i]$ is called the stego noise. The probability mass function of the stego noise is

$$\mathbf{p}[k] = \Pr\{\mathbf{r}[i] = k\} = \int_{k-\frac{1}{2}}^{k+\frac{1}{2}} f_\eta(x)\mathrm{d}x \quad \text{for any integers } i, k. \tag{7.14}$$

In theory, by adding realizations of the stego noise to the cover image we could embed $H(\mathbf{p})$ bits (the entropy of $\text{round}(\eta)$) at every pixel. Instead of trying to develop a practical method that achieves this capacity, which is not an easy task, we provide a simple suboptimal method.

It will be advantageous to work with bits represented using the pair $\{-1, 1\}$ rather than $\{0, 1\}$. Thus, from now till the end of this section, -1 and 1 represent binary 0 and 1, respectively. Next, we define a parity function π with the following antisymmetry property,

$$\pi(u + v, v) = -\pi(u, v) \tag{7.15}$$

for all integers u and v. The following function, for example, satisfies this requirement:

$$\pi(u, v) = \begin{cases} (-1)^u & \text{for } 2k|v| \leq u \leq 2k|v| + |v| - 1 \\ -(-1)^u & \text{for } (2k - 1)|v| \leq u \leq 2k|v| - 1 \\ 0 & \text{for } v = 0, \end{cases} \tag{7.16}$$

where k is an integer.

The embedding algorithm starts by generating a pseudo-random path through the image using the stego key and two independent stego noise sequences $\mathbf{r}[i]$ and $\mathbf{s}[i]$ with the probability mass function (7.14). The sender follows the embedding path and embeds one message bit, m, at the ith pixel, $\mathbf{x}[i]$, if and only if $\mathbf{r}[i] \neq \mathbf{s}[i]$. In this case, the sender can always embed one bit by adding either $\mathbf{r}[i]$ or $\mathbf{s}[i]$ to $\mathbf{x}[i]$ because due to the antisymmetry property of the parity function π (7.15)

$$\pi(\mathbf{x}[i] + \mathbf{s}[i], \mathbf{r}[i] - \mathbf{s}[i]) = -\pi(\mathbf{x}[i] + \mathbf{s}[i] + \mathbf{r}[i] - \mathbf{s}[i], \mathbf{r}[i] - \mathbf{s}[i]) \tag{7.17}$$
$$= -\pi(\mathbf{x}[i] + \mathbf{r}[i], \mathbf{r}[i] - \mathbf{s}[i]). \tag{7.18}$$

In the case when $\mathbf{r}[i] = \mathbf{s}[i]$, the sender does not embed any bit, replaces $\mathbf{x}[i]$ with $\mathbf{y}[i] = \mathbf{x}[i] + \mathbf{r}[i]$, and continues with embedding at the next pixel along the pseudo-random walk.

To complete the embedding algorithm, however, we need to resolve one more technicality of the embedding process. If, during embedding, $\mathbf{y}[i] = \mathbf{x}[i] + \mathbf{r}[i]$ gets out of its dynamic range, the sender has little choice but to slightly deviate from the stego noise model and instead add $\mathbf{r}'[i]$ so that $\pi(\mathbf{x}[i] + \mathbf{r}'[i], \mathbf{r}[i] - \mathbf{s}[i]) = m$ with $|\mathbf{r}'[i] - \mathbf{r}[i]|$ as small as possible.

The recipient first uses the stego key to generate the same stego noise sequences, $\mathbf{r}[i]$ and $\mathbf{s}[i]$, and the pseudo-random walk through the pixels. The message bits are read from the stego image pixels, $\mathbf{y}[i]$, as parities $m = \pi(\mathbf{y}[i], \mathbf{r}[i] - \mathbf{s}[i])$. No bit is extracted from pixel $\mathbf{y}[i]$ if $\mathbf{r}[i] = \mathbf{s}[i]$.

In summary, the sender starts by generating two independent stego noise sequences with the required probability mass function and then at each pixel attempts to embed one message bit by adding one of the two samples of the stego noise. This works due to the antisymmetry property of the parity function. If the two stego noise samples are the same, the embedding process does not embed any bits. The sender, however, still adds the noise to the image so that the stego image is, indeed, obtained by adding noise of a given pmf to the cover image as required.

Stochastic modulation embeds one bit at every pixel as long as $\mathbf{r}[i] \neq \mathbf{s}[i]$. Thus, the relative embedding capacity is $1 - \Pr\{\mathbf{r} = \mathbf{s}\} = 1 - \sum_k \mathbf{p}[k]^2$ bits per pixel, where the probabilities can be computed using (7.14). The random noise sources during image acquisition, such as the shot noise and the readout noise, are well modeled as Gaussian random variables with zero mean and variance σ^2. In this case, the relative embedding capacity per pixel is as shown in Figure 7.4. Because the stego noise sequences are more likely to be equal for small σ, the capacity increases with the noise variance. Also, note that the sender can communicate about 0.7 bits per pixel using quantized standard Gaussian noise $N(0,1)$.

Note that stochastic modulation is suboptimal in general because only $1 - \sum_k \mathbf{p}[k]^2$ bits are embedded at every pixel rather than $H(\mathbf{p})$ bits, the entropy of the stego noise (Exercise 7.4 shows that, indeed, $1 - \sum_k \mathbf{p}[k]^2 \leq H(\mathbf{p})$). Although it is not known how to construct steganographic systems reaching the theoretic embedding capacity for general stego noise, there exist capacity-reaching constructions for special instances of $\mathbf{p}$, such as when the stego noise amplitude is at most 1 (see Section 9.4.5).

No matter how plausible the heuristic behind stochastic modulation sounds, the fact is that it can be reliably detected using modern steganalysis methods based on feature extraction and machine learning. The main reason is that during image acquisition, the noise is injected *before* the signal is even quantized in the A/D converter and further processed. Adding the noise to the final TIFF or BMP image is not the same because this image already contains a multitude of complex dependences among neighboring pixels due to in-camera processing, such as demosaicking, color correction, and filtering. Thus, the stego noise should be superimposed on the raw sensor output rather than the final image. It is, however, not clear at this point how to embed bits in the pixel domain by modifying the raw sensor output. A possible solution is to apply coding methods for communication with non-shared selection channels, which is the topic of Chapter 9.

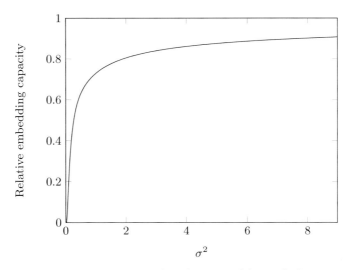

Figure 7.4 Relative embedding capacity of stochastic modulation (in bits per pixel) realized by adding white Gaussian noise of variance σ^2.

7.2.2   The question of optimal stego noise

For the ideal case of stochastic modulation capable of communicating $H(\mathbf{p})$ bits per pixel, it is natural to ask about the best stego noise that would embed the highest payload with the smallest embedding distortion. Stochastic modulation with such stego noise would minimize the embedding distortion, which could intuitively decrease the statistical detectability.

The properties of the optimal stego noise will obviously depend on how we measure the distortion between cover and stego images, $d(\mathbf{x}, \mathbf{y})$. The expected value of the embedding distortion per pixel, $E[d(x, y)]$, will be some function of the stego noise pmf $\mathbf{p}$, and will be denoted as $d(\mathbf{p}) = E[d(x, y)]$.

Imposing a bound on the embedding distortion, D, we wish to determine $\mathbf{p}_{\mathrm{opt}}$ that maximizes the relative embedding payload

$$\mathbf{p}_{\mathrm{opt}} = \arg \max_{d(\mathbf{p}) \leq D} H(\mathbf{p}). \tag{7.19}$$

This expression is recognized in information theory as the rate–distortion bound, which is a relationship that connects the communication rate and distortion. The relative payload for the optimal distribution, $H(\mathbf{p}_{\mathrm{opt}})$, depends on the distortion bound D and it is also a function of the distortion measure d. This is why we denote it as $H_d(D) = H(\mathbf{p}_{\mathrm{opt}})$. We can say that, given a bound on embedding distortion, D, the relative payload α that can be embedded with any instance of stochastic modulation must satisfy

$$\alpha \leq H_d(D). \tag{7.20}$$

Alternatively, the relative payload α can be embedded with distortion no smaller than $H_d^{-1}(\alpha)$, where H_d^{-1} is the inverse function to H_d. This translates into the following bound on the embedding efficiency (ratio of payload and distortion):

$$e \leq \frac{\alpha}{H_d^{-1}(\alpha)}. \tag{7.21}$$

We now find the stego noise distribution $\mathbf{p}_{\text{opt}}$ for a specific choice of the distortion measure, $d_\gamma(x,y) = |x-y|^\gamma$, as defined in (4.12). The expected distortion per pixel is

$$d_\gamma(\mathbf{p}) = \sum_k \mathbf{p}[k]|k|^\gamma \leq D. \tag{7.22}$$

We now show that the optimal stego noise distribution is the discrete generalized Gaussian

$$\mathbf{p}_{\text{opt}}[k] = \frac{e^{-\lambda|k|^\gamma}}{Z(\lambda)}, \tag{7.23}$$

where

$$Z(\lambda) = \sum_k e^{-\lambda|k|^\gamma} \tag{7.24}$$

is the normalization factor. To see this, we write for the entropy $H(\mathbf{p})$ of an arbitrary distribution $\mathbf{p}$ satisfying the bound (7.22)

$$H(\mathbf{p}) = \sum_k \mathbf{p}[k] \log \frac{1}{\mathbf{p}[k]} = \sum_k \mathbf{p}[k] \log \frac{1}{\mathbf{p}_{\text{opt}}[k]} + \sum_k \mathbf{p}[k] \log \frac{\mathbf{p}_{\text{opt}}[k]}{\mathbf{p}[k]} \tag{7.25}$$

$$\leq \log Z(\lambda) + \lambda \sum_k \mathbf{p}[k]|k|^\gamma \leq \log Z(\lambda) + \lambda D, \tag{7.26}$$

where we used the non-negativity of the KL divergence (Proposition B.5)

$$0 \leq D_{\text{KL}}(\mathbf{p}||\mathbf{p}_{\text{opt}}) = -\sum_k \mathbf{p}[k] \log \frac{\mathbf{p}_{\text{opt}}[k]}{\mathbf{p}[k]}. \tag{7.27}$$

From the property of the KL divergence, the equality is reached when $\mathbf{p} = \mathbf{p}_{\text{opt}}$, which proves the optimality of $\mathbf{p}_{\text{opt}}$ (7.23). The parameter λ is determined from the requirement $\sum_k \mathbf{p}[k]|k|^\gamma = D$.

Note that when the distortion is measured as energy, $\gamma = 2$, the bound (7.22) limits the variance of $\mathbf{p}$. In this case, the stego noise with maximal entropy is the discrete Gaussian, in compliance with the classical result from information theory that the highest-entropy noise among all variance-bounded distributions is the Gaussian distribution.

For steganography whose embedding changes are restricted to ± 1, the stego noise distribution satisfies $\mathbf{p}[k] = 0$ for $k \notin \{-1,0,1\}$ and $d_\gamma(\mathbf{p})$ becomes the change rate (4.17) for any $\gamma > 0$. In this case, the function $H_{d_\gamma}(x)$ can be determined analytically [233] (also see Exercise 7.5) to be

$$H_{d_\gamma}(x) = -x \log_2 x - (1-x) \log_2(1-x) + x, \tag{7.28}$$

the ternary entropy function (Section 8.6.2).

7.3   Steganalysis-aware steganography

Steganography is advanced through steganalysis. Thus, it is only natural to take into account existing attacks on steganographic techniques when designing a new one. In fact, OutGuess was originally designed as an advanced version of Jsteg resistant to the histogram attack (Chapter 5), rather than from the definition of steganographic security as presented in Chapter 6. Security with respect to known steganalysis is an obvious necessary condition that any new steganography aspiring to be secure must satisfy. Historically, steganalysis was often used as a guiding principle to avoid known pitfalls when designing the next generation of steganographic methods.

7.3.1 ± 1 embedding

It was recognized early on that the embedding operation of flipping the LSB creates many problems due to its asymmetry (even values are never decreased and odd values never increased during embedding). Flipping LSBs is an unnatural operation that introduces characteristic artifacts into the histogram. An obvious remedy is to use an embedding operation that is symmetrical. A trivial modification of LSB embedding is the so-called ± 1 embedding, also sometimes called LSB matching. This embedding algorithm embeds message bits as LSBs of cover elements; however, when an LSB needs to be changed, instead of flipping the LSB, the value is randomly increased or decreased, with the obvious exception that the values 0 and 255 are only increased or decreased, respectively. This has the effect of modifying the LSB but, at the same time, other bits may be modified as well. In fact, in the most extreme case, all bits may be modified, such as when the value $127 = (01111111)_2$ is changed to $128 = (10000000)_2$.

Note that the extraction algorithm of ± 1 embedding is the same as for LSB embedding – the message is read by extracting the LSBs of cover elements. The ± 1 embedding is much more difficult to attack than LSB embedding. While there exist astonishingly accurate attacks on LSB embedding (see Sample Pairs Analysis in Chapter 11), no attacks on ± 1 embedding with comparable accuracy currently exist (Sections 11.4 and 12.5 contain examples of a targeted and a blind attack).

7.3.2   F5 embedding algorithm

The effect of the embedding operation on the security of steganographic algorithms for the JPEG format is larger than on schemes that embed in the spatial domain. This is because there exist good models for the histogram of DCT coefficients (see Chapter 2). The F5 algorithm [241] was originally designed to overcome the histogram attack while still offering a large embedding capacity. The F5 contains two important ingredients – its embedding operation and matrix embedding. The operation of LSB flipping was replaced with decrementing the absolute value of the DCT coefficient by one. This preserves the natural shape of the DCT histogram, which looks after embedding as if the cover image was originally compressed using a lower quality factor. The second novel design element, the matrix embedding, is a coding scheme that decreases the number of embedding changes. For now, we do not consider matrix embedding in the algorithm description and instead postpone it to Chapter 8.

The F5 algorithm embeds message bits along a pseudo-random path determined from a user passphrase. The message bits are again encoded as LSBs of DCT coefficients along the path. If the coefficient's LSB needs to be changed, instead of flipping the LSB, the absolute value of the DCT coefficient is decreased by one. To avoid introducing easily detectable artifacts, the F5 skips over the DC terms and all coefficients equal to 0. In contrast to Jsteg, OutGuess, or Model-Based Steganography, F5 does embed into coefficients equal to 1.

Because the embedding operation decreases the absolute value of the coefficient by 1, it can happen that a coefficient originally equal to 1 or −1 is modified to zero (a phenomenon called "shrinkage"). Because the recipient will be reading the message bits from LSBs of non-zero AC DCT coefficients along the same path, the bit embedded during shrinkage would get lost.[3] Thus, if shrinkage occurs, the sender has to re-embed the same message bit, which, by the way, will always be a 0, at the next coefficient. However, by re-embedding these 0-bits, the sender will end up embedding a *biased* bit stream containing more zeros than ones. This would mean that odd coefficient values will be more likely to be changed than even-valued coefficients. Yet again, we run into an embedding asymmetry that introduces "staircase" artifacts into the histogram. There are at least two solutions to this problem. One is to embed the message bit m as the XOR of the coefficient LSB and some random sequence of bits also generated from the stego key. This way, shrinkage will be equally likely to occur when embedding a 1 or a 0. The implementation of F5 solves the problem differently by redefining the LSB for negative numbers:

$$\text{LSB}_{\text{F5}}(x) = \begin{cases} 1 - x \bmod 2 & \text{for } x < 0 \\ x \bmod 2 & \text{otherwise.} \end{cases} \quad (7.29)$$

Because in natural images the numbers of coefficients equal to 1 and −1 are approximately the same ($\mathbf{h}[1] \approx \mathbf{h}[-1]$), this simple measure will also cause shrinkage to occur when embedding both 0s and 1s with approximately equal probability. The pseudo-code for the embedding and extraction algorithm for F5 that does not employ matrix embedding is shown in Algorithms 7.1 and 7.2.

To calculate the embedding capacity of F5, realize that F5 does not embed into DCT coefficients equal to 0 and the DC term. Also, no bit is embedded during

---

[3] Realize that the recipient has no means of determining whether the DCT coefficient was originally zero or became zero due to embedding.

Algorithm 7.1 Embedding message $\mathbf{m} \in \{0,1\}^m$ in JPEG cover image $\mathbf{x} \in \mathcal{X}^n$ using the F5 algorithm (no matrix embedding employed).

```
// Initialize a PRNG using stego key (or passphrase)
// Input: message m ∈ {0,1}ᵐ, quantized JPEG DCT coefficients x ∈ 𝒳ⁿ
Path = Perm(n);
// Perm(n) is a pseudo-random permutation of {1,2,...,n}
y = x;
i = 1; j = 1; // i message index, j coefficient index
while (i ≤ m) & (j ≤ n) {
    if (x[Path[j]] ≠ 0) & (x[Path[j]] is not DC term) {
        if LSB_F5(x[Path[j]]) = m[i] {i = i + 1;}
        else {
            y[Path[j]] = x[Path[j]] − sign(x[Path[j]])
            if y[Path[j]] ≠ 0 {i = i + 1;}
        }
    }
    j = j + 1;
}
// y are stego DCT coefficients conveying i message bits
```

Algorithm 7.2 Extracting message $\mathbf{m}$ from a JPEG stego image $\mathbf{y} \in \mathcal{X}^n$ embedded using the F5 algorithm (no matrix embedding employed).

```
// Initialize a PRNG using stego key (or passphrase)
// Input: quantized JPEG DCT coefficients y ∈ 𝒳ⁿ
Path = Perm(n);
// Perm(n) is a pseudo-random permutation of {1,2,...,n}
i = 1; j = 1; // i message index, j coefficient index
while (j ≤ n) {
    if (y[Path[j]] ≠ 0) & (y[Path[j]] is not DC term) {
        m[i] = LSB_F5(y[Path[j]]) {i = i + 1;}
    }
    j = j + 1;
}
// read the header of extracted bits to find the message length and trun-
cate m
```

---

shrinkage, which will lead to loss of $(\mathbf{h}[1] + \mathbf{h}[-1])/2$ bits. Thus, given a JPEG file with in total n_{AC} AC DCT coefficients, the embedding capacity is $n_{AC} - \mathbf{h}[0] - (\mathbf{h}[-1] + \mathbf{h}[1])/2$, where $\mathbf{h}$ is the histogram of all AC DCT coefficients. For example, for a cover source formed by JPEG images with 80% quality, the average embedding capacity is about 0.75 bits per non-zero DCT coefficient, which is quite large.

We stress that the F5 algorithm does not preserve the histogram but preserves its crucial characteristics, such as its monotonicity and monotonicity of increments [241].

7.4   Minimal-impact steganography

Steganography by cover modification will inevitably introduce some embedding changes into the cover image. It is intuitively clear that the modifications may have different impact on statistical detectability depending on the local context, properties of the cover element being modified, and other factors. If we could quantify the contribution, $\rho[i]$, of modifying every cover element, $\mathbf{x}[i]$, to the overall statistical detectability, we would essentially convert the problem of maximizing the security into an optimization problem: "How to embed a given payload while minimizing the overall expected embedding impact?" This approach to steganography is rather general and it is also appealing because of its modular architecture. For example, progress in our understanding of the impact of embedding on security will only lead to updated values $\rho[i]$, while the optimization algorithm may stay the same.

Steganographic methods designed from the principle of minimal embedding impact may not minimize the KL divergence between cover and stego images for any chosen model of covers (see Exercise 7.6). As explained earlier in this chapter, our approach is heuristic, justified by the lack of accurate cover models. Indeed, schemes that painstakingly preserve some simple cover model may in fact be quite detectable because of the model misfit. The strategy proposed in this section is to abstract from a model and, instead of trying to preserve the cover model, accept in advance the fact that the steganography you will build will not be perfect and minimize the impact of embedding.

Let us assume that the impact of making an embedding change at pixel i can be captured using a scalar value $\rho[i] \geq 0$. Denoting by $\mathbf{x}$ and $\mathbf{y}$ the cover and stego image, the total embedding impact is defined as

$$d_\rho(\mathbf{x}, \mathbf{y}) = \sum_{i=1}^{n} \rho[i] \left(1 - \delta(\mathbf{x}[i] - \mathbf{y}[i])\right), \qquad (7.30)$$

where $\delta(x)$ is the Kronecker delta (2.23). We point out that (7.30) implicitly assumes that the embedding impact is additive because it is defined as a sum of impacts at individual pixels. In general, however, the embedding modifications could be interacting among themselves, reflecting the fact that making two changes to adjacent pixels might be more or less detectable than making the same changes to two pixels far apart from each other. A detectability measure that takes interaction among pixels into account would not be additive. If the density of embedding changes is low, however, the additivity assumption is plausible because the distances between modified pixels will generally be large and the embedding changes will not interfere much.

The reader should realize that $\rho[i]$ does not necessarily have to correspond to embedding distortion as defined in Chapter 4. For example, embedding changes in textured or noisy areas of the cover image are less likely to introduce detectable artifacts than changes in smooth areas. This could be captured by introducing weights, $\boldsymbol{\omega}[i] \geq 0$, for each cover element and defining

$$\rho[i] = \boldsymbol{\omega}[i]|\mathbf{x}[i] - \mathbf{y}[i]|^{\gamma}, \tag{7.31}$$

where γ is a non-negative parameter. If the embedding change is probabilistic and more than one value $\mathbf{y}[i]$ is possible, (7.31) is understood as the expected value. For example, in ± 1 embedding, $\mathbf{y}[i] = \mathbf{x}[i] + 1$ and $\mathbf{y}[i] = \mathbf{x}[i] - 1$ with equal probability and thus $E[|\mathbf{x}[i] - \mathbf{y}[i]|^{\gamma}] = 1$.

We now give a few examples of typical assignments ρ used in steganography. If $\boldsymbol{\omega}[i] = 1$ for all i and $|\mathbf{x}[i] - \mathbf{y}[i]| = 1$, d_ρ is the total number of embedding changes and (7.30) coincides with the distortion measure ϑ defined in Chapter 4. It is a reasonable measure of embedding impact when the magnitude of all embedding changes is 1. For $\boldsymbol{\omega}[i] = 1$ and $\gamma = 2$, d_ρ is the energy of modifications, while the choice $\boldsymbol{\omega}[i] = 1$ and $\gamma = 1$ gives the L_1 norm between $\mathbf{x}$ and $\mathbf{y}$ (see Chapter 4). Optimal-parity embedding as introduced in Section 5.2.4 could be interpreted as minimal-embedding-impact steganography with $\rho[i]$ equal to the isolation of the palette color at pixel i. As another example, if for some reason some cover elements are not to be modified under any circumstances, we can set $\boldsymbol{\omega}[i] = \infty$ for them. This measure of embedding impact will be used in Chapter 9.

The impact $\boldsymbol{\omega}[i]$ may also be determined from some side-information about the ith element. For example, let us assume that the cover is a color TIFF image sampled at 16 bits per channel (48 bits per pixel). The sender wishes to embed a message while decreasing the color depth to a true-color 8-bit per channel image while minimizing the *combined* quantization and embedding distortion. Let $\mathbf{z}[i]$ be the 16-bit color value and let $Q = 2^8$ be the quantization step for the color-depth reduction. The quantization error at the ith pixel is

$$\mathbf{e}[i] = Q\left|\frac{\mathbf{z}[i]}{Q} - \left[\frac{\mathbf{z}[i]}{Q}\right]\right|, \tag{7.32}$$

$0 \leq \mathbf{e}[i] \leq Q/2$. To embed a message bit as the LSB of the rounded value, the sender will need to round $\mathbf{z}[i]$ in the opposite direction, which would result in an increased quantization error of $Q - \mathbf{e}[i]$. Thus, the embedding distortion would be the difference between the two rounding errors $\rho[i] = Q - 2\mathbf{e}[i]$. Note that it is therefore advantageous for the sender to select for embedding those pixels with the smallest $\rho[i]$, which are exactly those 16-bit values $\mathbf{z}[i]$ that are close to the middle of quantization intervals, $\mathbf{e}[i] \approx Q/2$. In this case, the recipient cannot read the message because the selection channel is not available to him (he sees only the final quantized values). This problem can be solved using special coding techniques called wet paper codes explained in Chapter 9.

In general, it is a highly non-trivial problem to design steganographic schemes that embed a given payload while minimizing the total embedding impact. There exist many suboptimal schemes for special choices of the embedding impact. The whole of Chapter 8 is devoted to design of steganographic schemes when $\rho[i] = 1$ and $|\mathbf{x}[i] - \mathbf{y}[i]| = 1$, or in other words when the embedding impact is the number of embedding changes.

We have already encountered special cases of minimal-embedding-impact steganography in Chapter 5. The optimal parity assignment was designed to produce the smallest expected distortion when embedding in a palette image along a pseudo-random path. The suboptimal scheme for palette images called the embedding-while-dithering method also belongs to this category.

Since the design of steganographic schemes that attempt to minimize the embedding impact requires knowledge of certain coding techniques, we do not describe any specific instances of embedding schemes in this chapter. Instead, we postpone this topic to the next two chapters.

In the next section, we establish a fundamental performance bound for minimal-embedding-impact steganography. In particular, we derive a quantitative relationship between the maximal payload one can embed for a given bound on the embedding impact. Knowledge of this theoretical bound will give us an opportunity to evaluate the performance of suboptimal steganographic schemes and compare them.

7.4.1   Performance bound on minimal-impact embedding

This section is devoted to theoretical analysis of minimal-embedding-impact steganographic schemes. Our goal is to derive, for a given assignment $\rho[i]$, a relationship between the relative payload α and the minimal embedding impact needed to embed this payload using any steganographic method.

Let us assume that the sender wants to communicate $m = \alpha n$ bits in n pixels with assigned embedding-impact measure $\rho[i]$, $i = 1, \ldots, n$. For n-element objects, $\mathbf{x}, \mathbf{y}$, we define the modification pattern $\mathbf{s} \in \{0, 1\}^n$ as $\mathbf{s}[i] = 1$ when $\mathbf{x}[i] \neq \mathbf{y}[i]$ and $\mathbf{s}[i] = 0$ otherwise. Furthermore, let $d(\mathbf{s}) = d_\rho(\mathbf{x}, \mathbf{y})$ be the impact of making embedding changes at pixels with $\mathbf{s}[i] = 1$. Let us assume that the recipient also knows the cover $\mathbf{x}$. The sender then basically communicates the modification pattern $\mathbf{s}$. Assuming the sender selects each pattern $\mathbf{s}$ with probability $p(\mathbf{s})$, the amount of information that can be communicated is the entropy of $p(\mathbf{s})$,

$$H(p) = -\sum_{\mathbf{s}} p(\mathbf{s}) \log_2 p(\mathbf{s}). \tag{7.33}$$

Our problem is now reduced to finding the probability distribution $p(\mathbf{s})$ on the space of all possible flipping patterns $\mathbf{s}$ that minimizes the expected value of the embedding impact

$$\sum_{\mathbf{s}} d(\mathbf{s})p(\mathbf{s}) \tag{7.34}$$

subject to the constraints

$$H(p) = -\sum_{\mathbf{s}} p(\mathbf{s}) \log_2 p(\mathbf{s}) = m, \tag{7.35}$$

$$\sum_{\mathbf{s}} p(\mathbf{s}) = 1. \tag{7.36}$$

This problem can be solved using Lagrange multipliers. Let

$$F(p(\mathbf{s})) = \sum_{\mathbf{s}} p(\mathbf{s})d(\mathbf{s}) + a \left(m + \sum_{\mathbf{s}} p(\mathbf{s}) \log_2 p(\mathbf{s}) \right) + b \left(\sum_{\mathbf{s}} p(\mathbf{s}) - 1 \right). \tag{7.37}$$

Then,

$$\frac{\partial F}{\partial p(\mathbf{s})} = d(\mathbf{s}) + a \left(\log_2 p(\mathbf{s}) + \frac{1}{\log 2} \right) + b = 0 \tag{7.38}$$

if and only if $p(\mathbf{s}) = Ae^{-\lambda d(\mathbf{s})}$, where $A^{-1} = \sum_{\mathbf{s}} e^{-\lambda d(\mathbf{s})}$ and λ is determined from

$$-\sum_{\mathbf{s}} p(\mathbf{s}) \log_2 p(\mathbf{s}) = m. \tag{7.39}$$

Thus, the probabilities $p(\mathbf{s})$ follow an exponential distribution with respect to the embedding impact $d(\mathbf{s})$. Note that this result does not depend on the specific form of d and thus holds, for example, for non-additive measures as well.

If the embedding impact of the pattern $\mathbf{s}$ is an additive function of "singleton" patterns (patterns for which only one pixel is modified), then $d(\mathbf{s}) = \mathbf{s}[1]\boldsymbol{\rho}[1] + \cdots + \mathbf{s}[n]\boldsymbol{\rho}[n]$, and $p(\mathbf{s})$ accepts the form

$$p(\mathbf{s}) = Ae^{-\lambda \sum_{i=1}^{n} \mathbf{s}[i]\boldsymbol{\rho}[i]} = A \prod_{i=1}^{n} e^{-\lambda \mathbf{s}[i]\boldsymbol{\rho}[i]}, \tag{7.40}$$

$$A^{-1} = \sum_{\mathbf{s}} \prod_{i=1}^{n} e^{-\lambda \mathbf{s}[i]\boldsymbol{\rho}[i]} = \prod_{i=1}^{n} (1 + e^{-\lambda \boldsymbol{\rho}[i]}), \tag{7.41}$$

which further implies

$$p(\mathbf{s}) = \prod_{i=1}^{n} p(i, \mathbf{s}[i]), \tag{7.42}$$

where $p(i, 0)$ and $p(i, 1)$ are the probabilities that the ith pixel is not (is) modified during embedding,

$$p(i, 0) = \frac{1}{1 + e^{-\lambda \boldsymbol{\rho}[i]}}, \quad p(i, 1) = \frac{e^{-\lambda \boldsymbol{\rho}[i]}}{1 + e^{-\lambda \boldsymbol{\rho}[i]}}. \tag{7.43}$$

This means that the joint probability distribution $p(\mathbf{s})$ can be factorized and thus we need to know only the marginal probabilities $p(i, 1)$ that the ith pixel is modified. It also enables us to write for the entropy

$$H(p) = \sum_{i=1}^{n} H\left(p(i,1)\right),$$ (7.44)

where the function H applied to a scalar is the binary entropy function $H(x) = -x \log_2 x - (1-x) \log_2(1-x)$.

Note that in the special case when $\rho[i] = 1$, for all i, the embedding impact per pixel is the change rate $d/n = \beta$, and we obtain

$$m = \sum_{i=1}^{n} H\left(p(i,1)\right) = nH\left(\frac{e^{-\lambda}}{1+e^{-\lambda}}\right),$$ (7.45)

$$d = E\left(\sum_{i=1}^{n} p(i,1)\rho[i]\right) = \frac{ne^{-\lambda}}{1+e^{-\lambda}},$$ (7.46)

which gives the following relationship between the change rate and the relative message length $\alpha = m/n$:

$$\beta = H^{-1}(\alpha).$$ (7.47)

In Section 8.4.2 on matrix embedding, we rederive this bound in a different manner using purely combinatorial considerations.

We now derive the relationship between embedding capacity and impact in the limit of a large number of pixels, $n \to \infty$. Let us sort $\rho[i]$ from the smallest to the largest and normalize so that $\sum_i \rho[i] = 1$. Let ρ be a Riemann-integrable non-decreasing function on $[0,1]$ such that $\rho(i/n) = \rho[i]$. Then for $n \to \infty$, the average distortion per element

$$\frac{d}{n} = \frac{1}{n} \sum_{i=1}^{n} p(i,1)\rho[i] \to \int_0^1 p(x)\rho(x)\mathrm{d}x,$$ (7.48)

where $p(x) = e^{-\lambda\rho(x)}/(1+e^{-\lambda\rho(x)})$. By the same token,

$$\alpha = \frac{m}{n} = \frac{1}{n} \sum_{i=1}^{n} H\left(p(i,1)\right) \to \int_0^1 H\left(p(x)\right) \mathrm{d}x.$$ (7.49)

Finally, by direct calculation

$$\log 2 \int_0^1 H\left(p(x)\right) \mathrm{d}x = \lambda \int_0^1 \frac{\rho(x)e^{-\lambda\rho(x)}}{1+e^{-\lambda\rho(x)}}\mathrm{d}x + \int_0^1 \log\left(1+e^{-\lambda\rho(x)}\right)\mathrm{d}x$$ (7.50)

$$= \lambda \int_0^1 \frac{\left(\rho(x) + x\rho'(x)\right)e^{-\lambda\rho(x)}}{1+e^{-\lambda\rho(x)}}\mathrm{d}x + \log\left(1+e^{-\lambda\rho(1)}\right).$$

(7.51)

The second equality is obtained by integrating the second integral by parts. We derived the embedding-capacity–embedding-impact relationship in a parametric

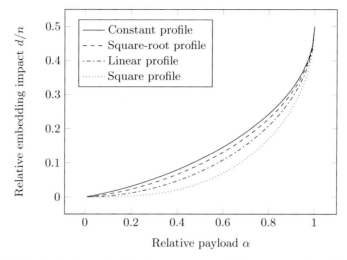

Figure 7.5 Minimal relative embedding impact versus relative message length for four embedding-impact profiles on $[0, 1]$. Constant: $\rho(x) = 1$, square-root profile: $\rho(x) = \sqrt{x}$, linear: $\rho(x) = x$, square: $\rho(x) = x^2$.

form

$$\frac{1}{n} d(\lambda) = G_\rho(\lambda), \tag{7.52}$$

$$\alpha(\lambda) = \frac{1}{\log 2} \left(\lambda F_\rho(\lambda) + \log \left(1 + e^{-\lambda \rho(1)} \right) \right), \tag{7.53}$$

where λ is a non-negative parameter and

$$G_\rho(\lambda) = \int_0^1 \frac{\rho(x) e^{-\lambda \rho(x)}}{1 + e^{-\lambda \rho(x)}} \, dx, \tag{7.54}$$

$$F_\rho(\lambda) = \int_0^1 \frac{(\rho(x) + x\rho'(x)) e^{-\lambda \rho(x)}}{1 + e^{-\lambda \rho(x)}} \, dx. \tag{7.55}$$

Figure 7.5 shows the embedding impact per pixel, d/n, as a function of relative payload α for four profiles $\rho(x)$. The square profile $\rho(x) = x^2$ has the smallest impact for a fixed payload among the four and the biggest difference occurs for small payloads. This is understandable because the square profile is the smallest among the four for small x.

Before closing this section, we provide one interesting result that essentially states that among a certain class of embedding operations for JPEG images, the embedding operation of F5 is optimal because, in some well-defined sense, it minimizes the embedding impact.

7.4.2   Optimality of F5 embedding operation

In this section, we show that the embedding operation of F5 (decreasing the absolute value of DCT coefficients) introduces the minimal embedding distortion among all operations from a certain class. Thus, one could also think of F5 as an instance of minimal-impact steganography.

As in Chapter 2, we denote by $\mathbf{d}[i]$ the DCT coefficients from the cover image after dividing by quantization steps but before rounding them to integers. Here, we use only a one-dimensional index assuming that the DCT coefficients are sorted in some way. While the range of $\mathbf{d}[i]$ depends on the implementation of the DCT transform, here we assume $\mathbf{d}[i]$ are real numbers. The DCT coefficients after rounding are denoted $\mathbf{D}[i]$. After applying a steganographic algorithm to the JPEG file, the quantized DCT coefficients $\mathbf{D}[i]$ are changed to $\mathbf{y}[i]$. The coefficients $\mathbf{D}[i]$ and $\mathbf{y}[i]$ are thus integers. We denote by $\mathbf{h}[k]$ the histogram of quantized AC DCT coefficients of the cover image. For $\gamma > 0$, the distortion due to rounding is

$$d_{\gamma,\text{round}} = \sum_{i=1}^{n} |\mathbf{d}[i] - \mathbf{D}[i]|^{\gamma} . \tag{7.56}$$

We take the total distortion due to quantization *and* embedding as the measure of embedding impact

$$d_{\gamma,\text{emb}} = \sum_{i=1}^{n} |\mathbf{d}[i] - \mathbf{y}[i]|^{\gamma} . \tag{7.57}$$

Additionally, we define the probabilistic ϵ-embedding operation that changes a DCT coefficient towards zero with probability $1 - \epsilon$ and away from zero with probability ϵ. In other words, $\mathbf{D}[i] = \mathbf{D}[i] - \text{sign}\,(\mathbf{D}[i])$ with probability $1 - \epsilon$ and $\mathbf{D}[i] = \mathbf{D}[i] + \text{sign}\,(\mathbf{D}[i])$ with probability ϵ. The F5 embedding operation is obtained for $\epsilon = 0$, while the ± 1 embedding corresponds to $\epsilon = \frac{1}{2}$.

Proposition 7.1. [Optimality of F5 embedding operation] *In the absence of any information about the unquantized DCT coefficients, the expected value of $d_{\gamma,\text{emb}}$ is minimized for $\epsilon = 0$, which is the embedding operation of F5, for any $\gamma > 0$.*

Proof. Viewing $\mathbf{d}[i]$ as instances of a random variable, let $f(x)$ be its probability distribution function. In Chapter 2, we learned that the histogram of quantized DCT coefficients in a JPEG file follows a distribution with a sharp peak at zero that is monotonically decreasing for positive coefficient values and increasing for negative coefficients. Thus, we assume that f is increasing on $(-\infty, 0)$ and decreasing on $(0, \infty)$ and therefore Lebesgue integrable. Let us inspect the embedding distortion for one value of the quantized coefficient $d \neq 0$, where, say, $d < 0$ (this means that f is increasing at d) under the assumption that a random fraction of β DCT coefficients are modified. The embedding operation will thus randomly change a fraction ϵ of $\beta \mathbf{h}[d]$ coefficients away from zero (increase their absolute value) and the fraction $1 - \epsilon$ towards zero (decrease their absolute value). Let us now express the increase of distortion due to embedding with respect to the distortion solely due to rounding $d_{\gamma,\text{emb}} - d_{\gamma,\text{round}}$ (follow Figure 7.6).

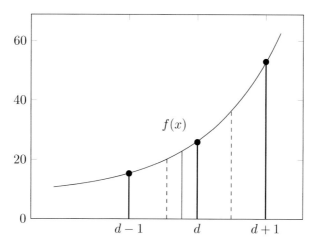

Figure 7.6 Illustrative example for derivations in the text.

The unquantized DCT coefficients that are quantized to d lie between $d - \frac{1}{2}$ and $d + \frac{1}{2}$ (the dashed lines in Figure 7.6). If an unquantized coefficient $x \in \left(d - \frac{1}{2}, d\right)$ is later changed to $d - 1$ during embedding (with probability ϵ), the increase in distortion is $\Delta_1(x) = (x - (d-1))^{\gamma} - (d-x)^{\gamma}$. If the coefficient is changed to $d + 1$ (with probability $1 - \epsilon$), the increase in distortion is $\Delta_2(x) = (d + 1 - x)^{\gamma} - (d - x)^{\gamma}$. For coefficients $x \in \left(d, d + \frac{1}{2}\right)$, the increase in distortion is $\Delta_3(x) = (x - (d-1))^{\gamma} - (x - d)^{\gamma}$, when d is changed to $d - 1$ (with probability ϵ), and $\Delta_4(x) = (d + 1 - x)^{\gamma} - (x - d)^{\gamma}$, when d is changed to $d + 1$ (with probability $1 - \epsilon$). Thus, the expected value of the distortion increase, $d(\epsilon) = d_{\gamma,\text{emb}} - d_{\gamma,\text{round}}$, is

$$d(\epsilon) = \int_{d-\frac{1}{2}}^{d} \left(\beta \epsilon f(x) \Delta_1(x) + \beta(1-\epsilon) f(x) \Delta_2(x)\right) \mathrm{d}x$$

$$+ \int_{d}^{d+\frac{1}{2}} \left(\beta \epsilon f(x) \Delta_3(x) + \beta(1-\epsilon) f(x) \Delta_4(x)\right) \mathrm{d}x, \qquad (7.58)$$

$$= \epsilon \beta \int_{d-\frac{1}{2}}^{d} f(x) \left(\Delta_1(x) - \Delta_2(x)\right) \mathrm{d}x$$

$$+ \epsilon \beta \int_{d}^{d+\frac{1}{2}} f(x) \left(\Delta_3(x) - \Delta_4(x)\right) \mathrm{d}x + C, \qquad (7.59)$$

where C does not depend on ϵ. We clearly have $\Delta_1(x) - \Delta_2(x) = \Delta_3(x) - \Delta_4(x) = (x - d + 1)^{\gamma} - (d + 1 - x)^{\gamma} = g(x)$. Moreover, $g(d + y) = -g(d - y)$ for

$y \in \left(-\frac{1}{2}, \frac{1}{2}\right)$ and $g(x) \geq 0$ on $\left(d, d + \frac{1}{2}\right)$. Thus, we can write for $d(\epsilon)$

$$d(\epsilon) = \epsilon\beta \int_0^{\frac{1}{2}} f(d-y)g(d-y)\mathrm{d}y + \epsilon\beta \int_0^{\frac{1}{2}} f(d+y)g(d+y)\mathrm{d}y + C \qquad (7.60)$$

$$= \epsilon\beta \int_0^{\frac{1}{2}} \left(f(d+y) - f(d-y)\right)g(d+y)\mathrm{d}y + C. \qquad (7.61)$$

Because f is increasing and $g(d+y) \geq 0$ on $\left(0, \frac{1}{2}\right)$, $d(\epsilon)$ is minimized when $\epsilon = 0$, which corresponds to the F5 embedding operation. $\qquad\square$

Summary

- In this chapter, we study methods for construction of practical steganographic schemes for digital images.
- There exist four heuristic principles that can be applied to decrease the KL divergence between cover and stego images and thus improve the steganographic security:
  - Model-preserving steganography
  - Making embedding mimic natural processing
  - Steganalysis-aware steganography
  - Minimum-embedding-impact steganography
- In model-preserving steganography, a simplified model of cover images is formulated and the embedding is forced to preserve that model.
- The most frequently adopted model is to consider the cover as a sequence of iid random variables. The model itself is either the sample distribution of cover elements or a parametric fit through the sample distribution. Steganography preserving the cover histogram is undetectable within the iid model.
- Model-preserving steganography can be attacked by identifying a quantity that is not preserved under embedding.
- Stochastic modulation is an example of embedding designed to mimic the image-acquisition process. The message is embedded by superimposing quantized iid noise with a given distribution.
- In steganalysis-aware steganography, the designer of the stego system focuses on making the impact of embedding undetectable using existing steganalysis schemes.
- Minimum-embedding-impact steganography starts by assigning to each cover element a scalar value expressing the impact of making an embedding change at that element. The designer then attempts to embed messages by minimizing the total embedding impact.
- Among all embedding operations that change a quantized DCT coefficient towards zero and away from zero with fixed probability, the operation of the F5 algorithm minimizes the combined distortion due to quantization and embedding.

Exercises

7.1 [Histogram after ±1 embedding] Let $\mathbf{h}[i]$, $i = 0, 1, \ldots, 255$, denote the histogram of an 8-bit cover grayscale image. Show that after embedding a message of relative length α using ±1 embedding, the stego image histogram is $\mathbf{h}_\alpha = \mathbf{A}\mathbf{h}$, where $\mathbf{A}$ is tri-diagonal with $1 - \alpha/2$ on the main diagonal and $\alpha/4$ on the two second main diagonals (with the exception of $\mathbf{A}[2, 1] = \mathbf{A}[254, 255] = \alpha/2$):

$$\mathbf{A} = \begin{pmatrix} 1-\frac{\alpha}{2} & \frac{\alpha}{4} & 0 & 0 & \cdots & 0 & 0 \\ \frac{\alpha}{2} & 1-\frac{\alpha}{2} & \frac{\alpha}{4} & 0 & \cdots & 0 & 0 \\ 0 & \frac{\alpha}{4} & 1-\frac{\alpha}{2} & \frac{\alpha}{4} & \cdots & 0 & 0 \\ \cdots & \cdots & \frac{\alpha}{4} & 1-\frac{\alpha}{2} & \frac{\alpha}{4} & \cdots & \cdots \\ 0 & 0 & \cdots & \frac{\alpha}{4} & 1-\frac{\alpha}{2} & \frac{\alpha}{4} & 0 \\ 0 & 0 & \cdots & 0 & \frac{\alpha}{4} & 1-\frac{\alpha}{2} & \frac{\alpha}{2} \\ 0 & 0 & \cdots & 0 & 0 & \frac{\alpha}{4} & 1-\frac{\alpha}{2} \end{pmatrix}. \tag{7.62}$$

7.2 [Impact of F5 on histogram] In a JPEG image, let $\mathbf{h}^{(kl)}[i]$ be the number of AC DCT coefficients corresponding to spatial frequency (k, l) that are equal in absolute value to i. Suppose that the F5 changes a total of n DCT coefficients. Thus, the probability that a randomly selected non-zero AC DCT coefficient is changed is $\beta = n/n_0$, where n_0 is the number of all non-zero AC DCT coefficients. Show that the expected value of the stego image histogram is

$$E\left[\mathbf{h}_\beta^{(kl)}[i]\right] = (1-\beta)\mathbf{h}^{(kl)}[i] + \beta\mathbf{h}^{(kl)}[i+1], \ i > 0, \tag{7.63}$$

$$E\left[\mathbf{h}_\beta^{(kl)}[0]\right] = \mathbf{h}^{(kl)}[0] + \beta\mathbf{h}^{(kl)}[1]. \tag{7.64}$$

7.3 [One-stego-sequence stochastic modulation] Consider the following version of stochastic modulation. The sender generates one stego noise sequence $\mathbf{r}[i] = \text{round}(\boldsymbol{\eta}[i])$, where $\boldsymbol{\eta}[i]$ follows an arbitrary distribution symmetrical about zero. Assume that the parity function satisfies

$$\pi(x+r, r) = -\pi(x-r, r) \in \{-1, 1\} \text{ for any } r \neq 0. \tag{7.65}$$

(Can you find one such π?) Ignoring the boundary effects for now, at each pixel $\mathbf{x}[i]$ where $\mathbf{r}[i] \neq 0$, the sender embeds one message bit $m \in \{-1, 1\}$ as

$$\mathbf{y}[i] = \mathbf{x}[i] + m \times \pi(\mathbf{x}[i] + \mathbf{r}[i], \mathbf{r}[i])\mathbf{r}[i]. \tag{7.66}$$

First, show that the effect of embedding is the same as adding to the cover image a random variable with the same distribution as $\mathbf{r}[i]$. Also show that the recipient can read one message bit as

$$m = \pi(\mathbf{y}[i], \mathbf{r}[i]) \tag{7.67}$$

whenever $\mathbf{r}[i] \neq 0$. Finally, if the same stego sequence is used in this simpler version of stochastic modulation and in the version described in Section 7.2.1, which version embeds at a lower distortion and why?

7.4 [Suboptimality of stochastic modulation] Prove that

$$1 - \sum_k \mathbf{p}[k]^2 \leq H(\mathbf{p}) \tag{7.68}$$

for any pmf $\mathbf{p}$. **Hint:** $\log \mathbf{p}[k] = \log\left(1 + (\mathbf{p}[k] - 1)\right) \leq \mathbf{p}[k] - 1$ by the log-inequality $\log(1 + x) \leq x, \forall x$.

7.5 [Stego noise with amplitude 1] Show that for stego noise sequences with amplitude 1 ($\mathbf{p}[k] = 0$ for $k \notin \{-1, 0, 1\}$), the entropy of the optimal distribution $H(\mathbf{p}_{\mathrm{opt}}) = H_{d_\gamma}(D) = -D \log_2 D - (1 - D)\log_2(1 - D) + D$ for any $\gamma > 0$. **Hint:** First show that for $\mathbf{p}[-1] + \mathbf{p}[1] = 2p$, a constant, $-\mathbf{p}[-1]\log_2 \mathbf{p}[-1] - \mathbf{p}[1]\log_2 \mathbf{p}[1]$ is maximal when $\mathbf{p}[-1] = \mathbf{p}[1] = p$. Thus, we need only to search for optimal distributions in the family of distributions $\mathbf{p} = (p, 1 - 2p, p)$ parametrized by a scalar parameter p. For such distributions, the distortion bound is $2p \leq D$ and the result is obtained by setting $p = D/2$.

7.6 [ϵ-embedding minimizing the KL divergence] The principle of minimal embedding impact does not have to produce steganographic schemes that minimize the KL divergence. Consider the following example. Let us model the quantized DCT coefficients in a JPEG file as iid instances of a random variable with distribution given by $\mathbf{h}$, the model of the histogram of quantized DCT coefficients from the cover. Let us assume that $\mathbf{h}$ is non-increasing for positive values, $\mathbf{h}[i] \geq \mathbf{h}[i + 1]$, and non-decreasing for negative values. Let us also assume that $\mathbf{h}[i] > 0$ for all i. A steganographic scheme that changes a fraction of β of non-zero DCT coefficients using the ϵ-embedding operation defined in Section 7.4.2 has the following impact on the histogram:

$$\mathbf{h}_\beta[i] = (1 - \beta)\mathbf{h}[i] + (1 - \epsilon)\beta\mathbf{h}[i + 1] + \epsilon\beta\mathbf{h}[i - 1] \quad \text{for } i > 1, \tag{7.69}$$

$$\mathbf{h}_\beta[i] = (1 - \beta)\mathbf{h}[i] + \epsilon\beta\mathbf{h}[i + 1] + (1 - \epsilon)\beta\mathbf{h}[i - 1] \quad \text{for } i < -1, \tag{7.70}$$

$$\mathbf{h}_\beta[1] = (1 - \beta)\mathbf{h}[1] + (1 - \epsilon)\beta\mathbf{h}[2], \tag{7.71}$$

$$\mathbf{h}_\beta[-1] = (1 - \beta)\mathbf{h}[-1] + (1 - \epsilon)\beta\mathbf{h}[-2], \tag{7.72}$$

$$\mathbf{h}_\beta[0] = \mathbf{h}[0] + (1 - \epsilon)\beta\mathbf{h}[1] + (1 - \epsilon)\beta\mathbf{h}[-1]. \tag{7.73}$$

Using the fact that KL divergence is locally quadratic for small β (see Proposition B.7), show that

$$\frac{2D_{\mathrm{KL}}(\mathbf{h}\|\mathbf{h}_\beta)}{\beta^2} = (1 - \epsilon)^2 \frac{(\mathbf{h}[1] + \mathbf{h}[-1])^2}{\mathbf{h}[0]} + \frac{((1 - \epsilon)\mathbf{h}[2] - \mathbf{h}[1])^2}{\mathbf{h}[1]}$$

$$+ \frac{((1 - \epsilon)\mathbf{h}[-2] - \mathbf{h}[-1])^2}{\mathbf{h}[-1]}$$

$$+ \sum_{i>1} \frac{((1 - \epsilon)\mathbf{h}[i + 1] + \epsilon\mathbf{h}[i - 1] - \mathbf{h}[i])^2}{\mathbf{h}[i]}$$

$$+ \sum_{i<-1} \frac{(\epsilon\mathbf{h}[i + 1] + (1 - \epsilon)\mathbf{h}[i - 1] - \mathbf{h}[i])^2}{\mathbf{h}[i]} + O(\beta). \tag{7.74}$$

By differentiating with respect to ϵ, show that the KL divergence is minimal when $\epsilon = \epsilon_1/\epsilon_2$, where

$$\epsilon_1 = \sum_{i<-1} \frac{(\mathbf{h}[i+1] - \mathbf{h}[i-1])(\mathbf{h}[i] - \mathbf{h}[i-1])}{\mathbf{h}[i]} - \frac{\mathbf{h}[-2]}{\mathbf{h}[-1]}(\mathbf{h}[-1] - \mathbf{h}[-2])$$

$$+ \frac{(\mathbf{h}[-1] + \mathbf{h}[1])^2}{\mathbf{h}[0]} - \frac{\mathbf{h}[2]}{\mathbf{h}[1]}(\mathbf{h}[1] - \mathbf{h}[2])$$

$$+ \sum_{i>1} \frac{(\mathbf{h}[i+1] - \mathbf{h}[i-1])(\mathbf{h}[i+1] - \mathbf{h}[i])}{\mathbf{h}[i]}, \tag{7.75}$$

$$\epsilon_2 = \sum_{i<-1} \frac{(\mathbf{h}[i+1] - \mathbf{h}[i-1])^2}{\mathbf{h}[i]} + \frac{\mathbf{h}[-2]^2}{\mathbf{h}[-1]} + \frac{(\mathbf{h}[-1] + \mathbf{h}[1])^2}{\mathbf{h}[0]}$$

$$+ \frac{\mathbf{h}[2]^2}{\mathbf{h}[1]} + \sum_{i>1} \frac{(\mathbf{h}[i+1] - \mathbf{h}[i-1])^2}{\mathbf{h}[i]}. \tag{7.76}$$

Use the monotonicity of $\mathbf{h}$ to show that $\epsilon_1 \leq \epsilon_2$ and thus $\epsilon \leq 1$. Contrast this result to Proposition 7.1.

7.7 [OutGuess does not preserve individual histograms] Even though OutGuess preserves the global histogram of all DCT coefficients in a JPEG file, it does not have to preserve histograms of individual DCT modes. Prove this statement by considering one LSB pair $\{2j, 2j+1\}$ for the global histogram, $\mathbf{h}$, and the same bin for the histogram, $\mathbf{h}^{(kl)}$, of a selected DCT mode (k, l). Assume that

$$\frac{\mathbf{h}[2j+1]}{\mathbf{h}[2j]} \neq \frac{\mathbf{h}^{(kl)}[2j+1]}{\mathbf{h}^{(kl)}[2j]}. \tag{7.77}$$

8   Matrix embedding

In the previous chapter, we learned that one of the general guiding principles for design of steganographic schemes is the principle of minimizing the embedding impact. The plausible assumption here is that it should be more difficult for Eve to detect Alice and Bob's clandestine activity if they leave behind smaller embedding distortion or "impact." This chapter introduces a very general methodology called matrix embedding using which the prisoners can minimize the total number of changes they need to carry out to embed their message and thus increase the embedding efficiency. Even though special cases of matrix embedding can be explained in an elementary fashion on an intuitive level, it is extremely empowering to formulate it within the framework of coding theory. This will require the reader to become familiar with some basic elements of the theory of linear codes. The effort is worth the results because the reader will be able to design more secure stegosystems, acquire a deeper understanding of the subject, and realize connections to an already well-developed research field. Moreover, according to the studies that appeared in [143, 95], matrix embedding is one of the most important design elements of practical stegosystems.

As discussed in Chapter 5, in LSB embedding or ± 1 embedding one pixel communicates exactly one message bit. This was the case of OutGuess as well as Jsteg. Assuming the message bits are random, each pixel is thus modified with probability $\frac{1}{2}$ because this is the probability that the LSB will not match the message bit. Thus, on average two bits are embedded using one change or, equivalently, the embedding efficiency is 2. If the message length is smaller than the embedding capacity of the cover image, it is possible to substantially increase the embedding efficiency and thus embed the same payload with fewer embedding changes.

To explain why this is possible at all, consider the following simple example. Let us assume that we wish to embed a message of relative length $\alpha = \frac{2}{3}$ using LSB embedding. Thus, given a cover image with n pixels, the message contains $n\alpha = 2n/3$ bits. This means that we need to embed 2 bits, $\mathbf{m}[1], \mathbf{m}[2]$, in a group of three pixels with grayscale values $\mathbf{g}[1], \mathbf{g}[2], \mathbf{g}[3]$. Classical LSB embedding would embed bit $\mathbf{m}[1]$ at $\mathbf{g}[1]$ and bit $\mathbf{m}[2]$ at $\mathbf{g}[2]$ while skipping $\mathbf{g}[3]$ and thus embed with embedding efficiency 2. We can, however, do better by embedding the bits as

$$m[1] = \mathrm{LSB}(g[1]) \oplus \mathrm{LSB}(g[2]), \tag{8.1}$$

$$m[2] = \mathrm{LSB}(g[2]) \oplus \mathrm{LSB}(g[3]), \tag{8.2}$$

where $\oplus$ is the exclusive or. If the cover values already satisfy both equations, no embedding changes are necessary. If the first equation is satisfied but not the second one, the sender can flip the LSB of $g[3]$. If the second equation is satisfied but not the first one, the sender should flip $g[1]$. If neither is satisfied, the sender will flip $g[2]$. Because the probability of each case is $\frac{1}{4}$, the expected number of changes is $0 \times \frac{1}{4} + 1 \times \frac{1}{4} + 1 \times \frac{1}{4} + 1 \times \frac{1}{4} = \frac{3}{4}$ and the embedding efficiency (in bits per change) is $e = \frac{2}{3/4} = \frac{8}{3} > 2$.

Note that in this scheme, we can no longer say which pixels convey the message bits. Both bits are communicated by the group *as a whole*. One can say that in matrix embedding the message is communicated by the embedding changes as well as their *position*. In fact, the message-extraction rule that the receiver uses is multiplication[1] of the vector of LSBs, $x = \mathrm{LSB}(g)$, by a matrix

$$m = \begin{pmatrix} 1 & 1 & 0 \\ 0 & 1 & 1 \end{pmatrix} x, \tag{8.3}$$

which gave this method its name – matrix embedding [19, 52, 98, 233].

Before presenting a general approach to matrix embedding based on coding theory, in Section 8.1 we generalize the simple trick of this introduction. Up until now, the material can be comfortably grasped by all readers without any background in coding theory. To prepare the reader for the material presented in the rest of this chapter, Section 8.2 contains a brief overview of the theory of binary linear codes. Readers not familiar with this subject are additionally urged to read Appendix C, which contains a more detailed tutorial. Section 8.3 contains the main result of this chapter, the matrix embedding theorem. The theorem gives us a general methodology to improve the embedding efficiency of steganographic methods using linear codes. We also revisit Section 8.1 and interpret the method from a new viewpoint. The theoretical limits of matrix embedding methods are the subject of Section 8.4. A matrix embedding approach suitable for hiding large payloads appears in Section 8.5. It also illustrates the usefulness of random codes. Section 8.6 deals with embedding methods realized using codes defined over larger alphabets (*q*-ary codes). Such codes can further increase the embedding efficiency at the cost of larger distortion. Finally, in Section 8.7 we explain an alternative approach to minimizing the number of embedding changes that is based on sum and difference covering sets of finite cyclic groups.

---

[1] All arithmetic operations are performed in the usual binary arithmetic (8.14) (also, see Appendix C for more details).

Plate 1   Fig 2.3

Plate 2   Fig 2.5

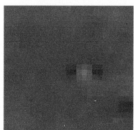

Plate 3 Fig 3.7

Plate 4 Fig 3.8

Plate 5 Fig 5.1

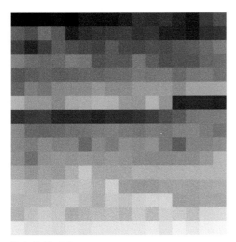

Plate 6 Fig 5.5

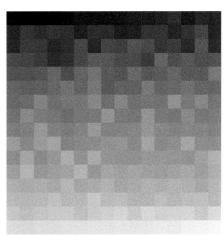

Plate 7 Fig 5.6

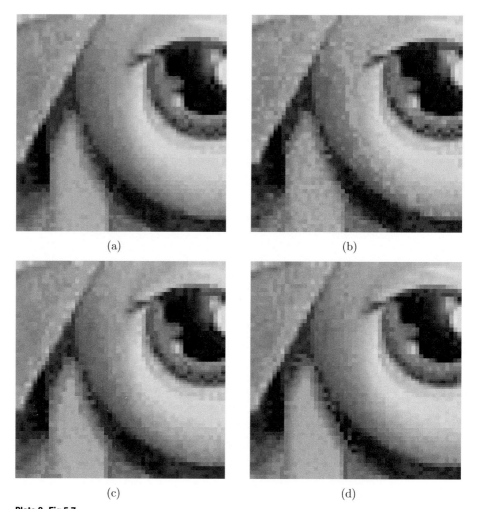

(a)   (b)

(c)   (d)

Plate 8   Fig 5.7

8.1   Matrix embedding using binary Hamming codes

In this section, we explain the matrix embedding method used in the F5 algorithm (Chapter 7), which was the first practical steganographic scheme to incorporate this embedding mechanism. The reader will recognize later that the method is based on binary Hamming codes.

Assume the sender wants to communicate a message with relative length $\alpha_p = p/(2^p - 1)$, $p \geq 0$, which means that p message bits, $\mathbf{m}[1], \ldots, \mathbf{m}[p]$, need to be embedded in $2^p - 1$ pixels. We denote by $\mathbf{x}$ the vector of LSBs of $2^p - 1$ pixels from the cover image (e.g., collected along a pseudo-random path if the embedding uses a stego key for random spread of the message bits). The sender and recipient share a $p \times (2^p - 1)$ binary matrix $\mathbf{H}$ that contains all non-zero binary vectors of length p as its columns. An example of such a matrix for $p = 3$ is

$$\mathbf{H} = \begin{pmatrix} 0\,0\,0\,1\,1\,1\,1 \\ 0\,1\,1\,0\,0\,1\,1 \\ 1\,0\,1\,0\,1\,0\,1 \end{pmatrix}. \tag{8.4}$$

As in the previous simple example, the sender modifies the pixel values so that the column vector of their LSBs, $\mathbf{y}$, satisfies

$$\mathbf{m} = \mathbf{Hy}. \tag{8.5}$$

We call the vector $\mathbf{Hy}$ the "syndrome" of $\mathbf{y}$. If by chance the syndrome of the cover pixels already communicates the correct message, $\mathbf{Hx} = \mathbf{m}$, which happens with probability $1/2^p$, the sender does not need to modify any of the $2^p - 1$ cover pixels, sets $\mathbf{y} = \mathbf{x}$, and proceeds to the next block of $2^p - 1$ pixels and embeds the next segment of p message bits.

When $\mathbf{Hx} \neq \mathbf{m}$, the sender looks up the difference $\mathbf{Hx} - \mathbf{m}$ as a column in $\mathbf{H}$ (there must be such a column because $\mathbf{H}$ contains all non-zero binary p-tuples). Let us say that it is the jth column and we write it as $\mathbf{H}[., j]$. By flipping the LSB of the jth pixel and keeping the remaining pixels unchanged,

$$\mathbf{y}[j] = 1 - \mathbf{x}[j], \tag{8.6}$$
$$\mathbf{y}[k] = \mathbf{x}[k], \quad k \neq j, \tag{8.7}$$

the syndrome of $\mathbf{y}$ now matches the message bits,

$$\mathbf{Hy} = \mathbf{m}. \tag{8.8}$$

This is because $\mathbf{Hy} = \mathbf{Hx} + \mathbf{H}(\mathbf{y} - \mathbf{x}) = \mathbf{Hx} - \mathbf{m} + \mathbf{H}[., j] + \mathbf{m} = \mathbf{m}$, because $\mathbf{Hx} - \mathbf{m} = \mathbf{H}[., j]$ and in binary arithmetic $\mathbf{z} + \mathbf{z} = 0$ for any $\mathbf{z}$.

To complete the description of the steganographic algorithm, the recipient follows the same path through the image as the sender and reads p message bits from the LSBs of each block of $2^p - 1$ pixels as the syndrome $\mathbf{m} = \mathbf{Hy}$.

Let us now compute the embedding efficiency of this embedding method. With probability $1/2^p$, the sender does not modify any of the $2^p - 1$ pixels. Furthermore, she makes exactly one change with probability $1 - 1/2^p$. The average number of embedding changes is thus $0 \times 1/2^p + 1 \times (1 - 1/2^p) = 1 - 1/2^p$ and the embedding efficiency is

$$e_p = \frac{p}{1 - 2^{-p}}. \tag{8.9}$$

The embedding efficiency and the relative payload $\alpha_p = p/(2^p - 1)$ for different values of p are shown in Table 8.1.

Table 8.1. Relative payload, α_p, and embedding efficiency, e_p, in bits per change for matrix embedding using binary Hamming codes $[2^p - 1, 2^p - 1 - p]$.

p	α_p	e_p
1	1.000	2.000
2	0.667	2.667
3	0.429	3.429
4	0.267	4.267
5	0.161	5.161
6	0.093	6.093
7	0.055	7.055
8	0.031	8.031
9	0.018	9.018

Note that with increasing p, the embedding efficiency, e_p, increases while the relative payload, α_p, decreases. The improvement over embedding each message bit at exactly one pixel is quite substantial. For example, a message of relative length 0.093 can be embedded with embedding efficiency 6.093, which means that a little over six bits are embedded with a single embedding change.

The motivational example from the beginning of this section is obtained for $p = 2$. Also, notice that the classical LSB embedding corresponds to $p = 1$, in which case the matrix $\mathbf{H}$ is 1×1 and the embedding efficiency is 2.

If the sender plans to communicate a message whose relative length α is not equal to any α_p, one needs to choose the largest p for which $\alpha_p \geq \alpha$ to make sure that the complete message can be embedded. This means that for $\alpha > \frac{2}{3}$, this matrix embedding method does not bring any improvement over classical embedding methods that embed one bit at each pixel.

The parameter p also needs to be communicated to the receiver, which can be arranged in many different ways. One possibility is to reserve a few pixels from the image using the stego key and embed the binary representation of p in LSBs of those pixels and then use the rest of the image to communicate the main payload.

Example 8.1: [Matrix embedding using binary Hamming codes] We now work out a simple example of how Hamming codes can be used in practice

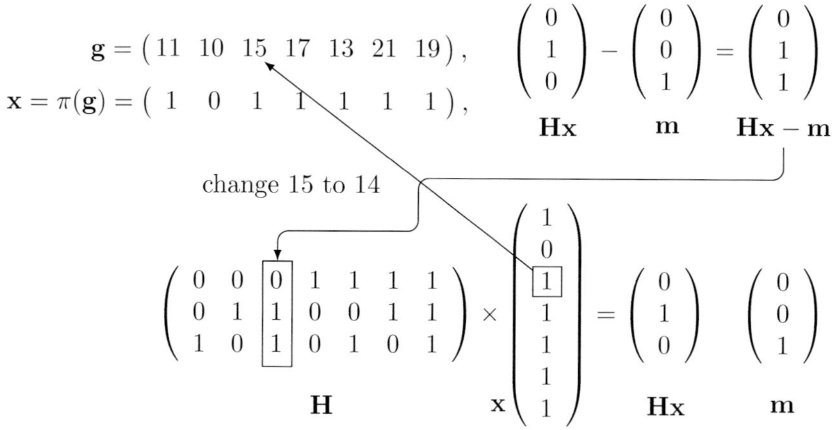

$$\mathbf{g} = \begin{pmatrix} 11 & 10 & 15 & 17 & 13 & 21 & 19 \end{pmatrix},$$

$$\mathbf{x} = \pi(\mathbf{g}) = \begin{pmatrix} 1 & 0 & 1 & 1 & 1 & 1 & 1 \end{pmatrix},$$

$$\begin{pmatrix} 0 \\ 1 \\ 0 \end{pmatrix} - \begin{pmatrix} 0 \\ 0 \\ 1 \end{pmatrix} = \begin{pmatrix} 0 \\ 1 \\ 1 \end{pmatrix}$$

$$\mathbf{Hx} \qquad \mathbf{m} \qquad \mathbf{Hx-m}$$

change 15 to 14

$$\begin{pmatrix} 0 & 0 & \boxed{0} & 1 & 1 & 1 & 1 \\ 0 & 1 & \boxed{1} & 0 & 0 & 1 & 1 \\ 1 & 0 & \boxed{1} & 0 & 1 & 0 & 1 \end{pmatrix} \times \begin{pmatrix} 1 \\ 0 \\ \boxed{1} \\ 1 \\ 1 \\ 1 \\ 1 \end{pmatrix} = \begin{pmatrix} 0 \\ 1 \\ 0 \end{pmatrix} \quad \begin{pmatrix} 0 \\ 0 \\ 1 \end{pmatrix}$$

$$\mathbf{H} \qquad\qquad \mathbf{x} \qquad\qquad \mathbf{Hx} \qquad \mathbf{m}$$

Figure 8.1 Example of embedding using binary Hamming codes.

to embed bits. Let us assume that we are embedding relative payload $\alpha = \frac{3}{7}$, which means we are embedding 3 bits in $2^3 - 1 = 7$ pixels. Given a block of pixel values $\mathbf{g} = (11, 10, 15, 17, 13, 21, 19)'$ and message $\mathbf{m} = (0, 0, 1)'$, Alice first converts the pixel values to bits, $\mathbf{x} = \mathbf{g} \bmod 2 = (1, 0, 1, 1, 1, 1, 1)'$, and computes the syndrome $\mathbf{Hx} = (0, 1, 0)'$, where $\mathbf{H}$ appears in (8.4). Then, she finds the triple $\mathbf{Hx} - \mathbf{m} = (0, 1, 1)'$ as a column in $\mathbf{H}$. It is the third column. Thus, to embed $\mathbf{m}$ in the block, she needs to flip the LSB of the third pixel. For example, she may change $\mathbf{g}[3] = 15$ to 14. Note that Alice could have achieved the same effect by changing the value to 16 as well.

There is a strong connection between matrix embedding methods and covering codes, which we now explore. The reader will realize that what appeared as a clever trick in this section will now fit into a much more general framework. Formulating the approach within the realm of coding theory will enable us to design more general and efficient matrix embedding methods as well as derive bounds on how much it is theoretically possible to minimize the number of embedding changes. To this end, the reader needs to become more familiar with the basic concepts from the theory of linear codes. Readers with a basic background in coding theory can skip the next section.

8.2   Binary linear codes

This section briefly introduces selected basic concepts and notation from the theory of binary linear codes that will be needed to understand the material in the rest of this chapter. The reader is encouraged to read Appendix C, which contains a brief tutorial on linear codes over finite fields. A good introduction to coding theory is [248].

We denote by $\mathbb{F}_2^n$ the vector space of all binary vectors, $\mathbf{x} = (\mathbf{x}[1], \ldots, \mathbf{x}[n])$, of length n where addition and multiplication by a scalar is performed elementwise,

$$\mathbf{x} + \mathbf{y} = (\mathbf{x}[1] + \mathbf{y}[1], \ldots, \mathbf{x}[n] + \mathbf{y}[n]), \tag{8.10}$$

$$b\mathbf{x} = (b\mathbf{x}[1], \ldots, b\mathbf{x}[n]), \tag{8.11}$$

for all $\mathbf{x}, \mathbf{y} \in \mathbb{F}_2^n$, $b \in \{0, 1\} = \mathbb{F}_2$. All operations in $\mathbb{F}_2$ are in the usual binary arithmetic

$$0 + 0 = 1 + 1 = 0, \tag{8.12}$$

$$0 + 1 = 1 + 0 = 0, \tag{8.13}$$

$$0 \cdot 1 = 1 \cdot 0 = 0, \tag{8.14}$$

$$1 \cdot 1 = 1, \tag{8.15}$$

$$0 \cdot 0 = 0. \tag{8.16}$$

A binary linear code $\mathcal{C}$ of length n is a vector subspace of $\mathbb{F}_2^n$. Its elements are called codewords. As a subspace, the code $\mathcal{C}$ is closed under linear combination of its elements, which means that $\forall \mathbf{x}, \mathbf{y} \in \mathcal{C}$ and $\forall b \in \{0, 1\}$, $\mathbf{x} + \mathbf{y} \in \mathcal{C}$ and $b\mathbf{x} \in \mathcal{C}$. The code has dimension k (and codimension $n - k$) if it has a basis consisting of $k \leq n$ linearly independent vectors. (We say that the code $\mathcal{C}$ is an $[n, k]$ code.) A code is completely described by its basis because all codewords can be obtained as linear combinations of the basis vectors. Writing the basis vectors as rows of a $k \times n$ matrix $\mathbf{G}$, we obtain its generator matrix.

Linear code can be alternatively described using its $(n - k) \times n$ parity-check matrix, $\mathbf{H}$, whose $n - k$ rows are linearly independent vectors that are orthogonal to $\mathcal{C}$, or

$$\mathbf{H}\mathbf{G}' = 0, \tag{8.17}$$

where the prime denotes transposition. The code can thus also be defined as

$$\mathcal{C} = \{\mathbf{c} \in \mathbb{F}_2^n | \mathbf{H}\mathbf{c} = 0\}. \tag{8.18}$$

The generator and parity-check matrices are not unique for the same reason that a vector subspace does not have a unique basis.

The space $\mathbb{F}_2^n$ can be endowed with a measure of distance called the Hamming distance defined as the number of places where $\mathbf{x}$ and $\mathbf{y}$ differ,

$$d_{\mathrm{H}}(\mathbf{x}, \mathbf{y}) = \left| \{i \in \{1, \ldots, n\} | \mathbf{x}[i] \neq \mathbf{y}[i]\} \right|. \tag{8.19}$$

In the context of steganography, $d_{\mathrm{H}}(\mathbf{x}, \mathbf{y})$ is the number of embedding changes. The Hamming weight of $\mathbf{x} \in \mathbb{F}_2^n$ is $w(\mathbf{x}) = d_{\mathrm{H}}(\mathbf{x}, \mathbf{0})$, which is the number of non-zero elements in $\mathbf{x}$.

The ball of radius r centered at $\mathbf{x}$ is the set

$$\mathcal{B}(\mathbf{x}, r) = \{\mathbf{y} \in \mathbb{F}_2^n | d_{\mathrm{H}}(\mathbf{x}, \mathbf{y}) \leq r\}, \tag{8.20}$$

and

$$V_2(r,n) = 1 + \binom{n}{1} + \binom{n}{2} + \cdots + \binom{n}{r} = \sum_{i=0}^{r} \binom{n}{i} \qquad (8.21)$$

is its volume (the number of vectors in $\mathcal{B}(\mathbf{x}, r)$).

One important characteristic of a code that will be relevant for applications in steganography is the covering radius

$$R = \max_{\mathbf{x} \in \mathbb{F}_2^n} d_H(\mathbf{x}, \mathcal{C}), \qquad (8.22)$$

where $d_H(\mathbf{x}, \mathcal{C}) = \min_{\mathbf{c} \in \mathcal{C}} d_H(\mathbf{x}, \mathbf{c})$ is the distance between $\mathbf{x}$ and the code. In other words, the covering radius is determined by the most distant point $\mathbf{x}$ from the code.

Recalling the matrix embedding method from Section 8.1, it should not be surprising that we define for each $\mathbf{x} \in \mathbb{F}_2^n$ its syndrome as $\mathbf{s} = \mathbf{H}\mathbf{x} \in \mathbb{F}_2^{n-k}$. For any syndrome $\mathbf{s} \in \mathbb{F}_2^{n-k}$, its coset $\mathcal{C}(\mathbf{s})$ is

$$\mathcal{C}(\mathbf{s}) = \{\mathbf{x} \in \mathbb{F}_2^n | \mathbf{H}\mathbf{x} = \mathbf{s}\}. \qquad (8.23)$$

Note that $\mathcal{C}(\mathbf{0}) = \mathcal{C}$ and $\mathcal{C}(\mathbf{s}_1) \cap \mathcal{C}(\mathbf{s}_2) = \emptyset$ for $\mathbf{s}_1 \neq \mathbf{s}_2$. The whole space $\mathbb{F}_2^n$ can thus be decomposed into 2^{n-k} cosets, each coset containing exactly 2^k elements,

$$\bigcup_{\mathbf{s} \in \mathbb{F}_2^{n-k}} \mathcal{C}(\mathbf{s}) = \mathbb{F}_2^n. \qquad (8.24)$$

Because $\mathcal{C}(\mathbf{s})$ is the set of all solutions to the equation $\mathbf{H}\mathbf{x} = \mathbf{s}$, from linear algebra the coset can be written as $\mathcal{C}(\mathbf{s}) = \{\mathbf{x} \in \mathbb{F}_2^n | \mathbf{x} = \tilde{\mathbf{x}} + \mathbf{c}, \mathbf{c} \in \mathcal{C}\} = \tilde{\mathbf{x}} + \mathcal{C}$, where $\tilde{\mathbf{x}}$ is an arbitrary member of $\mathcal{C}(\mathbf{s})$.

A coset leader $\mathbf{e}(\mathbf{s})$ is a member of the coset $\mathcal{C}(\mathbf{s})$ with the smallest Hamming weight. It is easy to see that the Hamming weight of any coset leader is at most R, the covering radius of $\mathcal{C}$. Take $\mathbf{x} \in \mathbb{F}_2^n$ arbitrary and calculate its syndrome $\mathbf{s} = \mathbf{H}\mathbf{x}$. Then,

$$R = \max_{\mathbf{z} \in \mathbb{F}_2^n} d_H(\mathbf{z}, \mathcal{C}) \geq d_H(\mathbf{x}, \mathcal{C}) = \min_{\mathbf{c} \in \mathcal{C}} w(\mathbf{x} - \mathbf{c}) = w(\mathbf{e}(\mathbf{s})) \qquad (8.25)$$

because when $\mathbf{c}$ goes through all codewords, $\mathbf{x} - \mathbf{c}$ goes through all members of the coset $\mathcal{C}(\mathbf{s})$. Note that the inequality is tight as there exists $\mathbf{z}$ such that $R = d_H(\mathbf{z}, \mathcal{C})$.

This result also implies that any syndrome, $\mathbf{s} \in \mathbb{F}_2^{n-k}$, can be obtained by adding at most R columns of $\mathbf{H}$. This is because $\mathcal{C}(\mathbf{s}) = \{\mathbf{x} | \mathbf{H}\mathbf{x} = \mathbf{s}\}$ and the weight of a coset leader is the smallest number of columns that need to be summed to obtain the coset syndrome $\mathbf{s}$. Thus, one method to determine the covering radius of a linear code is to first form its parity-check matrix and then find the smallest number of columns of $\mathbf{H}$ that can generate any syndrome.

8.3   Matrix embedding theorem

The matrix embedding theorem provides a recipe for how to turn any linear code into a matrix embedding method. The parameters of the code will determine the properties of the stegosystem, namely its payload and embedding efficiency. By making the connection between steganography and coding, we will be able to view the trick explained in Section 8.1 from a different angle and construct new useful embedding methods.

We will assume that Alice and Bob use a bit-assignment function $\pi : \mathcal{X} \rightarrow \{0,1\}$ that assigns a bit to each possible value of the cover element from $\mathcal{X}$. For example, $\mathcal{X} = \{0, \ldots, 255\}$ for 8-bit grayscale images and π could be the LSB of pixels (DCT coefficients), $\pi(x) = x \bmod 2$. Thus, a group of n pixels can be represented using a binary vector $\mathbf{x} \in \mathbb{F}_2^n$. The sole purpose of the embedding operation is to modify the bit assigned to the cover element, which could be achieved by flipping the LSB or adding ± 1 to the pixel value, etc. At this point, it is immaterial how the pixels are changed because we measure the embedding impact only as the number of embedding changes.

Recalling the definition from Section 4.3, the steganographic scheme is a pair of mappings Emb and Ext,

$$\text{Emb} : \mathbb{F}_2^n \times \mathcal{M} \rightarrow \mathbb{F}_2^n, \tag{8.26}$$
$$\text{Ext} : \mathbb{F}_2^n \rightarrow \mathcal{M}, \tag{8.27}$$

with the property

$$\text{Ext}\left(\text{Emb}(\mathbf{x}, \mathbf{m})\right) = \mathbf{m}, \quad \forall \mathbf{x} \in \mathbb{F}_2^n \text{ and } \mathbf{m} \in \mathcal{M}. \tag{8.28}$$

Here, $\mathbf{y} = \text{Emb}(\mathbf{x}, \mathbf{m})$ is the vector of bits extracted from the same block of n pixels in the stego image and $\mathcal{M}$ is the set of all messages that can be communicated. (The embedding capacity is $\log_2 |\mathcal{M}|$ bits.) Note that here we ignore the role of the stego key as it is, again, not important for our reasoning.

Let us suppose that it is possible to embed every message in $\mathcal{M}$ using at most R changes,

$$d_{\mathrm{H}}\left(\mathbf{x}, \text{Emb}(\mathbf{x}, \mathbf{m})\right) \leq R, \quad \forall \mathbf{x} \in \mathbb{F}_2^n, \mathbf{m} \in \mathcal{M}. \tag{8.29}$$

Because the embedding distortion is measured as the number of embedding changes, the embedding efficiency (in bits per embedding change) is

$$e = \frac{\log_2 |\mathcal{M}|}{R_{\mathrm{a}}}, \tag{8.30}$$

where R_{a} is the average number of embedding changes over all messages and covers

$$R_{\mathrm{a}} = E\left[d_{\mathrm{H}}\left(\mathbf{x}, \text{Emb}(\mathbf{x}, \mathbf{m})\right)\right]. \tag{8.31}$$

We also define the lower embedding efficiency

$$\underline{e} = \frac{\log_2 |\mathcal{M}|}{R}. \tag{8.32}$$

Obviously, since $R_a \le R$, we have $\underline{e} \le e$.

Having established the terminology and notation, we are now ready to formulate and prove the matrix embedding theorem.

Theorem 8.2. [Matrix embedding theorem] *An $[n, k]$ code with covering radius R can be used to construct a steganographic scheme that can embed $n - k$ bits in n pixels by making at most R embedding changes. The embedding efficiency is $e = (n - k)/R_a$, where $R_a = (1/2^n) \sum_{\mathbf{x} \in \mathbb{F}_2^n} d_H(\mathbf{x}, \mathcal{C})$ is the average distance to code.*

Proof. Let $\mathbf{H}$ be a parity-check matrix of the code, $\mathbf{x} \in \mathbb{F}_2^n$ the vector of LSBs of n pixels, and $\mathbf{m} \in \mathbb{F}_2^{n-k}$ the vector of message bits. Define the embedding mapping as

$$\mathbf{y} = \text{Emb}(\mathbf{x}, \mathbf{m}) \triangleq \mathbf{x} + \mathbf{e}(\mathbf{m} - \mathbf{Hx}), \tag{8.33}$$

where we remind the reader that $\mathbf{e}(\mathbf{m} - \mathbf{Hx})$ is a coset leader of the coset corresponding to the syndrome $\mathbf{m} - \mathbf{Hx}$. The corresponding extraction algorithm is

$$\mathbf{m} = \text{Ext}(\mathbf{y}) = \mathbf{Hy}. \tag{8.34}$$

In other words, the recipient reads the message by extracting the LSBs from the stego image, $\mathbf{y}$, and multiplying them by the parity-check matrix,

$$\mathbf{Hy} = \mathbf{Hx} + \mathbf{He} = \mathbf{Hx} + \mathbf{m} - \mathbf{Hx} = \mathbf{m}. \tag{8.35}$$

Because the message is being communicated as a syndrome of some linear code, steganography using matrix embedding is sometimes called syndrome coding. From (8.33), $d_H(\mathbf{x}, \mathbf{y}) = w(\mathbf{e}(\mathbf{m} - \mathbf{Hx}))$ and thus the number of embedding changes is at most R because the weight of every coset leader is at most R (8.25). If the message is a random bit stream, then all coset leaders are equally likely to participate in the embedding. Thus, the expected number of embedding changes is equal to the expected weight of a coset leader,

$$\frac{1}{2^{n-k}} \sum_{\mathbf{s} \in \mathbb{F}_2^{n-k}} w(\mathbf{e}(\mathbf{s})) = \frac{1}{2^n} \sum_{\mathbf{s} \in \mathbb{F}_2^{n-k}} 2^k w(\mathbf{e}(\mathbf{s})) \tag{8.36}$$

$$= \frac{1}{2^n} \sum_{\mathbf{s} \in \mathbb{F}_2^{n-k}} \sum_{\mathbf{x} \in \mathcal{C}(\mathbf{s})} d_H(\mathbf{x}, \mathcal{C}) = \frac{1}{2^n} \sum_{\mathbf{x} \in \mathbb{F}_2^n} d_H(\mathbf{x}, \mathcal{C}), \tag{8.37}$$

which is the average distance to code, R_a. In other words, we have just proved that $e = (n - k)/R_a$. The second equality follows from (8.25) expressing the fact that all coset members $\mathbf{x} \in \mathcal{C}(\mathbf{s})$ have the same distance to $\mathcal{C}$: $d_H(\mathbf{x}, \mathcal{C}) = w(\mathbf{e}(\mathbf{s}))$, equal to the weight of a coset leader of $\mathcal{C}(\mathbf{s})$. The fact that cosets partition the whole space implies the third and final equality.   □

8.3.1   Revisiting binary Hamming codes

We now revisit the example from Section 8.1 through the matrix embedding theorem. The embedding method used a parity-check matrix $\mathbf{H}$ whose columns were all non-zero binary p-tuples. (The matrix had $2^p - 1$ columns.) The length of the linear code was $n = 2^p - 1$ and its codimension p. These are binary Hamming codes, $\mathcal{H}_p$. Because the parity-check matrix contains all non-zero binary vectors of length p, any syndrome can be written as a trivial "sum" of one column from $\mathbf{H}$. Thus, the covering radius of the Hamming codes is $R = 1$.

The average distance to code, R_{a}, can be easily obtained by realizing that Hamming codes are perfect (see Appendix C) and the whole space $\mathbb{F}_2^n$ can be written as a union of balls with unit radius centered at the codewords,

$$\bigcup_{\mathbf{c} \in \mathcal{H}_p} \mathcal{B}(\mathbf{c}, 1) = \mathbb{F}_2^n. \tag{8.38}$$

Because the volume of each ball is exactly $1 + n = 2^p$ and all words with the exception of the center codeword have distance 1 from the code, the average distance to code is $R_{\mathrm{a}} = n/(n+1) = 1 - 2^{-p}$. Thus, the embedding efficiency of matrix embedding based on binary Hamming code $\mathcal{H}_p$ is $e_p = p/(1 - 2^{-p})$ in agreement with the result obtained in Section 8.1.

8.4   Theoretical bounds

The matrix embedding theorem tells us that linear codes can be used to construct steganographic schemes that impose fewer embedding changes than the simple paradigm of embedding one bit at every pixel. This is quite significant because it gives us the ability to communicate longer messages for a fixed distortion budget. It would be valuable to know the limits of this approach and determine the performance (embedding efficiency) of the best possible matrix embedding method and then attempt to reach this limit.

8.4.1   Bound on embedding efficiency for codes of fixed length

We now determine the largest possible embedding efficiency achievable using linear codes of fixed length n. To this end, we consider the embedding efficiency for codes capable of embedding relative payload α, or the class of $[n, n(1 - \alpha)]$ codes. An upper bound on embedding efficiency requires a lower bound on R and R_{a}. Because there are $\binom{n}{i}$ possible sums of i columns of the parity-check matrix $\mathbf{H}$, the number of cosets whose coset leaders have weight i is at most $\binom{n}{i}$. Thus, the covering radius R must be at least equal to R_n for which

$$\binom{n}{0} + \binom{n}{1} + \cdots + \binom{n}{R_n - 1} + \xi\binom{n}{R_n} = 2^{\alpha n}, \tag{8.39}$$

where $0 < \xi \leq 1$ is a real number. Note here that every $[n, n(1 - \alpha)]$ code has $2^{\alpha n}$ cosets.

Besides the lower bound on the covering radius, $R \geq R_n$, we obtain a lower bound for the average distance to code R_a,

$$R_a \geq \frac{\sum_{i=1}^{R_n - 1} i\binom{n}{i} + R_n \xi\binom{n}{R_n}}{2^{\alpha n}}, \tag{8.40}$$

and an upper bound on embedding efficiency

$$e = \frac{\alpha n}{R_a} \leq \frac{\alpha n 2^{\alpha n}}{\sum_{i=1}^{R_n - 1} i\binom{n}{i} + R_n \xi\binom{n}{R_n}}. \tag{8.41}$$

We remark that codes for which the number of coset leaders with Hamming weight i is exactly $\binom{n}{i}$ and $\xi = 1$ have the largest possible embedding efficiency within the class of codes of length n. Perfect codes have this property (see Appendix C). There are only four perfect codes: the trivial repetition code (see Exercises 8.1 and 8.6), Hamming codes, the binary Golay code, and the ternary Golay code [248].

The bounds derived above can be summarized in a proposition.

Proposition 8.3. [Bound on embedding efficiency, fixed code length]
Any matrix embedding scheme realized using a linear code of length n capable of embedding relative payload α must occasionally make at least R_n (8.39) embedding changes. The embedding efficiency is at most (8.41).

8.4.2   Bound on embedding efficiency for codes of increasing length

The second bound we derive will tell us the maximal embedding efficiency one can achieve using codes of arbitrary length for a fixed relative payload. We start by fixing the *change rate* $\beta = R/n$ and determine the largest relative payload α that one can embed by making at most βn changes as $n \to \infty$. We can limit ourselves to $\beta < \frac{1}{2}$ because we can always add the all-ones codeword to the code and thus make the code contain the repetition code whose covering radius is at most $n/2 \geq R$ (Exercise 8.1).

There are $\binom{n}{i}$ possible ways one can make i changes in n pixels. Thus, using at most $R = \beta n$ changes, we can embed at most

$$|\mathcal{M}| \leq \sum_{i=0}^{\beta n} \binom{n}{i} = V_2(\beta n, n) \tag{8.42}$$

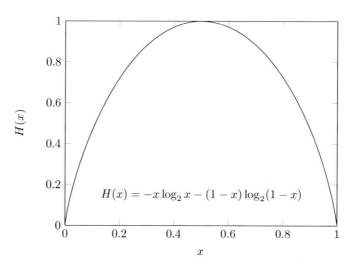

$$H(x) = -x \log_2 x - (1 - x) \log_2 (1 - x)$$

Figure 8.2 The binary entropy function.

messages if it is possible to arrange that every change communicates a different message. We now determine the asymptotic behavior of this sum. To obtain some insight, we will first use Stirling's formula

$$n! = \sqrt{2\pi n} \left(\frac{n}{e}\right)^n \left(1 + O\left(\frac{1}{n}\right)\right) \tag{8.43}$$

for the last (and the largest) term in the sum

$$\binom{n}{\beta n} = \frac{n!}{(n - \beta n)!(\beta n)!} \tag{8.44}$$

$$= \frac{\sqrt{2\pi n}(n/e)^n}{\sqrt{2\pi(n - \beta n)}((n - \beta n)/e)^{n-\beta n} \sqrt{2\pi \beta n}(\beta n/e)^{\beta n}}$$

$$\times \frac{1 + O\left(1/n\right)}{\left(1 + O\left(1/\beta n\right)\right)\left(1 + O\left(1/(n - \beta n)\right)\right)} \tag{8.45}$$

$$= \frac{\frac{1 + O(1/n)}{(1 + O(1/\beta n))(1 + O(1/(n-\beta n)))}}{\sqrt{2\pi n \beta(1 - \beta)}} (1 - \beta)^{-n(1-\beta)} \beta^{-n\beta} \tag{8.46}$$

$$= c(n) 2^{-n[(1-\beta)\log_2(1-\beta) - \beta \log_2 \beta]}, \tag{8.47}$$

where we denoted

$$c(n) = \frac{1}{\sqrt{2\pi n \beta(1 - \beta)}} \frac{1 + O\left(1/n\right)}{\left(1 + O\left(1/\beta n\right)\right)\left(1 + O\left(1/(n - \beta n)\right)\right)}. \tag{8.48}$$

Using the binary entropy function, $H(x) = -x \log_2 x - (1 - x) \log_2 (1 - x)$, this result can be rewritten as

$$\binom{n}{\beta n} = c(n) 2^{nH(\beta)}. \tag{8.49}$$

By taking the logarithm,

$$\log_2 \binom{n}{\beta n} = \log_2 c(n) + nH(\beta), \tag{8.50}$$

$$\frac{\log_2 \binom{n}{\beta n}}{nH(\beta)} = 1 + \frac{\log_2 c(n)}{nH(\beta)}. \tag{8.51}$$

Because

$$\frac{\log_2 c(n)}{nH(\beta)} = O\left(\frac{\log n}{n}\right), \tag{8.52}$$

we finally obtain

$$\lim_{n \to \infty} \frac{\log_2 \binom{n}{\beta n}}{nH(\beta)} = 1. \tag{8.53}$$

We have now proved that the largest term in $V_2(\beta n, n)$ behaves asymptotically as $2^{nH(\beta)}$ or $\log_2 V_2(\beta n, n) \gtrsim nH(\beta)$. We now derive an asymptotic upper bound, which will enable us to state that $\log_2 V_2(\beta n, n) \approx nH(\beta)$, where $f(n) \approx g(n)$ if $\lim_{n \to \infty} f(n)/g(n) = 1$. For this, we will need the tail inequality.

Lemma 8.4. [Tail inequality] *For any $\beta \le \frac{1}{2}$ and $n > 0$*

$$nH(\beta) \ge \log_2 V_2(\beta n, n). \tag{8.54}$$

Proof.

$$1 = (\beta + 1 - \beta)^n = \sum_{i=0}^{n} \binom{n}{i} \beta^i (1-\beta)^{n-i} \tag{8.55}$$

$$= (1-\beta)^n \sum_{i=0}^{n} \binom{n}{i} \left(\frac{\beta}{1-\beta}\right)^i \ge (1-\beta)^n \sum_{i=0}^{\beta n} \binom{n}{i} \left(\frac{\beta}{1-\beta}\right)^i \tag{8.56}$$

$$\ge (1-\beta)^n \sum_{i=0}^{\beta n} \binom{n}{i} \left(\frac{\beta}{1-\beta}\right)^{\beta n} = (1-\beta)^{n(1-\beta)} \beta^{\beta n} \sum_{i=0}^{\beta n} \binom{n}{i} \tag{8.57}$$

$$= 2^{n[(1-\beta)\log_2(1-\beta) + \beta \log_2 \beta]} V_2(\beta n, n), \tag{8.58}$$

which is the tail inequality. The second inequality holds because $\beta/(1-\beta) \le 1$ for $\beta \le \frac{1}{2}$. $\square$

Thus, using the tail inequality,

$$\log_2 \binom{n}{\beta n} \le \log_2 V_2(\beta n, n) \le nH(\beta) \tag{8.59}$$

and, because of (8.53), we have

$$\lim_{n \to \infty} \frac{\log_2 V_2(\beta n, n)}{nH(\beta)} = 1. \tag{8.60}$$

Therefore, the maximal number of bits one can embed using at most R changes is

$$m_{\mathrm{max}} \leq nH\left(\frac{R}{n}\right). \tag{8.61}$$

We can rewrite this to obtain a bound on the lower embedding efficiency, $\underline{e}$, for any relative payload α that can be embedded:

$$\alpha \leq \frac{m_{\mathrm{max}}}{n} \leq H\left(\frac{R}{n}\right), \tag{8.62}$$

$$H^{-1}(\alpha) \leq \frac{R}{n}, \tag{8.63}$$

$$\frac{n}{R} \leq \frac{1}{H^{-1}(\alpha)}, \tag{8.64}$$

$$\underline{e} = \frac{\alpha n}{R} \leq \frac{\alpha}{H^{-1}(\alpha)}. \tag{8.65}$$

The bound (8.65) on the lower embedding efficiency is also an asymptotic bound on the embedding efficiency e,

$$e \leq \frac{\alpha}{H^{-1}(\alpha)}. \tag{8.66}$$

This is because the relative average distance to code, R_{a}/n, which determines embedding efficiency, and the relative covering radius, R/n, for which we have a bound, are asymptotically identical as $n \to \infty$ [81].

Knowing the performance bounds, a practical problem for the steganographer is finding codes that could reach the theoretically optimal embedding efficiency with low computational complexity. Figure 8.3 shows the bound on embedding efficiency (8.66) and the performance of selected codes as a function of $1/\alpha$, where α is the relative payload. We can see that, although the Hamming codes have substantially higher embedding efficiency than the trivial paradigm of embedding one message bit at one pixel with $e = 2$, they are still far from the theoretical upper bound, indicating a space for improvement. A slightly better embedding efficiency can be obtained using BCH codes [208] and certain classes of non-linear codes [20].

An important result, which is beyond the scope of this book, states that the bound (8.66) is tight and asymptotically achievable using linear codes. In particular, the relative codimension $(n - k)/n$ of almost all random $[n, k]$ codes asymptotically achieves $H(R/n)$ for a fixed change rate $R/n < \frac{1}{2}$ and $n \to \infty$ (see, e.g., Theorem 12.3.5 in [46]). Thus, there exist embedding schemes based on linear codes whose embedding efficiency is asymptotically optimal.

Therefore, at least theoretically it should be possible to construct good matrix embedding schemes from random codes. However, while for structured codes, such as the Hamming codes, finding a coset leader was particularly simple, for

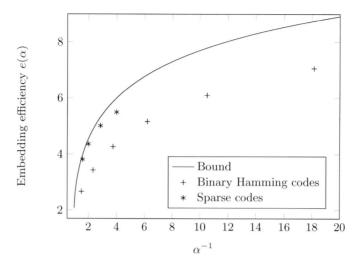

Figure 8.3 Embedding efficiency of binary Hamming codes (crosses) and the bound on embedding efficiency (8.66). The stars mark the embedding efficiency of sparse linear codes of length $n = 1,000$ and $n = 10,000$ as reported in [68].

general random codes it is an NP-complete problem whose complexity increases exponentially with n. In the special case of embedding large relative payloads, we can afford to keep the number of codewords small and find coset leaders using brute force (see Section 8.5). Another possibility is to use sparse linear codes where efficient algorithms based on message-passing exist [68]. In fact, such codes were shown to achieve embedding efficiency very close to the theoretical bound (8.66) (see Figure 8.3).

8.5   Matrix embedding for large relative payloads

The positive impact of matrix embedding on security is bigger when embedding large payloads because small payloads are less statistically detectable already. Unfortunately, matrix embedding using binary Hamming codes as introduced in Section 8.1 does not bring any improvement for payloads $\alpha > \frac{2}{3}$. In this section, we describe matrix embedding methods suitable when the relative payload is large, or $\alpha \approx 1$. The emphasis here is on the adjective "relative" because such payloads can in fact be quite small in absolute terms when compared with the embedding capacity. Take, for example, adaptive steganography where the sender constrains herself to a certain subset of the image right from the beginning to avoid introducing detectable artifacts or the perturbed quantization steganography as described in Chapter 9.

Recall that in matrix embedding, the stego vector of bits $\mathbf{y} = \mathbf{x} + \mathbf{e}(\mathbf{m} - \mathbf{Hx})$ is obtained by finding a coset leader for the syndrome $\mathbf{m} - \mathbf{Hx}$. And finding the coset leader is the most time-consuming part of the embedding algorithm. For large relative payloads $\alpha = (n-k)/n \approx 1$, the codes in matrix embedding $[n, n - \alpha n]$ have small dimension $k = n(1-\alpha)$, and thus a small number of codewords. This observation can be used to our advantage.

The method in this section will be based on random codes to show that even randomly generated codes can provide good performance, which is, after all, not that surprising because in the previous section we learned that random codes asymptotically reach the bound on embedding efficiency (8.66). Moreover, due to the small dimension, k, coset leaders can be found efficiently using look-up tables.

We assume that the sender wants to embed $n-k$ bits, $\mathbf{m} \in \{0,1\}^{n-k}$, in n pixels with bits $\mathbf{x} \in \{0,1\}^n$. The code parity-check matrix $\mathbf{H}$ is generated randomly (from a shared stego key) in its systematic form $\mathbf{H} = [\mathbf{I}_{n-k}, \mathbf{A}]$, where $\mathbf{I}_{n-k}$ is an $(n-k) \times (n-k)$ unity matrix and $\mathbf{A}$ is a random $(n-k) \times k$ matrix. The generator matrix is thus $\mathbf{G} = [\mathbf{A}', \mathbf{I}_k]$ (see Appendix C).

The algorithm has two steps:

1. Find any coset member $\tilde{\mathbf{y}} \in \mathcal{C}(\mathbf{m} - \mathbf{Hx})$ as the solution of $\mathbf{H}\tilde{\mathbf{y}} = \mathbf{m} - \mathbf{Hx}$. Because $\mathbf{H}$ is in its systematic form, for example $\tilde{\mathbf{y}} = (\mathbf{m} - \mathbf{Hx}, \mathbf{0})$, where $\mathbf{0} \in \{0,1\}^k$.

2. Find the coset leader $\mathbf{e}(\mathbf{m} - \mathbf{Hx})$. To do so, we need to find the codeword $\tilde{\mathbf{c}}$ closest to $\tilde{\mathbf{y}}$,

$$d_{\mathrm{H}}(\tilde{\mathbf{y}}, \tilde{\mathbf{c}}) = \min_{\mathbf{c} \in \mathcal{C}} d_{\mathrm{H}}(\tilde{\mathbf{y}}, \mathbf{c}) = \min_{\mathbf{c} \in \mathcal{C}} w(\tilde{\mathbf{y}} - \mathbf{c}) = w\left(\mathbf{e}(\mathbf{m} - \mathbf{Hx})\right), \qquad (8.67)$$

because the vector $\tilde{\mathbf{y}} - \mathbf{c}$ goes through all members of the coset. Thus, $\mathbf{e}(\mathbf{m} - \mathbf{Hx}) = \tilde{\mathbf{y}} - \tilde{\mathbf{c}}$, which completes the description of the embedding mapping Emb via the matrix embedding theorem. If the dimension of the code is small, the closest codeword can be quickly found by brute force or using precalculated look-up tables.

The relative embedding payload of these random $[n, k]$ codes is

$$\alpha_{k,n} = \frac{n-k}{n} = 1 - \frac{k}{n}. \qquad (8.68)$$

For a fixed code dimension k, $\alpha_{k,n} \to 1$ as n approaches ∞. Figure 8.4 shows the embedding efficiency of random codes of small dimension as a function of relative payload α. The points were obtained by averaging the embedding efficiency over 200 randomly generated codes. Note that the embedding efficiency is quite close to the bound (8.41) for codes of the same length n, indicating the fact that in the class of linear codes of fixed length, the random codes offer embedding efficiency close to the optimal value.

The reader is referred to [96] for examples of structured codes suitable for embedding large payloads. A simple ternary code that achieves very good performance for embedding large payloads is in Exercise 8.6 (and Figure 8.8). An alternative approach for efficient embedding of large payloads, which is not based

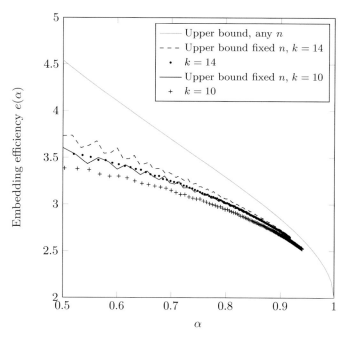

Figure 8.4 Embedding efficiency versus relative payload α for random codes of dimension $k = 10$ and 14. Also shown is the theoretical upper bound for codes of arbitrary length (8.66) and bounds for codes of a fixed length (8.41).

on syndrome coding, uses so-called Sum and Difference Covering Sets (SDCSs) of finite cyclic groups [158, 160, 174] (also see Section 8.7).

8.6   Steganography using q-ary symbols

All matrix embedding methods introduced in the previous sections used binary codes. The reason for this was that each cover element was represented with a bit through some bit-assignment function, π, such as its LSB. Fundamentally, there is no reason why we could not use q-ary representation for $q > 2$. Consider, for example, the ± 1 embedding where it is allowed to change a pixel value by ± 1. Each pixel can thus accept one of three possible values. By using a function that assigns a ternary symbol at each pixel, for example, $\pi(x) = x \bmod 3$, one ternary symbol can be embedded at each pixel rather than just one bit. Because one ternary symbol conveys $\log_2 3$ bits of information, the number of embedding changes would be decreased. Let us calculate the embedding efficiency of this ternary ± 1 embedding to see the improvement.

We already know that the embedding efficiency of classical (binary) ± 1 embedding is 2 because the probability that a pixel needs to be modified is $\frac{1}{2}$. Alternatively, two bits are embedded using one embedding change. When embedding a random stream of ternary symbols, the probability that a pixel will not have to be modified is $\frac{1}{3}$. In $\frac{2}{3}$ of the cases, the pixel value is modified by ± 1. Thus, the average number of embedding changes per pixel is $\frac{2}{3}$ and the payload per pixel is $\log_2 3$ bits, which leads to embedding efficiency $\log_2 3/(2/3) = 2.3774...$ bits per change. Thus, ternary ± 1 embedding has higher embedding efficiency than binary ± 1 embedding. Because the magnitude of embedding changes is the same in both cases, there is no reason not to use ternary ± 1 embedding over the binary embedding.

If ternary is better than binary, it is tempting to increase the magnitude of embedding changes even further and embed more bits per embedding change. For example, by allowing changes by up to ± 2, we would embed $\log_2 5$ bits per pixel and the embedding efficiency $\log_2 5/(4/5)$ would be even higher than for the ternary case. However, embedding changes of higher magnitude are also statistically more detectable than embedding changes of lower magnitude. Thus, it is not immediately clear that using pentary embedding would indeed lead to less detectable steganographic schemes. We revisit this intriguing topic in Section 8.6.3 after deriving the bound on embedding efficiency of q-ary codes.

As in the binary case, it is possible to construct matrix embedding schemes for q-ary linear codes and further increase the embedding efficiency. The only change is that we will now work with vectors of n q-ary symbols extracted from the cover/stego image rather than binary vectors. This translates to q-ary linear codes where the binary arithmetic is replaced with arithmetic in a finite field $\mathbb{F}_q$ (see Appendix C). Because the matrix embedding theorem can be proved in exactly the same manner, we provide only the formulation without the proof.

Theorem 8.5. [Matrix embedding theorem for q-ary codes] *An $[n,k]$ q-ary code with covering radius R can be used to construct a steganographic scheme that can embed $n-k$ q-ary symbols in n pixels by making at most R embedding changes. The embedding efficiency is $e = [(n-k)\log_2 q]/R_a$ bits per change, where $R_a = (1/q^n)\sum_{\mathbf{x}\in\mathbb{F}_q^n} d(\mathbf{x},\mathcal{C})$ is the average distance to code.*

8.6.1 q-ary Hamming codes

The binary Hamming codes from Sections 8.1 and 8.3.1 are an attractive choice for matrix embedding due to their simple implementation. In this section, we describe their generalized version that works over a finite field $\mathbb{F}_q$.

A q-ary Hamming code is defined using its parity-check matrix $\mathbf{H}$, which will now contain all different non-zero vectors $\mathbf{x}\in\mathbb{F}_q^p$ of length p (different *up to a multiplication by an element from the field*). For example, the parity-check matrix for the ternary Hamming code with $p=3$ is

$$\mathbf{H} = \begin{pmatrix} 1 & 0 & 0 & 0 & 1 & 1 & 1 & 0 & 2 & 1 & 2 & 1 & 1 \\ 0 & 1 & 0 & 1 & 0 & 1 & 1 & 1 & 0 & 2 & 1 & 2 & 1 \\ 0 & 0 & 1 & 1 & 1 & 0 & 1 & 2 & 1 & 0 & 1 & 1 & 2 \end{pmatrix}. \tag{8.69}$$

Thus, this is a $[13, 10]$ code with covering radius $R = 1$. The radius is 1 because we need only a linear combination of one column to obtain any syndrome. For example, the syndrome $(2, 2, 2)'$, which does not appear directly as a column in $\mathbf{H}$, can be obtained as a linear combination (multiple) of just one column, the seventh column.

In general, a q-ary Hamming code is a $\left[\frac{q^p - 1}{q - 1}, \frac{q^p - 1}{q - 1} - p\right]$ code with covering radius $R = 1$. The length is $n = (q^p - 1)/(q - 1)$ because there are $q^p - 1$ non-zero vectors of q-ary symbols of length p and each non-zero column appears in $q - 1$ versions (there are $q - 1$ non-zero multiples). Of course, the codimension is $n - k = p$.

Hamming codes can be used to construct steganographic schemes capable of embedding p q-ary symbols or $p \log_2 q$ bits in $(q^p - 1)/(q - 1)$ pixels by making on average $1 - q^{-p}$ changes (because the probability that the syndrome of the cover, $\mathbf{Hx} \in \mathbb{F}_q^p$, already matches the p message symbols is $1/q^p$). Thus, we have the corresponding relative payload α_p (in bits per pixel) and embedding efficiency e_p (in bits per change)

$$\alpha_p = \frac{p \log_2 q}{(q^p - 1)/(q - 1)}, \tag{8.70}$$

$$e_p = \frac{p \log_2 q}{1 - q^{-p}}. \tag{8.71}$$

The embedding efficiency of ternary Hamming codes is shown in Figure 8.5. Note that ternary Hamming codes offer larger embedding efficiency than binary Hamming codes.

Example 8.6: [±1 embedding using ternary Hamming codes with co-dimension p]

Assuming the parity-check matrix (8.69), let $\mathbf{g} = (14, 13, 13, 12, 12, 16, 18, 20, 19, 21, 23, 22, 24)'$ be the vector of grayscale values from a cover-image block of 13 pixels. The corresponding vector of ternary symbols is $\mathbf{x} = \mathbf{g} \bmod 3 = (2, 1, 1, 0, 0, 1, 0, 2, 1, 0, 2, 1, 0)'$, which has the following syndrome:

$$\mathbf{Hx} = \begin{pmatrix} 2 + 1 + 2 + 1 + 1 \\ 1 + 1 + 2 + 2 + 2 \\ 1 + 1 + 1 + 2 + 1 \end{pmatrix} \bmod 3 = \begin{pmatrix} 1 \\ 2 \\ 0 \end{pmatrix}. \tag{8.72}$$

Let us say that we wish to embed a ternary message $\mathbf{m} = (0, 1, 2)'$. We now need to find the element of the vector $\mathbf{x}$ that needs to be changed so that its syndrome is the desired message. Proceeding as in the matrix embedding theorem, we

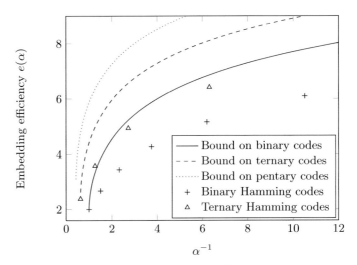

Figure 8.5 Embedding efficiency e as a function of α^{-1}, where α is the relative message length. The graph shows the bounds on embedding efficiency for binary (8.66) ($q = 2$, solid line), ternary ($q = 3$, dashed line), and pentary ($q = 5$, dotted line) codes (8.84), as well as the efficiency for binary ($+$) and ternary ($\triangle$) Hamming codes.

calculate[2]

$$\mathbf{m} - \mathbf{Hx} = \begin{pmatrix} 0 \\ 1 \\ 2 \end{pmatrix} - \begin{pmatrix} 1 \\ 2 \\ 0 \end{pmatrix} = \begin{pmatrix} 2 \\ 2 \\ 2 \end{pmatrix} = 2\mathbf{H}[.,7], \tag{8.73}$$

where $\mathbf{H}[.,7]$ is the seventh column of $\mathbf{H}$. Thus, $\mathbf{m} = \mathbf{Hx} + 2\mathbf{H}[.,7] = \mathbf{Hy}$, where $\mathbf{y}[i] = \mathbf{x}[i], i \neq 7$, and $\mathbf{y}[7] = \mathbf{x}[7] + 2 = 2$. Because $\mathbf{g}[7] = 18$, which is a ternary 0, we need to change $\mathbf{g}[7]$ from 18 to 17 because 17 is ternary 2 ($17 \bmod 3 = 2$). Note that changing $\mathbf{g}[7]$ to 20 would work as well, however at the expense of making an embedding change of magnitude 2 and thus this change would not be compatible with ± 1 embedding. To summarize, in order to embed the ternary message $(0, 1, 2)'$, the stego image pixel values will thus be $(14, 13, 13, 12, 12, 16, \mathbf{17}, 20, 19, 21, 23, 22, 24)'$.

8.6.2   Performance bounds for q-ary codes

Mimicking the flow of Section 8.4.1, we first determine the largest embedding efficiency of q-ary codes $[n, k]$ for a fixed n and k. Because there are $\binom{n}{i}(q-1)^i$ possible linear combinations of i columns of the parity-check matrix $\mathbf{H}$, the number of cosets whose coset leaders have weight i is at most $\binom{n}{i}(q-1)^i$. Thus,

---

[2] All arithmetic operations are modulo 3.

the covering radius R must be at least equal to R_n for which

$$\binom{n}{0} + \binom{n}{1}(q-1) + \cdots + \binom{n}{R_n-1}(q-1)^{R_n-1} + \xi\binom{n}{R_n}(q-1)^{R_n} = q^{n-k},$$
(8.74)

where $0 < \xi \leq 1$ is a real number. Note that an $[n,k]$ q-ary code has q^{n-k} cosets. This gives us a lower bound on the covering radius, $R \geq R_n$, as well as a lower bound for the average distance to code R_a (which is the average weight of a coset leader)

$$R_a \geq \frac{\sum_{i=1}^{R_n-1} i\binom{n}{i}(q-1)^i + R_n\xi\binom{n}{R_n}(q-1)^{R_n}}{q^{n-k}}.$$
(8.75)

A lower bound on R_a again provides an upper bound on embedding efficiency

$$e = \frac{(n-k)\log_2 q}{R_a} \leq \frac{(n-k)q^{n-k}\log_2 q}{\sum_{i=1}^{R_n-1} i\binom{n}{i}(q-1)^i + R_n\xi\binom{n}{R_n}(q-1)^{R_n}}.$$
(8.76)

This bound is reached by codes for which the number of coset leaders with Hamming weight i is exactly $\binom{n}{i}(q-1)^i$ and $\xi = 1$. The only such codes are perfect codes and they enjoy the largest possible embedding efficiency within the class of linear codes of length n.

Next, we derive a bound for codes of arbitrary length capable of communicating a fixed relative payload. The maximal number of messages $|\mathcal{M}|$ that can be communicated with change rate $\beta = R/n$ is bound from above by the number of possibilities one can make up to βn changes in n pixels,

$$|\mathcal{M}| \leq V_q(\beta n, n) = \sum_{i=0}^{\beta n} \binom{n}{i}(q-1)^i.$$
(8.77)

The asymptotic behavior of the sum is again determined by the largest last term:

$$\log_2 V_q(\beta n, n) \geq \log_2 \binom{n}{\beta n} + \beta n \log_2(q-1)$$
(8.78)

$$\approx n[H(\beta) + \beta \log_2(q-1)] = nH_q(\beta),$$
(8.79)

where we used the asymptotic expression (8.53) for $\log_2 \binom{n}{\beta n}$ and

$$H_q(x) = -x\log_2 x - (1-x)\log_2(1-x) + x\log_2(q-1)$$
(8.80)

is the q-ary entropy function shown in Figure 8.6 for $q = 3$.

The inequality (8.79) together with the equivalent of the tail inequality (see Exercise 8.3)

$$nH_q(\beta) \geq \log_2 V_q(\beta n, n)$$
(8.81)

proves that the relative payload that can be embedded using change rate β is bounded by

$$\alpha \leq \frac{\log_2 |\mathcal{M}|}{n} \approx H_q(\beta).$$
(8.82)

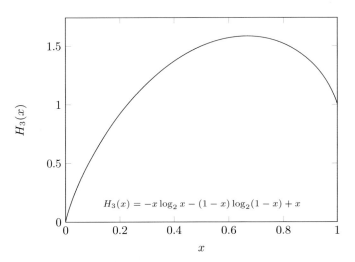

$$H_3(x) = -x \log_2 x - (1-x) \log_2 (1-x) + x$$

Figure 8.6 The ternary entropy function.

From here, we obtain the bound on the lower embedding efficiency

$$\underline{e} = \frac{\alpha}{\beta} \le \frac{\alpha}{H_q^{-1}(\alpha)}, \qquad (8.83)$$

which is also a bound on the embedding efficiency

$$e \le \frac{\alpha}{H_q^{-1}(\alpha)}. \qquad (8.84)$$

A matrix embedding scheme with the maximal embedding efficiency can thus embed relative payload α bpp by making on average $H_q^{-1}(\alpha)$ changes per pixel.

Figure 8.5 shows the upper bound (8.84) on embedding efficiency for $q = 2, 3, 5$ as a function of α^{-1}, where α is the relative payload in bpp. Note that the bounds start at the point $\alpha = \log_2 q$, $e = [q/(q-1)] \log_2 q$, which corresponds to embedding at the largest relative payload of $\log_2 q$ bpp. The same figure shows the benefit of using q-ary codes for a fixed relative payload α. For example, for $\alpha = 1$, the ternary ± 1 embedding can theoretically achieve embedding efficiency $e \simeq 4.4$, which is significantly higher than 2 – the maximal efficiency of binary codes at this relative message length. The embedding efficiency of binary and ternary Hamming codes for different values of p is shown with "+" and "△" symbols, respectively.

8.6.3   The question of optimal q

In the previous two sections, we learned that by encoding individual elements of the cover image into q-ary symbols rather than into bits, one can significantly increase the embedding efficiency of matrix embedding. However, in order to be able to modify the symbol assigned to each pixel to all q values from $\mathbb{F}_q$, we need to allow the following values of embedding changes:

$$\mathcal{D}_{\text{odd}} = \left\{ -\frac{q-1}{2}, -\frac{q-3}{2}, \dots, -1, 0, 1, \dots, \frac{q-1}{2} \right\} \tag{8.85}$$

for q odd, and

$$\mathcal{D}_{\text{even}} = \left\{ -\frac{q-2}{2}, -\frac{q-4}{2}, \dots, -1, 0, 1, \dots, \frac{q}{2} \right\} \tag{8.86}$$

for q even. Thus, for $q > 3$, the magnitude of modifications sometimes has to be larger than 1. The question is whether it is better to have fewer changes with larger magnitude or more changes with lower magnitude. In contrast to the case of binary or ternary codes when the magnitude of embedding changes was always 1, it is now no longer appropriate to measure embedding impact using the number of embedding changes because the changes have unequal magnitude. Instead, we measure embedding impact using the distortion measure as defined in Chapter 4,

$$d_\gamma(\mathbf{x}, \mathbf{y}) = \sum_{i=1}^{n} |\mathbf{x}[i] - \mathbf{y}[i]|^\gamma \tag{8.87}$$

for $\gamma \geq 1$.

Let us assume that we use optimal q-ary matrix embedding schemes with embedding efficiency reaching the bound (8.84). This means that one can embed relative payload α (bpp) by making on average $H_q^{-1}(\alpha)$ changes per pixel. Assuming that the message is encoded using symbols from $\mathbb{F}_q$ and forms a random stream, the magnitude of these changes is equally likely to reach any of the $q-1$ non-zero values in $\mathcal{D}$ and the expected impact per changed pixel is

$$\bar{d}_\gamma = \frac{1}{q-1} \sum_{d \in \mathcal{D}} |d|^\gamma, \tag{8.88}$$

where $\mathcal{D}$ stands for $\mathcal{D}_{\text{odd}}$ for q odd and $\mathcal{D}_{\text{even}}$ for q even. Thus, the expected embedding impact per pixel when embedding relative payload α using optimal q-ary matrix embedding is

$$\Delta(\alpha, q, \gamma) = \beta(\alpha)\bar{d}_\gamma = H_q^{-1}(\alpha)\frac{1}{q-1}\sum_{d \in \mathcal{D}} |d|^\gamma, \tag{8.89}$$

because $\beta(\alpha) = H_q^{-1}(\alpha)$ is the minimal change rate (8.82) for payload α embeddable using q-ary codes.

We can obtain some insight into the trade-off between the number of embedding changes and their magnitude by determining the value of q that minimizes $\Delta(\alpha, q, \gamma)$. Figure 8.7 shows $\Delta(\alpha, q, \gamma)$ for different values of q for $\alpha \in \{0.1, 0.2, \dots, 0.9\}$ and $\gamma = 1$. Note that $q = 3$ leads to the smallest embedding impact for all relative payloads. This statement holds true for any $\gamma \geq 1$ because $\Delta(\alpha, 2, \gamma) = \Delta(\alpha, 3, \gamma)$ for all γ (the magnitude of embedding changes is 1 in these cases) and $\Delta(\alpha, q, \gamma) \geq \Delta(\alpha, q, 1)$ for all $q > 3$. Thus, as long as the

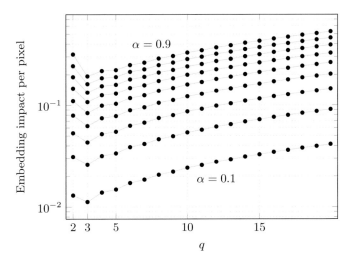

Figure 8.7 Expected embedding impact $\Delta(\alpha, q, 1)$ for various values of α and q. The minimal embedding impact is always obtained for $q = 3$.

embedding impact can be captured using d_γ, we can conclude that the most secure steganography is obtained for ternary codes and it does not pay off to make fewer embedding changes with larger magnitude (or use q-ary codes with $q > 3$). Of course, this conclusion hinges on the assumption that statistical detectability can be captured with a distortion measure. This is not, however, entirely clear. It is possible that for some combination of the embedding scheme and the cover-source model, the KL divergence between cover and stego objects will be lower for codes with $q > 3$. This issue is currently an open research problem.

8.7   Minimizing embedding impact using sum and difference covering set

The main theme of this chapter is improving the embedding efficiency of steganographic schemes by decreasing the number of embedding changes. So far, we have explored methods based on syndrome coding with linear codes (so-called matrix embedding). Although this approach is the one most developed today, there exists an alternative and quite elegant approach that originated from the theory of covering sets of cyclic groups. Moreover, it can be thought of as a generalization of ± 1 embedding to groups of multiple pixels. Additionally, by connecting a known problem in steganography with another branch of mathematics, steganography may benefit from future breakthroughs in this direction. In this section, we explain the main ideas and include appropriate references where the reader may find more detailed information. Following the spirit of this book, we first introduce a simple example that will later be generalized.

Table 8.2. Required modification of the cover pair $(\mathbf{x}[1], \mathbf{x}[2])$ depending on the value of $\mathrm{Ext}(\mathbf{x}[1], \mathbf{x}[2])$ (the first column) to embed a quaternary symbol $\mathbf{b} \in \mathbb{Z}_4 = \{0, 1, 2, 3\}$ (the first row).

$\mathbf{x} \backslash \mathbf{b}$	0	1	2	3
0	(0,0)	(1,0)	(0,1)	(−1,0)
1	(−1,0)	(0,0)	(1,0)	(0,1)
2	(0,−1)	(−1,0)	(0,0)	(1,0)
3	(1,0)	(0,−1)	(−1,0)	(0,0)

Consider a steganographic scheme whose embedding mechanism is limited to making ± 1 changes to each cover element. Let $\mathbf{x}[1]$, $\mathbf{x}[2]$ be two elements of the cover. We now show how one can embed a quaternary symbol[3] (or two message bits) into the pair $(\mathbf{x}[1], \mathbf{x}[2])$ by modifying at most one element by 1. The embedding will have the property that the symbol b can be extracted from the pair of stego elements $(\mathbf{y}[1], \mathbf{y}[2])$ using the following extraction mapping:

$$b = \mathrm{Ext}(\mathbf{y}[1], \mathbf{y}[2]) = (\mathbf{y}[1] + 2\mathbf{y}[2]) \bmod 4. \tag{8.90}$$

Table 8.2 shows that it is always possible to embed a quaternary symbol in each pair $(\mathbf{x}[1], \mathbf{x}[2])$ by modifying at most one element of the pair by ± 1. For example, when $\mathrm{Ext}(\mathbf{x}[1], \mathbf{x}[2]) = 2$ and we wish to embed symbol 0, we modify $\mathbf{x}[2] \rightarrow \mathbf{x}[2] - 1$. (In fact, in this case, we can achieve this same effect by modifying $\mathbf{x}[2] \rightarrow \mathbf{x}[2] + 1$, etc.) Here, we ignore the boundary issues when the modified element may get out of its dynamic range. If a random symbol stream is embedded using this method, we embed 2 bits by making a modification to one of the pixels with probability $\frac{12}{16} = \frac{3}{4}$. Thus, the embedding efficiency of this method is $e = \frac{2}{3/4} = \frac{8}{3} = 2.66\ldots$. This is a higher embedding efficiency than simply embedding each bit as the LSB of each pixel with efficiency 2. It is also higher than embedding a ternary symbol ($\log_2 3$ bits) using ± 1 embedding, which has embedding efficiency $\log_2 \frac{3}{2/3} = 2.377\ldots$.

This simple idea, which originally appeared in [174], can be greatly generalized [158, 160]. For any positive integer M, we will denote by $\mathbb{Z}_M = \{0, 1, \ldots, M - 1\}$ the finite cyclic group of order M where the group operation is addition modulo M. The group $(\mathbb{Z}_M, +)$ obviously satisfies the axioms of a group because it is closed with respect to the group operation (for any $a, b \in \mathbb{Z}_M$, the addition is defined as $a + b \bmod M \in \mathbb{Z}_M$), the operation is also associative, there exists an identity element, which is 1, and each element, a, has an inverse element, $M - a$.

Let $\mathcal{A} = \{\mathbf{a}[1], \ldots, \mathbf{a}[n]\}$ be a sequence of elements from $\mathbb{Z}_M$. The set $\mathcal{A}$ is called a Sum and Difference Covering Set (SDCS) with parameters (n, k, M) if

---

[3] A quaternary symbol is an element of the set $\{0, 1, 2, 3\} = \mathbb{Z}_4$.

for each $b \in \mathbb{Z}_M$ there exist $\mathbf{s}[i] \in \{0, 1, -1\}$, $i = 1, \ldots, n$, such that

$$\sum_{i=1}^{n} |\mathbf{s}[i]| \leq k, \tag{8.91}$$

$$\sum_{i=1}^{n} \mathbf{s}[i]\mathbf{a}[i] = b \text{ in } \mathbb{Z}_M. \tag{8.92}$$

In other words, it is possible to write every element of $\mathbb{Z}_M$ through addition or subtraction operations of at most k elements from $\mathcal{A}$.

The significance of such sets for steganography will become apparent on showing that an (n, k, M) SDCS can be used to construct a steganographic scheme that modifies each cover element by at most 1 and embeds $\log_2 M$ bits in n pixels by making at most k modifications. Consider an n-element cover (or a cover block), represented using the vector $\mathbf{x} = (\mathbf{x}[1], \ldots, \mathbf{x}[n]) \in \mathbb{Z}_M^n$, and a message symbol $b \in \mathbb{Z}_M$. Here, the cover representation is again obtained via some symbol-assignment function that maps cover elements to $\mathbb{Z}_M$. We define the embedding and extraction operation in the following manner:

$$\text{Emb}(\mathbf{x}, b) = (\mathbf{x}[1] + \mathbf{s}[1], \ldots, \mathbf{x}[n] + \mathbf{s}[n]) = \mathbf{y}, \tag{8.93}$$

$$\text{Ext}(\mathbf{y}) = \sum_{i=1}^{n} \mathbf{a}[i]\mathbf{y}[i], \tag{8.94}$$

where $\mathbf{s}[i]$ are such that $\sum_i |\mathbf{s}[i]| \leq k$ and $b = \sum_i \mathbf{a}[i]\mathbf{y}[i]$. All operations in the extraction function are understood as modulo M. To see the existence of such a vector $\mathbf{s}$, we first write the difference $w = b - \sum_{i=1}^{n} \mathbf{a}[i]\mathbf{x}[i]$ as an element of $\mathbb{Z}_M$,

$$w = \sum_{i=1}^{n} \mathbf{a}[i]\mathbf{s}[i], \tag{8.95}$$

with $\sum_{i=1}^{n} |\mathbf{s}[i]| \leq k$ because $\mathcal{A}$ is an (n, k, M) SDCS. To minimize the number of embedding operations, in the embedding function we should choose such a vector $\mathbf{s}$ with the minimal sum $\sum_{i=1}^{n} |\mathbf{s}[i]|$. If there exists more than one such vector, we choose randomly among them with uniform distribution. Then,

$$\sum_{i=1}^{n} \mathbf{a}[i]\mathbf{y}[i] = \sum_{i=1}^{n} \mathbf{a}[i](\mathbf{x}[i] + \mathbf{s}[i]) = b - w + w = b, \tag{8.96}$$

which proves that the extraction function indeed obtains the correct message symbol b. For practical implementations, the embedding function can be easily implemented using a look-up table tying each message symbol with the vector $\mathbf{s}$.

For further considerations, it will be convenient to denote $d_\mathcal{A}(b) = \sum_{i=1}^{n} |\mathbf{s}[i]|$ the minimal number of embedding modifications to embed symbol b. The embedding efficiency of the resulting steganographic scheme with SDCS $\mathcal{A}$ on covers with elements in $\mathbb{Z}_M$ is

$$e_\mathcal{A} = \frac{\log_2 M}{(1/M) \sum_{b=0}^{M-1} d_\mathcal{A}(b)}. \tag{8.97}$$

Example 8.7: Consider $\mathcal{A} = \{1, 2, 6\}$ and convince yourself that for $\sum_{i=1}^{3} |\mathbf{s}[i]| \leq$ 2 the sum $\sum_{i=1}^{3} \mathbf{s}[i]\mathbf{a}[i]$ can attain all values between -8 and 8. Out of these $M = 17$ values, $\sum_{i=1}^{3} |\mathbf{s}[i]| = 0$ for the value 0, $\sum_{i=1}^{3} |\mathbf{s}[i]| = 1$ for values from the set $\{-6, -2, -1, 1, 2, 6\}$, which is $\{11, 15, 16, 1, 2, 6\}$ in $\mathbb{Z}_{17}$, and $\sum_{i=1}^{3} |\mathbf{s}[i]| = 2$ for values in $\{-8, -7, -5, -4, -3, 3, 4, 5, 7, 8\}$. Thus, $\mathcal{A}$ is an SDCS with parameters $(3, 2, 17)$ and the embedding efficiency of the associated ± 1 embedding scheme is

$$e_{\mathcal{A}} = \frac{\log_2 17}{\frac{1}{17}(1 \times 0 + 6 \times 1 + 10 \times 2)} \approx 2.673. \tag{8.98}$$

This embedding scheme can embed $\log_2 17$ bits in $n = 3$ cover elements by making at most $k = 2$ modifications by ± 1.

If, for a given (n, k, M) SDCS, there does not exist an (n, k, M') SDCS with $M' > M$, we call the SDCS maximal (its associated steganographic method communicates the largest possible payload for a given choice of n and k). Because there are $V_3(n, k)$ possible ways one can make k or fewer changes by ± 1 to n cover elements, we obtain the following bound:

$$M \leq \sum_{i=0}^{k} 2^i \binom{n}{i}, \tag{8.99}$$

which is essentially the same bound as (8.77). For $k = 1$, the bound states $M \leq 1 + 2n$. In this case, it is possible to find the maximal SDCS with $M = 1 + 2n$, which is $\mathcal{A} = \{1, 2, \ldots, n\}$. To see that $\mathcal{A}$ is an $(n, 1, 1 + 2n)$ SDCS, the reader is encouraged to inspect Exercise 8.5, where the embedding scheme is formulated from a different perspective using rainbow coloring of lattices.

Finding maximal SDCSs is a rather difficult task. In fact, it is not easy to find SDCSs in general. Several parametric constructions of SDCSs are described in [160]. Table 8.3 contains examples of SDCSs useful for embedding large payloads, all found by a computer search. The embedding efficiency and relative embedding capacity of the associated embedding schemes are also displayed in Figure 8.8, where we compare the embedding efficiency of steganographic schemes constructed using SDCSs, Hamming codes, and the ternary repetition code from Exercise 8.6.

Curiously, the existence of an (n, k, M) SDCS does not imply the existence of SDCSs with (n, k, M'), $M' < M$. It is known, for example, that although there exists a $(9, 2, 132)$ SDCS, there are no SDCSs with parameters $(9, 2, x)$ for $x \in \{131, 129, 128, 127, 125\}$. The subject of finding SDCSs is related to other problems from discrete geometry, graph theory, and sum cover sets [73, 103, 105, 115]. Progress in these areas is likely to find applications in steganography via the considerations explained in this section.

Table 8.3. Examples of SDCSs and the relative payload, α, and embedding efficiency, e, of their associated steganographic schemes. Note that SDCSs with an even M cannot be enlarged to an odd M because we would lose the covering property. For example, for the SDCS $\{1,3,9,14\}$, $7 = -9 - 14 \bmod 30$ and 7 would not be covered in $\mathbb{Z}_{31}$.

(n,k,M)	SDCS	α	e
$(3,2,17)$	$\{1,2,6\}$	1.3625	2.6726
$(4,2,30)$	$\{1,3,9,14\}$	1.2267	2.9441
$(5,2,42)$	$\{1,2,7,14,18\}$	1.0785	3.1445
$(6,2,61)$	$\{1,2,5,11,19,27\}$	0.9885	3.3498
$(7,2,80)$	$\{1,22,26,30,34,36,39\}$	0.9031	3.5122
$(8,2,104)$	$\{2,4,6,13,16,34,39,40\}$	0.8376	3.6676
$(9,2,132)$	$\{2,11,33,34,44,50,55,58,62\}$	0.7827	3.8109
$(4,3,53)$	$\{1,2,6,18\}$	1.4320	2.5727
$(5,3,105)$	$\{1,3,14,36,42\}$	1.3428	2.7756
$(6,3,174)$	$\{1,3,9,21,51,86\}$	1.2405	2.9300
$(6,2,64)$	$\{1,2,4,12,21,28\}$	1	0.3021

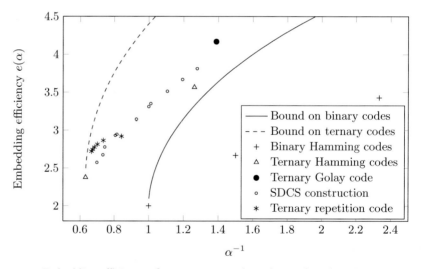

Figure 8.8 Embedding efficiency of ±1 steganographic schemes based on SDCS constructions from Table 8.3. For comparison, we include the embedding efficiency of ternary (codimension 1 and 2) and binary Hamming codes (codimension 1–3), the ternary Golay code, and the ternary repetition code from Exercise 8.6 for $n = 4, 7, 10, 13, 16$ (from the right).

Summary

- Matrix embedding (or syndrome coding) is a coding method that can increase the embedding efficiency of steganographic schemes.
- It can be applied only when the message length is smaller than the embedding capacity.

- The shorter the message, the larger the improvement due to matrix embedding.
- Matrix embedding is part of the theory of covering codes.
- Good matrix embedding schemes should have small average distance to code because it determines the embedding efficiency.
- Some of the simplest matrix embedding methods are based on Hamming codes.
- By assigning a q-ary symbol from a finite field to each cover element, it is possible to further increase embedding efficiency using q-ary codes.
- Measuring the embedding impact using distortion that takes into account the magnitude of modifications, ternary codes provide the minimal embedding impact. In particular, it does not pay off to make fewer embedding changes of magnitude larger than 1.
- It is possible to design steganographic schemes using sum and difference covering sets of finite cyclic groups. Current schemes find applications for embedding large payloads.

Table 8.4. Properties of optimal q-ary matrix embedding schemes when embedding into cover containing n pixels. The function $H_q(x)$ is the q-ary entropy function, $H_q(x) = -x \log_2 x - (1-x) \log_2(1-x) + x \log_2(q-1)$.

Maximal payload embeddable using up to R changes	$nH_q(R/n)$
Average number of embedding changes to embed m bits	$nH_q^{-1}(m/n)$
Maximal embedding efficiency to embed m bits	$\frac{m/n}{H_q^{-1}(m/n)}$

Exercises

8.1 [Binary repetition code] The binary repetition code of length n consists of two codewords $\mathcal{C} = \{(0,\ldots,0),(1,\ldots,1)\}$. Show that the covering radius of this code and its average distance to code are

$$R = \left\lceil \frac{n-1}{2} \right\rceil, \tag{8.100}$$

$$R_{\mathrm{a}} = \frac{n}{2}\left(1 - 2^{-n+1}\binom{n-1}{R}\right). \tag{8.101}$$

8.2 [Binary–ternary conversion] Write a computer program that converts a stream of q-ary symbols represented using integers $\{0,1,\ldots,q-1\}$ to a binary stream and vice versa. Make sure that the length of the binary stream is approximately $n \log_2 q$ for good encoding.

8.3 **[Tail inequality for $q > 2$]** Prove the tail inequality (8.81). **Hint:** First, you can assume that

$$0 \le \beta \le \frac{q-1}{q} \qquad (8.102)$$

because if we restrict our attention to codes containing the all-ones vector, and thus all its multiples $(0, \ldots, 0), (1, \ldots, 1), \ldots, (q-1, \ldots, q-1)$, no vector $\mathbf{x} \in \mathbb{F}_q^n$ can be further from these q codewords than $n(1 - 1/q)$ because the furthest vector contains q/n zeros, q/n ones, $\ldots$, and q/n symbols $q-1$. Then write

$$1 = (\beta + 1 - \beta)^n \ge (1 - \beta)^n \sum_{i=0}^{\beta n} \binom{n}{i} (q-1)^i \left(\frac{\beta}{1-\beta} \frac{1}{q-1} \right)^i, \qquad (8.103)$$

use the inequality $\beta/[(1-\beta)(q-1)] \le 1$, which follows from (8.102), and follow the same steps as in the proof of the tail inequality for $q = 2$.

8.4 **[Rainbow coloring]** Let $\mathcal{L}_d = \{(i_1, \ldots, i_d) | i_1 \in \mathbb{Z}, \ldots, i_d \in \mathbb{Z}\}$ be a d-dimensional integer lattice. For each lattice point, $(i_1, \ldots, i_d)$, we define its neighborhood as

$$\mathcal{N}(i_1, \ldots, i_d) = \left\{ (j_1, \ldots, j_d) \Big| \sum_{k=1}^{d} |i_k - j_k| = 1 \right\}. \qquad (8.104)$$

In other words, the neighborhood is formed by the lattice point itself and $2d$ other points that differ from the center point in exactly one coordinate by 1. Show that the following assignment of $2d + 1$ colors, c, $c \in \{0, \ldots, 2d+1\}$, to the lattice

$$\mathbf{c}(i_1, \ldots, i_d) = \sum_{k=1}^{d} k i_k \bmod (2d+1) \qquad (8.105)$$

has the property that the neighborhood of every lattice point contains exactly $2d + 1$ different colors.

8.5 **[± 1 embedding in groups]** One possibility to avoid modifying the pixels by more than 1, yet use the power of q-ary embedding, is to group pixels into disjoint subsets of d pixels. By modifying each pixel by ± 1, we obtain $2d$ possible modifications plus one case when no modifications are carried out. The rainbow coloring from the previous example will enable us to assign colors $((2d+1)$-ary symbols) to each pixel group and embed one $(2d+1)$-ary symbol by modifying at most one pixel by 1 (± 1 embedding). Show that the relative payload and embedding efficiency of this steganographic method are

$$\alpha_d = \frac{\log_2(2d+1)}{d}, \ e_d = \frac{\log_2(2d+1)}{1-(2d+1)^{-1}}. \qquad (8.106)$$

8.6 **[Large relative payload using ternary ± 1 embedding]** Let $\mathcal{C}$ be the ternary $[n, 1]$ repetition code (one-dimensional subspace of $\mathbb{F}_3^n$) with a ternary parity-check matrix $\mathbf{H} = [\mathbf{I}_{n-1}, \mathbf{u}]$, where $\mathbf{u}$ is the column vector of 2s. This code can be used to embed $(n-1)\log_2 3$ bits per n pixels, which gives relative payload $\alpha_n = (n-1)/n \log_2 3 \to \alpha_{\max} = \log_2 3$ bpp with increasing n. Thus this code is suitable for embedding large payloads close to the maximal relative payload $\alpha_{\max}$. If we denote the number of 0s, 1s, and 2s in an arbitrary vector of $\mathbb{F}_3^n$ by a, b, and c, respectively, then the average distance to $\mathcal{C}$ can be computed as

$$R_{\mathrm{a}} = \frac{1}{3^n} \sum (n - \max\{a, b, c\}) \binom{n}{a} \binom{n-a}{b}, \qquad (8.107)$$

where the sum extends over all triples $\{a, b, c\}$ of non-negative integers such that $a + b + c = n$. Thus, the embedding efficiency of this code is $e = [(n-1)\log_2 3]/R_{\mathrm{a}}$. For example, it is possible to embed $\alpha = 1.188$ bpp with embedding efficiency 2.918 bits per change for $n = 4$. The performance of this family of codes is shown in Figure 8.8. In this exercise, prove the expression for the average distance to code, R_{a}.

9 Non-shared selection channel

In Chapter 6, we learned that steganographic security can be measured with the Kullback–Leibler divergence between the distributions of cover and stego images. Four heuristic principles for minimizing the divergence were discussed in Chapter 7. One of them was the principle of minimal embedding impact, which starts with the assumption that each cover element, i, can be assigned a numerical value, $\rho[i]$, that expresses the contribution to the overall statistical detectability if that cover element was to be changed during embedding. If the values $\rho[i]$ are approximately the same across all cover elements, minimizing the embedding impact is equivalent to minimizing the number of embedding changes. The matrix embedding methods introduced in the previous chapter can be used to achieve this goal.

If $\rho[i]$ is highly non-uniform, Alice may attempt to restrict the embedding changes to a selection channel formed by those cover elements with small $\rho[i]$. Constraining the embedding process in this manner, however, brings a fundamental problem. Often, the values $\rho[i]$ are computed from the cover image or some side-information that is not available to Bob. Thus, Bob is generally unable to determine the same selection channel from the stego image and thus read the message. Channels that are not shared between the sender and the recipient are called non-shared selection channels. The main focus of this chapter is construction of methods that enable communication with non-shared selection channels.

We now give a few typical examples when non-shared selection channels arise. Imagine that Alice has a raw, never-compressed image and wants to embed information in its JPEG compressed form. Intuitively, the side-information in the form of the raw image should help her better conceal the embedding changes. When compressing the image, Alice can inspect the DCT coefficients after they have been divided by quantization steps but *before* they are rounded to integers and select for embedding those coefficients whose fractional part is close to 0.5. Such coefficients experience the *largest* quantization error during JPEG compression and the smallest combined error (rounding + embedding) if rounded to the "other value." For example, when rounding the coefficient -3.54, we can embed a bit by rounding it to -3 or to -4. The rounding distortion (rounding to -4) is 0.46. If embedding requires rounding to -3 instead, the combined rounding and embedding distortion is only slightly larger, 0.54. Selecting such coefficients for embedding, however, creates an obvious and seemingly insurmountable problem. Bob will not be able to tell which DCT coefficients in the *stego* JPEG file were used for embedding because the cover is not available to him and he cannot completely undo the loss due to rounding in JPEG compression.

As another simple example, consider adaptive steganography where the cover elements are chosen for embedding based on their neighborhood. Alice calculates for each pixel, i, in the cover image the variance, $\sigma^2[i]$, from all pixels in its local 3×3 neighborhood. Then, she sorts the variances $\sigma^2[i]$ from the largest to the smallest and embeds the payload of m bits using LSB embedding into the m pixels with the largest local variance. When Bob attempts to read the message, it may well happen that the m pixels with the largest local variance in the stego image will not be completely the same (or their order may not be the same) as those selected by Alice. Again, Bob is unable to read the message. The adaptive method for palette images presented in Section 5.2.5 is yet another example of this problem.

Non-shared selection channels in steganography are sometimes explained using the metaphor "writing on wet paper" [91]. Imagine that the cover image $\mathbf{x}$ was exposed to rain and some of it pixels got wet. Alice is allowed only to slightly modify the dry pixels (the selection channel) but not the wet pixels. During transmission, the stego image $\mathbf{y}$ dries out and thus Bob has no information about which pixels were dry. The question is how many bits can be communicated to Bob and how? This problem is recognized in information theory as writing in memory with defective cells [60, 108, 152, 231, 254]. A computer memory contains n cells out of which $n - k$ cells are permanently stuck at either 0 or 1. The device that writes data into the memory knows the locations and status of the stuck cells. The task is to write as many bits as possible into the memory (up to k) so that the reading device, that does not have any information about the stuck cells, can correctly read the data. Clearly, writing on wet paper is formally equivalent to writing in memory with stuck cells. (The stuck cells correspond to wet pixels.)

We now provide a simple argument [152] based on random binning that shows that asymptotically, as $n \to \infty$ for a fixed ratio k/n, it is possible to write all k bits into the memory. In other words, we can write as many bits in the memory as if the reading device knew the location of the stuck cells! This surprising fact also follows from the Gel'fand–Pinsker theorem for channels with random parameters known to the sender [99].

Select an arbitrary $\epsilon > 0$ and randomly assign all n-bit vectors to $2^{k-\epsilon n}$ disjoint bins. This assignment must be shared with the reading device. The index of the bin will be the message communicated. Because there are $2^{k-\epsilon n}$ bins, the message that can be communicated as the bin index has $k - \epsilon n$ bits. Given a specific message of $k - \epsilon n$ bits, find the bin $\mathcal{B}$ with index equal to the message. Then, find in $\mathcal{B}$ a word with $n - k$ cells stuck exactly as in the memory and write this word into the memory. The reading device will simply read an n-bit word from the memory, find the bin to which the word belongs, and from the shared codebook extract the bin index, which is the message. If we are unable to find a word in $\mathcal{B}$ that would be compatible with the defects, we declare a failure to write the message into the memory. We now show that the probability of failure is asymptotically negligible. Let us calculate the probability that we will not be able to find a defect-compatible word in bin $\mathcal{B}$. Because there are in total 2^k words compatible with the memory and we have $2^n/2^{k-\epsilon n} = 2^{n-k+\epsilon n}$ words in each bin, the probability that none of the 2^k compatible words will be in $\mathcal{B}$ is

$$\left(1 - \frac{2^k}{2^n}\right)^{2^{n-k+\epsilon n}} = \left(\left(1 - \frac{1}{2^{n-k}}\right)^{2^{n-k}}\right)^{2^{\epsilon n}} \rightarrow 0 \text{ as } n \rightarrow \infty \text{ for } \frac{k}{n} = const.,$$

(9.1)

because $\left(1 - 1/2^{n-k}\right)^{2^{n-k}} \rightarrow 1/e < 1$. Thus, for any $\epsilon > 0$ we can write $k - \epsilon n$ bits into the memory with probability that approaches 1 exponentially fast.

Although asymptotically optimal, the random-binning argument above is not a practical way to construct steganographic schemes with non-shared selection channels due to the enormous size of the codebook that needs to be shared. In Section 9.1, we describe an approach based on syndrome coding using random linear codes, which is quite suitable for applications in steganography. A practical and fast wet paper code can be obtained using random linear codes called LT codes (Section 9.2). To improve the embedding efficiency of wet paper codes, in Section 9.3 we present another random construction. The usefulness of having a practical solution for the non-shared selection channel is demonstrated in Section 9.4, where we list several fascinating and very diverse applications of wet paper codes.

Alternative approaches to writing on wet paper that are not discussed in this book are based on maximum distance separable codes, such as the Reed–Solomon codes [97], and on BCH codes [208].

9.1 Wet paper codes with syndrome coding

Steganographic schemes with non-shared selection channels can be realized using syndrome coding quite similar to matrix embedding in the sense that the message is communicated as the syndrome of stego image bits for some linear code. There are, however, some fundamental differences between coding for matrix embedding and for non-shared selection channels. These differences as well as similarities are commented upon throughout the text.

We assume that n cover elements are represented using a bit-assignment function, π, as a vector of n bits $\mathbf{x} \in \{0, 1\}^n$. For example, one can think of $\mathbf{x}$ as the vector of LSBs of pixels. The sender forms a selection channel of k changeable elements $\mathbf{x}[j]$, $j \in \mathcal{S} \subset \{1, \ldots, n\}$, $|\mathcal{S}| = k$, that can be modified during embedding. The remaining $n - k$ elements $\mathbf{x}[j]$, $j \notin \mathcal{S}$, are not to be changed. Using the writing-on-wet-paper metaphor, $\mathcal{S}$ contains indices of "dry" elements (functioning memory cells) while the rest of the elements are "wet" (stuck cells). The sender's goal is to communicate $m < k$ message bits $\mathbf{m} \in \{0,1\}^m$ to the recipient who has no information about the selection channel $\mathcal{S}$.

Approaching this problem using the paradigm of syndrome coding, the message is communicated to the recipient as a syndrome. To this end, the sender modifies some changeable elements in the cover image so that the bits assigned to the stego image, $\mathbf{y}$, satisfy

$$\mathbf{Dy} = \mathbf{m}, \qquad (9.2)$$

where $\mathbf{D}$ is an $m \times n$ binary matrix shared by the sender and the recipient. The recipient reads the message by multiplying the vector of stego image bits $\mathbf{y}$ by the matrix $\mathbf{D}$. While this appears identical to matrix embedding, there is one important difference because the sender is now allowed to modify only the changeable elements of $\mathbf{x}$.

Using the variable $\mathbf{v} = \mathbf{y} - \mathbf{x}$, (9.2) can be rewritten as

$$\mathbf{Dv} = \mathbf{m} - \mathbf{Dx}, \qquad (9.3)$$

where $\mathbf{v}[i] = 0$, for $i \notin \mathcal{S}$, and $\mathbf{v}[j]$, $j \in \mathcal{S}$, are to be determined. Because $\mathbf{v}[i] = 0$ for $i \notin \mathcal{S}$, the product, $\mathbf{Dv}$, on the left-hand side can be simplified. The sender can remove from $\mathbf{D}$ all $n - k$ columns corresponding to indices $i \notin \mathcal{S}$ and also remove from $\mathbf{v}$ all $n - k$ elements $\mathbf{v}[i]$, $i \notin \mathcal{S}$. To avoid introducing too many new symbols, we will keep the same symbol for the pruned vector $\mathbf{v}$ and write (9.3) as

$$\mathbf{Hv} = \mathbf{z}, \qquad (9.4)$$

where $\mathbf{H}$ is an $m \times k$ submatrix of $\mathbf{D}$ consisting of those columns of $\mathbf{D}$ with indices from $\mathcal{S}$. Note that $\mathbf{v} \in \{0,1\}^k$ is an unknown vector holding the embedding changes and $\mathbf{z} = \mathbf{m} - \mathbf{Dx} \in \{0,1\}^m$ is a known right-hand side. Equation (9.4) is a system of m linear equations for k unknowns $\mathbf{v}$. We now discuss several options for solving this system.

If the solution exists, it can be found using standard linear-algebra methods, such as Gaussian elimination. The solution will exist for any right-hand side $\mathbf{z}$ if the rank of $\mathbf{H}$ is m (or the rows of $\mathbf{H}$ are linearly independent), which means that we must have $m \leq k$ as the necessary condition. However, the complexity of Gaussian elimination will be prohibitively large, $O(km^2)$, because we cannot directly impose structure on $\mathbf{H}$ that would allow us to solve the system more efficiently. This is because $\mathbf{H}$ was obtained as a submatrix of a larger, user-selected matrix $\mathbf{D}$ through the selection channel over which the sender has no control because it is determined by the cover image or side-information as in the examples from the introduction to this chapter. In the next section, we introduce a class of sparse matrices for which fast algorithms for solving (9.4) are available. Before we do so, we make two more remarks.

If $\mathbf{D}$ is chosen randomly (e.g., generated from the stego key), the probability that $\mathbf{H}$ will be of full rank, $\text{rank}(\mathbf{H}) = m$, is $1 - O(2^{m-k})$ (see, for example, [31]) and thus approaches 1 exponentially fast with increasing $k - m$. This means that syndrome coding with random matrices is asymptotically capable of communicating the same number of bits as there are dry cover elements as if the recipient knew the selection channel. It is also another way to prove that the capacity of defective memory is asymptotically equal to the number of correctly functioning cells.

Another way of looking at the system (9.4) is to think of $\mathbf{H}$ as a parity-check matrix of some linear code of length k and codimension m, in which case solving for $\mathbf{v}$ requires finding a member of the coset $\mathcal{C}(\mathbf{z})$ (see Appendix C on coding or Section 8.2). Again, because it is $\mathbf{D}$ and not $\mathbf{H}$ over which the sender has complete control, we cannot easily impose that $\mathbf{H}$ be a parity-check matrix of some structured code that would enable us to efficiently find coset members. Note that by choosing $\mathbf{v}$ as a coset leader, rather than an arbitrary coset member, the sender would not only communicate through a non-shared selection channel but additionally minimize the number of embedding changes! However, the complexity of finding coset leaders for general codes is exponential in k and constitutes a much harder task (see Section 9.3 on improving the embedding efficiency of wet paper codes). Our focus for now will be on communicating through a non-shared selection channel, which means solving the system (9.4).

9.2   Matrix LT process

In the previous section, we showed that syndrome coding with matrix $\mathbf{D}$ shared between the sender and the recipient can be used to communicate secret messages using non-shared selection channels. While the recipient reads the message by performing a simple matrix multiplication $\mathbf{Dy}$, the sender needs to solve the linear system (9.4). In this section, we show how to perform this task with low complexity by choosing $\mathbf{D}$ from a special class of random sparse matrices.

The basic idea is to make $\mathbf{D}$ (and thus $\mathbf{H}$) sparse so that it can be put with high probability into upper-diagonal form simply by permuting its rows and columns. Imagine that matrix $\mathbf{H}$ has a column with exactly one 1 in, say, the j_1th row. The sender swaps this column with the first column and then swaps the first and j_1th rows, which brings the 1 to the upper left corner of $\mathbf{H}$. Note that at this stage, for the permuted matrix, $\mathbf{H}[1,1] = 1$ and $\mathbf{H}[j,1] = 0$, for $j > 1$. We now apply the same step again while ignoring the first column and the first row of the permuted matrix. Let us assume that we can find again a column with only one 1, say in the j_2th row,[1] and swap the column with the second column of $\mathbf{H}$ followed by swapping the second and j_2th rows. As a result, we will obtain a matrix with 1s on the first two elements of its main diagonal and 0s below them, $\mathbf{H}[1,1] = 1, \mathbf{H}[2,2] = 1, \mathbf{H}[j,1] = 0$ for $j > 1$, and $\mathbf{H}[j,2] = 0$ for $j > 2$. We continue this process, this time ignoring the first two columns and

---

[1] Since we are ignoring the first row, this column may have another 1 as its first element.

rows, and eventually stop after m steps. At the end of this process, the row and column permutations will produce a permuted matrix in an upper-diagonal form, $\mathbf{H}[i,i] = 1$ for $i = 1, \ldots, m$, $\mathbf{H}[j,i] = 0$ for $j > i$. Such a linear system can be efficiently solved using the standard back-substitution as in Gaussian elimination. (Note that the permutations preserve the low density of the matrix.) We call this permutation procedure a matrix LT process because it was originally invented for erasure-correcting codes called LT codes [164]. If, at some step during the permutation process, we cannot find a column with exactly one 1, we say that the matrix LT process has failed. The trick is to give $\mathbf{H}$ properties that will guarantee that the matrix LT process will successfully finish with a high probability.

If the Hamming weights of columns of $\mathbf{H}$ follow a probability distribution called the Robust Soliton Distribution (RSD) [164], the matrix LT process will not fail with high probability. Imposing this distribution on the columns of $\mathbf{D}$, the columns of $\mathbf{H}$ will inherit it, too, because $\mathbf{H}$ is a submatrix of $\mathbf{D}$ obtained by removing some of its columns. The RSD requires that the probability that a column in $\mathbf{D}$ has Hamming weight i, $1 \le i \le m$, be $(1/\eta)(\boldsymbol{\nu}[i] + \boldsymbol{\tau}[i])$, where

$$\boldsymbol{\nu}[i] = \begin{cases} 1/m & i = 1 \\ 1/[i(i-1)] & i = 2, \ldots, m, \end{cases} \tag{9.5}$$

$$\boldsymbol{\tau}[i] = \begin{cases} T/(im) & i = 1, \ldots, \lfloor m/T \rfloor - 1 \\ [T\log(T/\delta)]/m & i = \lfloor m/T \rfloor \\ 0 & i = \lfloor m/T \rfloor + 1, \ldots, m, \end{cases} \tag{9.6}$$

$\eta = \sum_{i=1}^{m}(\boldsymbol{\nu}[i] + \boldsymbol{\tau}[i])$, $T = c\log(m/\delta)\sqrt{m}$, δ and c are suitably chosen constants whose choice will be discussed later. An example of the RSD for $m = 100$ is shown in Figure 9.1. To generate a matrix with the number of ones in its columns following the RSD, we can first generate a sequence of integers $\mathbf{w}[1], \mathbf{w}[2], \ldots$ that follows the RSD. Then, the ith column of the matrix is generated by applying a random permutation to a column containing $\mathbf{w}[i]$ ones and $m - \mathbf{w}[i]$ zeros.

To obtain some insight into why this distribution looks the way it does, note that the columns with few ones are more frequent than denser columns to guarantee that the LT process will always find a column with just one 1. Without the rather mysterious spike at Hamming weight 18, however, the matrix would become too sparse to be of full rank. Thus, the spike ensures that the rank of the matrix is m. More rigorous analysis of the RSD appears in Chapter 50 of [171] and the exercises therein.

The analysis of LT codes [164] implies that when the Hamming weights of columns of $\mathbf{D}$ (and thus of $\mathbf{H}$) follow the RSD (9.5)–(9.6) the matrix LT process finishes successfully with probability $P_{\text{pass}} > 1 - \delta$ if the message length m and the number of changeable elements k satisfy

$$k \ge \theta m = \left(1 + O\left(\frac{\log^2(m/\delta)}{\sqrt{m}}\right)\right)m. \tag{9.7}$$

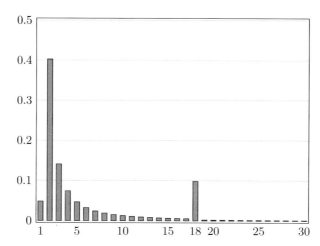

Figure 9.1 Robust soliton distribution for Hamming weights of columns of $\mathbf{D}$ for $\delta = 0.5$, $c = 0.1$, and $m = 100$.

Yet again, we see that asymptotically the sender can communicate k bits because $\theta \to 1$ as $m \to \infty$. The capacity loss due to the finite value of m is about 6% when $m = 10,000$, $c = 0.1$, and $\delta = 5$ ($\theta = 1.062$), while the probability that the LT process succeeds is about $P_{\text{pass}} \approx 0.75$. This probability increases and the capacity loss decreases with increasing message length (see Table 9.1).

We note at this point that in practice one can achieve low overhead and high probability of a successful pass through the LT process with $\delta > 1$, which is in contradiction with its probabilistic meaning. This is possible because the inequality $P_{\text{pass}} > 1 - \delta$ guaranteed by (9.7) is not tight.

Assuming the maximal-length message is sent ($m \approx k$), the average number of operations required to complete the LT process is [91]

$$O\left(n \log(m/\delta)\right) + O\left(m \log(m/\delta)\right) = O\left(n \log(m/\delta)\right), \tag{9.8}$$

which is significantly faster than Gaussian elimination. The first term arises from evaluating the product $\mathbf{Dx}$, while the second term is the complexity of the LT process. The gain in implementation efficiency over using simple Gaussian elimination is shown in Table 9.1.

9.2.1   Implementation

We now describe how the matrix LT process can be incorporated in a steganographic method. The sender starts by forming the matrix $\mathbf{D}$ with columns following the RSD. A stego key can be used to initialize the pseudo-random number generator. Applying the bit-assignment function to the cover image, the sender obtains the vector of bits $\mathbf{x}$ and computes the right-hand side $\mathbf{z} = \mathbf{m} - \mathbf{Dx}$. The matrix LT process is used to find the solution $\mathbf{v}$ to the linear system (9.4). Cover elements i with $\mathbf{v}[i] = 1$ should be modified to change their assigned bit.

Table 9.1. Running time (in seconds) for solving $m \times m$ and $m \times \theta m$ linear systems using Gaussian elimination and the matrix LT process, respectively ($c = 0.1$, $\delta = 5$); P_{pass} is the probability of a successful pass through the LT process. The experiments were performed on a single-processor Pentium PC with 3.4 GHz processor.

m	Gauss	LT	θ	P_{pass}
1,000	0.023	0.008	1.098	43%
10,000	17.4	0.177	1.062	75%
30,000	302	0.705	1.047	82%
100,000	9320	3.10	1.033	90%

The receiver forms the matrix $\mathbf{D}$, applies the bit-assignment function to image elements (obtaining vector $\mathbf{y}$), and finally extracts the message as the syndrome $\mathbf{m} = \mathbf{D}\mathbf{y}$. Note, however, that in order to do so the recipient needs to know the message length m because the RSD (and thus $\mathbf{D}$) depends on m as a parameter. (The remaining parameters c and δ can be public knowledge.) The message length m thus needs to be communicated to the receiver. For example, the sender can reserve a small portion of the cover image (e.g., determined from the stego key) where the parameter m will be communicated using a small matrix $\mathbf{D}_0$ with *uniform* distribution of 0s and 1s, instead of the RSD, and solve the system using Gaussian elimination. Because in typical applications m could be encoded using no more than 20 bits, the Gaussian elimination does not present a significant increase in complexity because solving a system of 20 equations should be fast. The payload of m bits is then communicated in the rest of the image using the matrix LT process whose matrix $\mathbf{D}$ follows the RSD.

To complete the algorithm description, we need to explain how the sender solves the problem with occasional failures of the matrix LT process. Again, among several different approaches that one can take, probably the simplest one is to make $\mathbf{D}$ dependent on the message length, m, for example by making the seed for the PRNG that generates the sequence of integers $\mathbf{w}[1], \ldots$ dependent on a combination of the stego key and message length. If a failure occurs, a dummy bit is appended to the message, and the matrix $\mathbf{D}$ is generated again followed by another run of the matrix LT process till a successful pass is obtained.

Algorithm 9.1 contains a pseudo-code for the matrix LT process to ease practical implementation. The input is a binary $m \times k$ matrix $\mathbf{H}$ and the right-hand side $\mathbf{z} \in \{0, 1\}^m$. The output is the solution $\mathbf{v}$ to the system $\mathbf{H}\mathbf{v} = \mathbf{z}$.

9.3   Wet paper codes with improved embedding efficiency

We already know that communication with non-shared selection channels using syndrome coding requires solving a linear system $\mathbf{H}\mathbf{v} = \mathbf{m} - \mathbf{D}\mathbf{x}$. For a random message, $\mathbf{m}$, the solution, $\mathbf{v}$, obtained using the matrix LT process will have

---

Algorithm 9.1 Matrix LT process for solving the linear system $\mathbf{Hv} = \mathbf{z}$.

$i = 1$ and $t = 0$
```
while (i ≤ m) & (∃i' ≥ i, ∑_{j>i} H[j, i'] = 1) {
```
    `swap rows` i `and` j_i; `//` where $\mathbf{H}[j_i, i'] = 1$
    `swap` $\mathbf{z}[i]$ `and` $\mathbf{z}[j_i]$;
    `swap columns` i `and` i';
    `swap` $\mathbf{v}[i]$ `and` $\mathbf{v}[i']$;
    $t = t + 1$;
    $\tau[t]$ `//` is the transposition $i \leftrightarrow i'$;
    $i = i + 1$;
```
}
```
`if` $i \leq m$ `declare failure and STOP`;
`//` $\mathbf{H}$ is now in upper-diagonal form
$\mathbf{v}[i] = 0$ `for` $m < i \leq k$;
`Use back-substitution to determine` $\mathbf{v}[i]$, $i \leq m$;
`//` Apply the transpositions τ to $\mathbf{v}$ in the reverse order
```
while t > 0 {
```
    $\mathbf{v} \leftarrow \tau[t](\mathbf{v})$;
    $t = t - 1$;
```
}
```
`//` The resulting $\mathbf{v}$ is the solution to the system $\mathbf{Hv} = \mathbf{z}$

---

on average 50% of ones and 50% of zeros. Since ones correspond to embedding changes, the message will be embedded with embedding efficiency 2. If the message is shorter than the maximal communicable message, $m < k$, there will be more than one solution. In the language of coding theory, the solutions will form a coset $\mathcal{C}(\mathbf{z}) = \{\mathbf{x} \in \{0,1\}^n | \mathbf{Hx} = \mathbf{z}\}$. If the sender selects the solution with the smallest number of ones (a coset leader) the embedding impact will be minimized or, equivalently, the embedding efficiency maximized. Unfortunately, the problem of finding a coset leader for general codes is NP-complete.

In this section, we describe a version of wet paper codes with improved embedding efficiency. It is a block-based scheme that embeds small message segments of p bits in each block using random codes of codimension p. For such codes, the problem of finding the solution $\mathbf{v}$ with the smallest number of ones can be solved simply using brute force.

Keeping the same notation, we assume there are k changeable pixels in a cover image consisting of n pixels and we wish to communicate $m < k$ message bits. The sender and receiver agree on a small integer p (e.g., $p \approx 20$) and using the stego key divide the cover image into $n_{\mathrm{B}} = m/p$ disjoint pseudo-random blocks, where each block will convey p message bits. Each block will thus contain $n/n_{\mathrm{B}} = pn/m$ cover elements (for simplicity we assume all quantities are integers). Since the blocks are formed pseudo-randomly, there will be on average

$(k/n) \times (pn/m) = pk/m = p/\alpha$ changeable pixels, where $\alpha = m/k, 0 \le \alpha \le 1$, is the relative payload.[2]

Using the stego key, the sender will generate a pseudo-random binary $p \times pn/m$ matrix $\mathbf{D}$ that will be used to embed p message bits in every block. Since the embedding efficiency is determined by the average distance to code (which should be as small as possible), the matrix $\mathbf{D}$ should not have any duplicate or zero columns. This can be guaranteed if the number of pixels in each block, n/n_B, satisfies $n/n_B < 2^p$ or, equivalently, $\alpha = m/n = pn_B/n > p/2^p$, which will not be satisfied for $p \approx 20$ only for extremely short payloads where detectability is not an issue anyway. Thus, the columns of $\mathbf{D}$ can be generated by drawing n/n_B integers from the set $\{1, 2, \ldots, 2^p - 1\}$ without replacement and writing them as binary vectors of length p.

As described in Section 9.1, in each block $\mathcal{B}$ the sender forms a binary submatrix $\mathbf{H}$ of $\mathbf{D}$ and computes the syndrome $\mathbf{z} = \mathbf{m} - \mathbf{Dx}$, where $\mathbf{m} \in \{0, 1\}^p$ is a segment of p message bits to be embedded at $\mathcal{B}$ and $\mathbf{x} \in \{0, 1\}^{pn/m}$ is the vector of bits of cover image pixels from $\mathcal{B}$. The submatrix $\mathbf{H}$ will in general be different in every block and will have exactly p rows and, on average, p/α columns. The sender now needs to find a coset leader, $\mathbf{v}$, of the coset $\mathcal{C}(\mathbf{z})$. To explain the method, we introduce the following concepts.

Let $\mathcal{U}_1 \subset \mathbb{F}_2^p$ be the set of all columns of $\mathbf{H}$ and $\mathcal{U}_{i+1} = \mathcal{U}_1 + \mathcal{U}_i - (\mathcal{U}_1 \cup \cdots \cup \mathcal{U}_i) - \{\mathbf{0}\}, i = 1, \ldots, p$, be a sequence of sets, where the sum of two sets is defined as $\mathcal{A} + \mathcal{B} = \{a + b | a \in \mathcal{A}, b \in \mathcal{B}\}$. Note that $\mathcal{U}_i = \emptyset$ for $i > R$, where R is the covering radius of $\mathbf{H}$. Also note that $\mathcal{U}_i$ is the set of syndromes that can be obtained by adding i columns of $\mathbf{H}$ but no less than i. Equivalently, $\mathcal{U}_i$ is the set of all coset leaders of weight i. For a given right-hand side $\mathbf{z}$, we could find a coset leader by generating the sets $\mathcal{U}_1, \mathcal{U}_2, \ldots$ and stop once $\mathbf{z} \in \mathcal{U}_r$. The problem is that the cardinality of these sets increases exponentially. We now describe a simple algorithm that enables us to find the coset leader with Hamming weight r by only generating the sets $\mathcal{U}_i$ for $i \le r/2$.

Let $\mathbf{z} = \mathbf{H}[., j_1] + \cdots + \mathbf{H}[., j_r]$, where $r \le R$ is the minimal number of columns of $\mathbf{H}$ adding up to $\mathbf{z}$. Note that $\mathbf{v}$ with zeros everywhere except for indices $j_1, \ldots, j_r$ is a coset leader. Since we work in binary arithmetic, $\mathbf{z} + \mathbf{H}[., j_1] + \cdots + \mathbf{H}[., j_{\lfloor r/2 \rfloor}] = \mathbf{H}[., j_{\lfloor r/2 \rfloor + 1}] + \cdots + \mathbf{H}[., j_r]$, which implies $(\mathbf{z} + \mathcal{U}_{\lfloor r/2 \rfloor}) \cap \mathcal{U}_{r - \lfloor r/2 \rfloor} \neq \emptyset$. This observation leads to Algorithm 9.2.

After the solution $\mathbf{v}$ has been found, the sender modifies those cover elements in the block for which $\mathbf{v}[i] = 1$. The modified block of pixels from the stego image is denoted $\mathbf{y}$, which completes the description of the embedding algorithm.

The recipient knows n from the stego image and knows p because it is public. Since the message length m is used in dividing the image into blocks, it needs to be communicated in the stego image as well, for example using the method

---

[2] Note that we measure the relative payload with respect to the number of changeable pixels, k, rather than the number of all pixels, n.

Algorithm 9.2 Meet-in-the-middle algorithm for finding coset leaders.

```
if (z ∈ U₁) {
    v[j₁] = 1; // because z = H[.,j₁] for some j₁
    set v[j] = 0 for all other j;
    return;
} else {
    l = r = 1;
}
while ((z + Uₗ) ∩ Uᵣ = ∅) {
    if (l = r) {
        r = r + 1;
        if (Uᵣ not yet generated) generate Uᵣ;
    } else {
        l = l + 1;
        if (Uₗ not yet generated) generate Uₗ;
    }
}
// any v ∈ (z + Uₗ) ∩ Uᵣ is a coset leader of weight l + r
```

---

described in Section 9.2.1. Knowing m, the recipient uses the secret stego key and partitions the rest of the stego image into the same disjoint blocks as the sender and extracts p message bits $\mathbf{m}$ from each block of pixels $\mathbf{y}$ as the syndrome $\mathbf{m} = \mathbf{D}\mathbf{y}$.

9.3.1   Implementation

To assess the memory requirements and complexity of Algorithm 9.2, we consider the worst case, when the sender needs to generate all sets $\mathcal{U}_1, \ldots, \mathcal{U}_{\lceil R/2 \rceil}$. The cardinalities of $\mathcal{U}_i$ exponentially increase with i, reach a maximum at around $i \approx R_{\mathrm{a}}$, the average distance to code, and then quickly fall off to zero for $i > R_{\mathrm{a}}$. We already know from Chapter 8 that with increasing length of the code (or increasing p), $R_{\mathrm{a}} \to R$. This means that the above algorithm avoids computing the largest of the sets $\mathcal{U}_i$. Nevertheless, we will still need to keep in memory all $\mathcal{U}_i$, $i = 1, \ldots, \lceil R/2 \rceil$ and the indices $j_1, \ldots, j_i$ for each element of $\mathcal{U}_i$. Because on average $|\mathcal{U}_1| = p/\alpha$, we have on average $|\mathcal{U}_i| \le \binom{p/\alpha}{i}$. Thus, the total memory requirements are bounded by $O\left(R/2 \times \binom{p/\alpha}{R/2}\right) \approx O\left(p 2^{(p/\alpha)H(R\alpha/2p)}\right) \approx O\left(p 2^{\kappa p}\right)$, where $\kappa = H(H^{-1}(\alpha)/2)/\alpha < 1$ and $H(x)$ is the binary entropy function. (For example, for $\alpha = \frac{1}{2}$, $\kappa = 0.61$.) Here, we used the asymptotic form for the binomial number (8.49) from Chapter 8, $\binom{n}{k} \approx 2^{nH(k/n)}$, and the fact that $R \approx (p/\alpha)H^{-1}(\alpha)$ is the expected number of embedding changes for large p (see Table 8.4).

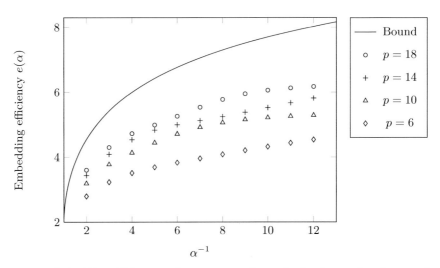

Figure 9.2 Embedding efficiency e of wet paper codes realized using random linear codes of codimension $p = 6, 10, 14, 18$ displayed as a function of $1/\alpha$. The solid curve is the asymptotic upper bound on embedding efficiency.

To obtain a bound on the computational complexity, note that we need to compute $\mathcal{U}_1 + \mathcal{U}_i$ for $i = 1, \ldots, R/2$. Thus, the computational complexity is bounded by $O\left(p/\alpha \times R/2 \times \binom{p/\alpha}{R/2}\right) \approx O\left(p^2 2^{\kappa p}\right)$. Because the complexity is exponential with p, the largest p for which this algorithm can be used with running times of the order of seconds on a single-processor PC with 3.4 GHz processor is about $p \approx 18$.

We make one comment on the solvability of (9.4) in each block. The equation $\mathbf{Hv} = \mathbf{z}$ will have a solution for all $\mathbf{z} \in \mathbb{F}_2^p$ if and only if $\text{rank}(\mathbf{H}) = p$. The probability of this is $1 - O\left(2^{p(1-1/\alpha)}\right)$ as this is the probability that a random binary matrix with dimension $p \times p/\alpha, \alpha = m/k$, will have full rank [30]. This probability quickly approaches 1 with decreasing message length m or with increasing p (for fixed m and k) because $k > m$.

For $k/m \approx 1$, the probability that $\text{rank}(\mathbf{H}) < p$ may become large enough to encounter a failure to embed all p bits in some blocks. For example, for $p = 18$ and $k/m = 2$ (or relative payload $\alpha = \frac{1}{2}$), $n = 10^6$, $k = 50,000$, the probability of failure is about 0.0043. The fact that the number of columns in $\mathbf{H}$ varies from block to block also contributes to failures. We note that the probability of failure quickly decreases with increasing k/m and is not an issue as long as $k/m > 3$ or $\alpha < \frac{1}{3}$. The failures can be dealt with by communicating to the receiver which blocks failed to hold all p bits. For details of this procedure, the reader is referred to [91].

9.3.2   Embedding efficiency

With increasing code length and fixed relative payload, matrix embedding schemes based on random linear codes asymptotically achieve the theoretical upper bound on embedding efficiency (see Chapter 8). Even though with $p = 18$ the codes' performance is still far from the limit, they lead to a significant improvement in embedding efficiency. Figure 9.2 shows the embedding efficiency as a function of the ratio $1/\alpha = k/m$ for a cover image with $n = 10^6$ pixels and $k = 50,000$ changeable pixels for $p = 6, 10, 14, 18$. The values were obtained by averaging over 100 embeddings of random messages in the same cover image with the same parameters k, n, and m.

Note that for a fixed p, the efficiency increases with shorter messages. Once the number of changeable pixels in each block exceeds 2^p, the embedding efficiency starts saturating at $p/(1 - 2^{-p})$, which is the value approached with decreasing payload α. This is because the p/α columns of $\mathbf{H}$ eventually cover the whole space $\mathbb{F}_2^p$ and thus we embed every non-zero syndrome $\mathbf{s} \neq \mathbf{0}$ using at most one embedding change.

We close this section with one more interesting observation. An observant reader will notice in Figure 9.2 that, for fixed p, the embedding efficiency increases with decreasing relative payload α in a curious non-uniform manner. In particular, while the codes with codimension $p = 14$ and $p = 18$ have almost identical embedding efficiency for $\alpha^{-1} = 5$, the difference becomes quite substantial for $\alpha^{-1} = 9$. In fact, it is even possible that for a fixed α, a higher embedding efficiency may be obtained for codes with lower p. In such a situation, it would be pointless to use codes with higher p as we would obtain worse performance and, at the same time, increase the computational complexity and memory requirements. A detailed analysis of this rather peculiar phenomenon, which is of importance to practitioners, appears in [93].

9.4   Sample applications

The methods for communication using non-shared selection channels introduced above are quite important tools for the steganographer because there exist numerous situations in steganography when non-shared channels arise. In this section, we demonstrate this for a rather diverse spectrum of applications ranging from minimum-impact steganography and public-key steganography to improved matrix embedding methods and an improved F5 algorithm (nsF5).

9.4.1   Minimal-embedding-impact steganography

According to the principle of minimal embedding impact, the steganographer first assigns to each cover element, i, a scalar value, $\rho[i]$, that expresses the increase in statistical detectability should the ith cover element be modified during embedding. If $\rho[i]$ are approximately uniform, minimizing the embedding impact is equivalent to minimizing the number of embedding changes, which was the topic of Chapter 8 on matrix embedding. If $\rho[i]$ vary greatly from element to element, the sender should embed the payload so that the sum of $\rho[i]$ over all modified cover elements is as small as possible.

One of the simplest strategies[3] the sender can adopt to communicate m message bits is to embed into m cover elements $\mathbf{x}[i_1], \ldots, \mathbf{x}[i_m]$ with the smallest $\rho[i]$, $\rho[i_k] \leq \rho[j]$ whenever $j \notin \mathcal{S}$, $\mathcal{S} = \{i_1, \ldots, i_m\}$. However, $\rho[i]$ are often determined by some side-information unavailable to the recipient, which means that we are facing a non-shared selection channel. Thus, the sender pronounces the elements from $\mathcal{S}$ changeable (or dry) and applies some of the methods explained in this chapter to communicate the payload to the recipient. We now provide details of a specific embedding method based on this strategy called perturbed-quantization steganography [89].

9.4.2   Perturbed quantization

Let us assume that the sender obtains the cover image through some process that ends with quantization, for example lossy compression, resampling, filtering, or the image-acquisition process itself. The sender's goal is to minimize the combined distortion due to processing and embedding. Let us assume that the input cover image (also called precover) is represented with a vector $\mathbf{X} \in \mathcal{Z}^N$, where $\mathcal{Z}$ is the range of its elements. For example, for a 16-bit grayscale image, $\mathcal{Z} = \{0, \ldots, 2^{16} - 1\}$. Here, we intentionally used the term "input cover image" and a capital letter to denote it because the cover against which security of the embedding changes should be evaluated will be obtained from $\mathbf{X}$ using a transformation F of the following form:

$$F = Q \circ T : \mathcal{Z}^N \to \mathcal{X}^n, \tag{9.9}$$

where $\mathcal{X}$ is the dynamic range of the transformed signal $\mathbf{x} = F(\mathbf{X})$ and the real-valued map $T : \mathcal{Z}^N \to \mathbb{R}^n$ is some form of processing. The circle stands for composition of mappings $Q \circ T(\mathbf{X}) = Q(T(\mathbf{X}))$. We denote by $T(\mathbf{X}) = \mathbf{u} \in \mathbb{R}^n$ the intermediate image. The map Q is an integer scalar quantizer with range $\mathcal{X}$ extended to work on vectors by coordinates: $Q(\mathbf{u}) = (Q(\mathbf{u}[1]), \ldots, Q(\mathbf{u}[n]))$. We stress that the cover image is the signal $\mathbf{x}$.

Following the principle of minimum embedding impact, we define $\rho[i]$ using the uniquely determined integer a, $a \leq \mathbf{u}[i] < a + 1$, as

$$\rho[i] = |\mathbf{u}[i] - a - (a + 1 - \mathbf{u}[i])| = 2\left|\mathbf{u}[i] - (a + 1/2)\right|. \tag{9.10}$$

---

[3] Exercises 9.3–9.6 investigate a more general strategy under the assumption that wet paper codes with the largest theoretical embedding efficiency are used.

In other words, $\boldsymbol{\rho}[i]$ is the increase in the quantization error when quantizing $\mathbf{u}[i]$ to the second closest value to $\mathbf{u}[i]$ instead of the closest one. If the sender uses a bit-assignment function that always assigns two different bits to integers a and $a+1$, the sender has the power to flip the bit assigned to $\mathbf{x}[i]$ by rounding to the second closest value.

We now give a few examples of mappings F that could be used for perturbed-quantization steganography.

Example 9.1: [Downsampling] For grayscale images in raster format, the transformation T maps an $M_1 \times N_1$ matrix of integers $\mathbf{X}[i,j]$ into an $m_1 \times n_1$ matrix of real numbers $\mathbf{u} = \mathbf{u}[r,s]$ using a resampling algorithm.

Example 9.2: [Decreasing the color depth by d bits] The transformation T maps an $M_1 \times N_1$ matrix of integers $\mathbf{X}[i,j]$ in the range $\mathcal{Z} = \{0, \dots, 2^{n_c} - 1\}$ into a matrix of real numbers $\mathbf{u}[i,j] = \mathbf{X}[i,j]/2^d$ of the same dimensions, $\mathbf{x}[i,j] \in \mathcal{X} = \{0, \dots, 2^{n_c - d} - 1\}$.

Example 9.3: [JPEG compression] For a grayscale image, the transformation T maps an $M_1 \times N_1$ matrix of integers $\mathbf{X}[i,j]$ into a matrix of DCT coefficients $\mathbf{u}[i,j]$ in a block-by-block manner (here we assume for simplicity that M_1 and N_1 are multiples of 8). In each 8×8 pixel block $\mathbf{B}$, the (k,l)th element of the transformed block in the DCT domain is $\mathrm{DCT}(\mathbf{B})[k,l]/\mathbf{Q}[k,l]$, where DCT is the two-dimensional DCT (2.20) and $\mathbf{Q}[k,l]$ is the (k,l)th element of the JPEG quantization matrix (see Section 2.3 for more details about the JPEG format).

Example 9.4: [Double JPEG compression]) Normally, the quantization error has uniform distribution, which limits the amount of changeable cover elements with small $\boldsymbol{\rho}[i]$. Their number can be artificially increased by repeated quantization. Imagine compressing a raw image using primary and secondary quantization matrices $\mathbf{Q}^{(1)}$ and $\mathbf{Q}^{(2)}$. When the quantization steps $\mathbf{Q}^{(1)}[k,l]$ and $\mathbf{Q}^{(2)}[k,l]$ satisfy $a\mathbf{Q}^{(1)}[k,l] = b\mathbf{Q}^{(2)}[k,l] + \frac{1}{2}\mathbf{Q}^{(2)}[k,l]$ for some integers a and b, it means that DCT coefficients $\mathbf{D}[k,l]$ that are equal to a after the first compression end up in the middle of quantization intervals during the second compression. Such coefficients are called contributing coefficients. Some combinations of the quality factors, such as $q_{\mathrm{f}}^{(1)} = 85$ and $q_{\mathrm{f}}^{(2)} = 70$, produce a large number of contributing coefficients and thus large embedding capacity. To embed n bits, the selection channel is formed by n contributing coefficients with the smallest impact (9.10). This algorithm is described in detail in the original publication [92].

We now provide some heuristic thoughts about the steganographic security of perturbed quantization. In order to mount an attack, the warden would have to find statistical evidence that some of the values $\mathbf{u}[j]$ were not quantized to their correct values. This, however, may not be easy in general for the following reasons. The sender is using side-information that is largely removed during quantization and is thus unavailable to the warden. Moreover, the rounding process at changeable elements is more influenced by noise naturally present in images than for the remaining elements.

The warden, however, might be able to model some regions in the image well enough (e.g., regions with a smooth gradient) and attempt to detect embedding changes in those regions only. Thus, the sender can (and should) exclude from the selection channel $\mathcal{S}$ those elements whose unquantized values can be predicted with better accuracy. This reasoning leads to two modifications of the perturbed-quantization (PQ) method from Example 9.4.

Example 9.5: [PQt] Texture-adaptive perturbed quantization narrows the selection channel only to contributing coefficients coming from blocks with the highest texture. The block texture $t(\mathbf{B})$ is computed from the singly compressed cover JPEG image with quality factor $q_{\mathrm{f}}^{(1)}$ decompressed to the spatial domain. The pixel block is divided into disjoint 2×2 blocks. For each 2×2 block, the difference between the highest and the lowest pixel value is calculated. The texture measure $t(\mathbf{B})$ is the sum of these differences over the whole block. To embed n bits, the selection channel is formed by n contributing coefficients from blocks with the highest measure of texture.

Example 9.6: [PQe] Energy-adaptive perturbed quantization narrows down the selection channel to contributing coefficients from blocks with the highest energy $e(\mathbf{B})$, which is calculated as a sum of squares of all quantized DCT coefficients in the block. To embed n bits, the selection channel is formed by n contributing coefficients selected from blocks with the highest energy.

In contrast to the distortion-based PQ method where the selection channel is formed by contributing DCT coefficients with the smallest rounding error, in PQt and PQe the contributing coefficients are selected on the basis of the block texture (or energy) rather than the rounding distortion. These two adaptive versions of perturbed quantization provide better security than the version from Example 9.4 (see [95] and the steganalysis results in Table 12.8).

9.4.3   MMx embedding algorithm

The MMx algorithm [141] is a steganographic algorithm for JPEG images with an embedding mechanism that is a combination of matrix embedding using binary $[2^p - 1, 2^p - 1 - p]$ Hamming codes and perturbed quantization. MMx minimizes the embedding impact by utilizing as side-information the unquantized version of the cover on its input (raw image before JPEG compression). It uses a clever trick to avoid having to use wet paper codes. Here, we only briefly explain the principle on a simple example of a scheme that uses $[7, 4]$ Hamming codes – it embeds 3 bits into 7 DCT coefficients $\mathbf{D}[i]$, $i = 1, \ldots, 7$, by making at most x embedding changes. We speak of an $(x, 3, 7)$ MMx method.

The sender first uses the non-rounded value of the ith DCT coefficient to derive the embedding impact, $\rho[i]$, at the ith DCT coefficient according to (9.10) with the only exception when either $a = 0$ or $a + 1 = 0$, or, in other words, when the value of the non-rounded DCT coefficient is in the interval $[-1, 1]$. Since the decoder will read message bits only from non-zero coefficients (as in F5), the sender never rounds such a coefficient to zero. Instead, she rounds it to 2 (if the coefficient is positive) or -2 (if the coefficient is negative). In this case, the embedding distortion will be increased to $1 + \rho[i]$. This choice removes the problem with shrinking a non-zero coefficient to zero at the expense of a larger embedding distortion.

During the actual embedding, the sender first tries to embed 3 bits in 7 DCT coefficients using at most one change as in matrix embedding using Hamming codes (Chapter 8). If the jth coefficient $\mathbf{D}[j]$ had to be rounded to the "other" side, the embedding impact is $\rho[j]$. The Hamming code uniquely determines the coefficient that needs to be modified. Chances are that the coefficient happens to have a large $\rho[j]$. The sender tries the embedding again, this time allowing two embedding changes (two coefficients to be rounded to the other side) and lists all pairs of columns $\mathbf{H}[., j'], \mathbf{H}[., j'']$ from the parity-check matrix $\mathbf{H}$ for which $\mathbf{H}[., j'] + \mathbf{H}[., j''] = \mathbf{H}[., j]$. For Hamming code $[7, 4]$ there will always be exactly three such pairs (in general, the number of pairs is equal to the code codimension p). For each pair, the sender calculates the embedding impact $\rho[j'] + \rho[j'']$. If one of these combined impacts is smaller than $\rho[j]$, the sender makes embedding changes at that coefficient pair instead to decrease the embedding impact. This embedding method can be extended by allowing up to x = 3 embedding changes to see whether the message can be embedded with an even smaller impact by modifying three pixels.

Depending on the number of allowed changes, we speak of an MMx algorithm, where x = $1, 2, 3, \ldots$ is the number of allowed embedding changes. In practice, only negligible improvement is typically obtained by using x > 3. For small payloads, the MMx method appears to resist blind steganalysis attacks the best out of all known steganographic methods for JPEG images [95] (as of late 2008). The reader is encouraged to inspect the results of blind steganalysis presented in Tables 12.7 and 12.8.

9.4.4   Public-key steganography

In this section, we explain how public-key cryptography can be used to construct a version of public-key steganography. Non-shared selection channels will again play an important role.

In public-key cryptography, there exist two keys – an encryption key E and a decryption key D. The encryption key is made public, while the decryption key is kept private. Public-key encryption schemes have the important property that, knowing the encryption key, it is computationally hard to derive the decryption key. Thus, giving everyone the ability to encrypt messages does not automatically give the ability to decrypt. When Alice wants to send an encrypted message, $\mathbf{m}$, to Bob, she can send him $E_\mathrm{B}(D_\mathrm{A}(\mathbf{m}))$, where E_B and D_A stand for the public (encryption) and private (decryption) key of Bob and Alice, respectively. Bob reads the message by $\mathbf{m} = E_\mathrm{A}(D_\mathrm{B}(E_\mathrm{B}(D_\mathrm{A}(\mathbf{m}))))$ because $D_\mathrm{B}E_\mathrm{B} = E_\mathrm{A}D_\mathrm{A} = \mathrm{Identity}$. Note that only Bob can read the message because only he has the private decryption key D_B. At the same time, he will know that it was Alice who sent the message as only she possesses the decryption key D_A.

The public-key scheme enables Alice and Bob to exchange secrets without previously agreeing on a secret key, which makes such schemes very useful in practice, such as in financial transactions over computer networks. The reader is referred to [207] to learn more about construction of such encryption schemes.

Public-key cryptography can be used to construct an equivalent paradigm for steganography [5]. Imagine that Alice uses a steganographic scheme with a public selection channel but encrypts her payload using a public-key encryption scheme. Upon receiving an image from Alice, Bob suspects steganography, extracts the payload, and decrypts it to see whether there is a secret message from Alice. Note that in this setup the encrypted message is publicly available to Eve (the warden) but, as long as the cryptosystem is strong, Eve will not be able to tell whether the extracted bit stream is a random sequence or ciphertext. The fact that the selection channel is public, however, may give Eve a starting point to mount a steganalytic attack. For example, she can compare statistical properties of pixels from the public selection channel with the remaining pixels and look for statistically significant deviations [90]. Without any doubts, public selection channels are a security weakness. This problem can be eliminated by using selection channels that are *completely random* implemented using wet paper codes with a public randomly generated matrix $\mathbf{D}$. This gives everyone the ability to read the message, as required, without giving any information about the selection channel.

9.4.5 $e + 1$ matrix embedding

In this section, we describe a clever method for using a binary code as if it were a ternary code with a correspondingly higher embedding efficiency. Let us assume that we have a matrix embedding scheme with bit-assignment function $\pi(x) = \mathrm{LSB}(x)$ based on a binary code $\mathcal{C}$ with embedding efficiency e bits per change. This means that the sender can embed m bits in n pixels on average by making m/e embedding changes. Let us denote the set of modified pixels as $\mathcal{S}$, $E[\|\mathcal{S}\|] = m/e$. If the sender modifies pixels in $\mathcal{S}$ by ± 1, rather than by flipping their LSBs, when making the embedding changes she has a choice to adjust the second LSB of every pixel in $\mathcal{S}$ to either 0 or 1 and thus embed an additional m/e bits in the second LSB plane. Since the recipient will not know the set $\mathcal{S}$, the sender has to use wet paper codes, this time with the second LSB as the bit-assignment function[4] and $\mathcal{S}$ as the set of changeable pixels.

The recipient first extracts m message bits from LSBs of the image using the parity-check matrix of the code $\mathcal{C}$ and then extracts an additional m/e bits from the second LSB plane using wet paper codes. Because a total of $m + m/e$ bits is embedded using m/e changes, the embedding efficiency of this scheme is

$$\frac{m + m/e}{m/e} = e + 1. \tag{9.11}$$

To summarize, if we allow the sender to modify the pixels by ± 1 rather than flip their LSBs, the embedding efficiency of the original binary matrix embedding scheme can be increased by 1. On the other hand, ± 1 embedding changes allow application of ternary codes that enjoy a higher bound on embedding efficiency. Thus, are we gaining anything using this scheme over ternary schemes? Quite surprisingly, it can be shown that the embedding efficiency of this $e + 1$ scheme is as high as what can be achieved using ternary codes with the additional advantage that there is no need to convert binary streams to ternary and vice versa.

We now prove that if the binary code $\mathcal{C}$ reaches its upper bound on embedding efficiency, $e = \alpha/H^{-1}(\alpha)$, the corresponding $e + 1$ matrix embedding scheme reaches the bound for ternary codes [258]. This interesting result tells us that if we have near-optimal binary matrix embedding, we can automatically construct near-optimal ternary codes! To see this, realize that for an optimal binary matrix embedding scheme, the relative payload embeddable using change rate β is $H(\beta)$ (see Table 8.4). The relative payload embeddable using the same change rate for the $e + 1$ scheme is $H(\beta) + \beta = H(\beta) + \beta \log_2(3 - 1) = H_3(\beta)$, which is exactly the ternary bound.

---

[4] $\pi_2(x) = \mathrm{LSB}(\lfloor x/2 \rfloor)$.

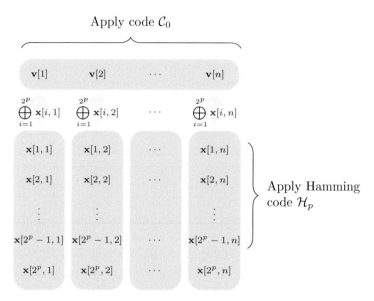

Figure 9.3 A block of $n2^p$ cover elements used in the ZZW code construction.

9.4.6   Extending matrix embedding using Hamming codes

Wet paper codes find numerous and sometimes quite unexpected applications in steganography. A nice example is the surprising ZZW construction [257], using which one can build new families of very efficient matrix embedding codes from existing codes. This construction is important because the new codes follow the upper bound on embedding efficiency.

Let $\mathcal{C}_0$ be a code (not necessarily linear) of length n that can embed m bits in n pixels using on average R_a changes. We will say that $\mathcal{C}_0$ is (R_a, n, m). The ZZW construction leads to a family of codes $\mathcal{C}_p$ that are $(R_\mathrm{a}, n2^p, m + pR_\mathrm{a})$, $p \geq 1$.

Consider a single cover block with $n2^p$ pixels. The following procedure, schematically depicted in Figure 9.3, would be repeated in each block if there is more than one block in the image. Divide the block into n disjoint subsets, each consisting of 2^p pixels. Denote by $\mathbf{x}[i, s]$, $i = 1, \ldots, 2^p$, $s = 1, \ldots, n$, the ith pixel in the sth subset. First, form the XOR of all bits from each subset s,

$$\mathbf{v}[s] = \bigoplus_{i=1}^{2^p} \mathbf{x}[i, s], \ s = 1, \ldots, n. \tag{9.12}$$

Then, using $\mathcal{C}_0$ embed m bits in $\mathbf{v}$ considering $\mathbf{v}$ as some fictitious cover image. This embedding will require changing r bits of $\mathbf{v}$ or, equivalently, changing the XOR of the corresponding subsets $s_1, s_2, \ldots, s_r \in \{1, \ldots, n\}$. On average, there will be R_a such subsets, or $E[r] = R_\mathrm{a}$. To change the XOR in (9.12), one pixel must be changed in each subset s_i. We will let this one change communicate an additional p bits through binary Hamming codes, which will give us the expected payload $m + pR_\mathrm{a}$ per $n2^p$ pixels for the code $\mathcal{C}_p$. However, because the receiver will not know which subsets communicate this additional payload (the receiver will not know the indices $s_1, \ldots, s_r$), the sender must use wet paper codes, which is the step described next.

Let $\mathbf{H}$ be the $p \times (2^p - 1)$ parity-check matrix of a binary Hamming code $\mathcal{H}_p$ (Section 8.3.1). Compute the syndrome of each subset as $\mathbf{s}^{(s)} = \mathbf{H}\mathbf{x}[., s] \in \{0, 1\}^p$, where $\mathbf{x}[., s] = (\mathbf{x}[1, s], \ldots, \mathbf{x}[2^p - 1, s])$ written as a column vector.[5] Concatenate all these syndromes to one column vector of np bits

$$\mathbf{s}^{(1)}, \mathbf{s}^{(2)}, \ldots, \mathbf{s}^{(n)}. \tag{9.13}$$

Now realize that due to the property of Hamming codes, for each s_i we can arrange that each syndrome $\mathbf{s}^{(s_i)}, i = 1, \ldots, r$, can be changed to an arbitrary syndrome by making at most one change to $\mathbf{x}[1, s], \ldots, \mathbf{x}[2^p - 1, s]$.

Label all p bits of syndromes coming from subsets $s_1, \ldots, s_r$ as dry (which makes in total pr dry bits) and all remaining bits in (9.13) as wet. If there is more than one block in the image, concatenate the vectors (9.13) from all blocks to form one long vector of Lnp bits, where L is the number of blocks. This vector will have $p(r_1 + \cdots + r_L)$ dry bits or on average $E[p(r_1 + \cdots + r_L)] = LR_{\mathrm{a}}p$ dry bits. Now form the random sparse matrix $\mathbf{D}$ with Lnp columns and $p(r_1 + \cdots + r_L)$ rows so that its columns follow the RSD as described in Section 9.2. Thus, using wet paper codes, we can communicate on average LpR_{a} message bits in the whole image (plus Lm bits embedded using $\mathcal{C}_0$ in each block). If the wet paper code dictates that a syndrome $\mathbf{s}^{(s)}$ be changed to $\mathbf{s}' \neq \mathbf{s}^{(s)}$, we can arrange for this by modifying exactly one bit in the corresponding vector of bits $\mathbf{x}[1, s], \ldots, \mathbf{x}[2^p - 1, s]$. If no change in the syndrome $\mathbf{s}^{(s)}$ is needed, all bits $\mathbf{x}[1, s], \ldots, \mathbf{x}[2^p - 1, s]$ must stay unchanged. But, because we still need to change the XOR of all bits $\mathbf{x}[1, s], \ldots, \mathbf{x}[2^p, s]$ in (9.12), we simply flip the 2^pth bit $\mathbf{x}[2^p, s]$ because this bit was put aside and does not participate in the syndrome calculation.

To summarize, we embed in each block of $n2^p$ cover elements $m + pR_{\mathrm{a}}$ bits using on average R_{a} changes. We can also say that the relative payload

$$\alpha_p = \frac{m + pR_{\mathrm{a}}}{n2^p} \tag{9.14}$$

can be embedded with embedding efficiency

$$e_p = \frac{m + pR_{\mathrm{a}}}{R_{\mathrm{a}}} = p + \frac{m}{R_{\mathrm{a}}}. \tag{9.15}$$

This newly constructed family of codes $\mathcal{C}_p$ has one important property. With increasing p, the embedding efficiency e_p follows the upper bound on embedding efficiency in the sense that the limit is finite [79],

$$\lim_{p \to \infty} e_p - \frac{\alpha_p}{H^{-1}(\alpha_p)} = \frac{1}{\log 2} - \frac{m}{R_{\mathrm{a}}} + \log_2 \frac{n}{R_{\mathrm{a}}} = \Delta(R_{\mathrm{a}}, n, m). \tag{9.16}$$

---

[5] Note that we are reserving the last element from each subset $\mathbf{x}[2^p, s]$ to be used later.

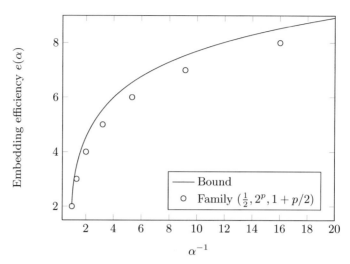

Figure 9.4 Embedding efficiency of codes from the family $\left(\frac{1}{2}, 2^p, 1 + p/2\right)$ for $p = 0, \ldots, 6$.

By inspecting this construction for the trivial embedding method that embeds 1 bit in 1 pixel using on average $\frac{1}{2}$ change, or the $\left(\frac{1}{2}, 1, 1\right)$ code, we discover something truly remarkable. The family of codes is $\left(\frac{1}{2}, 2^p, 1 + p/2\right)$ and its embedding efficiency for various values of p is shown in Figure 9.4. Surprisingly, this family outperforms all known matrix embedding schemes constructed from structured codes (both linear and non-linear) [257]. Extensions of codes that use random constructions [68] (also see Section 8.3 and Section 8.5) lead to even better code families.

9.4.7   Removing shrinkage from F5 algorithm (nsF5)

The F5 algorithm is a steganographic method for JPEG images. Its bit-assignment function is $\pi(x) = \text{LSB}(x)$ for $x \geq 0$ and $\pi(x) = 1 - \text{LSB}(x)$ for $x < 0$ while the embedding operation always decreases the absolute value of DCT coefficients.[6] The F5 algorithm also incorporates matrix embedding using binary Hamming codes.

When a DCT coefficient is changed to 0, which can happen only when the original coefficient value was 1 or -1, so-called shrinkage occurs. Because the recipient extracts the message only from non-zero coefficients, when shrinkage occurs, the sender keeps the embedding change and embeds the same payload again to prevent the recipient from losing a portion of the payload. This, however, decreases the embedding efficiency and the decrease is far from negligible because 1 and -1 are the most frequently occurring non-zero DCT coefficients.

---

[6] As explained in Chapter 7, this operation minimizes the total embedding impact in some well-defined sense.

The shrinkage presents a problem to the recipient only because he is unable to distinguish whether a DCT coefficient was changed to zero during embedding or was already equal to zero in the cover image. This is yet another example of a non-shared selection channel and, as such, it can be solved using wet paper codes. Because, coincidentally, the embedding efficiency of wet paper codes implemented using random linear codes with codimension $p = 18$ (Figure 9.2) is close to the embedding efficiency of binary Hamming codes (Figure 8.3), wet paper codes allow us to increase embedding efficiency by eliminating the adverse effect of shrinkage. This version of the F5 algorithm is called nsF5 (no-shrinkage F5).

We now provide a brief sketch of the implementation of both the embedding and the extraction algorithm. Let us assume that we want to embed a payload of m bits in a JPEG image consisting of n DCT coefficients out of which $n_{01} \geq m$ are non-zero and thus changeable. First, all coefficients (including zeros) are divided using the stego key into m/p randomly generated blocks. Each block will have exactly $n/(m/p) = np/m$ DCT coefficients out of which on average $n_{01}p/m$ will be non-zero and thus changeable. The relative payload is $\alpha = m/n_{01}$. The sender forms a random binary matrix $\mathbf{D}$ of dimensions $p \times np/m$ as described in Section 9.3. The sender now applies the method of Section 9.3 with random codes of codimension $p = 18$.

The recipient uses the stego key and divides all DCT coefficients in the stego image into the same blocks as the sender. He also generates the same random binary matrix $\mathbf{D}$ as the sender. A segment of p message bits is extracted as the syndrome $\mathbf{D}\mathbf{y}$ from the LSBs, $\mathbf{y}$, of all DCT coefficients in the block.

The improvement in embedding efficiency has a dramatic impact on the security of the F5 algorithm (see Table 12.8).

Summary

- When the placement of embedding changes is not shared with the recipient, we speak of a non-shared selection channel. Other synonyms are writing on wet paper or writing to memory with defective cells.
- There exist numerous situations in steganography when non-shared channels arise, such as in minimum-impact steganography, adaptive steganography, and public-key steganography.
- Communication using non-shared channels can be realized using syndrome codes also called wet paper codes.
- Syndrome codes with random matrices are capable of asymptotically communicating the maximum possible payload but have complexity cubic in the message length m.
- Sparse codes with robust soliton distribution of ones in their columns are also asymptotically optimal and can be implemented with complexity $O(n \log m)$, where n is the number of cover elements.

- Wet paper codes with improved embedding efficiency can be implemented using random linear codes with small codimension.

Exercises

9.1 [Writing in memory with one stuck cell] When the number of stuck cells is 1 or $n - 1$, it is easy to develop algorithms for writing in memory with defective cells. For $n - 1$ stuck cells, or $k = 1$, we can write one bit into the memory simply by adjusting it so that the message bit $\mathbf{m}[1]$ is equal to XOR of all bits in the memory.

The complementary case, when there is one stuck cell, is a little more complicated. Let $\mathbf{m}[i]$, $i = 1, \ldots, n - 1$, be the message to be written in the memory. If the stuck cell is the jth cell, $j > 1$, we write $\mathbf{m}$ into the memory in the following manner. If $\mathbf{m}[j - 1] = \mathbf{x}[j]$, we can write the message because the defect is compatible with the message. We write $(\mathbf{m}[1], \mathbf{m}[2], \ldots, \mathbf{m}[n - 1])$ into cells $(2, 3, \ldots, n)$ and we write 1 into $\mathbf{x}[1]$. If $\mathbf{m}[j - 1] \neq \mathbf{x}[j]$, we write the negation of $\mathbf{m}$, $1 - \mathbf{m}$, into cells $(2, 3, \ldots, n)$ and we write 0 into $\mathbf{x}[1]$.

If the stuck cell is the first cell, $j = 1$, we write the message into cells $(2, 3, \ldots, n)$ as is (if the stuck bit $\mathbf{x}[1] = 1$) and we write its negation into the same cells if the stuck bit $\mathbf{x}[1] = 0$. Convince yourself that the reading device can always read the message correctly using the following rule:

$$\text{if } \mathbf{x}[1] = 1, \text{ read the message as } (\mathbf{x}[2], \mathbf{x}[3], \ldots, \mathbf{x}[n]), \quad (9.17)$$

$$\text{if } \mathbf{x}[1] = 0, \text{ read the message as } (1 - \mathbf{x}[2], 1 - \mathbf{x}[3], \ldots, 1 - \mathbf{x}[n]). \quad (9.18)$$

9.2 [Rank of a random matrix] Show that the probability, $\mathbf{R}[k]$, that a randomly generated $k \times k$ binary matrix is of full rank is

$$\mathbf{R}[k] = \prod_{i=1}^{k} \left(1 - \frac{1}{2^i}\right) \to 0.2889\ldots, \text{ as } k \to \infty. \quad (9.19)$$

Note that here the rank should be computed in binary arithmetic. The rank of a binary matrix computed in binary arithmetic and the rank computed in real arithmetic may be different.

Hint: Use induction with respect to i for an $i \times k$ matrix.

9.3 [Minimum embedding impact I] Let the cover image have n pixels with embedding impact $\rho[i]$ and assume that the elements are already sorted so that $\rho[i]$ is non-decreasing, $\rho[i] \leq \rho[i + 1]$. In order to embed m bits, the sender marks pixels $1, \ldots, k$, $m \leq k$, as changeable and embeds the message into changeable pixels using binary wet paper codes with the best theoretically possible embedding efficiency $e = \alpha/H^{-1}(\alpha)$, where $\alpha = m/k$. Show that the value of k that minimizes the total embedding impact is

$$k_{\text{opt}} = \arg \min_{m \leq k \leq n} H^{-1}\left(\frac{m}{k}\right) \sum_{i=1}^{k} \rho[i]. \quad (9.20)$$

Hint: According to Table 8.4, the embedding will introduce $kH^{-1}(m/k)$ changes and the expected impact per changed pixel is $(1/k)\sum_{i=1}^{k}\rho[i]$.

9.4 **[Minimum embedding impact II]** Fix the ratio $\beta = m/n$ and consider (9.20) in the limit for $n \to \infty$. Assume that $\rho[i] = \rho(i/n)$ for some non-decreasing function (profile) ρ. Define $x = k/n$ and show that the optimization problem (9.20) becomes

$$x_{\text{opt}} = \frac{k_{\text{opt}}}{n} = \arg\min_{\beta \le x \le 1} H^{-1}\left(\frac{\beta}{x}\right) R(x), \tag{9.21}$$

where

$$R(x) = \int_0^x \rho(t)\mathrm{d}t. \tag{9.22}$$

Moreover, by differentiating show that x_{opt} is a solution of the following algebraic equation:

$$\frac{\rho(x)}{R(x)} = \frac{\beta}{\beta x + x^2 \log_2\left(1 - H^{-1}(\beta/x)\right)}. \tag{9.23}$$

Finally, show that $x = \beta$ is *never* a solution for any profile ρ. Put another way, it is always better to reserve more pixels as changeable than the m pixels with the smallest impact.
Hint:

$$\frac{\mathrm{d}H^{-1}(x)}{\mathrm{d}x} = \left(\log_2 \frac{1 - H^{-1}(x)}{H^{-1}(x)}\right)^{-1}. \tag{9.24}$$

9.5 **[Minimal embedding impact III]** Show that $x_{\text{opt}} = 1$ when $\beta \ge \beta_0$, where β_0 is the unique solution to

$$\frac{1}{R(1)} = \frac{\beta}{\beta^2 + \log_2\left(1 - H^{-1}(\beta)\right)}. \tag{9.25}$$

In other words, for any profile for a sufficiently large message it is always better to use *all* pixels ($k = n$).

9.6 **[Minimal embedding impact IV]** Investigate the profile $\rho(x) = x^{\gamma}$, $\gamma > 0$. Show that

$$x_{\text{opt}} = \begin{cases} c(\gamma)\beta & \text{for } \beta \le \beta_0 \\ 1 & \text{for } \beta > \beta_0, \end{cases} \tag{9.26}$$

where $c(\gamma)$ is the solution to

$$c \log\left(1 - H^{-1}\left(\frac{1}{c}\right)\right) = -\frac{\gamma}{\gamma + 1}. \tag{9.27}$$

In particular, the profile $\rho(x)$ for perturbed quantization is $\rho(x) = x$, because the quantization error is approximately uniformly distributed. Show that for this profile $c = 1.254$, obtaining thus the following rule of thumb: In perturbed-quantization steganography implemented using wet paper codes with optimal embedding efficiency, it is better to use $k \approx m + m/4$ rather than $k = m$.

10 Steganalysis

In the prisoners' problem, Alice and Bob are allowed to communicate but all messages they exchange are closely monitored by warden Eve looking for traces of secret data that may be hidden in the objects that Alice and Bob exchange. Eve's activity is called steganalysis and it is a complementary task to steganography. In theory, the steganalyst is successful in attacking the steganographic channel (i.e., the steganography has been broken) if she can distinguish between cover and stego objects with probability better than random guessing. Note that, in contrast to cryptanalysis, it is not necessary to be able to read the secret message to break a steganographic system. The important task of extracting the secret message from an image once it is known to contain secretly embedded data belongs to forensic steganalysis.

In Section 10.1, we take a look at various aspects of Eve's job depending on her knowledge about the steganographic channel. Then, in Section 10.2 we formulate steganalysis as a problem in statistical signal detection. If Eve knows the steganographic algorithm, she can accordingly target her activity to the specific stegosystem, in which case we speak of targeted steganalysis (Section 10.3). On the other hand, if Eve has no knowledge about the stegosystem the prisoners may be using, she is facing the significantly more difficult problem of blind steganalysis detailed in Section 10.4 and Chapter 12. She now has to be ready to discover traces of an arbitrary stegosystem. Both targeted and blind steganalysis work with one or more numerical features extracted from images and then classify them into two categories – cover and stego. For targeted steganalysis, these features are usually designed by analyzing specific traces of embedding, while in blind steganalysis the features' role is much more ambitious as their goal is to completely characterize cover images in some low-dimensional feature space. Blind steganalysis has numerous alternative applications, which are discussed in Section 10.5.

The reliability of steganalysis is strongly influenced by the source of covers. The prisoners may choose a source that will better mask their embedding and, on the other hand, they may also make fatal errors and choose covers so improperly that Eve will be able to detect even single-bit messages. The influence of covers on steganalysis is discussed in Section 10.6.

Although this chapter and this book focus on steganalysis based on analyzing statistical anomalies of pixel values that most steganographic algorithms leave behind, Eve may utilize other auxiliary information available to her. For example, she can inspect file headers or run brute-force attacks on the stego/encryption keys, hoping to reveal a meaningful message when she runs across a correct key. Such attacks are called system attacks (Section 10.7) and belong to forensic steganalysis (Section 10.8).

The main purpose of this chapter is to prepare the reader for Chapters 11 and 12, where specific steganalysis methods are described. Readers not familiar with the subject of statistical hypothesis testing would benefit from reading Appendix D before continuing.

10.1   Typical scenarios

Any information about the steganographic channel that is a priori available to Eve can aid her in mounting an attack. Recall from Chapter 4 that the steganographic channel consists of five basic elements:

- channel used to exchange data between Alice and Bob,
- cover source,
- message source,
- data-embedding and -extraction algorithms,
- source of stego keys driving the embedding/extraction algorithms.

In this book, with a few exceptions, we make an assumption that the warden is passive, which means that she passively observes the communication and does not interfere with it in any way. Thus, the physical channel used to exchange information is lossless and has no impact on steganalysis or steganography. The other four elements, the embedding algorithm and the source of covers, messages, and stego keys, however, are crucial for Eve. In general, the more information she has, the more successful she will be in detecting the presence of secretly embedded messages. Although there certainly exist many different situations with various levels of detail available to Eve about the individual elements of the steganographic channel, we highlight two typical and very different scenarios – the case of traffic monitoring and analysis of a seized computer.

An automatic traffic-monitoring device is an algorithm that analyzes every image passing through a certain network node. An example would be a program that inspects every image posted on a binary discussion group or a server monitoring all traffic through a specific Internet node. In this case, the warden has limited information about the cover source, the message source, or the stegosystem. This is the most difficult situation for Eve and one where Kerckhoffs' principle does not fully apply. Consequently, Eve needs steganalysis algorithms capable of detecting as wide a spectrum of steganographic schemes as possible.

If the set of parties communicating using steganography is narrowed down using some side-information, the diversity of the cover source may become lower, giving Eve a better chance to detect any steganographic activities. For example, when Eve already suspects that somebody may be using steganography, additional intelligence may be gathered that may provide some information about the cover source or the steganographic algorithm. Say, if Alice downloads a steganographic tool while being eavesdropped, Eve suddenly obtains prior information about the embedding algorithm, as well as the stego key space. Or, if Eve knows that Alice sends to Bob images obtained using her camera, Eve could purchase a camera of the same model and tailor her attack to this cover source.

Steganalysis may become significantly easier when a suspect's computer is seized and Eve's task is to steganalyze images on the hard disk. In this case, the stego tool may still reside on the computer or its traces may be recoverable even after it has been uninstalled (e.g., using Wetstone's Gargoyle `http://www.wetstonetech.com`). This gives her valuable prior information about the potential stego channel. In some situations, Eve may be able to find multiple versions of one image that are nearly identical. She can first investigate whether these slightly different versions are the result of steganography or some other natural process, such as compressing one image using two different JPEG compressors [95]. By comparing the images, she can learn about the nature of embedding changes and their placement (selection channel) and possibly conclude that the changes are due to embedding a secret message. Once Eve determines that steganography is taking place, the steganalysis is complete and she may continue with forensic steganalysis aimed at extracting the secret message itself or she may decide to interrupt the communication channel if it is within her competence.

10.2   Statistical steganalysis

In this section, you will learn that statistical steganalysis is a detection problem. In practice, it is typically achieved through some simplified model of the cover source obtained by representing images using a set of numerical features. Depending on the scope of the features, we recognize two major types of statistical steganalysis – targeted and blind. The following sections describe several basic strategies for selecting appropriate features for both targeted and blind steganalysis. Examples of specific targeted and blind methods are postponed to Chapters 11 and 12.

Before proceeding with the formulation of statistical steganalysis, we briefly review some basic facts from Chapter 6. The information-theoretic definition of steganographic security starts with the basic assumption that the cover source can be described by a probability distribution, P_c, on the space of all possible cover images, $\mathcal{C}$. The value $P_c(\mathcal{B}) = \int_{\mathcal{B}} P_c(\mathbf{x}) d\mathbf{x}$ is the probability of selecting cover $\mathbf{x} \in \mathcal{B} \subset \mathcal{C}$ for hiding a message. Assuming that a given stegosystem assumes on its input covers $\mathbf{x} \in \mathcal{C}, \mathbf{x} \sim P_c$, stego keys, and messages (both attaining values on their sets according to some distributions), the distribution of stego images is P_s.

10.2.1   Steganalysis as detection problem

Let us assume that Eve can collect sufficiently many cover images and estimate the pdf P_c. If her knowledge of the steganographic channel allows her to estimate the pdf P_s, or learn to distinguish between P_c and P_s using machine-learning tools, we speak of steganalysis of a *known stegosystem,* which is a detection problem that leads to simple hypothesis testing

$$H_0 : \quad \mathbf{x} \sim P_c, \qquad (10.1)$$
$$H_1 : \quad \mathbf{x} \sim P_s. \qquad (10.2)$$

Optimal Neyman–Pearson and Bayesian detectors can be derived for this problem using the likelihood-ratio test.

If Eve has no information about the stego system, the steganalysis problem becomes

$$H_0 : \quad \mathbf{x} \sim P_c, \qquad (10.3)$$
$$H_1 : \quad \mathbf{x} \nsim P_c, \qquad (10.4)$$

which is a composite hypothesis-testing problem and is in general much more complex.

Obviously, there are many other possibilities that fall in between these two formulations depending on the information about the steganographic channel available to Eve. For example, Eve may know the embedding algorithm but not the message source. Then, P_s will depend on an unknown parameter, the change rate β, and Eve faces the composite one-sided hypothesis-testing problem

$$H_0 : \quad \beta = 0, \qquad (10.5)$$
$$H_1 : \quad \beta > 0. \qquad (10.6)$$

Eve may employ appropriate tools, such as the generalized likelihood-ratio test, or convert this problem to simple hypothesis testing by considering β as a random variable and making an assumption about its distribution (the Bayesian approach).

10.2.2   Modeling images using features

Practical steganalysis of multimedia objects, such as digital images, cannot work with the full representation of images due to their large complexity and dimensionality. Instead, a simplified model whereby the detection problem becomes more tractable must be accepted. The models used in steganalysis are usually obtained by representing images using a set of numerical features. Each image, $\mathbf{x} \in \mathcal{C}$, is mapped to a d-dimensional feature vector $\mathbf{f} = (f_1(\mathbf{x}), \ldots, f_d(\mathbf{x})) \in \mathbb{R}^d$, where each $f_i : \mathcal{C} \to \mathbb{R}$. The random variables representing the cover source, $\mathbf{x} \sim P_c$, and the stego images, $\mathbf{y} \in P_s$, are thus transformed into the corresponding random variables $\mathbf{f}(\mathbf{x}) \sim p_c$ and $\mathbf{f}(\mathbf{y}) \sim p_s$ on $\mathbb{R}^d$. For accurate detection, the features need to be chosen so that the clusters of $\mathbf{f}(\mathbf{x})$ and $\mathbf{f}(\mathbf{y})$ have as little overlap as possible.

10.2.3   Optimal detectors

Any steganalysis algorithm is a detector, which can be described by a map F : $\mathbb{R}^d \rightarrow \{0, 1\}$, where $F(\mathbf{x}) = 0$ means that $\mathbf{x}$ is detected as cover, while $F(\mathbf{x}) = 1$ means that $\mathbf{x}$ is detected as stego. The set $\mathcal{R}_1 = \{\mathbf{x} \in \mathbb{R}^d | F(\mathbf{x}) = 1\}$ is called the critical region because the detector decides "stego" if and only if $\mathbf{x} \in \mathcal{R}_1$. The critical region fully describes the detector.

The detector may make two types of error – false alarms and missed detections. The probability of a false alarm, P_{FA}, is the probability that a random variable distributed according to p_{c} is detected as stego, while the probability of missed detection, P_{MD}, is the probability that a random variable distributed according to p_{s} is incorrectly detected as cover:

$$P_{\mathrm{FA}} = \Pr\{F(\mathbf{x}) = 1 | \mathbf{x} \sim p_{\mathrm{c}}\} = \int_{\mathcal{R}_1} p_{\mathrm{c}}(\mathbf{x}) \mathrm{d}\mathbf{x}, \tag{10.7}$$

$$P_{\mathrm{MD}} = \Pr\{F(\mathbf{x}) = 0 | \mathbf{x} \sim p_{\mathrm{s}}\} = 1 - \int_{\mathcal{R}_1} p_{\mathrm{s}}(\mathbf{x}) \mathrm{d}\mathbf{x}. \tag{10.8}$$

As derived in Section 6.1.1, the error probabilities of any detector must satisfy the following inequality:

$$(1 - P_{\mathrm{FA}}) \log \frac{1 - P_{\mathrm{FA}}}{P_{\mathrm{MD}}} + P_{\mathrm{FA}} \log \frac{P_{\mathrm{FA}}}{1 - P_{\mathrm{MD}}} \leq D_{\mathrm{KL}}(p_{\mathrm{c}} || p_{\mathrm{s}}) \leq D_{\mathrm{KL}}(P_{\mathrm{c}} || P_{\mathrm{s}}), \tag{10.9}$$

where D_{KL} is the Kullback–Leibler divergence between the two probability distributions. For secure steganography, $D_{\mathrm{KL}}(P_{\mathrm{c}} || P_{\mathrm{s}}) = 0$, and the only detector that Eve can build is one that randomly guesses between cover and stego, which means that Eve cannot construct an attack. For ϵ-secure stegosystems, $D_{\mathrm{KL}}(P_{\mathrm{c}} || P_{\mathrm{s}}) \leq \epsilon$, and the reliability of detection decreases with decreasing ϵ (see Figure 6.1).

In steganography, useful detectors must have a low probability of a false alarm. This is because images detected as potentially containing secret messages are likely to be subjected to further forensic analysis (see Section 10.8) to determine the steganographic program, the stego key, and eventually extract the secret message. This may require brute-force dictionary attacks that can be quite expensive and time-consuming. Thus, it is more valuable to have a detector with very low P_{FA} even though its probability of missed detection may be quite high (e.g., $P_{\mathrm{MD}} > 0.5$ or higher). Because steganographic communication is typically repetitive, even a detector with $P_{\mathrm{MD}} = 0.5$ is still quite useful as long as its false-alarm probability is small.

The hypothesis-testing problem in steganalysis is thus almost exclusively formulated using the Neyman–Pearson setting, where the goal is to construct a detector with the highest detection probability $P_{\mathrm{D}} = 1 - P_{\mathrm{MD}}$ while imposing a bound on the probability of false alarms, $P_{\text{FA}} \leq \epsilon_{\text{FA}}$. Even though it is possible to associate cost with both types of error, Bayesian detectors are typically not used in steganalysis, because the prior probabilities of encountering a cover or stego image can rarely be accurately estimated.

Given the bound on false alarms, ϵ_{FA}, the optimal Neyman–Pearson detector is the likelihood-ratio test

$$\text{Decide H}_1 \quad \text{when} \quad L(\mathbf{x}) = \frac{p_{\text{s}}(\mathbf{x})}{p_{\text{c}}(\mathbf{x})} > \gamma, \tag{10.10}$$

where $\gamma > 0$ is a threshold determined from the condition

$$P_{\text{FA}} = \int_{\mathcal{R}_1} p_{\text{c}}(\mathbf{x})\mathrm{d}\mathbf{x} = \epsilon_{\text{FA}}, \tag{10.11}$$

where

$$\mathcal{R}_1 = \{\mathbf{x} \in \mathbb{R}^d | L(\mathbf{x}) > \gamma\} \tag{10.12}$$

is the critical region of the detector. The ratio $L(\mathbf{x})$ is called the likelihood ratio.

Even though in theory Eve could construct optimal steganalysis algorithms by estimating the pdfs of cover/stego images and use the likelihood-ratio test, this can be done only for low-dimensional models of covers that may not describe covers well. Because the dimensionality of the feature spaces is often large, which is especially true for blind steganalysis, one can rarely estimate the underlying probability distributions accurately. Instead, the detection problem is viewed as a classification (pattern-recognition) problem and solved by training a classifier on a large database of cover and stego images. The classifier parameters are typically adjusted to obtain a low probability of false alarms.

10.2.4   Receiver operating characteristic (ROC)

For a given detector, the function $P_{\text{D}}(P_{\text{FA}})$ is called the Receiver-Operating-Characteristic (ROC) curve and it describes the performance of the detector. A few examples of ROC curves are given in Figure 10.1. The ROC curve of a poor steganalysis method is close to the diagonal line (Figure 10.1(b)), while good steganalysis methods have ROC curves that are close to the curve depicted in Figure 10.1(c).

Comparing detectors really means comparing their ROC curves. While some detectors may be unambiguously compared because their ROCs satisfy $P_{\text{D}}^{(1)}(P_{\text{FA}}) > P_{\text{D}}^{(2)}(P_{\text{FA}})$ for all $P_{\text{FA}} \in [0, 1]$ (e.g., Detector 1 is more reliable than Detector 2 for all P_{FA}), in general ROCs can intersect and then it is not clear which detector is better. An example of two curves that are hard to compare is in Figure 10.1(d). This problem can be avoided if we could extract a scalar measure of performance from each ROC because scalars can be unambiguously ordered. Among the different measures of performance that were proposed by various researchers, we name the following three:

- The area, ρ, between the ROC curve and the diagonal line normalized so that $\rho = 0$ when the ROC coincides with the diagonal line and $\rho \approx 1$ for ROC curves corresponding to nearly perfect detectors (Figure 10.1(c)). Mathematically,

$$\rho = 2 \int_0^1 P_D(x) dx - 1. \tag{10.13}$$

This quantity is sometimes called accuracy. A different but closely related measure is Area Under Curve (AUC) or Area Under ROC (AUR), which is simply the area under the ROC curve

$$\text{AUC} = \int_0^1 P_D(x) dx = \frac{1 + \rho}{2}. \tag{10.14}$$

- The minimal total average decision error under equal prior probabilities $(\Pr\{\mathbf{x} \text{ is stego}\} = \Pr\{\mathbf{x} \text{ is cover}\})$

$$P_E = \min_{P_{FA} \in [0,1]} \frac{1}{2} \left(P_{FA} + P_{MD}(P_{FA}) \right). \tag{10.15}$$

The minimum is reached at a point where the tangent to the ROC curve has slope $\frac{1}{2}$. In Figure 10.2, this point is marked with a circle.
- The false-alarm rate at probability of detection equal to $P_D = \frac{1}{2}$

$$P_D^{-1} \left(\frac{1}{2} \right). \tag{10.16}$$

This point is marked with a square in Figure 10.2.

None of these quantities is completely satisfactory because of the lack of fundamental reasoning behind them. The value $P_D^{-1}(1/2)$ is probably the most useful for steganalysis because, as explained above, what matters the most is the detector's false-alarm probability. The work of Ker [136] is an attempt to compare steganalysis detectors in a more fundamental manner in the limit as the change rate goes to zero.

In practice, the underlying distributions p_c and p_s are usually obtained experimentally in a sampled form. For one-dimensional distributions over the set of real numbers, p_s is usually a shifted version of p_c and both distributions are unimodal. In this case, the critical region is determined by a scalar threshold γ (see (10.10) and (10.12)). Thus, the ROC can be drawn either by fitting a parametric model through the sample distribution or directly by moving the threshold γ from $-\infty$ to $+\infty$ and computing the relative fraction of cover and stego features above γ (see Algorithm 10.1).

In the next section, we discuss methods for constructing the feature spaces. The specific form of the feature map $\mathbf{f}$ and the type of the hypothesis-testing problem depend on the steganalysis scenario and the steganographic channel.

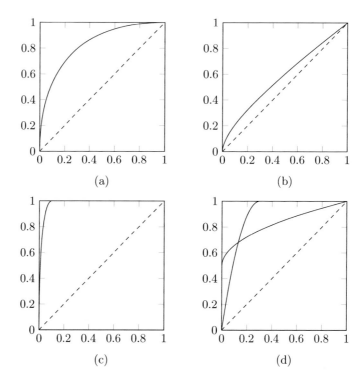

Figure 10.1 Examples of ROC curves. The x and y axes in all four graphs are the probability of false alarms, P_{FA}, and the probability of detection, P_D, respectively. (a) Example of an ROC curve; (b) ROC of a poor detector; (c) ROC of a very good detector; (d) two hard-to-compare ROCs.

Algorithm 10.1 Drawing an ROC curve for one-dimensional features $\mathbf{f}_s[i]$ (stego) and $\mathbf{f}_c[i]$ (cover), $i = 1, \ldots, k$, computed from k cover and k stego images.

```
f = sort(f_s ∪ f_c);
// f is the set of all features sorted to form a non-decreasing sequence
P_FA[0] = 1; P_D[0] = 1;
for i = 1 to 2k {
    if f[i] ∈ f_c  {P_FA[i] = P_FA[i − 1] − 1/k; P_D[i] = P_D[i − 1];}
    else        {P_D[i] = P_D[i − 1] − 1/k; P_FA[i] = P_FA[i − 1];}
    DrawLine((P_FA[i − 1], P_D[i − 1]), (P_FA[i], P_D[i]));
}
```

We distinguish two general cases: attacking a known steganographic method (or embedding operation) and attacking an unknown steganographic method. The corresponding approaches in steganalysis are called targeted and blind steganalysis.

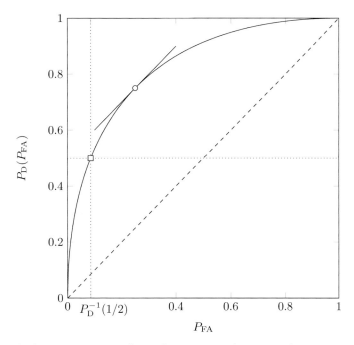

Figure 10.2 Scalar measures typically used to compare detectors. The square marks the point on the ROC curve corresponding to the criterion $P_\mathrm{D}^{-1}(1/2)$. The point marked with a circle corresponds to the minimal total decision error P_E.

10.3   Targeted steganalysis

The features in targeted steganalysis are constructed from knowledge of the embedding algorithm and are thus targeted to a specific embedding method or embedding operation (e.g., LSB embedding). On the other hand, features in blind steganalysis must be constructed in such a manner as to be able to detect every possible steganographic scheme, including future schemes. While a targeted steganalysis method can work very well with a single scalar feature, blind steganalysis methods often require larger sets of features and are usually implemented using machine-learning. In principle, however, both targeted and blind steganalysis can use multiple features and tools from machine-learning and pattern recognition. The main difference between them is the *scope* of their feature sets.

10.3.1   Features

We now describe several strategies for constructing features for targeted steganalysis and illustrate them with specific examples.

Because in targeted steganalysis the embedding mechanism of the stego system is known, it makes sense to choose as features the quantities that predictably change with embedding. While it is usually relatively easy to identify many such features, features that are especially useful are those that attain known values on either stego or cover images.

In the expressions below, β is the change rate (ratio between the number of embedding changes and the number of all elements in the cover image) and $\mathbf{f}_\beta$ is the feature computed from a stego image obtained by changing the ratio of β of its corresponding cover elements.

T1. **[Testing for stego artifacts]** Identify a feature that attains a specific known value, $\mathbf{f}_\beta$, on stego images and attains other, different values on cover images. Then, formulate a composite hypothesis-testing problem as

$$\mathrm{H}_0 :\quad \mathbf{f} = \mathbf{f}_\beta, \tag{10.17}$$

$$\mathrm{H}_1 :\quad \mathbf{f} \neq \mathbf{f}_\beta. \tag{10.18}$$

Note that here H_0 is the hypothesis that the image under investigation is stego, while H_1 stands for the hypothesis that it is cover.

T2. **[Known cover property]** Identify a feature $\mathbf{f}$ that predictably changes with embedding $\mathbf{f}_\beta = \Phi(\mathbf{f}_0; \beta)$ so that Φ can be inverted, $\mathbf{f}_0 = \Phi^{-1}(\mathbf{f}_\beta; \beta)$. Assuming $F(\mathbf{f}_0) = \mathbf{0}$ for some known function $F : \mathbb{R}^d \to \mathbb{R}^k$, estimate $\hat{\beta}$ from $F(\Phi^{-1}(\mathbf{f}_\beta; \hat{\beta})) = \mathbf{0}$ and test

$$\mathrm{H}_0 :\quad \hat{\beta} = 0, \tag{10.19}$$

$$\mathrm{H}_1 :\quad \hat{\beta} > 0. \tag{10.20}$$

Note that now we have the more typical case when the hypothesis H_0 stands for cover and H_1 for stego. Also, notice that a by-product of this approach is an estimate of the change rate, which can usually be easily related to the relative payload, α. For example, for methods that embed each message bit at one cover element, $\beta = \alpha/2$.

T3. **[Calibration]** In some cases, it is possible to estimate from the stego image what the value of a feature would be if it were computed from the cover image. This process is called calibration. Let $\mathbf{f}_\beta$ be the feature computed from the stego image and $\hat{\mathbf{f}}_0$ be the estimate of the cover feature. If the embedding allows expressing $\mathbf{f}_\beta$ as a function of $\mathbf{f}_0$ and β, $\mathbf{f}_\beta = \Phi(\mathbf{f}_0; \beta)$, we can again estimate β from $\mathbf{f}_\beta = \Phi(\hat{\mathbf{f}}_0; \hat{\beta})$ and again test

$$\mathrm{H}_0 :\quad \hat{\beta} = 0, \tag{10.21}$$

$$\mathrm{H}_1 :\quad \hat{\beta} > 0. \tag{10.22}$$

This method also provides an estimate of the change rate β (or the message payload α).

We now give examples of these strategies. Strategy T1 was pursued when deriving the histogram attack in Section 5.1.1. The attack was based on the observation that the histogram $\mathbf{h}$ of an 8-bit grayscale image fully embedded with LSB embedding must satisfy

$$\mathbf{h}[2k] \approx \mathbf{h}[2k+1], \quad k = 0, \ldots, 127. \tag{10.23}$$

Because the sum $\mathbf{h}[2k] + \mathbf{h}[2k+1]$ is invariant with respect to LSB embedding, we can take as a feature vector the histogram $\mathbf{h}$ and formulate the following composite hypothesis-testing problem:

$$H_0: \quad \mathbf{h}[2k] = \frac{\mathbf{h}[2k] + \mathbf{h}[2k+1]}{2}, \quad k = 0, \ldots, 127, \tag{10.24}$$

$$H_1: \quad \mathbf{h}[2k] \neq \frac{\mathbf{h}[2k] + \mathbf{h}[2k+1]}{2}, \quad k = 0, \ldots, 127, \tag{10.25}$$

which leads to the histogram attack from Section 5.1.1, where this problem was approached using Pearson's chi-square test.

As an example of Strategy T2, we revisit the attack on Jsteg from Section 5.1.2. There, we used some a priori knowledge about the cover image. Using the observation that histograms of cover JPEG images are approximately symmetrical, the following quantity was studied for steganalysis of Jsteg:

$$F(\mathbf{h}) = \sum_{k>0} \mathbf{h}[2k] + \sum_{k<0} \mathbf{h}[2k+1] - \sum_{k\geq0} \mathbf{h}[2k+1] - \sum_{k<0} \mathbf{h}[2k], \tag{10.26}$$

where $\mathbf{h}$ is the histogram of quantized DCT coefficients. Because the embedding mechanism is known, we were able to express the stego image histogram as a function of the relative message length $\alpha = 2\beta$ and the cover-image histogram $\mathbf{h}_0$, $\mathbf{h}_\alpha = \Phi(\mathbf{h}_0; \alpha)$, and invert the relationship $\mathbf{h}_0 = \Phi^{-1}(\mathbf{h}_\alpha; \alpha)$. The symmetry of the cover-image histogram gave us the equation $F(\mathbf{h}_0) = 0$, which was solved for the estimate $\hat{\alpha}$.

A specific example of Strategy T3 that uses calibration will be given in Chapter 11, where we describe a method for attacking the F5 algorithm. In this section, we explain only the basic principle of calibration.

Calibration attempts to estimate selected macroscopic quantities of the cover image from the stego image. This is, indeed, possible for JPEG images as the quantized DCT coefficients are robust with respect to steganographic embedding because the distortion is usually small. Calibration begins by decompressing the stego image to the spatial domain, then cropping the image by 4 columns and 4 rows, and then recompressing again using the same quantization matrix as that of the stego image. The resulting JPEG image is visually similar to the cover image and its quantized DCT coefficients are no longer influenced by steganographic embedding because the JPEG compression was performed on an 8×8 grid shifted by 4 pixels with respect to the grid in the cover image. Thus, the process of recompression on a shifted grid essentially erased the effect of embedding changes. Note that we cannot claim that the recompressed image is an approximation to the cover image because it is compressed on a shifted grid. Nevertheless, it is intuitively clear that macroscopic quantities, such as the histogram, should be approximately equal to those of the cover image. Note that geometrical transformations other than cropping can also be used, for example,

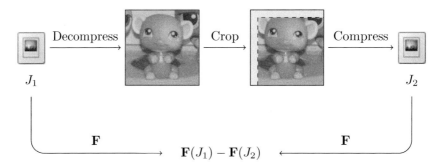

Figure 10.3 Calibration is used to estimate some macroscopic quantities of the cover image from the stego image.

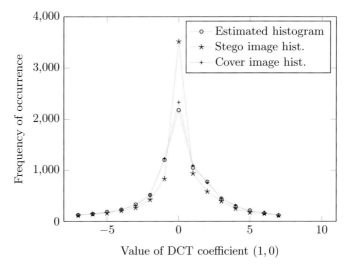

Figure 10.4 Histogram of the DCT coefficient for the spatial frequency $(1,0)$ for the cover image $(+)$, F5 fully embedded stego image $(\star)$, and the calibrated image $(\circ)$.

a slight rotation, resizing, or random warping as performed in the attack on watermarking schemes called Stirmark [151].

In Figure 10.4, we illustrate how accurately calibration works. The figure shows the histogram of DCT coefficients for the spatial frequency $(0,1)$ for the cover image $(+)$, a stego image fully embedded with the F5 algorithm $(\star)$, and the estimate of the cover-image histogram $(\circ)$ obtained using calibration. Clearly, calibration has provided a very close estimate to the original histogram.

In summary, the defining characteristic of targeted steganalysis is that the features are designed by analyzing the embedding mechanism of a specific stegosystem or a specific embedding operation (e.g., LSB embedding in the two cases above). In particular, in targeted schemes the features are designed without any ambitions to obtain an exhaustive representation of cover images in some lower-dimensional space.

10.3.2   Quantitative steganalysis

Many targeted steganalysis methods use as a detection statistic an estimate of the number of embedding changes (or the change rate). Steganalysis designed to estimate the change rate is called quantitative. Quantitative techniques are important in forensic steganalysis (see Section 10.8) because they give Eve additional, quite valuable forensic information. For example, when she detects messages in multiple images and the message-length estimates are clustered around multiples of some typical cipher block lengths, Eve can infer that the message is encrypted and even narrow down the possibilities for the encryption algorithm.

The change-rate estimate provided by quantitative steganalysis is subject to errors because the assumptions under which the estimator was derived were not satisfied. Let us take a closer look at Strategy T2 for targeted steganalysis. There are two different sources of error. The relationship $\mathbf{f}_\beta = \Phi(\mathbf{f}_0; \beta)$ is really an equality between expected values and should have been written more precisely as

$$E\left[\mathbf{f}_\beta\right] = \Phi(\mathbf{f}_0; \beta), \tag{10.27}$$

where the expected value is taken over embeddings inducing change rate β and over pseudo-random walks. The second source of error is the cover assumption $F(\mathbf{f}_0) = 0$. This is, again, an equality in expected value, this time over the covers

$$E[F(\mathbf{f}_0)] = 0. \tag{10.28}$$

The error of the output of many quantitative steganalyzers can be modeled as a realization of a random variable consisting of two factors: the within-image error and the between-image error [22, 23],

$$\hat{\beta} - \beta = \mathsf{w} + \mathsf{b}. \tag{10.29}$$

The within-image error, w, is caused by random correlations between the image and the message and the fact that the pseudo-random walk visits a different part of the image each time we embed. In Strategy T2, we can say more specifically that this error is due to the fact that the equality $\mathbf{f}_\beta = \Phi(\mathbf{f}_0; \beta)$ holds only in expectation. Imagine embedding with the same change rate, β, into the same image but each time along a different pseudo-random walk and with a different message. Repeating this experiment N_{w} times, the variations in the change-rate estimate are due to the within-image error. The within-image error distribution depends on the image content and on the relative payload itself. In a blue-sky picture, the variance of this error will be smaller than in an image containing a variety of different textures and objects. In general, the larger the payload the smaller is the contribution of the image content to this type of error. The within-image error is well modeled with a Gaussian distribution.

The between-image error is a random variable whose distribution is tied to properties of natural images. In Strategy T2, this error is due to the failure of the covers to satisfy $F(\mathbf{f}_0) = 0$. Because the within-image error is Gaussian, we can isolate the realization of the between-image error for a given image by simply averaging all N_w estimates. The distribution of the between-image error is then sampled by repeating this experiment over N_b images. Student's t-distribution is often a good fit for the between-image error.

As an example, we now look more closely at the quantitative steganalyzer for Jsteg (Section 5.1.2). There, Figure 5.4 shows the histograms of estimated payload for five payloads across $N_b = 954$ grayscale images. The histograms show the mixture of both types of error. To separate them, we perform the following experiment with the same database of images. For a fixed change rate $\beta = 0.2$, each image was embedded with Jsteg $N_w = 200$ times with different messages and stego keys (which determine the pseudo-random walk). By running the quantitative steganalyzer (5.25), we obtain a matrix of change-rate estimates $\hat{\beta}[i,j]$, $i = 1, \ldots, N_w$, $j = 1, \ldots, N_b$. The distribution of the between-image error is obtained by taking the average change-rate estimate over N_w embeddings for each image,

$$ \mathbf{b}[j] = \frac{1}{N_w} \sum_{i=1}^{N_w} \hat{\beta}[i,j] - \beta. \tag{10.30} $$

The sample pdf of the between-image error is shown in Figure 10.5. The figure also includes the log–log empirical cdf plot showing the tail probability $\Pr\{\mathbf{b} > x\}$ as a function of x and the corresponding Gaussian fit using a thin line (see Appendix A for more details about the plot). It is apparent that this error exhibits thick tails and the Gaussian model is not a good fit. The plot, however, seems to indicate that Student's t-distribution might be a reasonably good fit because the tail of the experimental data becomes linear for large x in agreement with the model ($\Pr\{\mathsf{x} > x\} \approx x^{-\nu}$ for a random variable x following Student's t-distribution with ν degrees of freedom). The inter-quartile range (IQR) for the between-image error is $[0.198, 0.218]$, suggesting thus that the right tail is thicker than the left tail.

Figure 10.6 shows the within-image error and the log–log empirical cdf plot for the right tail obtained for one randomly chosen image from the database based on $N_w = 10{,}000$ embeddings. The within-image error appears to be well modeled with a Gaussian distribution. Again, this property seems to be generic among other quantitative steganalyzers [23].

We close this section with a few notes about some interesting recent developments in quantitative steganalysis. If the statistical distributions of random variables involved in Strategy T2 can be derived or if at least reasonable assumptions about them can be made, it becomes possible to derive more accurate quantitative estimators using the maximum-likelihood principle,

$$ \hat{\beta} = \arg\max_{\beta} \Pr\{\mathbf{f}_\beta | \beta\}. \tag{10.31} $$

An example of this approach is [133], which is also briefly mentioned in Section 11.1.3.

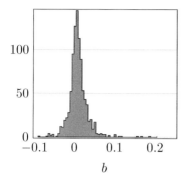 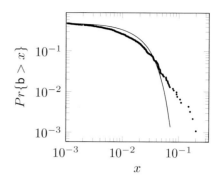

Figure 10.5 Distribution of the between-image error and its log–log empirical cdf plot for the quantitative steganalyzer for Jsteg (5.25). The thin line is a Gaussian fit.

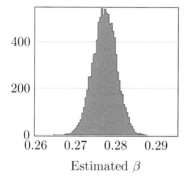 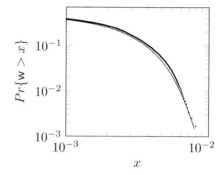

Figure 10.6 Distribution of the within-image error and its log–log empirical cdf plot for one image for the quantitative steganalyzer for Jsteg (5.25). The thin line is a Gaussian fit.

An interesting possibility to construct quantitative steganalyzers from blind steganalyzers was proposed in [195] (also see Section 12.4.1). It turns out that it is possible to learn the relationship between the feature vector used by blind steganalyzers (see the next section) and the change rate using regression methods and obtain as a result a very accurate quantitative steganalyzer for essentially every steganographic method that is detectable using that feature vector. An important advantage of this cookie-cutter approach to quantitative steganalysis is that it can be applied even when the embedding mechanism is completely unknown as long as one can obtain a large set of stego images embedded with messages of known size.

10.4   Blind steganalysis

The goal of blind steganalysis is to detect any steganographic method irrespective of its embedding mechanism. As in targeted steganalysis, we cannot work with the full representation of images and instead transform them to a lower-dimensional feature space, where the distributions of cover and stego images are p_c and p_s. Ideally, we would want the feature space to be complete [144] in the sense that for any steganographic scheme

$$D_\text{KL}(P_\text{s}||P_\text{c}) > \epsilon \quad \Rightarrow \quad D_\text{KL}(p_\text{s}||p_\text{c}) > 0 \qquad (10.32)$$

so that we do not lose on our ability to distinguish between cover and stego images by representing the images with their features. In practice, we may be satisfied with a weaker property, namely the requirement that it be hard to practically construct a stego scheme with $D_\text{KL}(p_\text{s}||p_\text{c}) = 0$.

Next, we outline several general strategies for selecting good features for blind steganalysis and then review the options available for constructing detectors.

10.4.1   Features

The impact of embedding can be considered as adding noise of certain specific properties. Thus, many features are designed to be sensitive to adding noise while at the same time being insensitive to the image content.

B1. **[Noise moments]** Transform the image to some domain, such as the Fourier or wavelet domain, where it is easier to separate image content and noise. Compute some statistical characteristics of the noise component (such as statistical moments of the sample distributions of transform coefficients). By working with the noise residual instead of the image, we essentially improve the signal-to-noise ratio (here, signal is the stego noise and noise is the cover image itself) and thus improve the features' sensitivity to embedding changes, while decreasing their undesirable dependence on image content.

B2. **[Calibrated features]** Identify a feature, $\mathbf{f}$, that is likely to predictably change with embedding and calibrate it. In other words, compute the difference

$$\mathbf{f}_\beta - \hat{\mathbf{f}}_0. \qquad (10.33)$$

This process is graphically shown in Figure 10.3. The purpose of calibration is two-fold. It makes the feature approximately zero-mean on the set of covers

$$E_{p_\text{c}}[\mathbf{f}_\beta(\mathsf{x}) - \hat{\mathbf{f}}_0(\mathsf{x})] \approx 0 \qquad (10.34)$$

and it decreases its variance,

$$\text{Var}_{p_\text{c}}[\mathbf{f}_\beta(\mathsf{x}) - \hat{\mathbf{f}}_0(\mathsf{x})]. \qquad (10.35)$$

For these features to work best, it is advisable to construct them in the same domain as where the embedding occurs. For example, when designing features for detection of steganographic schemes that embed data in quantized DCT coefficients of JPEG images, compute the features directly from the quantized DCT coefficients. This makes sense because the embedding changes are "lumped" in that domain, while in a different domain, such as the Fourier domain, the effect of embedding changes is more spread out.

B3. **[Targeted features]** Many features for blind steganalysis originated in targeted steganalysis. In fact, it is quite reasonable to include in the feature set the features that can reliably detect specific steganographic schemes because, this way, the blind steganalyzer will likely detect these steganographic schemes well.

B4. **[Known properties of covers]** If the covers are known to satisfy some a priori statistical properties, such as the symmetry of the DCT histogram, they can and should be taken into consideration for design of features. As an example, consider the histograms $\mathbf{h}^{(kl)}[i]$ of DCT coefficients for a specific spatial frequency (k, l), $k, l = 0, \ldots, 7$. Since these histograms are known to follow the generalized Gaussian distribution, a potentially useful feature is the square error between the histogram and its parametric generalized Gaussian fit

$$\sum_i \left(\mathbf{h}^{(kl)}[i] - g(i; \hat{\mu}, \hat{\alpha}, \hat{\beta}) \right)^2, \tag{10.36}$$

where $g(x; \mu, \alpha, \beta) = [\alpha/(2\beta\Gamma(1/\beta))]e^{-\left|\frac{x-\mu}{\beta}\right|^\alpha}$ is the generalized Gaussian pdf and $\hat{\mu}, \hat{\alpha}, \hat{\beta}$ are its mean, shape, and width parameters estimated from $\mathbf{h}^{(kl)}[i]$. One might as well take the estimated parameters directly as features or use their calibrated versions. Alternatively, non-parametric models may be used as well (e.g., sample distributions of DCT coefficients or their groups).

Specific examples of features that can be used for blind steganalysis are given in Chapter 12.

10.4.2   Classification

After selecting the feature set, Eve has at least two options to construct her detector. One possibility is to mathematically describe the probability distribution of cover-image features, for example, by fitting a parametric model, $\hat{p}_c$, to the sample distribution of cover features and test

$$H_0 : \quad \mathbf{x} \sim \hat{p}_c, \tag{10.37}$$
$$H_1 : \quad \mathbf{x} \not\sim \hat{p}_c. \tag{10.38}$$

The second option is to use a large database of images and embed them using every known steganographic method[1] with uniformly distributed change rates β (or payloads distributed according to some distribution if the change-rate distribution is available) and then fit another distribution, $\hat{p}_s$, through the ex-

---

[1] To be more precise, the selection of stego images in Eve's training set should reflect the probability with which they occur in the steganographic channel. These probabilities are, however, unknown in general.

perimentally obtained data. Eve now faces a simple hypothesis-testing problem,

$$H_0 : \quad \mathbf{x} \sim \hat{p}_c, \qquad (10.39)$$

$$H_1 : \quad \mathbf{x} \sim \hat{p}_s. \qquad (10.40)$$

Even with a complete feature set, however, approaching the above detection problems using the likelihood-ratio test (10.10) is typically not feasible. This is because the feature spaces that aspire to be complete in the above practical sense are still relatively high-dimensional (with dimensionality of the order of 10^3 or higher) to obtain accurate parametric or non-parametric models of p_c. Thus, in practice the detection is formulated as classification. After all, Eve is interested only in detecting stego images, which is a simpler problem than estimating the underlying distributions. She trains a classifier on features $\mathbf{f}(\mathbf{x})$ for $\mathbf{x}$ drawn from a sufficiently large database of cover and stego images to recognize both classes.

Eve can construct her classifier in two different manners. The first option is to train a *cover-versus-all-stego* binary classifier on two classes: cover images and stego images produced by a sufficiently large number of stego algorithms and an appropriate distribution of message payloads. The hope is that if the classifier is trained on all possible archetypes of embedding operations, it should be able to generalize to previously unseen schemes. With this approach, there is always the possibility that in the future some new steganographic algorithm may produce stego images whose features will be incompatible with the distribution $\hat{p}_s$, in which case such images may be misclassified as cover.

Alternatively, Eve can train a classifier that can recognize cover images in the feature space and marks everything that does not resemble a cover image as potentially stego. Mathematically, Eve needs to specify the null-hypothesis region $\mathcal{R}_0$ containing features of cover images ($\mathcal{R}_0$ is the complement of the critical region $\mathcal{R}_1$). This can be done, for example, by covering the support of $\hat{p}_c$ with hyperspheres [167, 168]. An alternative approach using a one-class neighbor machine is explained in Section 12.3. This one-class approach to blind steganalysis has several important advantages. First, the classifier training is simplified because only cover images are used. Also, the classifier does not need to be retrained when new embedding methods appear. The potential problem is that the database has to be very large and *diverse*. We emphasize the adjective diverse because we certainly do not wish to misidentify processed covers (e.g., sharpened images) as containing stego just because the classifier has not been trained on them.

In general, the modular structure of blind detectors makes them very flexible and gives them the ability to evolve with progress in steganography and in machine-learning. One can obviously easily exchange the machine-learning engine, add more features, or expand the training database. Note that all these actions require retraining the classifier.

10.5   Alternative use of blind steganalyzers

Even though the main purpose of blind steganalysis is to detect steganography, there are many other important applications of this approach to steganalysis. In this section, we review these applications, postponing specific examples to Chapter 12.

10.5.1   Targeted steganalysis

A blind steganalyzer can be used to construct a targeted attack on any steganographic scheme simply by training the blind steganalyzer on a narrower training set consisting of cover images and the corresponding stego images embedded by a specific steganographic scheme. If there already exist targeted attacks on the scheme, it is advisable to augment the feature set by the features used in the targeted attacks to further improve the steganalyzer accuracy. This approach to targeted steganalysis often produces the most reliable steganalysis. For example, the blind steganalyzer for JPEG images described in Chapter 12 is more accurate in detecting F5 and OutGuess than the first targeted attacks on both schemes (see the targeted attack on F5 in Chapter 11 and the attack on OutGuess in [87]).

Also, in this application dimensionality-reduction methods, such the method of [173], could be applied to decrease the dimensionality of the feature space and provide a simpler and perhaps more accurate targeted detector. Moreover, it is possible to use the quantitative response of the detector, such as the distance between the stego image feature and the cluster of cover features, to derive an estimate of the number of embedding changes (see [195] and Section 12.4.1).

10.5.2   Multi-classification

An important advantage of blind steganalysis is that the position of the stego image feature in the feature space provides additional information about the embedding algorithm. In fact, the features embedded by different steganographic algorithms form clusters and thus it is possible to classify stego images into known steganographic schemes instead of the binary classification between cover and stego images. A binary classifier can be extended to classify into $k > 2$ classes by building $\binom{k}{2}$ binary classifiers distinguishing between every pair of classes and then fusing the results using a classical voting system. This method for multi-classification is known as the Max–Wins principle [114].

Multi-classification into known steganographic methods is the first step towards forensic steganalysis, whose goal is to identify the embedding algorithm [189] and the secret stego key [94], and eventually extract the embedded message.

10.5.3   Steganography design

Blind steganalyzers can also be used as an oracle for design of steganographic algorithms. The security of a new algorithm can be readily tested by constructing a targeted attack using the approach outlined in Section 10.5.1. The steganalysis results thus provide immediate feedback and guidance to Alice. Depending on how the features are constructed, Alice may be able to identify the features that contribute to successful detection the most and then modify the steganographic algorithm to decrease the impact of embedding on those features. In fact, this approach to steganography has become standard today and the majority of research articles proposing new embedding algorithms report steganalysis results using blind steganalyzers (e.g., [217]).

10.5.4   Benchmarking

The feature set in a blind steganalyzer capable of detecting known steganographic schemes is likely to be a good low-dimensional model of covers. This suggests using the feature space as a simplified model of covers for benchmarking steganographic schemes by evaluating the KL divergence between the features of covers and stego objects calculated from some fixed large database. Since the KL divergence is generally hard to estimate accurately in high-dimensional spaces, alternative statistics, such as the two-sample statistic called maximum mean discrepancy, could be used instead [104, 191]. The third possibility is to benchmark for small payloads only and use for benchmarking the Fisher information evaluated in the feature space as explained in Section 6.1.2.

It is important to realize that benchmarking steganography in this manner is with respect to the feature model and the image database. It is possible that two steganographic techniques might rank differently using a different feature set (model) or image database [37] (also see the discussion in the next section). The problem of fair benchmarking is an active area of research.

10.6   Influence of cover source on steganalysis

The properties of the cover-image source have a major influence on the accuracy of steganalysis. In general, the more "spread out" the cover-image pdf P_c is, the easier it is for Alice to hide messages without increasing the KL divergence $D_{\mathrm{KL}}(P_c \| P_s)$ and the more difficult it is to detect them for Eve. For this reason, one should avoid using covers with little redundancy, such as images with a low number of colors represented in palette image formats, because there the spatial distribution of colors is more predictable. Among the most important attributes of the cover source that influence steganalysis accuracy, we name the color depth, image content, image size, and previous processing. Experimental evaluation of the influence of the cover source on the reliability of blind and targeted steganalysis has been the subject of [22, 23, 102, 129, 130, 139].

Images with a higher level of noise or complex texture have a more spread-out distribution than images that were compressed using lossy compression or denoised. Scans of films or analog photographs are especially difficult for steganalysis because high-resolution scans of photographs resolve the individual grains in the photographic material, and this graininess manifests itself as high-frequency noise. It is also generally easier to detect steganographic changes in color images than in grayscale images, because color images provide more data for statistical analysis and because Eve can utilize strong correlations between color channels.

The image size has an important influence on steganalysis as well. Intuitively, it should be more difficult to detect a fixed relative payload in smaller images than in larger images because features computed from a shorter statistical sample are inherently more noisy. This intuitive observation is analyzed in more detail in Chapter 13 on steganographic capacity. The effect of image size on reliability of steganalysis also means that JPEG covers with a low quality factor are harder to steganalyze reliably because the size of the cover is determined by the number of non-zero coefficients, which decreases with decreasing quality factor.

Image processing may play a decisive role in steganalysis. Processing that is of low-pass character (denoising, blurring, and even lossy JPEG compression to some degree) generally suppresses the noise naturally present in the image, which makes the stego noise more detectable. This is especially true for spatial-domain steganalysis. In fact, it is possible that a certain steganalysis technique can have very good performance on one image database and, at the same time, be almost useless on a different source of images. Thus, it is absolutely vital to test new steganalysis techniques on as large and as diverse a source of covers as possible.

Sometimes, it may not be apparent at first sight that a certain processing may introduce artifacts that may heavily influence steganalysis. For example, a transformation of grayscales, such as contrast adjustment, histogram equalization, or gamma correction, generally does not influence the image noise component in any significant manner. However, it may introduce characteristic spikes and zeros into the histogram [222]. This unusual artifact is caused by discretization of the grayscale transformation to force it to map integers to integers. The spikes and valleys in the histogram that would otherwise not be present in the image can aid steganalysis methods that use the histogram for their reasoning. A good example is the superior performance for detection of ± 1 embedding of ALE (it uses Amplitudes of Local Extrema of the histogram) steganalysis [37, 38] on the database of images supplied with the image-editing software Corel Draw. The images happen to have been processed using a grayscale transformation, which makes ALE perform very well.

JPEG images are less sensitive to processing that occurred prior to compression. This is because the compression has a tendency to suppress many artifacts that are otherwise strikingly present in the uncompressed images. JPEG images, however, present challenges of their own. Because the statistical distribution of DCT coefficients in a JPEG file is significantly influenced by the quality factor, for some steganographic schemes separate steganalyzers may need to be built for each quality factor to improve their accuracy. This is a complication as JPEG format allows customized quantization tables and it would not be practical to construct steganalyzers for each possible matrix. Another complication for steganalysis of JPEG images is repeated JPEG compression with different quality factors because it leads to a phenomenon called "double compression" that may drastically change the statistical distribution of DCT coefficients. When a singly compressed JPEG file is decompressed to the spatial domain and then recompressed with the same 8×8 grid of pixel blocks, the coefficients experience double quantization, which may lead to an unusual-looking histogram. Take, for example, the case when DCT coefficients from a fixed DCT mode were originally quantized with the primary quantization step 7 and then quantized with the secondary step 4. Note that no coefficients quantize to 4 after the second compression and the histogram will have a zero at the first multiple of 4. Coefficients with value 7 after the first compression now quantize to 8. Interestingly, some portion of the coefficients equal to 14 after the first compression now quantize to 12 and a portion of them to 16. The histogram of the DCT mode in the doubly compressed image will thus exhibit unusual patterns not typically found in JPEG images (see Figure 10.7). In other words, the resonance between the primary and secondary quantization steps may leave a characteristic imprint in the histogram of individual DCT modes. Double-compression artifacts may be mistaken by some steganalysis methods as an impact of steganography.

The process of calibration introduced in Section 10.3 is especially vulnerable to double compression because, during calibration, the effects of both compressions are suppressed and the steganalyst does not estimate the doubly compressed cover but instead the cover singly compressed by the second quality factor. One possible approach here is to estimate the primary quantization matrix, such as by using methods explained in [192, 196], and then calibrate by mimicking both compression processes. The need to do this significantly complicates steganalysis as one now needs to build steganalyzers for all possible combinations of primary and secondary quality factors [193].

There are some pathological situations when a certain combination of covers and steganographic system is so unfortunate that it becomes possible to detect even a single modification! Consider images obtained by decompressing JPEG images of a fixed quality to the spatial domain. This image source is significantly different (and much smaller) than the set of all uncompressed images because JPEG is a many-to-one mapping. This is because, during JPEG compression, many slightly different 8×8 blocks of pixels are mapped to the same 8×8 block of quantized DCT coefficients. Thus, after embedding a message, e.g., using LSB or ± 1 embedding in the spatial domain, it is very likely that no 8×8 block of quantized DCT coefficients can, when decompressed, produce the pixel values in the modified spatial block. At the same time, because the steganographic changes

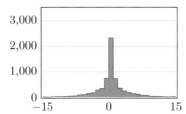 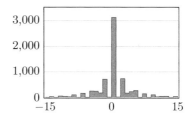

Figure 10.7 Histogram of luminance DCT coefficients for spatial frequency $(1,1)$ for the image shown in Figure 5.1 compressed with quality factor 70 (left), and the same for the image first compressed with quality factor $q_f^{(1)} = 85$ and then with $q_f^{(2)} = 70$ (right). The quantization steps for the primary and secondary compression for this DCT mode were $\mathbf{Q}^{(1)}[1,1] = 7$, $\mathbf{Q}^{(2)}[1,1] = 4$.

are small, after transforming the block back to the DCT domain, the coefficients will still exhibit traces of quantization due to the previous JPEG compression. In fact, if the number of embedding changes is sufficiently small (say one or two embedding changes per block), it is possible to recover the original block of cover-image pixels simply by a brute-force search for the modified pixels. This way, Eve can not only detect the presence of a secret message with high probability but also identify which pixels have been modified! Steganalysis based on this idea is called JPEG compatibility steganalysis [85].

10.7   System attacks

When the steganographic algorithm F5 for JPEG images was introduced, most researchers focused on attacking the impact of the embedding changes using statistical steganalysis. It was only later that Niels Provos pointed out[2] that the JPEG compressor in F5 implementation always inserted the following JPEG comment into the header of the stego image "JPEG Encoder Copyright 1998, James R. Weeks and BioElectroMech," which is rarely present in JPEG images produced by common image-editing software. This comment thus serves as a relatively reliable detector of JPEG images produced by F5. This is an example of a system attack. Johnson [119, 121, 122] and [142] give examples of other unintentional fingerprints left in stego images by various stego products.

Although these weaknesses are not as interesting from the mathematical point of view, it is very important to know about them because they can markedly simplify steganalysis or provide valuable side-information about the steganographic channel.

The size of the stego key space can also be used to attack a steganographic scheme even though the embedding changes it introduces are otherwise statistically undetectable. The stego key usually determines a pseudo-random path

---

[2] Personal communication by Andreas Westfeld.

Algorithm 10.2 System attack on stego image **y** by trying all possible keys. The stego key is found once a meaningful message is extracted.

```
// Input: stego image y
while (Keys left) {
   k = NextKey();
   m = Ext(y, k);
   if (m meaningful) {
     output('Image is stego.');
     output('Message = ', m);
     output('Key     = ', k);
     STOP;
   }
}
output('Image is cover.');
```

---

through the image where the message bits are embedded. A weak stego key or a small stego key space create a security weakness that can be used by Eve to mount the system attack shown in Algorithm 10.2.

Depending on the size of the stego key space, Eve can go through all stego keys or use a dictionary attack. For each key tried, Eve extracts an alleged message. Once she obtains a legible message, she will know that Alice and Bob use steganography and she will have the correct stego key. At this point, a malicious warden can choose to impersonate either party or simply block the communication.

The attack above will not work if the message is encrypted prior to embedding, because the warden cannot reliably distinguish between a random bit stream and an encrypted message. However, encrypting the message using a strong encryption scheme with a secure key still does not mean that the stego key does not have to be strong because Eve can determine the stego key by other means than inspecting the message. She can still run through all stego keys as above; however, this time she will be checking the statistical properties of the pixels along the pseudo-random path rather then the extracted message bits. The statistical properties of pixels along the true embedding path should be different than statistical properties of a randomly chosen path, as long as the image is not fully embedded.[3] To see this, assume that the steganographic scheme embeds the payload by changing n_0 pixels while visiting $k < n$ pixels along some embedding path in an n-pixel cover image. The change rate for the first k pixels along the embedding path will be n_0/k, while the change rate when following a random path through the image is only $n_0/n < n_0/k$. Thus, a sudden increase of this ratio is indicative of the fact that a correct stego key was used.

---

[3] Fully embedded images can most likely be reliably detected using other methods.

For example, for simple LSB embedding, $n_0/k \approx 1/2$ while $n_0/n = \alpha/2 < 1/2$, where α is the relative payload. In this case, the warden can apply the histogram attack from Chapter 5 as her detector. More examples of this type of system attack can be found in [94].

10.8   Forensic steganalysis

The goal of steganalysis is to detect the presence of secret messages. In the classical prisoners' problem, once Eve finds out that Alice and Bob communicate using steganography, she decides to block the communication channel. In practice, however, Eve may not have the resources or authority to do so or may not even want to because blocking the channel would only alert Alice and Bob that they are being subjected to eavesdropping. Instead, Eve may try to determine the steganographic algorithm and the stego key, and eventually extract the message. Such activities belong to forensic steganalysis, which can be loosely defined as a collection of tasks needed to identify individuals who are communicating in secrecy, the stegosystem they are using, its parameters (the stego key), and the message itself. We list these tasks below, with the caveat that in any given situation some of these tasks do not have to be carried out because the information may already be a priori available.

1. Identification of web sites, Internet nodes, or computers that should be analyzed for steganography.
2. Development of algorithms that can distinguish stego images from cover images.
3. Identification of the embedding mechanism, e.g., LSB embedding, ± 1 embedding, embedding in the frequency domain, embedding in the image palette, sequential, random, or content-adaptive embedding, etc.
4. Determining the steganographic software.
5. Searching for the stego key and extracting the embedded data.
6. Deciphering the extracted data and obtaining the secret message (cryptanalysis).

The power of steganography is that it not only provides privacy in the sense that no one can read the exchanged messages, but also hides the very presence of secret communication. Thus, the primary problem in steganography detection is to decide what communication to monitor in the first place. Since steganalysis algorithms may be expensive and slow to run, focusing on the right channel is of paramount importance.

Second, the communication through the monitored channel will be inspected for the presence of secret messages using steganalysis methods. Once an image has been detected as containing a secret message, it is further analyzed to determine the steganographic method.

The warden can continue by trying to recover some attributes of the embedded message and properties of the stego algorithm. For example, some detection algorithms may provide Eve with an estimate of the number of embedding changes. In this case, she can approximately infer the length of the secret message. If the approximate location of the embedding changes can be determined, this may point to a class of stego algorithms. For example, Eve can use the histogram attack, described in Chapter 5, to determine whether the message has been sequentially embedded. If the prisoners reuse the stego key, then Eve may have multiple stego images embedded with the same key, which can help her determine the embedding path [137, 139] and narrow down the class of possible stego methods.

The character of the embedding changes also leaks information about the embedding mechanism. If Eve can determine that the LSBs of pixels were modified, she can then focus on methods that embed into LSBs. Eventually, Eve may guess which stego method has been used and attempt to determine the stego key and extract the embedded message. If the message is encrypted, Eve then needs to perform cryptanalysis on the extracted bit stream.

In her endeavor, Eve can mount different types of attacks depending on the information available to her [120]. The most common case, which we have already considered, is the *stego-image-only-attack* in which Eve has only the stego image. However, in some situations, Eve may have additional information available that may aid in her effort. For example, in a criminal case the suspect's computer may be available, with several versions of the same image on the hard disk. This will allow Eve to directly infer the location, number, and character of embedding modifications. This scenario is known as a *known-cover attack*.

If the steganographic algorithm is known to Eve, her options further increase as she can now mount two more attacks – *the known-stego-method attack* and *the known-message attack*. Eve can, for example, embed the message with various keys and identify the correct key by comparing the locations of embedding changes in the resulting stego image with those in the stego image under investigation.

Summary

- Steganalysis is the complementary task to steganography. Its goal is to detect the presence of secret messages.
- Steganography is broken when the mere presence of a secret message can be proved. In particular, it is not necessary to read the message to break a stegosystem.
- Activities directed towards extracting the message belong to forensic steganalysis.
- Steganalysis can be formulated as the detection problem using a variety of hypothesis-testing scenarios.

- There are two major types of steganalysis attacks – statistical and system attacks.
- System attacks use some weakness in the implementation or protocol. Statistical attacks try to distinguish cover and stego images by computing statistical quantities from images.
- All statistical steganalysis methods work by representing images in some feature space where a detector is constructed.
- If the features are designed to detect a specific stegosystem, we speak of targeted steganalysis.
- Features designed to attack an arbitrary stegosystem lead to blind steganalysis algorithms.
- The features for targeted schemes are designed by
  - identifying quantities that predictably change with embedding,
  - estimating these quantities for the cover image from the stego image (calibration), or
  - finding a function of such quantities that attains a known value on covers.
- Features for blind steganalysis are usually constructed in a heuristic manner to be sensitive to typical steganographic changes and insensitive to image content. Calibration can be used to achieve this goal. The features can be computed in the spatial domain, frequency domain, or wavelet domain.
- The goal of quantitative steganalysis is to estimate the embedded payload (or, more accurately, the number of embedding changes).
- The error of the estimate from quantitative steganalyzers has two components – a within-image error and a between-image error. The within-image error depends on the image content and the payload. It is well modeled using a Gaussian distribution. The between-image error is the estimator bias for each image caused by the properties of natural images. It is well modeled using Student's t-distribution.

Exercises

10.1 [Gauss–Gauss ROC] Consider a scalar feature with $p_c = N(0, \sigma^2)$ and $p_s = N(\mu, \sigma^2)$, $\mu > 0$. Prove the following expressions for the ROC curve and the three measures of detector performance from the text:

$$P_D(P_{FA}) = Q\left(Q^{-1}(P_{FA}) - \frac{\mu}{\sigma}\right),$$
(10.41)

$$P_D^{-1}\left(\frac{1}{2}\right) = Q\left(\frac{\mu}{\sigma}\right),$$
(10.42)

$$\rho = 1 - 2Q\left(\frac{\mu}{\sigma\sqrt{2}}\right),$$
(10.43)

$$P_E = Q\left(\frac{\mu}{2\sigma}\right),$$
(10.44)

where $Q(x)$ is the complementary cumulative distribution function of a standard normal variable $N(0, 1)$,

$$Q(x) = \int_x^\infty \frac{1}{\sqrt{2\pi}} e^{-\frac{t^2}{2}} \, dt. \tag{10.45}$$

Hint: When computing ρ,

$$\frac{\rho + 1}{2} = \int_0^1 Q\left(Q^{-1}(x) - \frac{\mu}{\sigma}\right) dx = -\int_{-\infty}^\infty Q\left(y - \frac{\mu}{\sigma}\right) Q'(y) dy \tag{10.46}$$

$$= \int_{-\infty}^\infty \int_{y-\frac{\mu}{\sigma}}^\infty \frac{1}{2\pi} e^{-\frac{x^2+y^2}{2}} \, dx dy = \iint_{x-y \geq -\frac{\mu}{\sigma}} \frac{1}{2\pi} e^{-\frac{x^2+y^2}{2}} \, dx dy \tag{10.47}$$

and make a substitution

$$r = \frac{1}{\sqrt{2}}(x + y), \tag{10.48}$$

$$s = \frac{1}{\sqrt{2}}(x - y). \tag{10.49}$$

11 Selected targeted attacks

Steganalysis is the activity directed towards detecting the presence of secret messages. Due to their complexity and dimensionality, digital images are typically analyzed in a low-dimensional feature space. If the features are selected wisely, cover images and stego images will form clusters in the feature space with minimal overlap. If the warden knows the details of the embedding mechanism, she can use this side-information and design the features accordingly. This strategy is recognized as targeted steganalysis. The histogram attack and the attack on Jsteg from Chapter 5 are two examples of targeted attacks.

Three general strategies for constructing features for targeted steganalysis were described in the previous chapter. This chapter presents specific examples of four targeted attacks on steganography in images stored in raster, palette, and JPEG formats. The first attack, called Sample Pairs Analysis, detects LSB embedding in the spatial domain by considering pairs of neighboring pixels. It is one of the most accurate methods for steganalysis of LSB embedding known today. Section 11.1 contains a detailed derivation of this attack as well as several of its variants formulated within the framework of structural steganalysis. The Pairs Analysis attack is the subject of Section 11.2. It was designed to detect steganographic schemes that embed messages in LSBs of color indices to a preordered palette. The EzStego algorithm from Chapter 5 is an example of this embedding method. Pairs Analysis is based on an entirely different principle than Sample Pairs Analysis because it uses information from pixels that can be very distant. The third attack, presented in Section 11.3, is targeted steganalysis of the F5 algorithm that demonstrates the use of calibration as introduced in Section 10.3. The last attack concerns detection of ± 1 embedding in the spatial domain and is detailed in Section 11.4. These steganalysis methods were chosen to illustrate the basic principles on which many targeted attacks are based.

11.1   Sample Pairs Analysis

A large number of steganographic applications today use LSB embedding for hiding messages. The likely reasons for its popularity are ease of implementation, speed, and large embedding capacity. Thus, reliable methods for detection of this embedding paradigm are of great interest.

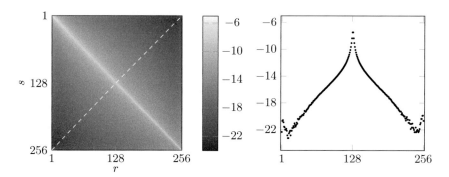

Figure 11.1 Left: Logarithm of the normalized adjacency histogram of horizontally neighboring pixel pairs (r, s) from 6000 never-compressed raw digital-camera images (see the description of Database RAW in Section 11.1.1). Right: A cross-section along the minor diagonal of the adjacency histogram.

In Chapter 5, we introduced an attack on LSB embedding called the histogram attack. It was based on an observation that flipping LSBs has a tendency to even out the histogram counts in LSB pairs (pairs of intensities differing by their LSBs $\{2k, 2k + 1\}, k = 0, \ldots, 127$). The histogram attack worked reasonably well when the stego image was fully embedded or when the image was only partially embedded but along a known embedding path, such as sequentially by rows or columns. When the message is scattered along a pseudo-random path, the histogram attack will not work unless the majority (i.e., 99%) of all pixels were used for embedding.

Steganalysis methods that use only a first-order statistic, such as histograms of pixels, cannot capture the relationship among neighboring pixels. By utilizing the fact that neighboring pixels in images exhibit strong correlations, it is possible to construct much more reliable and accurate steganalysis methods. The first method that used local correlations among neighboring pixels to accurately detect LSB embedding was the heuristic RS Analysis [84]. The simplest case of RS Analysis, Sample Pairs Analysis (SPA), was later rederived [62] in a way that enabled multiple extensions and improvements [61, 63, 127, 128, 131, 133, 134, 135, 163]. Together, these contributions revolutionized steganalysis of LSB embedding and provided extremely accurate methods capable of reliably detecting payloads as small as 0.03 bpp in some cover sources. Next, we describe in detail the simplest version of SPA as it appeared in [62]. Later in this chapter, SPA is reformulated within a more general framework that provides deeper insight into its inner workings and gives birth to several important generalizations and improvements.

From Section 5.1.1, we already know that LSB embedding predictably changes the image histogram. The problem is that the variety of cover-image histograms is so high that no reasonable assumption can be made about them. The situation, however, is quite different when one looks at *pairs* of neighboring pixels. Let $\mathcal{P}$ be the set of all pairs of, say horizontally, neighboring pixels. Due to local correlations among pixels of natural images, we are more likely to see a pair of neighboring pixels (r, s) than a pair (r', s') whenever $|r - s| < |r' - s'|$. In other words, the larger the difference $|r - s|$ is, the less probably will such a pair occur in a natural image.[1] Thus, unlike the histogram of pixels, the histogram of pixel pairs has a predictable shape (see Figure 11.1). Let us now take a look at a pixel pair (r, s) with $r < s$. The pair can accept four different forms,

$$(r, s) \in \{(2i, 2j), (2i, 2j + 1), (2i + 1, 2j), (2i + 1, 2j + 1)\}. \tag{11.1}$$

Let $\mathbf{h}[r, s]$ be the number of horizontally adjacent pixel pairs (r, s) in the image. Because $r < s$, we expect the counts for the pair $(2i, 2j + 1)$ to be the lowest among the four pairs, and the counts for $(2i + 1, 2j)$ to be the highest. LSB embedding changes any pair into another with probability that depends on the change rate β. Because the set of four pairs (11.1) is obviously closed under LSB embedding, the count of pixel pairs in the stego image, $\mathbf{h}_\beta$, is a convex combination of the cover counts for all four pairs. For example, if the payload is embedded pseudo-randomly in the image,

$$\begin{aligned}
\mathbf{h}_\beta[2i + 1, 2j] &= \beta\,(1 - \beta)\,\mathbf{h}[2i, 2j] + \beta^2 \mathbf{h}[2i, 2j + 1] \\
&\quad + (1 - \beta)^2\,\mathbf{h}[2i + 1, 2j] \\
&\quad + \beta\,(1 - \beta)\,\mathbf{h}[2i + 1, 2j + 1], \tag{11.2} \\
\mathbf{h}_\beta[2i, 2j + 1] &= \beta\,(1 - \beta)\,\mathbf{h}[2i, 2j] + (1 - \beta)^2\,\mathbf{h}[2i, 2j + 1] \\
&\quad + \beta\,(1 - \beta)\,\mathbf{h}[2i + 1, 2j + 1] \\
&\quad + \beta^2 \mathbf{h}[2i + 1, 2j]. \tag{11.3}
\end{aligned}$$

Similar expressions can be obtained for the other two counts. The important observation here is that $\mathbf{h}_\beta[2i + 1, 2j]$ will decrease with β because three out of four terms in (11.2) are smaller than $\mathbf{h}[2i + 1, 2j]$. By a similar argument, $\mathbf{h}_\beta[2i, 2j + 1]$ will increase with β. Eventually, when $\beta = 0.5$, the expected values of all four counts will be the same. We will put the pair (r, s) into set $\mathcal{X}$ whenever $r < s$ and s is even, and we will include the pair in set $\mathcal{Y}$ whenever $r < s$ and s is odd. With LSB embedding, the counts of pixel pairs in $\mathcal{X}$ will decrease, while the counts of pairs in $\mathcal{Y}$ will increase.

A similar analysis can be carried out for pairs (r, s) for which $r > s$. There, the situation is complementary in the sense that the counts of pairs $(2i + 1, 2j)$ will increase, while the counts of $(2i, 2j + 1)$ will decrease. Thus, we include in $\mathcal{X}$ pairs with $r > s$, s odd, while $(r, s) \in \mathcal{Y}$ when $r > s$ and s even. This way, the cardinality of $\mathcal{X}$ will decrease with LSB embedding, while the cardinality of $\mathcal{Y}$ will increase, for all pairs (r, s), $r \neq s$.

---

[1] Because the order of pixels in the pair matters, we will denote the pair in round brackets (r, s), rather than curly brackets $\{r, s\}$.

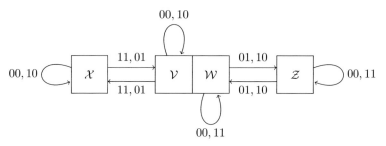

Figure 11.2 Transitions between primary sets, $\mathcal{X}, \mathcal{V}, \mathcal{W}, \mathcal{Z}$, under LSB flipping. Note that $\mathcal{Y} = \mathcal{V} \cup \mathcal{W}$.

By now, we have partitioned the set of all pairs of horizontally neighboring pixels $\mathcal{P}$ into three sets:

$$\mathcal{X} = \{(r, s) \in \mathcal{P} | (s \text{ is even and } r < s) \text{ or } (s \text{ is odd and } r > s)\}, \qquad (11.4)$$

$$\mathcal{Y} = \{(r, s) \in \mathcal{P} | (s \text{ is even and } r > s) \text{ or } (s \text{ is odd and } r < s)\}, \qquad (11.5)$$

$$\mathcal{Z} = \{(r, s) \in \mathcal{P} | r = s\}. \qquad (11.6)$$

The symmetry of the definitions of the sets $\mathcal{X}$ and $\mathcal{Y}$ indicates that for cover images the cardinalities of $\mathcal{X}$ and $\mathcal{Y}$ should be the same,

$$|\mathcal{X}| = |\mathcal{Y}|, \qquad (11.7)$$

because in natural images it should be equally likely to have $r > s$ or $r < s$ independently of whether s is even or odd. However, we now know that the equality will be broken by flipping LSBs. Thus, with embedding the difference $|\mathcal{X}| - |\mathcal{Y}|$ will no longer be zero. Clearly, we identified a quantity that predictably changes with embedding and whose value is known on cover images. All we now need to do is to quantify this important observation.

To this end, we further partition $\mathcal{Y}$ into two subsets $\mathcal{W}$ and $\mathcal{V}$, $\mathcal{V} = \mathcal{Y} - \mathcal{W}$, where

$$\mathcal{W} = \{(r, s) \in \mathcal{P} | r = 2k, s = 2k + 1 \text{ or } r = 2k + 1, s = 2k\}. \qquad (11.8)$$

In other words, $\mathcal{W} \cup \mathcal{Z}$ is the set of all pairs from $\mathcal{P}$ that belong to one LSB pair $\{2k, 2k + 1\}$.

The sets $\mathcal{X}, \mathcal{W}, \mathcal{V}$, and $\mathcal{Z}$ are called primary sets. Note that $\mathcal{P} = \mathcal{X} \cup \mathcal{W} \cup \mathcal{V} \cup \mathcal{Z}$.

We now analyze what happens to a given pixel pair (r, s) under LSB embedding. There are four possibilities:

1. both r and s stay unmodified (modification pattern 00)
2. only r is modified (modification pattern 10)
3. only s is modified (modification pattern 01)
4. both r and s are modified (modification pattern 11).

LSB embedding may cause a given pixel pair to move from its primary set to another primary set. The transitions of pairs between the primary sets are depicted in Figure 11.2. When an arrow points from set $\mathcal{A}$ to set $\mathcal{B}$, it means that a pixel pair originally in $\mathcal{A}$ moves to $\mathcal{B}$ if modified by the modification pattern associated with the arrow.

For each modification pattern $\Omega \in \{00, 10, 01, 11\}$ and any subset $\mathcal{A} \subset \mathcal{P}$, we denote by $\phi(\Omega, \mathcal{A})$ the expected fraction of pixel pairs in $\mathcal{A}$ modified with pattern Ω. Under the assumption that the message bits are embedded along a random path through the image, each pixel in the image is equally likely to be modified. Thus, the expected fraction of pixels modified with a specific modification pattern $\Omega \in \{00, 10, 01, 11\}$ is the same for every primary set $\mathcal{A} \in \{\mathcal{X}, \mathcal{V}, \mathcal{W}, \mathcal{Z}\}$, $\phi(\Omega, \mathcal{X}) = \cdots = \phi(\Omega, \mathcal{Z}) = \phi(\Omega)$. With change rate β, we obtain the following transition probabilities:

$$\phi(00) = (1 - \beta)^2, \tag{11.9}$$
$$\phi(01) = \phi(10) = \beta(1 - \beta), \tag{11.10}$$
$$\phi(11) = \beta^2. \tag{11.11}$$

Together with the transition diagram of Figure 11.2, we can now express the expected cardinalities of the primary sets for the stego image as functions of the change rate β and the cardinalities of the cover image. Denoting the primary sets after embedding with a prime, we obtain

$$E[|\mathcal{X}'|] = (1 - \beta)|\mathcal{X}| + \beta|\mathcal{V}|, \tag{11.12}$$
$$E[|\mathcal{V}'|] = (1 - \beta)|\mathcal{V}| + \beta|\mathcal{X}|, \tag{11.13}$$
$$E[|\mathcal{W}'|] = (1 - 2\beta + 2\beta^2)|\mathcal{W}| + 2\beta(1 - \beta)|\mathcal{Z}|. \tag{11.14}$$

From now on, we drop the expectations and assume that the cardinalities of the primary sets from the stego image will be close to their expectations.

Following Strategy T2 from Chapter 10, our goal is to derive an equation for the unknown quantity β using only the cardinalities of primed sets because they can be calculated from the stego image. Equations (11.12) and (11.13) imply that

$$|\mathcal{X}'| - |\mathcal{V}'| = (|\mathcal{X}| - |\mathcal{V}|)(1 - 2\beta). \tag{11.15}$$

Because $|\mathcal{X}| = |\mathcal{Y}|$, we have $|\mathcal{X}| = |\mathcal{V}| + |\mathcal{W}|$, and from (11.15)

$$|\mathcal{X}'| - |\mathcal{V}'| = |\mathcal{W}|(1 - 2\beta). \tag{11.16}$$

The transition diagram shows that the embedding process does not modify the union $\mathcal{W} \cup \mathcal{Z}$. Denoting $\kappa = |\mathcal{W}| + |\mathcal{Z}| = |\mathcal{W}'| + |\mathcal{Z}'|$, on replacing $|\mathcal{Z}|$ with $\kappa - |\mathcal{W}|$, equation (11.14) becomes

$$|\mathcal{W}'| = |\mathcal{W}|(1 - 2\beta)^2 + 2\beta(1 - \beta)\kappa. \tag{11.17}$$

Algorithm 11.1 Sample Pairs Analysis for estimating the change rate from a stego image. The constant γ is a threshold on the test statistic $\hat{\beta}$ set to achieve $P_{\text{FA}} < \epsilon_{\text{FA}}$, where ϵ_{FA} is a bound on the false-alarm rate.

```
// Input M × N image x
// Form pixel pairs
P = {(x[i,j], x[i,j+1])|i = 1,..., M, j = 1,..., N − 1}
x = y = 0; κ = 0;
for k = 1 to M(N − 1) {
    (r, s) ← kth pair from P
    if (s even & r < s) or (s odd & r > s){x = x + 1;}
    if (s even & r > s) or (s odd & r < s){y = y + 1;}
    if (⌊s/2⌋ = ⌊r/2⌋){κ = κ + 1;}
}
if κ = 0 {output('SPA failed because κ = 0'); STOP}
a = 2κ; b = 2 (2x − M(N − 1)); c = y − x;
β± = Re((−b ± √(b² − 4ac))/(2a));
β̂ = min(β+, β−);
if β̂ > γ {
    output('Image is stego');
    output('Estimated change rate = ', β̂);
}
```

Eliminating $|\mathcal{W}|$ from (11.16) and (11.17) leads to

$$|\mathcal{W}'| = (|\mathcal{X}'| - |\mathcal{V}'|)(1 - 2\beta) + 2\beta(1 - \beta)\kappa. \qquad (11.18)$$

Since $|\mathcal{X}'| + |\mathcal{Y}'| + |\mathcal{Z}'| = |\mathcal{X}'| + |\mathcal{V}'| + |\mathcal{W}'| + |\mathcal{Z}'| = |\mathcal{P}|$, (11.18) is equivalent to

$$2\kappa\beta^2 + 2(2|\mathcal{X}'| - |\mathcal{P}|)\beta + |\mathcal{Y}'| - |\mathcal{X}'| = 0, \qquad (11.19)$$

which is a quadratic equation for the unknown change rate β. This equation can be directly solved for β because all coefficients in this equation can be evaluated from the stego image (recall that $\kappa = |\mathcal{W}'| + |\mathcal{Z}'|$). The final estimate of the change rate $\hat{\beta}$ is obtained as the smaller root of this quadratic equation. Note that when $\kappa = 0$, $\mathcal{W} \cup \mathcal{Z} = \emptyset$ and thus $|\mathcal{X}| = |\mathcal{X}'| = |\mathcal{Y}| = |\mathcal{Y}'| = |\mathcal{P}|/2$. In this case (11.19) becomes a useless identity and we cannot estimate β. However, since κ is the number of pixel pairs where both values belong to the same LSB pair, this will happen with very small probability for natural images.

11.1.1   Experimental verification of SPA

In this section, we test SPA on images from two databases to demonstrate its accuracy and illustrate how much the accuracy may depend on the cover-image source.

Database RAW consists of 2567 raw, never-compressed, 24-bit color images of different dimensions (all larger than 1 megapixel). The images were acquired using 22 different digital cameras ranging from low-cost point-and-shoot cameras to semi-professional SLR cameras. For the tests, the images were converted to 8-bit grayscale.

Database SCAN consists of 3000 high-resolution never-compressed 1500×2100 scans of films in the 32-bit CMYK TIFF format downloaded from the NRCS Photo Gallery (`http://photogallery.nrcs.usda.gov`). For the tests, the images were converted to 8-bit grayscale and downsampled to 640×480 using bicubic resizing.

For each database, the cover images were embedded using LSB embedding with random bit streams of three different relative payloads $\alpha = 0.05, 0.25$, and 0.5 bpp, which correspond to expected change rates $\beta = 0.025, 0.125$, and 0.25. Then, SPA was applied to all cover as well as stego images. The median estimated change rate and its sample median absolute error (MAE^2) are shown in Table 11.1. Notice that the estimator produced markedly better results on digital-camera images than on scans. For scans, the estimator exhibits a positive bias and a significantly larger MAE compared with the result from the database of digital-camera images. This is due to the fact that the scans were much more noisy than the digital-camera images because when scanning at high dpi the scan captures the film grain, which casts a characteristic stochastic texture resembling noise.

To further render the difference between the results on these two databases, Figures 11.3 and 11.4 show the histogram of the estimated change rate for raw digital-camera images and film scans. Besides the obvious fact that the results for scans are much more scattered, note that the distribution is non-Gaussian, asymmetric, and has thick tails. These properties seem to be characteristic for change-rate estimators in general. The estimation error is a mixture of two errors – the within-image error and the between-image error (see Section 10.3.2). In this particular experiment, since the embeddings were done by fixing the payload rather than the change rate, there is a third source of error caused by the fact that, when embedding a fixed payload α, the actual change rate will not be exactly $\beta = \alpha/2$ but will fluctuate around this expected value, following a binomial distribution [22].

11.1.2   Constructing a detector of LSB embedding using SPA

We now use the data obtained in the experiments to illustrate how one may proceed with constructing an LSB detector in practice. The reader will see how important for steganalysis the type of side-information about the steganographic

---

^{2} MAE is a more robust statistic than variance. Moreover, since we know from Section 10.3.2 that the error of quantitative steganalyzers often has thick tails matching Student's t-distribution, the variance may not even exist. Also, see discussion in Section A.1.

Table 11.1. Median estimated change rate and its median absolute error (MAE) obtained using Sample Pairs Analysis for raw digital-camera images and film scans.

	Median/MAE of $\hat{\beta}$			
	Cover ($\beta = 0$)	$\beta = 0.025$	$\beta = 0.125$	$\beta = 0.25$
RAW	0.0015/0.0076	0.0264/0.0074	0.1261/0.0062	0.2507/0.0050
SCAN	0.0130/0.0331	0.0373/0.0316	0.1347/0.0254	0.2575/0.0177

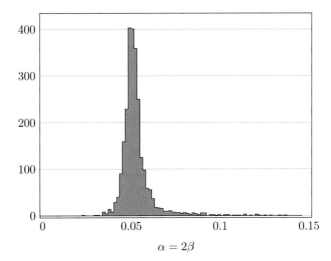

$$\alpha = 2\beta$$

Figure 11.3 Histogram of estimated message length for images from Database RAW embedded with $\alpha = 0.05$ bpp.

channel available to Eve is. Let us assume that Eve knows that Alice and Bob always embed relative payload $\alpha = 0.05$ (change rate $\beta \approx 0.025$) using LSB embedding in randomly selected pixels. If, additionally, Eve knows that Alice and Bob use digital-camera images, she uses the data from experiments on Database RAW and draws an ROC using Algorithm 10.1. To avoid introducing any systematic errors due to a low number of images in the database, she divides the database into two disjoint subsets $\mathcal{D} = \mathcal{D}_1 \cup \mathcal{D}_2$ and takes the cover features (the features are the estimates $\hat{\beta}$) only from images in $\mathcal{D}_1$ and stego features only from images in $\mathcal{D}_2$.

The ROC describes a class of detectors that differ by their threshold γ,

$$\text{Decide "stego" if } \hat{\beta} > \gamma. \tag{11.20}$$

Different choices of γ correspond to different points on the ROC curve (see Figure 11.5). From the graph, we have $P_{\mathrm{D}}^{-1}\left(\frac{1}{2}\right) \approx 0.009$ (see equation (10.16)). Actually, Eve can select a point on the ROC curve (circle in Figure 11.5) that will give her a higher $P_{\mathrm{D}} \approx 0.68$ at the same $P_{\mathrm{FA}} = P_{\mathrm{D}}^{-1}\left(\frac{1}{2}\right)$. Also note that the ROC curve is not concave. Since we know that ROCs for optimal detectors

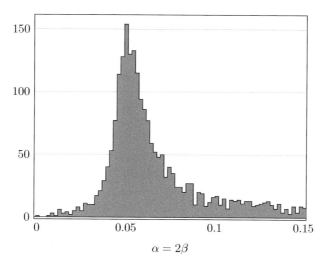

$$\alpha = 2\beta$$

Figure 11.4 Histogram of estimated message length for images from Database SCAN embedded with $\alpha = 0.05$ bpp.

must be concave (see Appendix D), we can further improve the detector's ROC by connecting the points $(0,0)$ and $(u_1, P_D(u_1))$ (the origin and the square in Figure 11.5), which corresponds to the following family of detectors F_u in the interval $u \in [0, u_1]$:

$$F_u(\mathbf{x}) = \begin{cases} F_{u_1}(\mathbf{x}) & \text{with probability } u/u_1 \\ F_0(\mathbf{x}) & \text{with probability } 1 - u/u_1. \end{cases} \qquad (11.21)$$

Note that this detector would decrease the false-alarm rate at $P_D = 0.5$ from roughly 0.009 to 0.006. By imposing a bound on the false-alarm rate, Eve can now select the threshold and use the appropriate detector in her eavesdropping.

For comparison, we show in Figure 11.6 the ROC curve for detecting LSB embedding at relative payload $\alpha = 0.05$ in raw scans of film. Note that the performance of the detector is markedly worse than for digital-camera images. The ROC can be made concave using the same procedure as above.

We note that if Eve were facing a different steganographic channel where the change rate follows a known distribution f_β, she could construct a detector in the same way as above with one difference: the change rate in stego images, $\mathcal{D}_2$, would have to follow f_β instead of being fixed at 0.025. If Eve has no prior information about the payload, she can still obtain the distribution of $\hat{\beta}$ on cover images, fit a parametric model, and fix the threshold on the basis of the bound on false alarms. This time, however, Eve will not be able to determine the probability of correctly detecting a stego image as stego as this depends on the distribution of payloads.

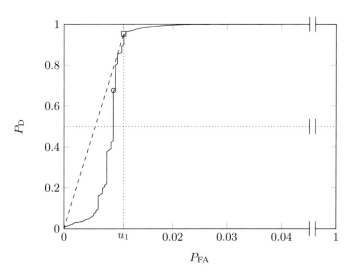

Figure 11.5 ROC for the SPA detector distinguishing cover digital-camera images (Database RAW) from the same images embedded using LSB embedding in randomly selected pixels with $\alpha = 0.05$. Because the ROC very quickly reaches $P_{\mathrm{D}} = 1$, we display the curve only in the range $P_{\mathrm{FA}} \leq 0.05$. The circle corresponds to the point with $P_{\mathrm{D}} = 0.68$ with $P_{\mathrm{FA}} < 0.01$. The line connecting the origin and the square symbol is a concave hull of the ROC that can be obtained using the detector at $u = u_1$ and the detector at $u = 0$.

11.1.3   SPA from the point of view of structural steganalysis

Sample Pairs Analysis can be interpreted within a more general framework called structural steganalysis [63, 128, 133, 134]. This reformulation is appealing because it provides deeper insight into the inner workings of SPA and allows several important generalizations that lead to more accurate steganalysis. In this section, we describe the framework as it appeared in [128] and then discuss possible avenues that can be taken while referring the reader to the literature.

Similar to SPA, we start by dividing the image into pairs of pixels and then divide the pairs into three types of trace sets, $\mathcal{C}, \mathcal{E}, \mathcal{O}$, parametrized by integer index i,

$$\mathcal{C}_i = \{(r, s) \in \mathcal{P} \,|\, \lfloor r/2 \rfloor - \lfloor s/2 \rfloor = i \}, \tag{11.22}$$

$$\mathcal{E}_i = \{(r, s) \in \mathcal{P} \,|\, r - s = i,\ s \text{ even} \}, \tag{11.23}$$

$$\mathcal{O}_i = \{(r, s) \in \mathcal{P} \,|\, r - s = i,\ s \text{ odd} \}. \tag{11.24}$$

The set $\mathcal{C}_i$ contains all pairs whose values differ by i after right-shifting their binary representation (dividing by 2 and rounding down). There are four possibilities for $(r, s) \in \mathcal{C}_i$: $r - s = 2i$, which covers two cases when either both r and s are even or both are odd, or $r - s = 2i - 1$ (if r is even and s odd), or $r - s = 2i + 1$ (if r is odd and s even). Thus, each trace set $\mathcal{C}_i$ can be written as

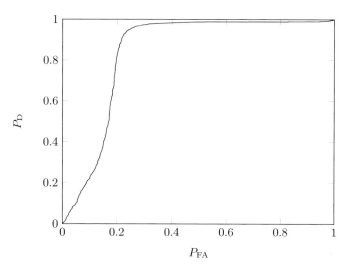

Figure 11.6 ROC curve for the SPA detector of relative payload $\alpha = 0.05$ in raw scans of film (Database SCAN). When comparing this figure with Figure 11.5, note the range of the x axis.

a disjoint union of four trace subsets

$$C_i = \mathcal{E}_{2i} \cup \mathcal{O}_{2i-1} \cup \mathcal{E}_{2i+1} \cup \mathcal{O}_{2i}. \tag{11.25}$$

Note that trace sets C_i are invariant with respect to LSB embedding because the value of $\lfloor r/2 \rfloor$ does not depend on the LSB of r. The four trace subsets of C_i, however, are in general not invariant with respect to LSB embedding. The transition diagram between the trace subsets, including the probabilities of transition, is shown in Figure 11.7. As an example, we explain the transitions from $\mathcal{E}_{2i}$. All pairs (r, s) from $\mathcal{E}_{2i}$ have both r and s even. The probability that a pair $(r, s) \in \mathcal{E}_{2i}$ ends up again in $\mathcal{E}_{2i}$ is the probability that neither r nor s gets flipped, which is $(1 - \beta)^2$. The probability of the transition $\mathcal{E}_{2i} \to \mathcal{O}_{2i}$ is the probability that both get flipped, which is β^2. The probabilities of the remaining two transitions, $\mathcal{E}_{2i} \to \mathcal{E}_{2i+1}$ and $\mathcal{E}_{2i} \to \mathcal{O}_{2i-1}$, are equal to the probability that one pixel gets flipped but not the other, which is $\beta(1 - \beta)$.

The cardinality of each trace set after changing a random portion of β of LSBs is a random variable with the following expected values derived from the transition diagram:

$$\begin{pmatrix} E[|\mathcal{E}'_{2i}|] \\ E[|\mathcal{O}'_{2i-1}|] \\ E[|\mathcal{E}'_{2i+1}|] \\ E[|\mathcal{O}'_{2i}|] \end{pmatrix} = \begin{pmatrix} b^2 & ab & ab & a^2 \\ ab & b^2 & a^2 & ab \\ ab & a^2 & b^2 & ab \\ a^2 & ab & ab & b^2 \end{pmatrix} \begin{pmatrix} |\mathcal{E}_{2i}| \\ |\mathcal{O}_{2i-1}| \\ |\mathcal{E}_{2i+1}| \\ |\mathcal{O}_{2i}| \end{pmatrix}, \tag{11.26}$$

where $a = \beta$, $b = 1 - \beta$. Here, we again use the prime to denote the sets obtained from the stego image. For any $0 \le \beta < \frac{1}{2}$, the matrix is invertible and we can

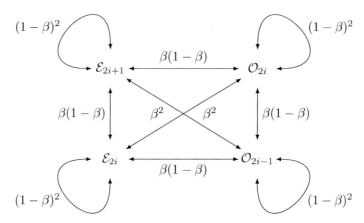

Figure 11.7 Diagram of transitions between trace sets from $\mathcal{C}_i$.

express the cardinalities of cover-image trace sets as

$$
\begin{pmatrix}
|\mathcal{E}_{2i}| \\
|\mathcal{O}_{2i-1}| \\
|\mathcal{E}_{2i+1}| \\
|\mathcal{O}_{2i}|
\end{pmatrix}
= \frac{1}{(b-a)^2}
\begin{pmatrix}
b^2 & -ab & -ab & a^2 \\
-ab & b^2 & a^2 & -ab \\
-ab & a^2 & b^2 & -ab \\
a^2 & -ab & -ab & b^2
\end{pmatrix}
\begin{pmatrix}
E[|\mathcal{E}'_{2i}|] \\
E[|\mathcal{O}'_{2i-1}|] \\
E[|\mathcal{E}'_{2i+1}|] \\
E[|\mathcal{O}'_{2i}|]
\end{pmatrix}.
\tag{11.27}
$$

Thus, in theory, if we knew the change rate, we should be able to recover the original cardinalities of cover trace sets by substituting the trace-set cardinalities of the stego image, arguing that they should be close to their expected values as long as the number of pixels in each trace set is large. Alternatively, if we succeed in finding a condition that the trace-set cardinalities of covers must satisfy, we obtain equation(s) for the unknown change rate β. In other words, we are again following Strategy T2 from Chapter 10. By the same reasoning as in the derivation of SPA, we expect to see in covers approximately the same number of pairs with $r - s = j$ independently of whether s is even or odd, or $|\mathcal{E}_j| \approx |\mathcal{O}_j|$. LSB embedding violates this condition only for odd values of j. Indeed, the condition $|\mathcal{E}_{2i}| = |\mathcal{O}_{2i}|$ implies $E[|\mathcal{E}'_{2i}|] = E[|\mathcal{O}'_{2i}|]$ as can be easily verified from the first and fourth equations from (11.26). The condition

$$
|\mathcal{E}_{2i+1}| = |\mathcal{O}_{2i+1}|
\tag{11.28}
$$

leads to the following quadratic equation for β obtained from the third equation from (11.27) and the second equation from (11.27) written for $\mathcal{O}_{2i+1}$ rather than $\mathcal{O}_{2i-1}$:

$$
\begin{aligned}
&- \beta(1-\beta)|\mathcal{E}'_{2i}| + \beta^2|\mathcal{O}'_{2i-1}| + (1-\beta)^2|\mathcal{E}'_{2i+1}| - \beta(1-\beta)|\mathcal{O}'_{2i}| \\
&= -\beta(1-\beta)|\mathcal{E}'_{2i+2}| + (1-\beta)^2|\mathcal{O}'_{2i+1}| + \beta^2|\mathcal{E}'_{2i+3}| - \beta(1-\beta)|\mathcal{O}'_{2i+2}|,
\end{aligned}
\tag{11.29}
$$

which simplifies to

$$\beta^2 \left(|\mathcal{C}_i| - |\mathcal{C}_{i+1}| \right)$$
$$+ \beta \left(|\mathcal{E}'_{2i+2}| + |\mathcal{O}'_{2i+2}| - 2|\mathcal{E}'_{2i+1}| + 2|\mathcal{O}'_{2i+1}| - |\mathcal{E}'_{2i}| - |\mathcal{O}'_{2i}| \right)$$
$$+ \left(|\mathcal{E}'_{2i+1}| - |\mathcal{O}'_{2i+1}| \right) = 0 \tag{11.30}$$

Here, we used the fact that $|\mathcal{C}_i| = |\mathcal{C}'_i|$ and $|\mathcal{C}_i| = |\mathcal{E}'_{2i}| + |\mathcal{O}'_{2i-1}| + |\mathcal{E}'_{2i+1}| + |\mathcal{O}'_{2i}|$, which follows from (11.25). The smaller root of this quadratic equation is the change-rate estimate obtained from the ith trace set $\mathcal{C}_i$.

At this point, we have multiple choices regarding how to aggregate the available equations to estimate the change rate:

1. Sum all equations (11.30) for all indices i (or for some limited range, such as $|i| \leq 50$) and solve the resulting single quadratic equation. This option is essentially the step taken in the generalized version of SPA as it appeared in [63].
2. Solve (11.30) for some small values of $|i|$, e.g., $|i| \leq 2$, obtaining individual estimates $\hat{\beta}_i$, and estimate the final change rate as $\hat{\beta} = \min_i \hat{\beta}_i$. This choice has been shown to provide more stable results compared with SPA [127].
3. Solve (11.28) in the least-square sense [128, 163]

$$\hat{\beta} = \arg\min_{\beta} \sum_i \left(|\mathcal{E}_{2i+1}| - |\mathcal{O}_{2i+1}| \right)^2, \tag{11.31}$$

where we substitute from (11.27) for the cardinalities of cover trace sets, replacing the expected values with the observed cardinalities of stego image trace sets.

The above formulation of SPA allows direct extensions to groups of more than two pixels (see the triples analysis and quadruples analysis by Ker [128, 131], which was reported to provide more accurate change-rate estimates, especially for short messages). An alternative extension of SPA to groups of multiple pixels appeared in [61].

The least-square estimator (11.31) can be interpreted as a maximum-likelihood estimator under the assumption that the differences between the cardinalities of cover trace sets, $|\mathcal{E}_{2i+1}| - |\mathcal{O}_{2i+1}|$, are independent realizations of a Gaussian random variable. In this sense, the least-square estimator is a step in the right direction because it views the cover image as an entity unknown to the steganalyzer and postulates a statistical model for it. The iid Gaussian assumption is, however, clearly false because the cardinalities of trace sets decrease with increasing value of $|i|$ (trace sets with large values of $|i|$ are sparsely populated) and thus exhibit larger relative variations. It is not immediately clear what assumptions can be imposed on the cover trace sets to derive a "more proper" maximum-likelihood variant of SPA. An interesting solution was proposed by Ker [133], who introduced the concept of a precover consisting of pixel pairs with precisely $|\mathcal{E}_{2i+1}| + |\mathcal{O}_{2i+1}|$ pairs of pixels differing by $2i + 1$. A specific cover is then obtained by randomly (uniformly) associating each pair with either $\mathcal{E}_{2i+1}$ or $\mathcal{O}_{2i+1}$.

This assumption allows one to model the differences $|\mathcal{E}_{2i+1}| - |\mathcal{O}_{2i+1}|$ as random variables with a binomial distribution (or their Gaussian approximation) and derive an appropriate maximum-likelihood estimator of the change rate. This ML version of SPA has been shown to provide improved estimates for low embedding rates.

We close this section with a brief mention of other related approaches to detection of LSB embedding in the spatial domain. Methods based on statistics of differences between pairs of neighboring pixels include [88, 155, 256]. Approaches that use a classical signal-detection framework appeared in [40] and [55]. A qualitatively different method called the WS method [83] (Weighted Stego image) that uses local image estimators is presented in Exercises 11.3 and 11.4. An improved version of this method has been shown to produce some of the most accurate results for detection of LSB embedding in the spatial domain [138]. The WS method for the JPEG domain appeared in [21]. Another advantage of the WS method is that it is less sensitive than SPA to the assumption that the message is embedded along a pseudo-random path and thus gives better results when the embedding path is non-random or adaptive. Structural steganalysis has also been extended to detection of embedding in two LSBs in [135, 253].

11.2   Pairs Analysis

Although Pairs Analysis can detect LSB embedding in grayscale and color images in general, it was originally developed for quantitative steganalysis of methods that embed messages in indices to a sorted palette using LSB embedding. The EzStego algorithm of Chapter 5 is a typical example of such schemes. Prior to embedding, EzStego first sorts the palette colors according to their luminance and then reindexes the image data accordingly so that the visual appearance of the image does not change. Then, the usual LSB embedding in indices is applied. Besides EzStego, early versions of Steganos (http://steganos.com) and Hide&Seek (ftp://ftp.funet.fi/pub/crypt/steganography/hdsk41.zip) also employed a similar method.

Before describing Pairs Analysis, we introduce notation and analyze the impact of EzStego embedding on cover images. Let $\mathbf{c}[i] = (\mathbf{r}[i], \mathbf{g}[i], \mathbf{b}[i]), i = 0, \ldots, 255$ be the 256 colors from the sorted palette. During LSB embedding, each color can be changed only into the other color from the same color pair $\{\mathbf{c}[2k], \mathbf{c}[2k+1]\}$, $k = 0, \ldots, 127$. For a fixed k, we extract the colors $\mathbf{c}[2k]$ and $\mathbf{c}[2k+1]$ from the whole image, for example by scanning it by rows (Figure 11.8). This sequence of colors can be converted to a binary vector of the same length by associating a "0" with $\mathbf{c}[2k]$ and a "1" with $\mathbf{c}[2k+1]$. This binary vector will be called a color cut for the pair $\{\mathbf{c}[2k], \mathbf{c}[2k+1]\}$ and will be denoted $\mathbf{Z}(\mathbf{c}[2k], \mathbf{c}[2k+1])$. Because palette images have a small number of colors and natural images contain macroscopic structure, $\mathbf{Z}$ is more likely to exhibit long runs of zeros or ones rather than some random pattern. The embedding process will disturb this structure

$$Z(\,4\,,\,5\,) = (\,0\;1\;1\;1\;0\;0\;1\;0\,)$$

Figure 11.8 An example of a color cut. The pixels shaded in gray represent two colors from the same LSB pair. Crossed-out pixels in white represent the remaining colors.

and increase the entropy of $\mathbf{Z}$. Finally, when the maximal-length message has been embedded in the cover image (1 bit per pixel), $\mathbf{Z}$ will be a random binary sequence and the entropy of $\mathbf{Z}$ will be maximal.

Let us now take a look at what happens during embedding to color cuts for the "shifted" color pairs $\{\mathbf{c}[2k-1], \mathbf{c}[2k]\}$, $k = 1, \ldots, 127$. During embedding, the colors $\mathbf{c}[2k-2]$ and $\mathbf{c}[2k-1]$ are exchanged for each other and so are the colors $\mathbf{c}[2k]$ and $\mathbf{c}[2k+1]$. Even after embedding the maximal message (each pixel modified with probability $\frac{1}{2}$), the color cut $\mathbf{Z}(\mathbf{c}[2k-1], \mathbf{c}[2k])$ will still show some residual structure. To see this, imagine a binary sequence $\mathbf{W}$ that was formed from the cover image by scanning it by rows and associating a "0" with the colors $\mathbf{c}[2k-2]$ and $\mathbf{c}[2k-1]$ and a "1" with the colors $\mathbf{c}[2k]$ and $\mathbf{c}[2k+1]$. Convince yourself that, after embedding a maximal pseudo-random message in the image, the color cut $\mathbf{Z}(\mathbf{c}[2k-1], \mathbf{c}[2k])$ is the same as starting with the sequence $\mathbf{W}$ and skipping each element of $\mathbf{W}$ with probability $\frac{1}{2}$. Because $\mathbf{W}$ showed structure in the cover image, most likely long runs of 0s and 1s, we see that randomly chosen subsequences of $\mathbf{W}$ will show some residual structure as well.

We are now ready to describe the steganalysis method. Denoting the concatenation of bit strings with the symbol "&," we first concatenate all color cuts $\mathbf{Z}(\mathbf{c}[2k], \mathbf{c}[2k+1])$ into one vector

$$\mathbf{Z} = \mathbf{Z}(\mathbf{c}[0], \mathbf{c}[1]) \& \ldots \& \mathbf{Z}(\mathbf{c}[254], \mathbf{c}[255]), \tag{11.32}$$

and all color cuts for shifted pairs $\mathbf{Z}(\mathbf{c}[2k-1], \mathbf{c}[2k])$ into

$$\mathbf{Z}' = \mathbf{Z}(\mathbf{c}[1], \mathbf{c}[2]) \& \ldots \& \mathbf{Z}(\mathbf{c}[253], \mathbf{c}[254]) \& \mathbf{Z}(\mathbf{c}[255], \mathbf{c}[0]). \tag{11.33}$$

Next, we define a simple measure of structure in a binary vector as the number of homogeneous bit pairs 00 and 11 in the vector. For example, a vector of n 1s will have $n-1$ homogeneous bit pairs. We denote by $R(\beta)$ the expected relative[3] number of homogeneous bit pairs in $\mathbf{Z}$ after flipping the LSBs of indices of a fraction β of randomly chosen pixels, $0 \leq \beta \leq 1$. We recognize β as the change rate. Similarly, let $R'(\beta)$ be the expected relative number of homogeneous bit pairs in $\mathbf{Z}'$. For $\beta < \frac{1}{2}$, this change rate corresponds to relative payload $\alpha = 2\beta$.

Exercise 11.1 shows that $R(x)$ is a quadratic polynomial with its vertex at $x = \frac{1}{2}$ and $R\left(\frac{1}{2}\right) = (n-1)/(2n) \approx \frac{1}{2}$ (see Figure 11.9). The value $R(\beta)$ is known

---

[3] By relative, we mean the number of pairs normalized by n.

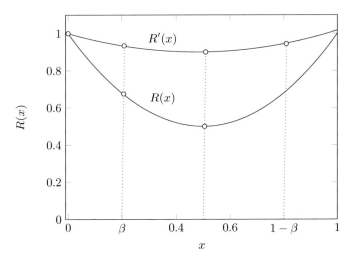

Figure 11.9 Expected number of homogeneous pairs $R(x)$ and $R'(x)$ in color cuts $\mathbf{Z}$ and $\mathbf{Z}'$ as a function of the change rate x. The circles correspond to y values that can be obtained from the stego image with unknown change rate β. The values $R(0)$ and $R'(0)$ are not known but satisfy $R(0) = R'(0)$.

as it can be calculated from the stego image. Note that $R(\beta) = R(1 - \beta)$ is also known.

The value of $R'\left(\frac{1}{2}\right)$ can be derived from $\mathbf{Z}'$ (see Exercise 11.2), while $R'(\beta)$ and $R'(1 - \beta)$ can be calculated from the stego image and the stego image with all colors flipped, respectively. Modeling $R'(x)$ as a second-degree polynomial, the difference $D(x) = R(x) - R'(x) = Ax^2 + Bx + C$ is also a second-degree polynomial.

Finally, we accept one additional assumption,

$$R(0) = R'(0), \tag{11.34}$$

which says that the number of homogeneous pairs in $\mathbf{Z}$ and $\mathbf{Z}'$ must be the same if no message has been embedded. This is, indeed, intuitive because there is no reason why the color cuts of pairs and shifted pairs in the cover image should have different structures.

In summary, we know the four values

$$D(0) = R(0) - R'(0) = 0, \tag{11.35}$$
$$D(1/2) = R(1/2) - R'(1/2), \tag{11.36}$$
$$D(\beta) = R(\beta) - R'(\beta), \tag{11.37}$$
$$D(1 - \beta) = R(1 - \beta) - R'(1 - \beta), \tag{11.38}$$

which gives us four equations for four unknowns A, B, C, β:

$$C = 0, \tag{11.39}$$
$$4D(1/2) = A + 2B, \tag{11.40}$$
$$D(\beta) = A\beta^2 + B\beta, \tag{11.41}$$
$$D(1 - \beta) = A(1 - \beta)^2 + B(1 - \beta). \tag{11.42}$$

It can easily be verified that

$$\beta\left(D(\beta) - D(1 - \beta) + 4D(1/2)\right) = 2\beta^2 A + 2\beta^2 B + \beta B \tag{11.43}$$
$$= D(\beta) + \beta^2 4D(1/2), \tag{11.44}$$

which is a quadratic equation for β,

$$4D(1/2)\beta^2 - \left(D(\beta) - D(1 - \beta) + 4D(1/2)\right)\beta + D(\beta) = 0. \tag{11.45}$$

The smaller of the two roots is our approximation to the unknown change rate β. The pseudo-code for Pairs Analysis is shown in Algorithm 11.2.

11.2.1   Experimental verification of Pairs Analysis

The performance of Pairs Analysis is demonstrated using experiments on a database of 180 color GIF images. The images were originally stored in a high-quality JPEG format and came from four different digital cameras. For the test, the images were resampled to 800×600 pixels using Corel Photo-Paint 9 (with the anti-alias option) and converted to palette images with the following options: optimized palette, ordered dithering. All images were embedded using the EzStego algorithm with pseudo-random message spread with relative payload $\alpha = 0, 0.2, 0.4, 0.6, 0.8, 1$ and then processed using Pairs Analysis. The results are shown in Figure 11.11.

Pairs Analysis can be further improved by scanning the image along a space-filling curve rather than in a row-by-row manner [243]. The Hilbert scan (Figure 11.10) is an example of a scanning order that becomes in a limit a space-filling curve [184]. Because this scanning order is more likely to capture uniform segments in the image than the simple raster order, the corresponding color cuts have a higher number of homogeneous pairs, which translates into slightly higher detection accuracy.

11.3   Targeted attack on F5 using calibration

In this section, we illustrate Strategy T3 for design of targeted steganalysis (from Section 10.3) by describing an attack on the F5 algorithm (see the description of the algorithm in Section 7.3.2). From Exercise 7.2, we know that the histogram of absolute values of DCT coefficients $\mathbf{h}_\beta^{(kl)}$ corresponding to DCT mode (k, l) in

---

Algorithm 11.2 Pairs Analysis. γ is a threshold on the test statistic $\hat{\beta}$ set to achieve $P_{\text{FA}} < \epsilon_{\text{FA}}$, where ϵ_{FA} is the bound on the false-alarm rate.

```
// Arrange all pixels by scanning the image along
// a continuous path into a 1d vector v[i], i = 1, ..., M × N
v_flip = LSBflip(v);   // flip LSBs of all stego elements
Z = {∅}; Z' = {∅}; Z'' = {∅};
for k = 0 to 127  Z = Z&ColorCut(v, 2k, 2k + 1);
for k = 0 to 126  {
   Z' = Z'&ColorCut(v, 2k + 1, 2k + 2);
   Z'' = Z''&ColorCut(v_flip, 2k + 1, 2k + 2);
}
Z' = Z'&ColorCut(v, 255, 1); Z'' = Z''&ColorCut(v_flip, 255, 1);
D(β) = CountHomog(Z)-CountHomog(Z');
D(1 - β) = CountHomog(Z)-CountHomog(Z'');
D(1/2) = 1/2 - R'(1/2); // compute R'(1/2) from Exercise 11.2
a = 4D(1/2); b = D(1 - β) - D(β) - a; c = D(β);
β± = Re((-b ± √(b² - 4ac))/(2a));  β̂ = min(β+, β−);
if β̂ > γ {
   output('Image is stego');
   output('Estimated change rate = ', β̂);
}
function: y = CountHomog(b)
   y = 0; n_b = length(b);
   for i = 1 to n_b {
      y = y + b[i]b[i − 1] + (1 − b[i])(1 − b[i − 1]);
   }
   y = y/n_b;
function: Z = ColorCut(x, 2k, 2k + 1)
   j = 0; n_x = length(x);
   for i = 1 to n_x {
      if (x[i] = 2k or x[i] = 2k + 1) {
         Z[j] = x[i] mod 2; j = j + 1;
      }
   }
```

---

an F5-embedded stego image satisfies

$$E\left[\mathbf{h}_\beta^{(kl)}[i]\right] = (1 - \beta)\mathbf{h}^{(kl)}[i] + \beta\mathbf{h}^{(kl)}[i + 1], \quad i > 0, \tag{11.46}$$

$$E\left[\mathbf{h}_\beta^{(kl)}[0]\right] = \mathbf{h}^{(kl)}[0] + \beta\mathbf{h}^{(kl)}[1], \tag{11.47}$$

where β is the embedding change rate. Because F5 preserves the symmetry of the histograms, we cannot use the same approach as in attacking Jsteg. Instead,

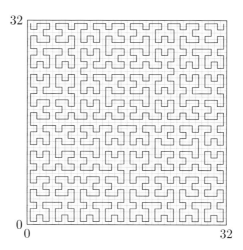

Figure 11.10 Pixel-scanning order along a Hilbert curve for a 32×32 image.

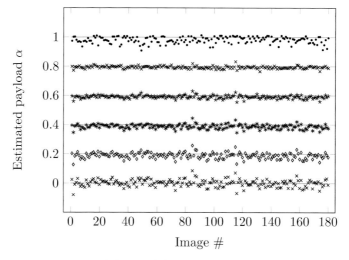

Figure 11.11 Estimated payload ($\alpha = 2\beta$) from 180 GIF images embedded using EzStego with relative payloads $\alpha = 0, 0.2, 0.4, 0.6, 0.8, 1$. The straight lines mark the true expected change rate.

the histogram of absolute values of DCT coefficients for the cover image, $\mathbf{h}_0^{(kl)}$, is estimated from the stego image using calibration as described in Section 10.3.

Denoting the estimated histograms of absolute values of DCT coefficients corresponding to spatial frequency (k, l) as $\hat{\mathbf{h}}_0^{(kl)}$, the change rate can be estimated from equations (11.46)–(11.47), where we substitute $\hat{\mathbf{h}}_0^{(kl)}$ for the cover image histograms and replace the expected values with the sample values

$$\mathbf{h}_\beta^{(kl)}[i] = (1 - \beta)\hat{\mathbf{h}}_0^{(kl)}[i] + \beta\hat{\mathbf{h}}_0^{(kl)}[i + 1], \quad i > 0, \tag{11.48}$$

$$\mathbf{h}_\beta^{(kl)}[0] = \hat{\mathbf{h}}_0^{(kl)}[0] + \beta\hat{\mathbf{h}}_0^{(kl)}[1]. \tag{11.49}$$

This is a system of linear equations for various values of (k, l) and i for just one unknown – the change rate β. Not all values (k, l) and i, however, should be used. The histograms are in general less populated for higher spatial frequencies and higher values of i. Again, we have several possibilities for how to aggregate the equations to obtain the change-rate estimate (see Section 11.1.3). For brevity, here we present only the approach proposed in [86].

The steganalyst is advised to obtain three least-square estimates $\hat{\beta}_{01}, \hat{\beta}_{10}, \hat{\beta}_{11}$ from histograms with $(k, l) \in \{(0, 1), (1, 0), (1, 1)\}$ and $i = 0, 1$,

$$\hat{\beta}_{kl} = \arg\min_\beta \left(\mathbf{h}_\beta^{(kl)}[0] - \hat{\mathbf{h}}_0^{(kl)}[0] - \beta\hat{\mathbf{h}}_0^{(kl)}[1]\right)^2$$

$$+ \left(\mathbf{h}_\beta^{(kl)}[1] - (1 - \beta)\hat{\mathbf{h}}_0^{(kl)}[1] - \beta\hat{\mathbf{h}}_0^{(kl)}[2]\right)^2, \tag{11.50}$$

which can be solved analytically because the function that is minimized is a quadratic polynomial in β,

$$\hat{\beta}_{kl} = \frac{\hat{\mathbf{h}}_0^{(kl)}[1]\left(\mathbf{h}_\beta^{(kl)}[0] - \hat{\mathbf{h}}_0^{(kl)}[0]\right) + \left(\mathbf{h}_\beta^{(kl)}[1] - \hat{\mathbf{h}}_0^{(kl)}[1]\right)\left(\hat{\mathbf{h}}_0^{(kl)}[2] - \hat{\mathbf{h}}_0^{(kl)}[1]\right)}{\left(\hat{\mathbf{h}}_0^{(kl)}[1]\right)^2 + \left(\hat{\mathbf{h}}_0^{(kl)}[2] - \hat{\mathbf{h}}_0^{(kl)}[1]\right)^2}. \tag{11.51}$$

The final estimate $\hat{\beta}$ is obtained as

$$\hat{\beta} = \frac{\hat{\beta}_{01} + \hat{\beta}_{10} + \hat{\beta}_{11}}{3}. \tag{11.52}$$

The performance of this estimator is evaluated in [87].

11.4   Targeted attacks on ± 1 embedding

The main reason why LSB embedding in the spatial domain can be detected very reliably is the non-symmetrical character of this embedding operation. By symmetrizing the embedding operation to "add or subtract 1 at random" (± 1 embedding) instead of flipping the LSB, a majority of accurate attacks on LSB embedding is thwarted. In this section, we show some targeted attacks on ± 1 embedding, which are, in fact, applicable to more general embedding paradigms in the spatial domain as long as the impact of embedding the message can be described as adding iid noise to the image (e.g., stochastic modulation from Section 7.2.1).

The first attempts to construct steganalytic methods for detection of embedding by noise adding appeared in [39, 40, 242]. We now describe the approach proposed in [107] and its extension [129, 130].

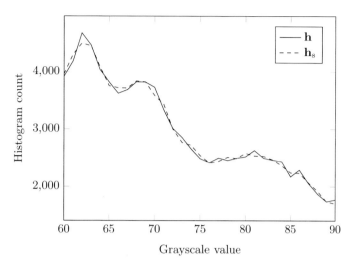

Figure 11.12 Adding stego noise to an image smoothens its histogram.

Any steganographic scheme that embeds messages by adding independent noise to the cover image will smooth the image histogram (see Figure 11.12). This is because the stego image can be viewed as a sum of two independent random variables – the cover image and the stego noise. Since the probability mass function of the sum of two independent random variables is the convolution of their probability mass functions, the stego image histogram, $\mathbf{h}_\mathrm{s}$, is a low-pass version of the cover-image histogram, $\mathbf{h}$,

$$\mathbf{h}_\mathrm{s} = \mathbf{h} \star \mathbf{f}, \tag{11.53}$$

or

$$\mathbf{h}_\mathrm{s}[i] = \sum_j \mathbf{h}[j]\mathbf{f}[i - j], \tag{11.54}$$

where the indices i, j run over the index set determined by the number of colors in the histogram (e.g., for 8-bit grayscale images, $i, j \in \{0, \ldots, 255\}$). In (11.53), $\mathbf{f}$ is the probability mass function of the stego noise. The specific form of this convolution for ±1 embedding has been derived in Exercise 7.1.

This observation gives us an idea for deriving useful features for steganalysis by analyzing the histogram smoothness. Due to the low-pass character of the convolution, $\mathbf{h}_\mathrm{s}$ will be smoother than $\mathbf{h}$ and thus its energy will be concentrated in lower frequencies. This can be captured by switching to the Fourier representation of the histograms and the noise pmf. For clarity, in this section we will denote Fourier-transformed quantities of all variables with the corresponding capital letters. The Discrete Fourier Transform (DFT) of an N-dimensional vector $\mathbf{x}$ is defined as

$$\mathbf{X}[k] = \sum_{j=0}^{N-1} \mathbf{x}[j] e^{-\mathrm{i}\frac{2\pi jk}{N}}, \tag{11.55}$$

where i in (11.55) stands for the imaginary unit ($\mathrm{i}^2 = -1$). The Fourier transform of the stego-image histogram is obtained as an elementwise multiplication of the cover-image histogram and the noise pmf,

$$\mathbf{H}_{\mathrm{s}}[k] = \mathbf{H}[k]\mathbf{F}[k] \text{ for each } k. \tag{11.56}$$

The function $\mathbf{H}_{\mathrm{s}}$ is called the Histogram Characteristic Function (HCF) of the stego image. At this point, a numerical quantity is needed that can be computed from the HCF and that would evaluate the location of the energy in the spectrum. Because the absolute value of the DFT is symmetrical about the midpoint value $k = N/2$, a reasonable measure of the energy distribution is the Center Of Gravity (COG) of $|\mathbf{H}|$ computed for indices $k = 0, \dots, N/2 - 1$,

$$\mathrm{COG}(\mathbf{H}) = \frac{\sum_{k=0}^{N/2-1} k|\mathbf{H}[k]|}{\sum_{k=0}^{N/2-1} |\mathbf{H}[k]|}. \tag{11.57}$$

It can be shown using the Chebyshev sum-inequality (see Exercise 11.5) that as long as $|\mathbf{F}[k]|$ is non-increasing,[4]

$$\mathrm{COG}(\mathbf{H}_{\mathrm{s}}) \leq \mathrm{COG}(\mathbf{H}), \tag{11.58}$$

which can be intuitively expected because the stego image histogram is smoother and thus the energy of the HCF will shift towards lower frequencies (see Figure 11.13).

For steganalysis of color images, we will use the three-dimensional color histogram, $\mathbf{h}[j_1, j_2, j_3]$, which denotes the number of pixels with their RGB color (j_1, j_2, j_3). Furthermore, the one-dimensional DFT (11.55) is replaced with its three-dimensional version

$$\mathbf{H}[k_1, k_2, k_3] = \sum_{j_1,j_2,j_3=0}^{N-1} \mathbf{h}[j_1, j_2, j_3] e^{-\mathrm{i}\frac{2\pi(j_1 k_1 + j_2 k_2 + j_3 k_3)}{N}}. \tag{11.59}$$

Due to the symmetry of the three-dimensional DFT, we constrain ourselves to the first octant, $k_1, k_2, k_3 \geq 0$, and compute the three-dimensional COG with its mth coordinate, $m = 1, 2, 3$,

$$\mathrm{COG}(\mathbf{H})[m] = \frac{\sum_{k_1,k_2,k_3=0}^{\frac{N}{2}-1} k_m |\mathbf{H}[k_1, k_2, k_3]|}{\sum_{k_1,k_2,k_3=0}^{\frac{N}{2}-1} |\mathbf{H}[k_1, k_2, k_3]|}. \tag{11.60}$$

The center of gravity (11.57) and (11.60) can be used directly as a feature for steganalysis. For color images with a low level of noise, such as decompressed

---

[4] Many known noise distributions, such as the Gaussian or Laplacian distributions, have monotonically decreasing $|\mathbf{F}[k]|$.

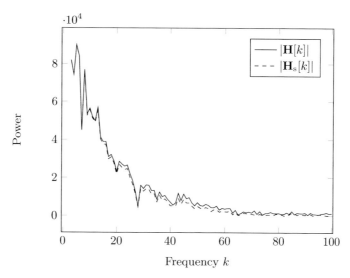

Figure 11.13 Stego image HCF falls off to zero faster because the stego image histogram is smoother.

JPEGs or professionally enhanced images, it is possible to identify a fixed threshold separating the COG of cover and fully embedded stego images relatively well [107]. Figure 11.14 (top) shows the COG of 186 cover and stego images embedded with change rate $\beta = 0.25$ with ± 1 embedding (the change rate is computed with respect to $3n$, where n is the number of pixels in the image). The images for this test came from the test database of 954 images described in Section 5.1.2. They were taken by three different cameras in an uncompressed format at their native resolution and then converted to true-color BMP images and JPEG compressed with quality factor 75. All images were decompressed back to the spatial domain and cropped to their central 800×600 region. To obtain a scalar value that can be easily compared, the figure shows the average

$$\frac{\mathrm{COG}[1] + \mathrm{COG}[2] + \mathrm{COG}[3]}{3}. \tag{11.61}$$

Since the three-dimensional COGs lie very close to the axis of the first octant, the averaging does not affect the separability of the statistic between cover and stego images in any significant manner.

Figure 11.14 (bottom) shows the result of the exact same experiment performed directly on the raw images rather than their JPEG compressed form. Observe that, even though the embedding does decrease the value of the COG, it is no longer possible to separate cover and stego images because the COG of covers varies too much. Surprisingly, the steganalysis of images from a Kodak DC290 camera still works. This is because this camera seems to suppress the noise in images as part of its in-camera processing chain. Thus, subsequent steganographic embedding is more detectable.

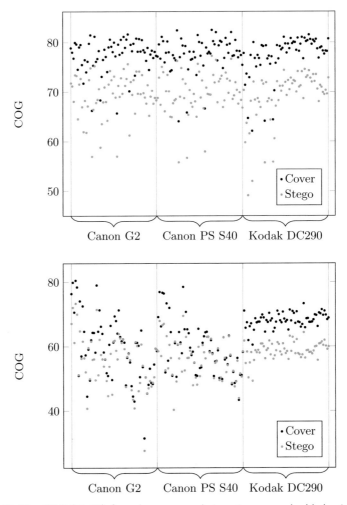

Figure 11.14 The COG (11.61) for color cover and stego images embedded using ± 1 embedding in the spatial domain with change rate $\beta = 0.25$. The bottom graph was generated from raw, never-compressed color images from three different digital cameras, while the top graph corresponds to their JPEG compressed versions (quality factor 75). All images were cropped to their central 800×600 portion.

In general, the performance of steganalysis based on COG of the HCF varies greatly with the cover source and it is not possible to identify a universal threshold that would separate cover and stego images well. The detection becomes even worse for grayscale images. We obviously need a way to calibrate the COG to remove its dependence on cover source and content.

It has been proposed in [130] to estimate the COG of the cover image by resizing the stego image to half its size by averaging values in groups of 2×2 pixels. It is hoped that the resizing will produce just the right amount of blurring to enable approximate estimation of the COG of the cover image HCF. Given a stego image $\mathbf{y}[i,j], i = 1, \ldots, M, j = 1, \ldots, N$, the downsampled stego image $\hat{\mathbf{y}}[k,l], k = 1, \ldots, \lfloor M/2 \rfloor, l = 1, \ldots, \lfloor N/2 \rfloor$, is obtained as

$$\hat{\mathbf{y}}[k,l] = \frac{1}{4}\left(\mathbf{y}[2k,2l] + \mathbf{y}[2k+1,2l] + \mathbf{y}[2k,2l+1] + \mathbf{y}[2k+1,2l+1]\right). \quad (11.62)$$

Denoting by $\hat{\mathbf{H}}$ and $\mathrm{COG}(\hat{\mathbf{H}})$ the HCF and its COG for the resized image, $\hat{\mathbf{y}}$, the ratio

$$\frac{\mathrm{COG}(\mathbf{H})}{\mathrm{COG}(\hat{\mathbf{H}})} \quad (11.63)$$

is taken as the calibrated COG feature for steganalysis. For certain cover sources, the calibrated COGs for cover and stego images are better separated than $\mathrm{COG}(\mathbf{H_s})$ and $\mathrm{COG}(\mathbf{H})$ because dividing by the estimated COG decreases image-to-image variations [130].

For grayscale images, the performance can be improved also by working with a better model than the pixels' sample pmf. It is particularly appealing to work with a higher-order statistical model. Denoting the set of all pairs of horizontally adjacent pixels in the image as $\mathcal{P}$, we represent the image with an adjacency histogram, also called a co-occurrence matrix,

$$\mathbf{t}[i,j] = \{(x,y) \in \mathcal{P} | x = i, y = j\}. \quad (11.64)$$

In other words, $\mathbf{t}[i,j]$ is defined as the number of horizontally neighboring pixel pairs with grayscales i and j. This matrix is sparser than the histogram (because there are 256^2 possible pairs) and thus reacts more sensitively to embedding changes. Because of local correlations present in natural images, the adjacency histogram has the largest values on the diagonal and then it quickly falls off (see Figure 11.1). The HCF of the adjacency histogram is now two-dimensional, $\mathbf{T}[k,l]$, and can be used for steganalysis in the same way as the HCF above. The author in [130] used an alternative scalar quantity defined as

$$\mathrm{COG}(\mathbf{T}) = \frac{\sum_{k,l=0}^{\frac{N}{2}-1}(k+l)|\mathbf{T}[k,l]|}{\sum_{k,l=0}^{\frac{N}{2}-1}|\mathbf{T}[k,l]|}. \quad (11.65)$$

Figure 11.15 shows the improvement in detection of ± 1 embedding in grayscale decompressed JPEG images by computing the COG from the adjacency histogram (11.65) rather than from the histogram itself (11.57). The experiment was performed on the same set of images coming from three digital cameras and cropped to their central 800×600 portion. Note that while the COG computed from the HCF does not have any significant distinguishing power (top), the COG of the adjacency histogram seems to work reasonably well on this image set (bottom). The reader is referred to the original publication [129, 130] for a more detailed discussion of experimental results of the above methods when tested on various cover sources. Further extension of this approach appears in [159], where the authors proposed to work with differences of neighboring pixels rather than the pixel values directly. Overall, the performance of steganalysis based on

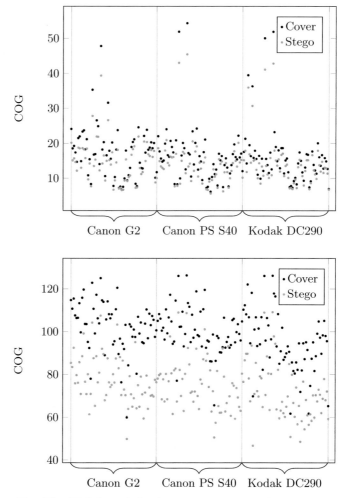

Figure 11.15 Top: The COG (11.57) for the same images as in Figure 11.14 converted to grayscale and JPEG compressed using quality factor 75. The embedding change rate was $\beta = 0.25$ (w.r.t. the number of pixels). Bottom: The COG of cover and stego images of the adjacency histogram (11.65) computed for the same grayscale JPEG images.

HCF appears to vary considerably across different image sources. This is true in general for most spatial-domain steganalyzers, including the blind constructions explained in Section 12.5 [37, 139].

Other methods for detection of noise adding and ±1 embedding include methods based on signal estimation [55, 220, 249], histogram artifacts [37, 38], blind steganalyzers [10, 252], and blind steganalysis methods described in Section 12.5.

Summary

- LSB embedding in the spatial domain can be reliably detected even at low relative payloads due to the asymmetry of the embedding operation.
- Sample Pairs Analysis is an example of an attack on LSB embedding in pseudo-randomly spread pixels. It works by analyzing the embedding impact on subsets of pairs of neighboring pixels.
- Structural steganalysis is a reformulation of SPA using the concept of trace sets. This formulation makes it possible to derive detectors that can incorporate statistical assumptions about the cover image and provide a convenient framework for generalizing SPA to work with groups of more than two pixels.
- Pairs Analysis is another method for detection of LSB embedding that is based on a different principle by considering the spatial distribution of colors from each LSB pair in the entire image. It is especially suitable for attacking LSB embedding in palette image formats.
- The steganographic algorithm F5 can be attacked using calibration by first quantifying the relationship between the histograms of stego and cover images and then estimating the cover-image histogram using calibration.
- ± 1 embedding in the spatial domain can be detected using the center of gravity of the histogram characteristic function (absolute value of the Fourier transform of the histogram) as the feature. This is because adding an independent stego noise to the cover image smooths the histogram. The accuracy of this method can be improved by considering the adjacency histogram and by applying calibration (resampling the stego image).

Exercises

11.1 [Pairs analysis I] Prove that the expected value of the relative number of homogeneous pairs $R(\beta)$ for color cut $\mathbf{Z}$ for change rate $\beta \in [0, 1]$ is a parabola with its minimum at $\frac{1}{2}$. In particular, $R\left(\frac{1}{2}\right) = (n-1)/(2n) \approx \frac{1}{2}$ for large n, where n is the length of the color cut (number of pixels in the image). **Hint:** Write the color cut $\mathbf{Z}$ of the cover image as a concatenation of r segments consisting of consecutive runs of $k_1, \ldots, k_r$ 0s or 1s, $k_1 + \cdots + k_r = n$. Thus, $R(0) = \sum_{i=1}^{r} (k_i - 1) = n - r$. After changing the LSB of a random portion of β pixels, the probability that a homogeneous pair of consecutive bits will stay homogeneous is $\beta^2 + (1 - \beta)^2$. The expected number of homogeneous pairs in the ith segment is thus $(\beta^2 + (1 - \beta)^2)(k_i - 1) + 2\beta(1 - \beta)$, where the last term comes from the right end of the segment (an additional pair will be formed at the boundary if the last bit in the segment flips and the first bit of the next segment does not flip, or vice versa). This last term is missing from the last segment.

11.2 [Pairs analysis II] Let $\mathbf{Z}' = \{b[i]\}_{i=1}^{n}$ be the color cut for the shifted color pairs and $R'(\beta)$ be the number of homogeneous pairs after flipping a portion

β of pixels. Prove that the expected value of $R'\left(\frac{1}{2}\right)$ is

$$E\left[R'\left(\frac{1}{2}\right)\right] = \sum_{k=1}^{n-1} 2^{-k} h_k, \qquad (11.66)$$

where h_k is the number of homogeneous pairs in the sequence of pairs

$$(\mathbf{b}[1], \mathbf{b}[1+k]), (\mathbf{b}[2], \mathbf{b}[2+k]), (\mathbf{b}[3], \mathbf{b}[3+k]), \ldots, (\mathbf{b}[n-k], \mathbf{b}[n]). \qquad (11.67)$$

Hint: Let $\mathbf{W} = \mathbf{W}_1 \& \cdots \& \mathbf{W}_{128}$ be the concatenation of binary sequences $\mathbf{W}_j$ formed from the stego image by scanning it by rows and associating a 0 with the colors $\mathbf{c}[2j-2]$ and $\mathbf{c}[2j-1]$ and a 1 with the colors $\mathbf{c}[2j]$ and $\mathbf{c}[2j+1]$. The expected form of the color cut $\mathbf{Z}'(\mathbf{c}[2j-1], \mathbf{c}[2j])$ after embedding a maximal message in the cover image is the same as starting with the sequence $\mathbf{W}_j$ and skipping each element of $\mathbf{W}_j$ with probability $\frac{1}{2}$. Imagine you are going through $\mathbf{Z}'$ while skipping each element with probability $\frac{1}{2}$. Then, the probability of skipping exactly $k-1$ elements in a row is 2^{-k}, $k = 1, 2, \ldots$. Because there are h_k homogeneous pairs in the sequence of pairs $\mathbf{b}[1]\mathbf{b}[1+k]$, $\mathbf{b}[2]\mathbf{b}[2+k]$, $\mathbf{b}[3]\mathbf{b}[3+k]$, $\ldots$, $\mathbf{b}[n-k]\mathbf{b}[n]$, the expected number of homogeneous pairs separated by $k-1$ elements is $2^{-k}h_k$. The formula for the expected value is obtained by summing these contributions from $k=1$ to the maximal separation satisfying $k-1 = n-2$.

11.3 [Weighted stego image] In this (and the next) exercise, you will derive another attack on LSB embedding in the spatial domain called a Weighted Stego (WS) attack. Let $\mathbf{x}[i]$, $i = 1, \ldots, n$ be the pixel values from an 8-bit grayscale cover image containing n pixels. The value of $\mathbf{x}[i]$ after flipping its LSB will be denoted with a bar,

$$\bar{\mathbf{x}}[i] \triangleq \text{LSBflip}(\mathbf{x}[i]) = \mathbf{x}[i] + 1 - 2(\mathbf{x}[i] \bmod 2). \qquad (11.68)$$

Let $\mathbf{y}[i]$ denote the stego image after flipping the fraction β of pixels along a pseudo-random path (this corresponds to embedding an unbiased binary message of relative message length 2β). Let $\mathbf{w}_\theta[i]$ be the "weighted" stego image,

$$\mathbf{w}_\theta[i] = \mathbf{y}[i] + \theta(\bar{\mathbf{y}}[i] - \mathbf{y}[i]), \ 0 \le \theta \le 1, \qquad (11.69)$$

with weight θ. Let

$$D(\theta) = \sum_{i=1}^{n} (\mathbf{w}_\theta[i] - \mathbf{x}[i])^2 \qquad (11.70)$$

be the sum of squares of the differences between the pixels of the weighted stego image and the cover image. Show that $D(\theta)$ is minimal for $\theta = \beta$. In other words,

$$\beta = \arg\min_{\theta} D(\theta) = \arg\min_{\theta} \sum_{i=1}^{n} (\mathbf{w}_\theta[i] - \mathbf{x}[i])^2. \qquad (11.71)$$

Hint: Substitute (11.69) into (11.71) and divide the sum over pixels for which $\mathbf{x}[i] = \mathbf{y}[i]$ (unmodified pixels) and $\mathbf{y}[i] = \bar{\mathbf{x}}[i]$ (flipped pixels). Then simplify and differentiate with respect to θ to find the minimum of the polynomial quadratic in θ.

11.4 **[WS attack on LSB embedding]** Equation (11.71) cannot be used directly for estimating the change rate β from the stego image because the cover-image pixels $\mathbf{x}[i]$ are not known to the steganalyst. To turn the result of the previous exercise into an attack, replace $\mathbf{x}[i]$ with its estimate $\hat{\mathbf{x}}[i]$ obtained from its neighboring values and find the minimum value of (11.70) by differentiating it with respect to θ and solving $D'(\theta) = 0$ for θ. You should obtain the following estimator of the change rate:

$$\hat{\beta} = \frac{1}{n} \sum_{i=1}^{n} (\mathbf{y}[i] - \bar{\mathbf{y}}[i])(\mathbf{y}[i] - \hat{\mathbf{x}}[i]). \tag{11.72}$$

This estimator can be further improved by introducing non-negative local weights $\mathbf{w}[i]$,

$$\hat{\beta} = \sum_{i=1}^{n} \mathbf{w}[i](\mathbf{y}[i] - \bar{\mathbf{y}}[i])(\mathbf{y}[i] - \hat{\mathbf{x}}[i]), \tag{11.73}$$

where $\sum_i \mathbf{w}[i] = 1$. The purpose of the weights is to give more emphasis to those pixels where we expect $\hat{\mathbf{x}}[i]$ to be a more accurate estimate of the cover pixel and less emphasis to those pixels that are likely to be poorly estimated. Because our ability to estimate the cover image is better in smooth regions, the weights should be inversely proportional to some local measure of texture. A good, albeit empirical, choice is $\mathbf{w}[i] = 1/(5 + \hat{\sigma}^2[i])$, where $\hat{\sigma}^2[i]$ is the sample pixel variance estimated from a 3×3 neighborhood of pixel i.
Implement this estimator for

$$\hat{\mathbf{x}}[i, j] = \frac{1}{4} \left(\mathbf{y}[i-1, j] + \mathbf{y}[i+1, j] + \mathbf{y}[i, j-1] + \mathbf{y}[i, j+1] \right) \tag{11.74}$$

and test its performance on images. More details about this estimator can be found in [83, 138]. This attack can be adapted to work for JPEG images as well [21, 244].

11.5 **[Monotonicity of COG]** Prove (11.58) using the Chebyshev sum-inequality [176] valid for any non-decreasing sequence $\mathbf{a}[i]$, non-increasing sequence $\mathbf{b}[i]$, and a non-negative sequence $\mathbf{p}[i]$,

$$\sum_{i=0}^{N-1} \mathbf{p}[i] \sum_{i=0}^{N-1} \mathbf{p}[i]\mathbf{a}[i]\mathbf{b}[i] \leq \sum_{i=0}^{N-1} \mathbf{p}[i]\mathbf{a}[i] \sum_{i=0}^{N-1} \mathbf{p}[i]\mathbf{b}[i]. \tag{11.75}$$

Hint: Substitute into the Chebyshev inequality

$$\mathbf{p}[i] = |\mathbf{H}[i]|, \tag{11.76}$$
$$\mathbf{a}[i] = i, \tag{11.77}$$
$$\mathbf{b}[i] = |\mathbf{F}[i]|. \tag{11.78}$$

11.6 [**Attack on MBS**] Consider Model-Based Steganography for JPEG images as explained in Chapter 7 and mount the following targeted attack that follows Strategy T1 from Section 10.3. The histogram of stego images follows the model obtainable from both cover and stego images. The model is a generalized Cauchy distribution and thus symmetrical. Real images, however, rarely have perfectly symmetrical histograms. Thus, it might be possible to construct a targeted attack by testing using the chi-square test if the image histogram follows the model.

12 Blind steganalysis

The goal of steganalysis is to detect the presence of secretly embedded messages. Depending on how much information the warden has about the steganographic channel she is trying to attack, the detection problem can accept many different forms. In the previous chapter, we dealt with the situation when the warden knows the steganographic method that Alice and Bob might be using. With this knowledge, Eve can tailor her steganalysis to the particular steganographic channel using several strategies outlined in Section 10.3. If Eve has no information about the steganographic method, she needs blind steganalysis capable of detecting as wide a spectrum of steganographic methods as possible. Design and implementation of practical blind steganalysis detectors is the subject of this chapter.

The first and most fundamental step for Eve is to accept a model of cover images and represent each image using a vector of features. In contrast to targeted steganalysis, where a single feature (e.g., an estimate of message length) was often enough to construct an accurate detector, blind steganalysis by definition requires many features. This is because the role of features in blind steganalysis is significantly more fundamental – in theory they need to capture all possible patterns natural images follow so that every embedding method the prisoners can devise disturbs at least some of the features. In Section 10.4, we loosely formulated this requirement as *completeness* of the feature space and outlined possible strategies for constructing good features.

The second step in building a blind steganalyzer is selecting a classification tool. There are many choices, including neural networks, clustering algorithms, support vector machines, and other tools of soft computing, pattern recognition, and data mining. It seems that the method of choice today is Support Vector Machines (SVMs) due to their ease of implementation and tuning as well as superior performance. In Appendix E, we provide a brief introduction to the theory of SVMs and their implementation and training. Readers not familiar with this approach to machine-learning should read this appendix and become familiar with the approach at least on a conceptual level.

The third and final step is the actual training of the classifier and setting its parameters to satisfy desired performance criteria, such as making sure that the false-alarm rate for known steganography algorithms is below a certain bound. An essential part of training is the database of images. Ideally, it should reflect the properties of the cover source. For example, if Eve knows that Alice likes to send to Bob images taken with her camera, the warden may purchase a camera of the same model and use it to create the training database. If the warden has little or no information about the cover source, which would be the case of an automatic traffic-monitoring device, constructing a blind steganalyzer is much harder. There are significant differences in how difficult it is to detect steganographic embedding in various cover sources. For example, as already mentioned in Chapter 11, scans of film or analog photographs are typically very noisy and contain characteristic microscopic structure due to the grains present in film. Because this structure is stochastic in nature, it complicates detection of embedding changes. Thus, tuning a blind steganalyzer to produce a low false-positive rate across all cover sources, without being overly restrictive for "well-behaved" cover sources, such as low-noise digital-camera images, may be quite a difficult task. This is a serious problem that is not easily resolved [37]. One possible avenue Eve can take is to first classify the digital image into several categories, such as scan, digital-camera image, raw, JPEG, computer graphics, etc., and then send the image to a classifier that was trained separately for each cover category. Creating such a system requires large computational and storage resources because each database should contain images that were also processed using commonly used image-processing operations, such as denoising, filtering, recoloring, resizing, rotation, cropping, etc. In general, the larger the database, the more accurate and reliable the steganalyzer will be.

Many different blind steganalysis methods have been proposed in the literature [10, 11, 12, 67, 78, 102, 167, 168, 189, 190, 210, 225, 236, 252]. Even though, in principle, a blind detector can be used to detect steganography in any image format, one can expect that features computed in the same domain as where the embedding is realized would be the most sensitive to embedding because in this domain the changes to individual cover elements are lumped and independent. Therefore, we divide the description of blind steganalysis according to the embedding domain. In the next section, we give a specific example of a feature set for blind steganalysis of stegosystems that embed messages by manipulating quantized DCT coefficients. Using this feature set, in Sections 12.2 and 12.3 we present the details of a specific implementation of a blind steganalyzer using SVMs and a one-class neighbor machine. Both steganalyzers are tested regarding how well they can detect stego images and how they generalize to previously unseen steganographic methods. An example of using blind steganalysis for construction of targeted attacks is included in Section 12.4. Blind steganalysis of stegosystems that embed messages in the spatial domain is included in Section 12.5.

12.1   Features for steganalysis of JPEG images

In this section, we give a specific example of features [190] for construction of blind steganalyzers for steganography that embeds messages by modifying quantized DCT coefficients of a JPEG file.

For simplicity, we will consider only grayscale JPEG images represented using an array of quantized DCT coefficients $\mathbf{D}[k, l, b]$ and the JPEG quantization matrix $\mathbf{Q}[k, l]$. Here, $\mathbf{D}[k, l, b]$, $k, l = 0, \ldots, 7$, $b \in \{1, \ldots, N_{\mathrm{B}}\}$, is the (k, l)th DCT coefficient in the bth 8×8 block and N_{B} is the number of all 8×8 blocks in the image. We will assume that the blocks are ordered in a row-wise manner, meaning that $b = 1$ is the block that was originated by transforming the block of 8×8 pixels in the upper left corner, $b = \lceil N/8 \rceil$ corresponds to the last block in the first row of blocks, and $b = \lceil M/8 \rceil \times \lceil N/8 \rceil = N_{\mathrm{B}}$ points to the block in the lower right corner. Here, we assumed that the image has $M \times N$ pixels.

Some features can be described more easily with an alternative data structure obtained by rearranging $\mathbf{D}$ into an $8 \lceil M/8 \rceil \times 8 \lceil N/8 \rceil$ matrix by replacing each 8×8 pixel block with the corresponding block of DCT coefficients. We denote the rearranged coefficients again with the same letter, hoping that this will not cause confusion because each representation can be easily recognized from the number of indices of $\mathbf{D}$. To remove any potential source of ambiguity here, we provide a formal description:

$$\mathbf{D}[8i_{\mathrm{B}} + k + 1, 8j_{\mathrm{B}} + l + 1] = \mathbf{D}\left[k, l, \lceil N/8 \rceil i_{\mathrm{B}} + j_{\mathrm{B}} + 1\right], \tag{12.1}$$

$$i_{\mathrm{B}} = 0, \ldots, \lceil M/8 \rceil - 1, \; j_{\mathrm{B}} = 0, \ldots, \lceil N/8 \rceil - 1. \tag{12.2}$$

The reader is referred to Chapter 2 for more details about the JPEG format.

The features should capture all relationships that exist among DCT coefficients. A good approach is to consider the coefficients as realizations of a random variable that follows a certain statistical model and choose as features the model parameters estimated from the data. Unfortunately, the great diversity of natural images prevents us from finding one well-fitting model and it is thus necessary to build the features from several models in order to obtain good steganalysis results.

Additionally, the features are required to be sensitive to typical steganographic embedding changes and not depend on the image content so that one can easily separate the cluster of cover and stego image features. To satisfy this requirement, the features are calibrated as explained in Chapter 10 (see Figure 10.3). In calibration, the stego JPEG image J_1 is decompressed to the spatial domain, cropped by 4 pixels in both directions, and recompressed with the same quantization table as J_1 to obtain J_2. The calibrated form of feature f is thus the difference

$$f(J_1) - f(J_2). \tag{12.3}$$

12.1.1   First-order statistics

The first set of features is derived from the assumption that DCT coefficients are realizations of an iid random variable. This means that their complete statistical description can be captured using their probability mass function. The features will thus be formed by the sample pmf computed from the DCT coefficients.

The sample pmf, which we will also call the normalized histogram, of all $64 \times N_B$ luminance DCT coefficients is a D-dimensional vector

$$\mathbf{H}[r] = \frac{1}{64 \times N_B} \sum_{k,l=0}^{7} \sum_{b=1}^{N_B} \delta \left(r - \mathbf{D}[k,l,b] \right), \tag{12.4}$$

where $r = L, \ldots, R$, $L = \min_{k,l,b} \mathbf{D}[k,l,b]$, $R = \max_{k,l,b} \mathbf{D}[k,l,b]$, and $D = R - L + 1$. Here, $\delta(x)$ is the Kronecker delta (2.23).

Because the distribution of DCT coefficients varies for different modes, it is possible to consider the coefficients as 64 parallel iid channels, each corresponding to one DCT mode. Thus, further useful features will be provided by the normalized histogram of individual DCT modes. For a fixed DCT mode (k,l), let $\mathbf{h}^{(kl)}[r]$, $r = L, \ldots, R$, denote the D-dimensional vector representing the histogram of values $\mathbf{D}[k,l,b]$, $b = 1, \ldots, N_B$,

$$\mathbf{h}^{(kl)}[r] = \frac{1}{N_B} \sum_{b=1}^{N_B} \delta \left(r - \mathbf{D}[k,l,b] \right). \tag{12.5}$$

Additionally, for a fixed integer $r \in \{L, \ldots, R\}$ we define the so-called dual histogram as an 8×8 matrix

$$\mathbf{g}^{(r)}[k,l] = \frac{1}{N_B(r)} \sum_{b=1}^{N_B} \delta \left(r - \mathbf{D}[k,l,b] \right). \tag{12.6}$$

In words, $\mathbf{g}^{(r)}[k,l]$ is how many times the value r occurs as the (k,l)th DCT coefficient in all N_B blocks and $N_B(r) = \sum_{k,l} \sum_{b=1}^{N_B} \delta \left(r - \mathbf{D}[k,l,b] \right)$ is the normalization constant. The dual histogram captures the distribution of a given coefficient value r among different DCT modes. Note that if a steganographic method preserves all individual histograms, it also preserves all dual histograms and vice versa.

If we were to take the complete vectors (12.4), (12.5), and matrices (12.6) as features, the dimensionality of the feature space would be too large. Because DCT coefficients typically follow a distribution with a sharp spike at $r = 0$ (see Chapter 2), the sample pmf can be accurately estimated only around zero while its values for larger values of r exhibit fluctuations that are of little value. The same holds true of the individual DCT modes. The most populated are low-frequency modes with small $k + l$. Thus, as the first set of features, we select the first-order statistics shown in Table 12.1. We remind the reader that the features for blind steganalysis are not used directly in this form but are calibrated using (12.3).

Table 12.1. Non-calibrated features formed by 165 first-order statistics of DCT coefficients from a JPEG image.

Feature name	Feature	Index range	Features
Global histogram	$\mathbf{H}[r]$	$-5 \leq r \leq 5$	11
AC histograms	$\mathbf{h}^{(kl)}[r]$	$-5 \leq r \leq 5,\ 0 < k+l \leq 2$	11×5
Dual histograms	$\mathbf{g}^{(r)}[k,l]$	$-5 \leq r \leq 5,\ 0 < k+l \leq 3$	11×9

12.1.2   Inter-block features

The statistical models for DCT coefficients that were used in the previous section do not capture the fact that natural images exhibit dependences over distances larger than the block size. Consequently, coefficients from neighboring blocks are not independent. The relationship among coefficients $\mathbf{D}[k,l,b]$ and $\mathbf{D}[k,l,b+1]$ from neighboring blocks can be captured using the joint probability distribution.

The features in this section are more easily described by rearranging the DCT coefficients into the two-dimensional array $\mathbf{D}[i,j]$, $i = 1, \ldots, 8\lceil M/8 \rceil$, $j = 1, \ldots, 8\lceil N/8 \rceil$ defined in (12.1) obtained simply by replacing 8×8 blocks of pixels with blocks of corresponding DCT coefficients.

The first feature set is defined as the sum of the sample joint probability matrices in the horizontal and vertical directions,

$$
\mathbf{C}[s,t] = \frac{\displaystyle\sum_{i=1}^{8\lceil \frac{M}{8} \rceil - 8} \sum_{j=1}^{8\lceil \frac{N}{8} \rceil} \delta\left(s - \mathbf{D}[i,j]\right)\delta\left(t - \mathbf{D}[i+8,j]\right)}{64\left(\lceil M/8 \rceil - 1\right)\lceil N/8 \rceil}
$$
$$
+ \frac{\displaystyle\sum_{i=1}^{8\lceil \frac{M}{8} \rceil} \sum_{j=1}^{8\lceil \frac{N}{8} \rceil - 8} \delta\left(s - \mathbf{D}[i,j]\right)\delta\left(t - \mathbf{D}[i,j+8]\right)}{64\lceil M/8 \rceil\left(\lceil N/8 \rceil - 1\right)}. \tag{12.7}
$$

This matrix is called the co-occurrence matrix as it describes the distribution of pairs of neighboring DCT coefficients. The matrix $\mathbf{C}$ usually has a sharp maximum at $(s,t) = (0,0)$ and then quickly falls off. This is why we select as features only the values $\mathbf{C}[s,t]$ for $-2 \leq s,t \leq 2$.

The next (scalar) feature captures the fact that most steganographic techniques in some sense add entropy to the array of quantized DCT coefficients and thus increase the differences between dependent coefficients across blocks. The

Table 12.2. Non-calibrated features formed by 28 higher-order inter-block statistics of DCT coefficients from a JPEG image.

Feature name	Feature	Index range	Features
Co-occurrence matrix	$\mathbf{C}[s,t]$	$-2 \le s, t \le 2$	25
Variation	V		1
Blockiness	B_1, B_2		2

dependences are measured using a quantity known in mathematics as variation,

$$V = \frac{\sum_{i=1}^{8\lceil \frac{M}{8} \rceil - 8} \sum_{j=1}^{8\lceil \frac{N}{8} \rceil} \left| \mathbf{D}[i,j] - \mathbf{D}[i+8,j] \right|}{64 \left(\lceil M/8 \rceil - 1 \right) \lceil N/8 \rceil}$$

$$+ \frac{\sum_{i=1}^{8\lceil \frac{M}{8} \rceil} \sum_{j=1}^{8\lceil \frac{N}{8} \rceil - 8} \left| \mathbf{D}[i,j] - \mathbf{D}[i,j+8] \right|}{64 \lceil M/8 \rceil \left(\lceil N/8 \rceil - 1 \right)}. \tag{12.8}$$

An integral measure of dependences among coefficients from neighboring blocks is the blockiness defined as the sum of discontinuities along the 8×8 block boundaries in the *spatial domain*. Embedding changes are likely to increase the blockiness rather than decrease it. We define two blockiness measures for $\gamma = 1$ and $\gamma = 2$:

$$B_\gamma = \frac{\sum_{i=1}^{\lfloor \frac{M-1}{8} \rfloor} \sum_{j=1}^{N} \left| \mathbf{x}[8i,j] - \mathbf{x}[8i+1,j] \right|^\gamma}{N \lfloor (M-1)/8 \rfloor + M \lfloor (N-1)/8 \rfloor}$$

$$+ \frac{\sum_{i=1}^{M} \sum_{j=1}^{\lfloor \frac{N-1}{8} \rfloor} \left| \mathbf{x}[i,8j] - \mathbf{x}[i,8j+1] \right|^\gamma}{N \lfloor (M-1)/8 \rfloor + M \lfloor (N-1)/8 \rfloor}. \tag{12.9}$$

Here, M and N are the image height and width in pixels and $\mathbf{x}[i,j], i = 1, \ldots, M, j = 1, \ldots, N$, are grayscale values of the decompressed JPEG image. These two features are the only features computed from the spatial representation of the JPEG image.

The higher-order functionals measuring inter-block dependences among DCT coefficients are summarized in Table 12.2.

12.1.3   Intra-block features

DCT coefficients within one 8×8 block exhibit weak (intra-block) dependences that cannot be captured with the features introduced so far [210]. In particular, for a fixed index b, we need to describe the relationship among coefficients that are adjacent in the horizontal direction, $\mathbf{D}[k, l-1, b]$, $\mathbf{D}[k, l, b]$, and $\mathbf{D}[k, l+1, b]$, as well as for the vertical and both diagonal directions. To this end, it will be useful to represent DCT coefficients again using the matrix $\mathbf{D}[i,j]$.

Let $\mathbf{A}[i,j] = |\mathbf{D}[i,j]|$ be the matrix of absolute values of DCT coefficients in the image. Instead of modeling directly the intra-block dependences among DCT coefficients, which are quite weak, we will model the differences among them because the differences will be more sensitive to embedding changes. Thus, we form four difference arrays along four directions: horizontal, vertical, diagonal, and minor diagonal (further denoted as $\mathbf{A}_\mathrm{h}[i,j]$, $\mathbf{A}_\mathrm{v}[i,j]$, $\mathbf{A}_\mathrm{d}[i,j]$, and $\mathbf{A}_\mathrm{m}[i,j]$ respectively)

$$\mathbf{A}_\mathrm{h}[i,j] = \mathbf{A}[i,j] - \mathbf{A}[i,j+1], \tag{12.10}$$

$$\mathbf{A}_\mathrm{v}[i,j] = \mathbf{A}[i,j] - \mathbf{A}[i+1,j], \tag{12.11}$$

$$\mathbf{A}_\mathrm{d}[i,j] = \mathbf{A}[i,j] - \mathbf{A}[i+1,j+1], \tag{12.12}$$

$$\mathbf{A}_\mathrm{m}[i,j] = \mathbf{A}[i+1,j] - \mathbf{A}[i,j+1]. \tag{12.13}$$

Note that $\mathbf{A}_\mathrm{h}$ is an $M \times (N-1)$ matrix, $\mathbf{A}_\mathrm{v}$ is $(M-1) \times N$, and $\mathbf{A}_\mathrm{d}$ and $\mathbf{A}_\mathrm{m}$ are $(M-1) \times (N-1)$. Viewing the individual elements of these matrices as realizations of Markov variables, we compute four sample transition probability matrices $\mathbf{M}_\mathrm{h}, \mathbf{M}_\mathrm{v}, \mathbf{M}_\mathrm{d}, \mathbf{M}_\mathrm{m}$:

$$\mathbf{M}_\mathrm{h}[s,t] = \frac{\sum_{u=1}^{M} \sum_{v=1}^{N-2} \delta(\mathbf{A}_\mathrm{h}[u,v]-s)\delta(\mathbf{A}_\mathrm{h}[u,v+1]-t)}{\sum_{u=1}^{M} \sum_{v=1}^{N-2} \delta(\mathbf{A}_\mathrm{h}[u,v]-s)}, \tag{12.14}$$

$$\mathbf{M}_\mathrm{v}[s,t] = \frac{\sum_{u=1}^{M-2} \sum_{v=1}^{N} \delta(\mathbf{A}_\mathrm{v}[u,v]-s)\delta(\mathbf{A}_\mathrm{v}[u+1,v]-t)}{\sum_{u=1}^{M-2} \sum_{v=1}^{N} \delta(\mathbf{A}_\mathrm{v}[u,v]-s)}, \tag{12.15}$$

$$\mathbf{M}_\mathrm{d}[s,t] = \frac{\sum_{u=1}^{M-2} \sum_{v=1}^{N-2} \delta(\mathbf{A}_\mathrm{d}[u,v]-s)\delta(\mathbf{A}_\mathrm{d}[u+1,v+1]-t)}{\sum_{u=1}^{M-2} \sum_{v=1}^{N-2} \delta(\mathbf{A}_\mathrm{d}[u,v]-s)}, \tag{12.16}$$

$$\mathbf{M}_\mathrm{m}[s,t] = \frac{\sum_{u=1}^{M-2} \sum_{v=1}^{N-2} \delta(\mathbf{A}_\mathrm{m}[u+1,v]-s)\delta(\mathbf{A}_\mathrm{m}[u,v+1]-t)}{\sum_{u=1}^{M-2} \sum_{v=1}^{N-2} \delta(\mathbf{A}_\mathrm{m}[u+1,v]-s)}. \tag{12.17}$$

Since the range of differences between absolute values of neighboring DCT coefficients could be quite large, if the matrices $\mathbf{M}_\mathrm{h}, \mathbf{M}_\mathrm{v}, \mathbf{M}_\mathrm{d}, \mathbf{M}_\mathrm{m}$ were taken directly as features, the dimensionality of the feature space would be impractically large. Thus, we use only the central portion of the matrices, $-4 \le s,t \le 4$ with the note that the values in the difference arrays $\mathbf{A}_\mathrm{h}[i,j]$, $\mathbf{A}_\mathrm{v}[i,j]$, $\mathbf{A}_\mathrm{d}[i,j]$, and $\mathbf{A}_\mathrm{m}[i,j]$ larger than 4 are set to 4 and values smaller than -4 are set to -4 prior to calculating $\mathbf{M}_\mathrm{h}, \mathbf{M}_\mathrm{v}, \mathbf{M}_\mathrm{d}, \mathbf{M}_\mathrm{m}$. To further reduce the features' dimensionality, all four matrices are averaged,

$$\mathbf{M} = \frac{1}{4}\left(\mathbf{M}_\mathrm{h} + \mathbf{M}_\mathrm{v} + \mathbf{M}_\mathrm{d} + \mathbf{M}_\mathrm{m}\right), \tag{12.18}$$

which gives a total of $9 \times 9 = 81$ features (Table 12.3).

The final feature set summarized in Tables 12.1–12.3 contains $165 + 28 + 81 = 274$ calibrated features: 165 first-order statistics, 28 inter-block statistics, and 81

Table 12.3. Non-calibrated features formed by 81 higher-order intra-block statistics of DCT coefficients from a JPEG image.

Feature name	Feature	Index range	Features
Average Markov matrix	$\mathbf{M}[s,t]$	$-4 \leq s, t \leq 4$	81

intra-block features. The following sections demonstrate several applications that use this feature set.

12.2   Blind steganalysis of JPEG images (cover-versus-all-stego)

In blind steganalysis, images are first mapped to some low-dimensional feature space where they can be classified into two categories (cover and stego) using standard machine-learning tools. In general, there are countless ways to represent an image using a low-dimensional feature. The previous section showed one particular 274-dimensional representation suitable for JPEG images. This feature set is now used to construct a blind steganalyzer for JPEG images. The idea is to train a binary classifier to distinguish between two classes of images – the class of cover images and the class of stego images embedded using multiple steganographic methods with a mixture of payloads. We call such a steganalyzer a "cover-versus-all-stego classifier." The author hopes that by considering details of a specific construction, the reader will be exposed to some typical issues that arise when designing a blind steganalyzer using this approach and will be able to apply the acquired knowledge for other feature spaces and machine-learning tools. We also note that the cover-versus-all-stego approach is not the only possible way to construct blind steganalyzers. In Section 12.3, the reader will learn about an alternative approach in which a classifier is trained to recognize only the class of cover images.

First, we describe the database of images used for training and testing as well as the set of steganographic techniques used to produce stego images for training. Then, we show how to implement the steganalyzer using support vector machines. Finally, the steganalyzer is subjected to tests to give the reader a sense of the level of performance that can be achieved in practice.

12.2.1   Image database

Because the database used to construct and test the blind steganalyzer may substantially influence the resulting performance, it is always necessary to describe it in detail. In particular, one should include the number of images, their size, their origin, and details of processing they have been subjected to. In our example, the images were all created from 6004 source raw images taken by 22 different digital cameras with sizes ranging from 1.4 to 6 megapixels with an average size of 3.2 megapixels. All images were originally acquired in the raw format and then converted to grayscale and saved as 75% quality JPEG. The database was divided into two disjoint sets $\mathcal{D}_1$ and $\mathcal{D}_2$. The first set with 3500 images was used only for training the SVM classifier while the second set with 2504 images was used to evaluate the steganalyzer performance. No image appeared in both databases that would be taken by the same camera in order to make the evaluation of the steganalyzer more realistic.

12.2.2 Algorithms

When training a blind steganalyzer, we need to present it with stego images from as many (diverse) steganographic methods as possible to give it the ability to generalize to previously unseen stego images. The tacit assumption we are making here is that the steganalyzer will be able to recognize stego images embedded with an *unknown* steganographic scheme because the features will occupy a location in the feature space that is more compatible with stego images rather than cover images. As will be seen in Section 12.2.6, this assumption is not always satisfied and alternative approaches to blind steganalysis need to be explored (Section 12.3).

All stego images for training were prepared from the training database $\mathcal{D}_1$ using six steganographic techniques – JP Hide&Seek, F5, Model-Based Steganography without (MBS1) and with (MBS2) deblocking, Steghide, and OutGuess. The algorithms for F5, Model-Based Steganography, and OutGuess were described in Chapter 7. JP Hide&Seek is a more sophisticated version of Jsteg and its embedding mechanism mostly modifies LSBs of DCT coefficients. The source code is available from `http://linux01.gwdg.de/~alatham/stego.html`. Steghide is another algorithm that preserves the global first-order statistics of DCT coefficients but using a different mechanism than OutGuess. The embedding is always done by swapping coefficients rather than modifying their LSBs, which means that no correction phase is needed. Steghide is described in [110]. These selected six algorithms form the warden's knowledge base about steganography from which she constructs her detector.

12.2.3 Training database of stego images

In practice, the warden will rarely have any information about the distribution of payloads. Thus, a reasonable strategy is to select the least informative distribution, such as uniform distribution of payload. In our example, with the exception of MBS2, for the remaining five algorithms one third of the training database $\mathcal{D}_1$ was embedded with random messages of relative length 1 (full embedding capacity), another third with payload 0.5, and the last third with 0.25.[1] Thus, each image was embedded with five algorithms with one payload out of three. The stego images for MBS2 were embedded with an even mixture of relative payloads 0.3 and 0.15 of the embedding capacity of MBS1. This measure was necessary because MBS2 often fails to embed longer messages. Thus, the training database contained a total of 6×3500 stego images embedded with an even mixture of relative payloads $1, 0.5$, and 0.25 (with the exception of MBS2).

For practical applications, the number of training stego images produced by each embedding algorithm should reflect the a priori probabilities of encountering stego images generated by each steganographic algorithm. In the absence of any prior information, again one can use uniform distribution and assume that we are equally likely to encounter a stego image from any of the stego algorithms. Thus, the training database of stego images was formed by randomly selecting 3500 stego images from all 6×3500 stego images.

12.2.4   Training

The training consists of computing the feature vectors for each cover image from $\mathcal{D}_1$ and for each stego image from the training database created in the previous section. There are many tools one can use for classification purposes. Here, we describe an approach based on soft-margin weighted support vector machines (C-SVMs) with Gaussian kernel.[2] The kernel width γ and the penalization parameter C are typically determined by a grid-search on a multiplicative grid, such as

$$(C, \gamma) \in \left\{ (2^i, 2^j) | i \in \{-3, \ldots, 9\}, j \in \{-5, \ldots, 3\} \right\}, \tag{12.19}$$

to determine the values of the parameters leading to a false-alarm rate below the threshold required by each particular application. To obtain a more robust estimate of these two parameters, one typically uses multiple cross-validation. For example, in five-fold cross-validation, the training database is randomly divided into five mutually disjoint parts, four of which are used for training and the remaining fifth part for testing the classifier performance to see whether the false-alarm rate is below the required threshold, such as $P_{\mathrm{FA}} \leq 0.01$ on the fifth validation part.

---

[1] This means that a larger number of bits was embedded with stego algorithms with higher embedding capacity, which might correspond to how users would use the algorithms in practice. Consequently, one cannot use such results to fairly compare different stego algorithms.

[2] The reader is now encouraged to browse through Appendix E to become more familiar with SVMs.

Table 12.4. Probability of detection P_{D} when presenting the blind steganalyzer with stego images embedded by "known" algorithms on which the steganalyzer was trained. The false-alarm rate on the testing set of 2504 cover images was 1.04%.

Algorithm [bpnc]	P_{D}
F5 [1.0]	99.96%
F5 [0.5]	99.60%
F5 [0.25]	90.73%
JP HS [1.0]	99.84%
JP HS [0.5]	98.28%
JP HS [0.25]	73.52%
MBS1 [1.0]	99.96%
MBS1 [0.5]	99.80%
MBS1 [0.3]	98.88%
MBS1 [0.15]	71.19%
MBS2 [0.3]	99.12%
MBS2 [0.15]	77.92%
OutGuess [1.0]	99.96%
OutGuess [0.5]	99.96%
OutGuess [0.25]	98.12%
Steghide [1.0]	99.96%
Steghide [0.5]	99.84%
Steghide [0.25]	96.37%
Cover	98.96%

12.2.5 Testing on known algorithms

The blind steganalyzer is first tested on stego images embedded with the same six algorithms on which it was trained. The stego images were prepared from database $\mathcal{D}_2$ in the same manner as for training. One third was embedded with relative payload $1, 0.5$, and 0.25, again with the same exception for MBS2. From this set of 6×2504 stego images, 2504 images were randomly selected for testing.

The results of the test are shown in Table 12.4. The steganalyzer can detect fully embedded images with all algorithms with probability better than 99%. To determine the steganalyzer's false-alarm rate, it was also presented with 2504 cover images from $\mathcal{D}_2$ out of which 1.04% were detected as stego. This is in good agreement with the design false-alarm rate of $\epsilon_{\mathrm{FA}} = 1\%$ used to determine the SVM parameters γ and C during training. The table also confirms our intuition that with decreasing payload, the missed-detection rate increases. Overall, we can state that the steganalyzer is very successful in detecting known steganographic algorithms embedded with payloads larger than 25% of each algorithm's capacity.

In the next section, the same blind steganalyzer is tested on stego images produced by previously unseen steganographic methods.

Table 12.5. Probability of detection of four "unknown" steganographic algorithms for various payloads expressed in bits per non-zero AC DCT coefficient (bpnc), for –F5 and MMx, and in percentages of embedding capacity for Jsteg.

Algorithm [bpnc]	P_{D}
–F5 [1.0]	99.08%
–F5 [0.5]	99.60%
–F5 [0.25]	98.48%
MM2 [0.66]	99.64%
MM2 [0.42]	99.20%
MM2 [0.26]	53.67%
MM3 [0.66]	99.72%
MM3 [0.42]	99.32%
MM3 [0.26]	58.51%
Jsteg [1.00]	42.41%
Jsteg [0.50]	42.43%
Jsteg [0.25]	42.05%
Cover	98.96%

12.2.6   Testing on unknown algorithms

As explained in the introduction of Section 12.2, the cover-versus-all-stego steganalyzer is supposed to detect all steganographic methods. The hope is that when the detector is presented with a new steganographic algorithm, it will be able to generalize and correctly classify the image as containing stego. We remind the reader that our blind detector was trained on six steganographic algorithms that represent the knowledge base of the warden. In order to test how well the detector can recognize steganographic algorithms on which it was not trained, we intentionally did not use in the training phase the following algorithms: Jsteg (Chapter 5), –F5, and two MMx algorithms (Chapter 9). The –F5 algorithm works in exactly the same manner as F5 but the embedding operation is reversed (the absolute value of the DCT coefficient is always increased rather than decreased). Reversing the embedding operation has the benefit of removing shrinkage from F5, which enables easier implementation and increases the algorithm embedding efficiency.

The payloads embedded using MM2 and MM3 (Chapter 9) were 0.66, 0.42, and 0.26 bpnc. These values correspond to relative payloads $\alpha = \frac{2}{3}, \frac{3}{7}$, and $\frac{4}{15}$ bpnc for binary Hamming codes $[2^p - 1, 2^p - p - 1]$ for $p = 2, 3, 4$ (see Table 8.1). Because the embedding capacity of both –F5 and MMx is equal to the number of non-zero DCT coefficients, payloads expressed in bpnc also express the relative payload size with respect to the maximal embedding capacity of both algorithms. Finally, for Jsteg the images were embedded with relative payloads 1.0, 0.5, and 0.25. The quality factor for all stego images was again set to 75, the quality factor of all images from the training set.

The results shown in Table 12.5 demonstrate that the blind detector can generalize to –F5 and MMx. Even though the embedding mechanism of –F5 is very different from those of the six algorithms on which the blind steganalyzer was trained, images produced by –F5 are reliably detected.

Quite surprisingly, however, the images embedded by Jsteg are detected the least reliably despite the fact that Jsteg is a relatively poor steganographic algorithm that is easily detectable using a variety of targeted attacks, such as the one explained in Section 5.1.2 or [21, 156, 157]. Jsteg introduces severe artifacts into the histogram of DCT coefficients and the steganalyzer has not seen such artifacts before. Because it was tuned to a low probability of false alarm, it conservatively assigns such images to the cover class. This analysis underlies the need to train the classifier on as diverse set of steganographic algorithms as possible to give it the ability to generalize.

An alternative approach to blind steganalysis that is less prone to such catastrophic failures but gives an overall smaller accuracy on known algorithms is the construction based on a one-class detector. In the next section, we describe one simple approach to such one-class steganalyzers implemented using a one-class neighbor machine.

12.3   Blind steganalysis of JPEG images (one-class neighbor machine)

The cover-versus-all-stego blind steganalyzer described in the previous section failed to correctly classify stego images embedded by Jsteg (this algorithm was kept "unknown" when training the classifier to test its ability to generalize to previously unseen stego methods). It appears that blind steganalyzers trained on examples of both cover and stego images will always be prone to such failures simply because the SVM (or some other machine-learning method) will by definition learn the boundary between the classes and this boundary may be a poor choice for stego images that occupy a completely different portion of the feature space (images that do not look like covers or like known stego images).

A reasonable alternative to resolve this problem is to train a classifier to recognize only cover images so that anything that fails to appear as cover will be classified as stego. This problem is recognized in the machine-learning literature as novelty detection. In this section, we describe a very simple one-class steganalyzer and demonstrate its performance on experiments. Despite its simplicity, its accuracy is commanding, which makes this approach to blind steganalysis quite promising. The reader is referred to [194] for a more detailed treatment as well as detailed comparative study of various one-class steganalyzers.

For the description of a One-Class Neighbor Machine (OC-NM) [181], we need the notion of a sparsity measure. Let us assume that we have l cover images from which we compute l d-dimensional features. We will denote the jth component of the ith feature as $\mathbf{f}[i,j]$, $i = 1, \ldots, l$, $j = 1, \ldots, d$. The array $\mathbf{f}$ is our training set. The function $S_{\mathbf{f}} : \mathbb{R}^d \to \mathbb{R}$, which depends on $\mathbf{f}$, is a sparsity measure if and only if

$$\forall \mathbf{x}, \mathbf{y} \in \mathbb{R}^d, \quad p_c(\mathbf{x}) > p_c(\mathbf{y}) \;\Rightarrow\; S_{\mathbf{f}}(\mathbf{x}) < S_{\mathbf{f}}(\mathbf{y}), \tag{12.20}$$

where p_c is the distribution of cover features. In other words, the sparsity measure characterizes the closeness of $\mathbf{x}$ to the training set. The OC-NM works by identifying a threshold γ so that all features $\mathbf{x}$ with $S_{\mathbf{f}}(\mathbf{x}) > \gamma$ are classified as stego.

The training of an OC-NM is simple because we need only to find the threshold γ. It begins with calculating the sparsity of all training samples $\mathbf{m}[i] = S_{\mathbf{f}}(\mathbf{f}[i,.])$, $1 \le i \le l$, and ordering them so that $\mathbf{m}[1] \ge \mathbf{m}[2] \ge \ldots \ge \mathbf{m}[l]$. By setting $\gamma = \mathbf{m}[P_{\mathrm{FA}} l]$, we ensure that a fraction of exactly P_{FA} training features are classified as stego. Assuming the features $\mathbf{f}[i,.]$ are iid samples drawn according to the pdf p_c, with $l \to \infty$ the OC-NMs were shown to converge to a detector with the required probability of false alarm [181].

Note that there is a key difference between utilizing the training features in OC-NM and in classifiers based on SVMs. While SVMs use only a fraction of them during classification (support vectors defining the hyperplane), OC-NMs use all training features, which shows the relation to classifiers of the nearest-neighbor type.

The original publication on OC-NMs [181] presents several types of sparsity measures. In this book, we adopted the one based on the so-called Hilbert kernel density estimator

$$S_{\mathbf{f}}(\mathbf{x}) = \log \left(\frac{1}{\sum_{i=1}^{l} 1/(\|\mathbf{x} - \mathbf{f}[i,.]\|_2)^{hd}} \right), \tag{12.21}$$

where $\|\mathbf{x}\|_2$ is the Euclidean norm of $\mathbf{x}$. The parameter h in (12.21) controls the smoothness of the sparsity measure. Intuitively, when $\mathbf{x}$ "is surrounded" by the training features, the sparsity measure will be small (negative). Points that are farther away from the training features will lead to larger values of $S_{\mathbf{f}}$.

12.3.1   Training and testing

An important advantage of OC-NMs is the simplicity of their training. In our case, the training of an OC-NM classifier simply consists of computing the set of features $\mathbf{f}$ from 3500 cover JPEG images from $\mathcal{D}_1$. The ability of the detector to recognize stego images will be estimated on 2504 JPEG images from $\mathcal{D}_2$ not used during training. The description of the databases of cover images $\mathcal{D}_1$ and $\mathcal{D}_2$, as well as the process of creating the stego images, can be found in Section 12.2.

The detection accuracy of the OC-NM seems to vary very little with the sparsity parameter. All experimental results reported in Tables 12.6 and 12.7 were obtained with the sparsity parameter h set to 0.01.

Table 12.6. Detection accuracy of OC-NM on F5, MBS1, MBS2, JP HS, OutGuess, and Steghide. For comparison, the detection accuracy of the cover-versus-all-stego classifier described in Section 12.2.5 is repeated in this table. We emphasize that the cover-versus-all-stego classifier was trained on a mixture of stego images embedded by the same six algorithms.

Algorithm [bpnc]	OC-NM	Cover-versus-all-stego
F5 [1.0]	98.96%	99.96%
F5 [0.5]	20.10%	99.60%
F5 [0.25]	2.40%	90.73%
JP HS [1.0]	99.52%	99.84%
JP HS [0.5]	41.73%	98.28%
JP HS [0.25]	19.04%	73.52%
MBS1 [1.0]	99.92%	99.96%
MBS1 [0.5]	29.50%	99.80%
MBS1 [0.3]	4.27%	98.88%
MBS1 [0.15]	1.76%	71.19%
MBS2 [0.3]	32.47%	99.12%
MBS2 [0.15]	2.88%	77.92%
OutGuess [1.0]	100.00%	99.96%
OutGuess [0.5]	57.51%	99.96%
OutGuess [0.25]	5.19%	98.12%
Steghide [1.0]	99.44%	99.96%
Steghide [0.5]	16.61%	99.84%
Steghide [0.25]	2.84%	96.37%
Cover	98.64%	98.96%

Table 12.6 shows the percentage of correctly classified stego images embedded with six different steganographic algorithms and various payloads. We also reprint the detection accuracy of the cover-versus-all-stego classifier from Section 12.2 for comparison. As one could expect, the binary cover-versus-all-stego classifier has a better performance because it was trained on a mixture of stego images embedded using the same algorithms. The advantage of the OC-NM becomes apparent when testing on algorithms unseen by the binary classifier (Table 12.7). Here, the difference in performance is only marginal for –F5 and MMx. The biggest difference occurs for Jsteg images, which are reliably detected as stego by the OC-NM and almost completely missed by the cover-versus-all-stego classifier. The last row of the table shows the false-alarm rates for both methods, which are fairly similar.

12.4   Blind steganalysis for targeted attacks

By training a blind steganalyzer on the set of stego images embedded using a specific steganographic algorithm, we essentially obtain a cookie-cutter approach for constructing targeted attacks. The targeted steganalyzer built in this way will

Table 12.7. Detection accuracy of OC-NM on –F5, MMx, and Jsteg. For comparison, the detection accuracy of the cover-versus-all-stego classifier described in Section 12.2.6 is repeated here. We note that the cover-versus-all-stego classifier was not trained on stego images produced by these three algorithms.

Algorithm [bpnc]	OC-NM	Cover-versus-all-stego
–F5 [1.0]	100.00%	99.08%
–F5 [0.5]	100.00%	99.60%
–F5 [0.25]	93.93%	98.48%
MM2 [0.66]	100.00%	99.64%
MM2 [0.42]	99.92%	99.20%
MM2 [0.26]	17.69%	53.67%
MM3 [0.66]	100.00%	99.72%
MM3 [0.42]	99.92%	99.32%
MM3 [0.26]	15.14%	58.51%
Jsteg [1.0]	100%	42.41%
Jsteg [0.75]	100%	42.33%
Jsteg [0.5]	100%	42.37%
Jsteg [0.2]	98.09%	42.09%
Jsteg [0.1]	65.36%	32.98%
Jsteg [0.05]	40.27%	5.99%
Cover	98.64%	98.96%

naturally have a better performance than the blind classifiers from Sections 12.2 and 12.3 because the distribution of stego images will be less spread out. This section provides the reader with an example of how accurate the targeted attacks are for the following steganographic algorithms: F5, –F5, nsF5, Model-Based Steganography without deblocking (MBS1), JP Hide&Seek, Steghide, MMx, and three versions of perturbed quantization in double-compressed JPEG images as described in Chapter 9 (PQ, PQe, PQt).

As in the previous section, all classifiers were implemented as soft-margin support vector machines with Gaussian kernel trained on 3500 cover images and 3500 stego images[3] with an even mixture of *short* payloads 0.05, 0.1, 0.15, and 0.2 bpnc. The detection performance was evaluated on 2504 cover images from database $\mathcal{D}_2$ and their stego versions embedded with the same payload mixture as in the training set. For this experiment, the quality factor of all JPEG images was set to 70. For the three perturbed quantization methods, the primary and secondary quality factors were 85 and 70, respectively. The cover images for these three methods were JPEG images doubly compressed with quality factors 85 and 70.

Table 12.8 shows the detection results obtained using the performance criteria P_{E} and $P_{\mathrm{D}}^{-1}\left(\frac{1}{2}\right)$ (see Section 10.2.4).

---

[3] The same database $\mathcal{D}_1$ was used as in Section 12.2.1.

Table 12.8. Performance of targeted steganalyzers obtained by training a binary classifier for JPEG images on covers and a mixture of stego images embedded with relative payloads $0.05, 0.1, 0.15$, and 0.2 bpnc. The results are from the testing set. Two performance criteria are shown; the minimal average error probability P_E and false-alarm probability at 50% detection, $P_\mathrm{D}^{-1}(1/2)$. The abbreviations of stego algorithms are explained in the text.

Algorithm	P_E	$P_\mathrm{D}^{-1}(1/2)$
F5	7.5%	0.00%
–F5	4.4%	0.00%
nsF5	14.9%	1.04%
JP Hide&Seek	3.2%	0.04%
MBS1	2.8%	0.00%
MM2	18.4%	0.96%
MM3	19.5%	1.24%
Steghide	1.4%	0.00%
PQ	6.4%	0.12%
PQe	27.4%	10.08%
PQt	26.9%	10.84%

Because the payloads embedded using each algorithm were the same, this table tells us how detectable different steganographic methods are. Note that even though the tested payloads are relatively short, very reliable targeted attacks are possible on OutGuess, F5, –F5, Model-Based Steganography, Steghide, and JP Hide&Seek. The improved version of F5 without shrinkage (nsF5) is the best tested method that does not use any side-information at the sender (the uncompressed JPEG image). As expected, the methods that do use this side-information (versions of PQ and MMx) are among the least detectable.

To summarize, this section demonstrates a general approach to targeted steganalysis that does not require knowledge of the embedding mechanism. The basic idea is to train a binary classifier to recognize the class of cover images and stego images embedded with the particular steganographic method.

12.4.1 Quantitative blind attacks

In the previous section, we learned that targeted attacks can be constructed by training a blind steganalyzer on a set of cover and stego images embedded by a specific stegosystem. Such detectors are binary classifiers returning a "yes" or "no" answer when presented with an image. In this section, we explain how to use the feature set described in Section 12.1 for construction of quantitative attacks capable of estimating the change rate (payload). Intuitively, because the features sensitively react to embedding changes, the difference between the feature vectors of the cover and stego image increases with increasing change rate. Thus, one could potentially extract more information from the *position* of the feature vector and build a change-rate estimator using regression.

Let us assume that we have l stego images embedded with a mixture of change rates $\beta[i]$. Denoting their feature vectors $\mathbf{f}[i,j]$, $i = 1,\ldots,l$, $j = 1,\ldots,d$, we will seek the estimate of the change rate as a linear combination of the features

$$\hat{\beta}[i] = \mathbf{f}[i,.]\boldsymbol{\theta}, \tag{12.22}$$

where $\boldsymbol{\theta} \in \mathbb{R}^d$ is a column vector. The unknown vector parameter $\boldsymbol{\theta}$ is determined by minimizing the total square error

$$\sum_{i=1}^{l} (\mathbf{f}[i,.]\boldsymbol{\theta} - \beta[i])^2 . \tag{12.23}$$

This least-square problem has a standard solution in the form (see Section D.8)

$$\hat{\boldsymbol{\theta}} = (\mathbf{f}'\mathbf{f})^{-1}\mathbf{f}'\boldsymbol{\beta}. \tag{12.24}$$

Change-rate estimators constructed in this fashion typically exhibit very good accuracy, which generally depends on how well the features react to embedding. To give the reader a sense of the performance, we include a sample of the results published in the paper that pioneered this approach to quantitative steganalysis [195]. The results were obtained for the same database of JPEG images as in the previous sections enlarged by an additional 3159 images, bringing the total number of images to 9163. All images were originally acquired in raw format and then, for the purpose of this test, compressed with 80% quality JPEG. Seven steganographic algorithms were used to embed all images from the database with a uniform mixture of relative payloads. To be more precise, the length of the message was chosen randomly between zero and the maximum embedding capacity for each algorithm and each image. Matrix embedding was turned off for F5 to better control the change rate. The database was then divided into two halves, each containing $l \approx 4600$ images. The first half was used to construct the linear regressor (find $\boldsymbol{\theta}$), while the other half was used for testing the performance.

The feature set $\mathbf{f}$ was formed by 274 calibrated DCT features described in Section 12.1 augmented with the number of non-zero DCT coefficients as an additional 275th feature (thus, $d = 275$). All 275 features were normalized to have zero mean and unit variance on the training set.

Figure 12.1 shows the scatter plot of the estimated change rate versus true change rate for the F5 algorithm and Model-Based Steganography. Table 12.9 displays the sample Median Absolute Error (MAE) of the estimator and its bias (the median of the error computed from all estimates) for seven steganographic algorithms.

This approach to quantitative steganalysis has one overwhelming advantage over the strategies presented in Section 10.3 and methods described in Chapter 11. Here, all that is needed to construct the quantitative attack is a set of stego images embedded with known change rates. In particular, the steganalyst does not need any information about the embedding algorithm. As long as

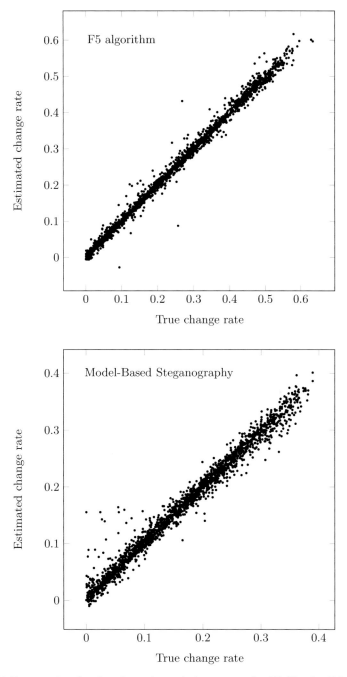

Figure 12.1 Scatter plot showing the estimated change rate for F5 (Section 7.3.2) and Model-Based Steganography (Section 7.1.2) versus the true change rate. All estimates were made on images from the testing set.

Table 12.9. Median absolute error (MAE) and bias for the quantitative change-rate estimator built from a 275-dimensional feature set implemented using a linear least-square fit (results for the testing set).

Algorithm	MAE	Bias
F5	8.39×10^{-3}	-5.29×10^{-4}
Jsteg	8.38×10^{-3}	-5.29×10^{-4}
MB1	9.07×10^{-3}	3.86×10^{-5}
MMX	3.25×10^{-3}	1.58×10^{-4}
PQ	5.69×10^{-2}	-2.89×10^{-3}
OutGuess	2.53×10^{-3}	1.51×10^{-4}
Steghide	3.23×10^{-3}	2.60×10^{-4}

the blind steganalyzer successfully detects the steganographic method, it can be turned into a quantitative steganalyzer using regression!

The reader is referred to [195] for more details about this approach to quantitative steganalysis. This reference also contains an alternative, slightly more accurate implementation using methods of support vector regression [215].

12.5   Blind steganalysis in the spatial domain

The blind steganalyzer for JPEG images explained in the previous sections used features that were computed directly from DCT coefficients. This brought two important advantages – the features were sensitive to embedding and, in combination with calibration, they were less sensitive to image content. The calibration worked because quantized DCT coefficients are robust with respect to small (embedding) changes when the quantization is performed on a desynchronized 8×8 grid. Unfortunately, this principle is unavailable in the spatial domain. However, it is possible to calibrate and simultaneously increase the features' sensitivity to embedding by calculating the features from the stego image *noise residual* $\mathbf{r} = \mathbf{y} - F(\mathbf{y})$ obtained using a denoising filter F. Working with the noise residual instead of the whole image has two important advantages:

1. The cover image content is suppressed and thus the features exhibit less variation from image to image.
2. The SNR between the stego noise and the cover image is increased, which leads to increased sensitivity of the features to embedding.

As an example of this approach to blind steganalysis in the spatial domain, we now describe the Wavelet Absolute Moments (WAM) feature set originally published in [102] and then show the results when applying this feature set to steganalysis of ±1 embedding. It is worth mentioning that many other features proposed for blind steganalysis [11, 67, 252] were intuitively constructed using the same principle. The role of the denoising filter is replaced with local content predictors [67, 252] or image-quality metrics [11]. All methods implicitly make use of the difference between the stego image and its low-pass-filtered version. The goal is, however, the same – to remove the image content and make the features more sensitive to stego noise.

12.5.1 Noise features

The features' definition stems from the inner mechanism of the wavelet-based denoising filter described in [175]. For the sake of simplicity, we will assume that the image under investigation, $\mathbf{y}$, is a grayscale 512×512 image. The image is first transformed using a single-level 8-tap (decimated) Daubechies wavelet transform W. The wavelet transform is a linear transformation that provides decomposition of the image localized both in frequency and in space. The transformed image consists of four 256×256 arrays of real numbers, $\mathbf{L}, \mathbf{H}, \mathbf{V}, \mathbf{D}$, called low-frequency, horizontal, vertical, and diagonal subbands (see Figure 12.2). This transformation can be calculated, for example, using the Matlab Wavelet Toolbox with the command [L,H,V,D]=dwt2(y,'db8'); where y is either an array of integers in the range $\{0, \dots, 255\}$, for an 8-bit grayscale image, or one color channel in a true-color image. Since the steganographic modifications of most methods constitute a high-frequency signal, most of the energy of the stego noise is in the high spatial frequencies. Thus, we will work only with the following three subbands: $\mathbf{H}[i, j], \mathbf{D}[i, j], \mathbf{V}[i, j], i, j = 1, \dots, 256.$[4]

Each subband is viewed as a sum of two random variables – the signal due to high-pass filtering of the noise-free image (mostly the edges) and the noise. The image is modeled as a non-stationary Gaussian signal $N(0, \sigma^2[i, j])$ and the noise as a stationary Gaussian signal $N(0, \sigma_n^2)$ with a known variance σ_n^2 (typically $\sigma_n^2 \approx 2$ gives good results for an 8-bit grayscale image). The variance $\sigma^2[i, j]$ is estimated from a local neighborhood of the wavelet coefficient $\mathbf{H}[i, j]$,

$$\hat{\sigma}^2[i, j] = \min(\sigma_3^2[i, j], \sigma_5^2[i, j], \sigma_7^2[i, j], \sigma_9^2[i, j]), \tag{12.25}$$

where $\sigma_w^2[i, j]$ is the sample variance estimated from a $w \times w$ square neighborhood,

$$\sigma_w^2[i, j] = \max\left(0, \frac{1}{w^2} \sum_{k,l=-\lfloor \frac{w}{2} \rfloor}^{\lfloor \frac{w}{2} \rfloor} (\mathbf{H}[i + k, j + l])^2 - \sigma_n^2\right). \tag{12.26}$$

---

[4] This choice is justifiable for detection of steganography that imposes independent modifications to cover-image pixels, such as ± 1 embedding. Steganalysis of stego noise with different spectral characteristics (e.g., with its energy shifted towards medium spatial frequencies as in [36, 76]) may benefit from analyzing subbands from higher-level wavelet decomposition.

Figure 12.2 Example of a grayscale 512×512 image and its wavelet transform using an 8-tap Daubechies wavelet. The H, V, and D subbands contain the high-frequency noise components of the image along the horizontal, vertical, and diagonal directions.

The noise residual in the horizontal subband, $\mathbf{r_H}$, is obtained by subtracting from it a Wiener-filtered version of the subband,

$$\mathbf{r_H}[i,j] = \mathbf{H}[i,j] - \frac{\hat{\sigma}^2[i,j]}{\hat{\sigma}^2[i,j] + \sigma_n^2}\mathbf{H}[i,j]. \qquad (12.27)$$

The noise residuals $\mathbf{r_D}$ and $\mathbf{r_V}$ can be obtained using similar formulas.

The steganographic features are calculated for each subband separately as the first nine central absolute moments $\boldsymbol{\mu}_c[k], k = 1, \ldots, 9$, of the noise residual

$$\boldsymbol{\mu}_c[k] = C \sum_{i,j=1}^{256} |\mathbf{r_H}[i,j] - \bar{\mathbf{r}}_\mathbf{H}|^k, \qquad (12.28)$$

where $\bar{\mathbf{r}}_\mathbf{H} = C \sum_{i,j} \mathbf{r_H}[i,j]$ is the sample mean of $\mathbf{r_H}$ and $C = \frac{1}{256^2}$ is a normalization constant. Since there are three subbands, there will be a total of 27 features for a grayscale image and 3×27 features for a color image.

Note that the amount of data samples (256^2) in each subband does not allow estimating all moments accurately. In particular, only the sample moments

Table 12.10. Total error probability P_E and $P_\mathrm{D}^{-1}(1/2)$ for detectors constructed to detect ± 1 embedding in raw images (RAW), their JPEG versions (JPEG80), and scans of film (SCAN). The whole ROCs are displayed in Figure 12.3.

	P_E	$P_\mathrm{D}^{-1}(1/2)$
RAW	16.6%	1.4%
JPEG80	1.7%	0%
SCAN	22.8%	6.1%

$\boldsymbol{\mu}_\mathrm{c}[k]$ for $k \leq 4$ are relatively accurate estimates of the true moments. However, even though the higher-order sample moments are not accurate estimates of the moments, they can still be valuable as features in steganalysis as long as they sensitively react to embedding changes.

12.5.2   Experimental evaluation

We demonstrate the performance of the blind steganalysis method based on 27 noise moments by steganalyzing ± 1 embedding in the spatial domain of grayscale images. The results will be reported separately on three different image sources, which will enable us to demonstrate the influence of the cover source on statistical detectability as already discussed in Section 12.2. Two of the databases are the RAW and SCAN sets described in Section 11.1.1. The third database, JPEG80, contains images from RAW compressed with 80% quality JPEG (the images were decompressed to the spatial domain before embedding).

Each database was randomly split in half, obtaining two image sets $\mathcal{D}_1$ and $\mathcal{D}_2$. Images from $\mathcal{D}_1$ were used as covers for training and were embedded with relative payload randomly chosen from the set $\{0.1, 0.25, 0.5\}$ bpp (for RAW and JPEG80) and from $\{0.25, 0.5, 0.75, 1\}$ bpp for SCAN. The set $\mathcal{D}_2$ was used to build the testing set in the same manner.

The detectors were soft-margin weighted support vector machines with Gaussian kernels as in Section 12.2. A separate detector was trained for each database.

The detection performance of the blind steganalyzer on ± 1 embedding for each database is shown in Table 12.10 using the total probability of error, P_E, and false-alarm rate at 50% detection, $P_\mathrm{D}^{-1}(1/2)$. Figure 12.3 shows the whole ROC curves drawn by changing the threshold b^* as described in Section E.5.5.

This experiment demonstrates that the features computed from the noise residual in the wavelet domain enable reliable detection of ± 1 embedding in the spatial domain. Note that the detection is significantly more reliable for digital-camera images than for scans. This is attributed to the much higher noise level of scans as discussed in Section 11.1.1. Note the difference between detection performance on uncompressed camera images and decompressed JPEGs. This is because the lossy compression acts as a low-pass filter and thus it is easier to filter out the steganographic changes.

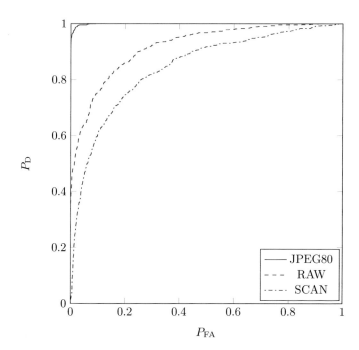

Figure 12.3 ROC for detection of ± 1 embedding in images from the training databases RAW, JPEG80, and SCAN. The stego images contained an equal mixture of images embedded with relative payloads $0.1, 0.25$, and 0.50 bpp for RAW and JPEG80, and 0.25, 0.5, 0.75, and 1.0 bpp for SCAN.

Summary

- The goal of blind steganalysis is to detect an arbitrary steganographic method.
- It works by first selecting a set of numerical features that can be computed from an image. The feature set plays the role of a low-dimensional model of cover images. Then, a classifier is trained on many examples of cover and stego image features to recognize cover and stego images in the feature space.
- The performance of a blind steganalyzer is typically evaluated on a separate testing set using the ROC curve or selected numerical characteristics, such as the total probability error, P_E, or the false-alarm rate at 50% detection rate, $P_\mathrm{D}^{-1}\left(\frac{1}{2}\right)$.
- A reliable blind steganalysis classifier for JPEG images can be constructed from various first-order and higher-order statistics of quantized DCT coefficients. Calibration further improves the classifier performance because it makes the features more sensitive to embedding while removing the influence of the cover content.
- A blind steganalyzer can be built either as a binary cover-versus-all-stego classifier or as a one-class classifier. In the first case, the two classes are formed by cover images and stego images embedded by as many steganographic methods as possible with varying payload. The alternative approach is to train a one-

class steganalyzer that recognizes only covers and marks features incompatible with the training set as stego (or suspicious).

- The equivalent of calibration in the spatial domain is the principle of computing the features from the image noise residual obtained using a denoising filter.
- An example of a blind steganalysis method in the spatial domain is shown. It is based on computing higher-order moments of the image noise residual in high-frequency wavelet subbands.
- Blind steganalysis can also be used to construct targeted and quantitative attacks and to classify stego images into known steganographic programs.

13 Steganographic capacity

Intuition tells us that steganographic capacity should perhaps be defined as the largest payload that Alice can embed in her cover image using a specific embedding method without introducing artifacts detectable by Eve. After all, knowledge of this secure payload appears to be fundamental for the prisoners to maintain the security of communication. Unfortunately, determining the secure payload for digital images is very difficult even for the simplest steganographic methods, such as LSB embedding. The reason is the lack of accurate statistical models for real images. Moreover, it is even a valid question whether capacity can be meaningfully defined for an individual image and a specific steganographic method. Indeed, capacity of noisy communication channels depends only on the channel and not on any specific communication scheme.

This chapter has two sections, each devoted to a different capacity concept. In Section 13.1, we study the steganographic capacity of perfectly secure stegosystems. Here, we are interested in the maximal relative payload (or rate) that can be securely embedded in the limit as the number of pixels in the image approaches infinity. Capacity defined in this way is a function of only the physical communication channel and the cover source rather than the steganographic scheme itself. It is the maximal relative payload that Alice can communicate if she uses the best possible stegosystem. The significant advantage of this definition is that we can leverage upon powerful tools and constructions previously developed for study of robust watermarking systems. This apparatus also enables simultaneous study of stegosystems with distortion-limited embedder and distortion-limited active warden (Section 6.3). For some simple models of covers, it is also possible to compute the capacity and design capacity-reaching stegosystems. However, the steganographic capacity derived from these models is often overly optimistic because simple models do not describe real images well. Also, because of the model mismatch, it is hard to relate specific numerical results to real cover sources. Despite these objections, the theoretical results provide deeper insight into the problem as well as the possibility of progressive improvement to include more realistic cover models.

Section 13.2 deals with a more practical issue, which is the largest payload that can be embedded in covers of a certain size using a specific embedding method. To distinguish this concept from the capacity as defined in Section 13.1, we use the term "secure payload" instead. The section starts with a bold pragmatic assumption that it is unlikely that a perfectly secure stegosystem can ever be built for real covers due to their overwhelming complexity. Under the assumption that the steganographic system is not perfectly secure, the secure payload is defined in absolute terms as the critical size of payload for which the KL divergence between covers and stego images reaches a certain fixed value. The secure payload defined in this way is bounded by the square root of the number of pixels, n, which means that the safe communication rate approaches zero as $1/\sqrt{n}$. The result is supported in Section 13.2.2 by experiments carried out for a selected embedding algorithm and a chosen blind steganalyzer.

Before starting with the technical arguments, we note that Alice may attempt to determine the secure payload experimentally using current best blind steganalyzers on a large database of cover images from the investigated cover source. By analyzing images embedded with different payloads, she can estimate the secure payload as the critical size of payload at which her steganalyzer starts making random guesses (e.g., $P_{\mathrm{E}} \geq 0.5 - \delta$ for some selected δ; see (10.15) for the definition of P_{E}). Secure payload defined this way informs the prisoners about the longest message that can be safely embedded given the current state of the art in steganalysis. The obvious disadvantage is that the estimate is now tied to a specific steganalyzer. Alice can thus obtain only an upper bound on the true size of the secure payload, which may dramatically change with further progress in steganalysis. The reader is referred to [102, 95, 217] for some recent results.

13.1   Steganographic capacity of perfectly secure stegosystems

In this section, we define steganographic capacity of perfectly secure stegosystems with a distortion-limited embedder and an active distortion-limited warden. The results for the passive-warden scenario and unlimited distortion of the embedder can be obtained as corollaries. The capacity is defined as the largest rate (relative payload) over all possible steganographic schemes satisfying the distortion constraints and the steganographic constraint (the distributions of cover and stego objects should coincide).

The material in this section deviates from the rest of this book in the sense that it requires advanced knowledge of information theory that goes well beyond the prerequisites of this book. Proofs of the statements in this section can be found in the original publications [49, 180, 237]. The reader is referred to [50] for a comprehensive text on information theory.

We start by reminding the reader of the concept of a steganographic channel as defined in Section 6.3. A steganographic channel consists of the source of covers, $\mathbf{x} \in \mathcal{X}^n$, where $\mathcal{X}$ is a finite set (e.g., for grayscale images, $\mathcal{X} = \{0, \ldots, 255\}$), secret keys, $\mathbf{k} \in \mathcal{K}$, and the source of messages, $\mathbf{m} \in \mathcal{M}$, with corresponding distributions on their sets, P_{c}, P_{k}, P_{m}, the embedding and extraction algorithms,

$\mathrm{Emb}_n : \mathcal{X}^n \times \mathcal{K} \times \mathcal{M} \to \mathcal{X}^n$, $\mathrm{Ext}_n : \mathcal{X}^n \times \mathcal{K} \to \mathcal{M}$,[1] and the physical communication channel described by a matrix of conditional probabilities $A(\tilde{\mathbf{y}}|\mathbf{y})$ that a stego object $\mathbf{y}$ sent by Alice is received in its noisy form as $\tilde{\mathbf{y}}$ by Bob. The attack channel is considered to be memoryless, or

$$A(\tilde{\mathbf{y}}|\mathbf{y}) = \prod_{i=1}^{n} A(\tilde{\mathbf{y}}[i]|\mathbf{y}[i]). \tag{13.1}$$

Furthermore, we assume that the expected embedding distortion imposed by Alice and the channel distortion per cover element are bounded,

$$E\left[d(\mathsf{x},\mathsf{y})\right] = \sum_{\mathbf{x}} P_{\mathrm{c}}(\mathbf{x}) \sum_{\mathbf{k}} P_{\mathrm{k}}(\mathbf{k}) \sum_{\mathbf{m}} P_{\mathrm{m}}(\mathbf{m}) d\left(\mathbf{x}, \mathrm{Emb}_n(\mathbf{x},\mathbf{k},\mathbf{m})\right) \le D_1, \tag{13.2}$$

$$\sum_{\mathbf{y},\tilde{\mathbf{y}}} P_{\mathrm{c}}(\mathbf{y}) A(\tilde{\mathbf{y}}|\mathbf{y}) d(\tilde{\mathbf{y}},\mathbf{y}) \le D_2, \tag{13.3}$$

where $d(.,.) : \mathcal{X}^n \times \mathcal{X}^n \to [0,\infty)$ is a measure of distortion per cover element. The requirement of perfect security is $P_{\mathrm{c}} = P_{\mathrm{s}}$ or, equivalently, $D_{\mathrm{KL}}(P_{\mathrm{c}}||P_{\mathrm{s}}) = 0$, where P_{s} is the distribution of stego objects $\mathbf{y} = \mathrm{Emb}_n(\mathbf{x},\mathbf{k},\mathbf{m})$.

In this section, we allow the embedding mapping to fail to embed a message in the sense that $\mathrm{Ext}_n\left(\mathrm{Emb}_n(\mathbf{x},\mathbf{k},\mathbf{m}),\mathbf{k}\right) \neq \mathbf{m}$ in some cases, as long as the probability of this occurring is small with growing length of covers, n. To be more precise, we say that rate R is *achievable* if there exists a sequence of embedding and extraction mappings $(\mathrm{Emb}_n)_{n=1}^{\infty}$, $(\mathrm{Ext}_n)_{n=1}^{\infty}$, such that

$$\sup_{A} \Pr\left\{\mathrm{Ext}_n\left(\mathrm{Emb}_n(\mathbf{x},\mathbf{k},\mathbf{m})\right) \neq \mathbf{m}\right\} \to 0 \tag{13.4}$$

as $n \to \infty$ and $|\mathcal{M}| \ge 2^{nR}$. The random variable $\mathrm{Emb}_n(\mathbf{c},\mathbf{k},\mathbf{m})$ is required to comply with the distortion constraint (13.2). The supremum is taken over all noisy channels $A(\tilde{\mathbf{y}}|\mathbf{y})$ constrained by (13.3).

The *steganographic capacity* is defined as the supremum of all achievable rates. It will be denoted as $C_{\mathrm{steg}}(D_1, D_2)$ as it clearly depends on the distortion bounds. Note that C_{steg} is not a concept defined for one specific embedding scheme! Instead, it is a function of only the cover source and the distortion bounds.

We also define the concept of a *covert channel* as the conditional probabilities $Q(\mathbf{y}, u|\mathbf{x}) = \Pr\{\mathsf{y} = \mathbf{y}, \mathsf{u} = u|\mathsf{x} = \mathbf{x}\}$, where u is an auxiliary random variable on a finite but arbitrarily large alphabet $\mathcal{U}$. This random variable is used to define a random codebook shared between Alice and Bob to realize the steganographic communication.

---

[1] Here, we use the subscript n to stress the fact that Emb_n and Ext_n act on n-element covers.

We say that the covert channel Q is *feasible* if

$$\sum_{\mathbf{y},u,\mathbf{x}} Q(\mathbf{y},u|\mathbf{x})P_c(\mathbf{x})d\left(Q(\mathbf{x},\mathbf{y})\right) \leq D_1, \tag{13.5}$$

$$\sum_{u,\mathbf{x}} Q(\mathbf{y},u|\mathbf{x})P_c(\mathbf{x}) = P_c(\mathbf{y}). \tag{13.6}$$

Thus, a feasible covert channel complies with the embedding-distortion constraint and perfect steganographic security.

The next theorem gives an expression for the steganographic capacity C_{steg} for the most general case of an active warden.

Theorem 13.1. [Steganographic capacity]

$$C_{\text{steg}}(D_1, D_2) = \sup_Q \inf_A \{I(\mathsf{u};\tilde{\mathsf{y}}) - I(\mathsf{u};\mathsf{x})\}, \tag{13.7}$$

where $(\mathsf{u},\mathsf{x}) \to \mathsf{y} \to \tilde{\mathsf{y}}$ *forms a Markov chain.*

In the theorem, the supremum is taken over all feasible channels Q and the infimum over all attack channels A satisfying the distortion constraint.

For the passive-warden scenario, $A(\tilde{\mathbf{y}}|\mathbf{y}) = 1$ when $\tilde{\mathbf{y}} = \mathbf{y}$, and the capacity result simplifies to the following corollaries.

Corollary 13.2. [Passive warden] *For the passive-warden case* $(D_2 = 0)$,

$$C_{\text{steg}}(D_1, 0) = \sup_Q H(\mathsf{y}|\mathsf{x}). \tag{13.8}$$

Corollary 13.3. [Passive warden, unlimited embedding distortion] *For unlimited embedding distortion,* $D_1 = \infty$, *and a passive warden*

$$C_{\text{steg}}(\infty, 0) = H(\mathsf{x}). \tag{13.9}$$

The capacity can be calculated for some simple cover sources, such as the source of uniformly distributed random bits, $\mathcal{X} = \{0,1\}$, when covers $\mathbf{x}$ are Bernoulli sequences $B\left(\frac{1}{2}\right)$ (see Appendix A for the definition) and the distortion is the Hamming distance d_{H} [180]. It has also been established for iid Gaussian sources [49].

13.1.1   Capacity for some simple models of covers

The capacity for the Bernoulli $B\left(\frac{1}{2}\right)$ cover source has been derived in [17, 197] without the steganographic constraint $P_c = P_s$. For the passive-warden scenario,

$$C_{\text{steg}}(D_1, 0) = \begin{cases} H(D_1) & \text{if } 0 \leq D_1 \leq \frac{1}{2} \\ 1 & \text{if } D_1 \geq \frac{1}{2} \end{cases} \tag{13.10}$$

and for the active warden

$$C_{\text{steg}} = \begin{cases} (D_1/d')\,(H(d') - H(D_2)) & \text{if } 0 \leq D_1 \leq d' \\ H(D_1) - H(D_2) & \text{if } d' \leq D_1 \leq \frac{1}{2} \\ 1 - H(D_2) & \text{if } D_1 \geq \frac{1}{2}, \end{cases} \qquad (13.11)$$

where $d' = 1 - 2^{-H(D_2)}$ and $H(x)$ is the binary entropy function (Figure 8.2). In this special case, the steganography constraint $P_c = P_s$ does not influence the steganographic capacity because the capacity-reaching distribution is $P_s = P_c$. It is possible to construct steganographic schemes that reach the capacity using random linear codes [254].

More realistic cover sources formed by iid Gaussian sequences were studied in [49]. The authors computed a lower bound on the capacity of perfectly secure stegosystems for such sources and studied the increase in capacity for ϵ-secure systems. They also described a practical QIM[2] lattice-based construction for high-capacity secure stegosystems and contrasted their work with results obtained for digital watermarking that do not include the steganographic constraint $P_c = P_s$. The authors of [237] describe codes for construction of high-capacity stegosystems and show how their approach can be generalized from iid sources to Markov sources and sources over continuous alphabets.

Finally, we note that the work [180] also investigated steganographic capacity for an alternative definition of an almost sure distortion of the attack channel in which the channel distortion constraint is replaced with $\Pr\{d(\mathbf{x}, \mathbf{y}) \leq D_2\} = 0$,[3] and showed that for this form of the active warden, the steganographic capacity is also given by Theorem 13.1.

13.2   Secure payload of imperfect stegosystems

In the previous section, we introduced a communication-theoretic definition of steganographic capacity as the maximal achievable rate of stegosystems complying with the required distortion constraints in the limit when the number of elements in the cover goes to infinity. We also learned that for sufficiently simple models of covers, it is possible to compute the capacity and even construct capacity-reaching perfectly secure stegosystems or ϵ-secure stegosystems with positive rate, depending on the cover model. However, the problem of determining the maximal size of the secure payload for a specific steganographic scheme and for realistic covers remains open. Due to our ignorance about the cover model, all steganographic techniques designed for real digital multimedia are likely to be statistically detectable in the sense that their KL divergence will

---

[2] A version of Quantization Index Modulation (QIM) appears in Chapter 6, otherwise, the reader is referred to [51].

[3] This distortion constraint was for the first time used in [219].

be unbounded with increasing number of elements n in the cover. This indicates that for such imperfect schemes the secure relative payload will converge to zero. In this section, we establish the so-called *Square-Root Law* (SRL) for imperfect stegosystems which says that for a fixed ϵ-secure stegosystem the absolute secure payload grows only as the square root of n and thus the communication rate approaches zero as $O(1/\sqrt{n})$.

The fact that the secure payload of practical steganographic schemes digital-media covers might be sublinear in the number of cover elements was first suspected by Anderson [3] in 1996:

"Thanks to the Central limit theorem, the more covertext we give the warden, the better he will be able to estimate its statistics, and so the smaller the rate at which [the steganographer] will be able to tweak bits safely. The rate might even tend to zero..."

The first insight into the problem was obtained from analysis of batch steganography and pooled steganalysis [132]. In batch steganography, Alice and Bob try to minimize the chances of being caught by dividing the payload into chunks (not necessarily uniformly) and embedding each chunk in a different image. The warden is aware of this fact and thus pools her results from testing all images sent by both prisoners. One of the main results obtained in this study is the fact that if there exists a detector for the stegosystem, the secure payload grows only as the square root of the number of communicated covers. This result could be interpreted as the square-root law for a single image by dividing it into smaller blocks. In the next section, we describe a different proof of this law for covers modeled as sequences of iid variables and for steganographic algorithms that do not preserve the cover model. In Section 13.2.2, the SRL is experimentally verified using a blind steganalyzer in the DCT domain for the embedding operation of F5 (Section 7.3.2).

13.2.1   The SRL of imperfect steganography

Before proceeding with the technical result, we note that we will study only stegosystems for which the embedding and extraction mappings satisfy

$$\text{Ext}\left(\text{Emb}(\mathbf{x}, \mathbf{k}, \mathbf{m}), \mathbf{k}\right) = \mathbf{m}, \quad \forall \mathbf{x} \in \mathcal{C}, \mathbf{k} \in \mathcal{K}, \mathbf{m} \in \mathcal{M}. \tag{13.12}$$

In other words, it is assumed that every message can be embedded in every cover. If the steganographic method has this property of homogeneity, one could define the secure payload as $\log_2 |\mathcal{M}|$ for the largest set $\mathcal{M}$ for which the KL divergence between cover and stego images is below a fixed threshold $\epsilon > 0$.

Let us assume that the cover source produces sequences of n iid realizations of a discrete random variable x with probability mass function $\mathbf{p}_0[i] > 0$ for all i, where i is an index whose range is determined by the range of cover elements (for example, $i \in \{0, \ldots, 255\}$ for an 8-bit grayscale image). Let us further assume that after embedding using change rate β, the stego image can be described as an iid sequence with pmf

$$\mathbf{p}_\beta[i] = \mathbf{p}_0[i] + \beta\mathbf{r}[i], \qquad (13.13)$$

where $\mathbf{r}[i]$ are constants[4] independent of β, such that $\mathbf{r}[k] \neq 0$ for at least one k. Before discussing the plausibility of assuming that the impact of embedding on the cover source is in the form of equation (13.13), notice that for imperfect steganography there must indeed exist at least one $\mathbf{r}[k] \neq 0$, otherwise the steganographic method would be undetectable within our cover-source model.

Assumption (13.13) holds for many practical stegosystems and digital-media covers. In general, it is true whenever the embedding visits individual cover elements and applies an independent embedding operation to each visited cover element. This is true, for example, for steganographic schemes whose embedding impact can be modeled as adding to x an independent random variable ξ with pmf $\mathbf{u}$ that is linear in β. This is because the distribution of the stego image elements y = x + ξ is the convolution $\mathbf{u} \star \mathbf{p}$ or $\mathbf{p}_\beta[i] = \sum_j \mathbf{u}[i-j]\mathbf{p}_0[j]$.

Examples of practical steganographic schemes that do satisfy (13.13) include LSB embedding (see equation (5.10)), ± 1 embedding (see Section 7.3.1 and Exercise 7.1), stochastic modulation (Section 7.2.1), and the embedding operation of F5, –F5, nsF5, and MMx algorithms (see Exercise 7.2 for F5, Section 9.4.7 for nsF5, and Section 9.4.3 for MMx). Even though the relationship (13.13) may not be valid for steganography that preserves the first-order statistics (histogram) of the cover, such as OutGuess, a similar dependence[5] may exist for some proper subsets of the cover or for local pairs of cover elements (since most stego methods disturb higher-order statistics). For example, OutGuess preserves only the global histogram but not the histograms of individual DCT modes. Thus, assumption (13.13) applies to virtually all practical steganographic schemes for some proper model of the cover.

Theorem 13.4. [The SRL of imperfect steganography] Under assumption (13.13),

1. *If the absolute number of embedding changes $\beta(n)n$ increases faster than $\sqrt{n}$ in the sense that $\lim_{n\to\infty} \beta(n)n/\sqrt{n} = \infty$, then for sufficiently large n there exist steganalysis detectors with arbitrarily small probability of false alarms and missed detection.*
2. *If the absolute number of embedding changes increases slower than $\sqrt{n}$, $\lim_{n\to\infty} \beta(n)n/\sqrt{n} = 0$, then the steganographic algorithm can be made ϵ-secure for any $\epsilon > 0$ for sufficiently large n.*

---

[4] Note that since $\mathbf{p}_\beta$ is a pmf, we must have $\sum_i \mathbf{r}[i] = 0$.

[5] When $\mathbf{p}_\beta$ is the histogram of pairs (groups) of pixels, the dependence $\mathbf{p}_\beta$ on β may not be linear, but the arguments of this section apply as long as $\partial\mathbf{p}_\beta[k]/\partial\beta\big|_{\beta=0} \neq 0$ for some (multi-dimensional) index k.

3. *Finally,   when $\lim_{n\to\infty} \beta(n)n/\sqrt{n} = C$,   the   stegosystem   is $(C^2/2)\sum_i \mathbf{r}^2[i]/\mathbf{p}_0[i]$-secure for sufficiently large n.*

Proof. Before proving the theorem, we clarify one technical issue. The theorem is formulated with respect to the number of embedding changes rather than payload. This is quite understandable because statistical detectability is influenced by the impact of embedding and not the payload. However, for steganographic schemes that exhibit a linear relationship between the number of changes and payload, such as schemes that do not employ matrix embedding, the theorem can be easily seen in terms of payload.

Proof of Part 1: Under Kerckhoffs' principle, the warden knows the distribution of cover elements $\mathbf{p}_0$. Her task is to distinguish whether she is observing realizations of a random variable with pmf $\mathbf{p}_\beta$ with $\beta = 0$ or with $\beta > 0$. To show the first part of the SRL, the warden needs only to construct a sufficiently good detector rather than the best possible one. In particular, it is sufficient to constrain ourselves to the index k for which $\mathbf{r}[k] \neq 0$, say $\mathbf{r}[k] > 0$. Let $(1/n)\mathbf{h}_\beta[i]$ be the random variable corresponding to the value of the ith bin of the normalized histogram of the stego image embedded with change rate β. We will investigate the following scaled variable:

$$\nu_{\beta,n} = \sqrt{n}\left(\frac{1}{n}\mathbf{h}_\beta[k] - \mathbf{p}_0[k]\right). \tag{13.14}$$

From the assumption of independence of stego elements, $\nu_{\beta,n}$ is a binomial random variable with expected value and variance

$$E\left[\nu_{\beta,n}\right] = \beta\sqrt{n}\mathbf{r}[k], \tag{13.15}$$

$$\mathrm{Var}\left[\nu_{\beta,n}\right] = \mathbf{p}_\beta[k]\left(1 - \mathbf{p}_\beta[k]\right). \tag{13.16}$$

With $n \to \infty$, the distribution of $\nu_{\beta,n}$ converges to a Gaussian distribution with the same mean and variance. Thus, from now on, we will make the following simplifying assumption:

$$\nu_{\beta,n} \text{ is Gaussian.} \tag{13.17}$$

This step will simplify further arguments and, it is hoped, highlight the main idea without cluttering it with technicalities. The interested reader is referred to Exercise 13.1 to see how the arguments can be carried out without assumption (13.17).

The variance of $\nu_{\beta,n}$ can be bounded

$$\mathrm{Var}[\nu_{\beta,n}] = \sigma_{\beta,n}^2 \leq \frac{1}{4} \tag{13.18}$$

for all $\beta \in [0, 1]$ because $x(1 - x) \leq \frac{1}{4}$.

The difference between the mean values,

$$E[\nu_{\beta,n}] - E[\nu_{0,n}] = E[\nu_{\beta,n}] = \beta\sqrt{n}\mathbf{r}[k], \tag{13.19}$$

tends to ∞ with $n \to \infty$ when $\beta\sqrt{n} \to \infty$. Note that for $\beta = 0$, $\nu_{0,n}$ is Gaussian $N(0, \sigma_{0,n}^2)$ with $\sigma_{0,n}^2 \leq \frac{1}{4}$.

Consider the following detector:

$$\nu_{\beta,n} > T \text{ decide stego } (\beta > 0), \tag{13.20}$$
$$\nu_{\beta,n} \leq T \text{ decide cover } (\beta = 0), \tag{13.21}$$

where T is a fixed threshold. We will now show that T can be chosen to make the detector probability of false alarms and missed detections satisfy

$$P_{\text{FA}} \leq \epsilon_{\text{FA}}, \tag{13.22}$$
$$P_{\text{MD}} \leq \epsilon_{\text{MD}} \tag{13.23}$$

for arbitrary $\epsilon_{\text{FA}}, \epsilon_{\text{MD}} > 0$ for sufficiently large n. The threshold $T(\epsilon_{\text{FA}})$ will be determined from the requirement that the right tail, $x \geq T(\epsilon_{\text{FA}})$, for the Gaussian variable $N(0, \sigma_{0,n}^2)$ is less than or equal to ϵ_{FA}. This can be conveniently written using the tail probability function $Q(x) = (1/\sqrt{2\pi}) \int_x^\infty e^{-\frac{t^2}{2}} dt$ for a standard normal random variable (see Appendix A) as[6]

$$\epsilon_{\text{FA}} \geq Q\left(\frac{T(\epsilon_{\text{FA}})}{1/2}\right) \geq Q\left(\frac{T(\epsilon_{\text{FA}})}{\sigma_{0,n}}\right) \Rightarrow \frac{1}{2}Q^{-1}(\epsilon_{\text{FA}}) \leq T(\epsilon_{\text{FA}}), \tag{13.24}$$

or it is sufficient to take $T = \frac{1}{2}Q^{-1}(\epsilon_{\text{FA}})$.

Because of the growing difference between the means (13.19), we can find n large enough that the left tail, $x \leq T(\epsilon_{\text{FA}})$, of the Gaussian variable $\nu_{\beta,n}$ is less than or equal to ϵ_{MD},

$$\epsilon_{\text{MD}} \geq Q\left(\frac{\beta\sqrt{n}\mathbf{r}[k] - T(\epsilon_{\text{FA}})}{1/2}\right) \geq Q\left(\frac{\beta\sqrt{n}\mathbf{r}[k] - T(\epsilon_{\text{FA}})}{\sigma_{\beta,n}}\right) \tag{13.25}$$

or

$$\frac{1}{2}Q^{-1}(\epsilon_{\text{MD}}) \leq \beta\sqrt{n}\mathbf{r}[k] - T(\epsilon_{\text{FA}}), \tag{13.26}$$

which can be satisfied for sufficiently large $n \geq n_{\text{MD}}$ because $\beta\sqrt{n} \to \infty$.

Proof of Part 2: We now show the second part of the SRL. In particular, we prove that if $\beta\sqrt{n} \to 0$, the steganography is ϵ-secure for any $\epsilon > 0$ for sufficiently large n.

Let $\mathbf{x}$ and $\mathbf{y}$ be n-dimensional random variables representing the elements of cover and stego images. Due to the independence assumption about cover

[6] Note that $Q^{-1}(x)$ is a decreasing function.

elements and the embedding operation

$$D_{\mathrm{KL}}(\mathbf{x}\|\mathbf{y}) = nD_{\mathrm{KL}}(\mathbf{p}_0\|\mathbf{p}_\beta) = n\sum_i \mathbf{p}_0[i]\log\frac{\mathbf{p}_0[i]}{\mathbf{p}_\beta[i]} \tag{13.27}$$

$$= -n\sum_i \mathbf{p}_0[i]\log\left(1 + \frac{\beta\mathbf{r}[i]}{\mathbf{p}_0[i]}\right) \tag{13.28}$$

$$= -n\beta\sum_i \mathbf{r}[i] + \frac{n\beta^2}{2}\sum_i \frac{\mathbf{r}^2[i]}{\mathbf{p}_0[i]} - n\sum_i \mathbf{p}_0[i]\frac{\theta^3[i]}{3}, \tag{13.29}$$

where $\theta[i] \in (0, \beta\mathbf{r}[i]/\mathbf{p}_0[i])$. We used Taylor expansion of $\log(1+x) = x - \frac{1}{2}x^2 + \frac{1}{3}\theta^3$ and the Lagrange form of the remainder. Since $\sum_i \mathbf{r}[i] = 0$, we readily obtain that the KL divergence is locally quadratic in β because it is assumed that at least one $\mathbf{r}[k] > 0$ and $\mathbf{p}_0[i] > 0$ for all i. Under the assumption that $\beta\sqrt{n} \to 0$, the quadratic term converges to zero. Because $\theta^3[i] < \beta^3\mathbf{r}^3[i]/\mathbf{p}_0^3[i]$, we have for the Lagrange remainder

$$\left|n\sum_i \mathbf{p}_0[i]\frac{\theta^3[i]}{3}\right| < n\beta^2\beta\left|\frac{1}{3}\sum_i \frac{\mathbf{r}^3[i]}{\mathbf{p}_0^2[i]}\right| \to 0. \tag{13.30}$$

This establishes the result that $D_{\mathrm{KL}}(\mathbf{x}\|\mathbf{y})$ can be made arbitrarily small for sufficiently large n. Putting this another way, when the number of embedding changes is smaller than the square root of the number of cover elements n, the steganographic scheme becomes asymptotically perfectly secure.

Proof of Part 3: In the third case when $\beta\sqrt{n} \to C$, the cubic term also goes to zero because $\beta \to 0$. The quadratic term can be bounded

$$\frac{n\beta^2}{2}\sum_i \frac{\mathbf{r}^2[i]}{\mathbf{p}_0[i]} \le \frac{C^2}{2}\sum_i \frac{\mathbf{r}^2[i]}{\mathbf{p}_0[i]}. \tag{13.31}$$

$\square$

Having established the SRL of imperfect steganography, we now discuss its implications. Put in simple words, the law states that the secure payload of a wide class of practical embedding methods is proportional to the square root of the number of cover elements that can be used for embedding. For the steganographer, this means that an r-times larger cover image can hold only $\sqrt{r}$-times larger payload at the same level of statistical detectability (the same KL divergence). The SRL also means that it is easier for the warden to detect the same relative payload in larger images than in smaller ones.

The SRL has additional important implications for steganalysis. For example, when evaluating the statistical detectability of a steganographic method for a *fixed relative payload*, we will necessarily obtain better detection on a database of large images than on small images. To make comparisons meaningful, we should fix the "root rate," defined as the number of bits per square root of the number of cover elements, rather than the rate. This would also alleviate issues with interpreting steganalysis results on a database of images with varying size.

For schemes that do use matrix embedding, due to the non-linear relationship between the change rate and the relative payload, the critical size of the payload does not necessarily scale with $\sqrt{n}$. This is because βn changes may communicate payload up to $nH(\beta)$ bits, where $H(x)$ is the binary entropy function. Because for $x \to 0$, $H(x) \approx x \log(1/x)$, we can say that the critical size of the payload picks up an additional logarithmic factor and scales as $n\beta \log(1/\beta)$, which for $\beta \approx 1/\sqrt{n}$ behaves as $\sqrt{n} \log n$.

On the more fundamental level, the SRL essentially tells us that the conventional approach to steganography is highly suboptimal because we already know from Section 6.2 and Section 13.1 that when the cover source is known, the secure payload is linear in the number of cover elements (the rate is positive). The reader should also compare this result with the capacity of noisy communication channels [50], where the number of message bits that can be sent without any errors increases *linearly* with the number of communicated symbols.

Finally, we note that the law can be generalized in several directions. It has been established for a more general class of Markov covers and steganography realized by applying a mutually independent embedding operation at each cover element [71]. Also, the linear form of $\mathbf{p}_\beta$ with respect to β in assumption (13.13) can be relaxed. All that is really necessary to carry out essentially the same steps in the proof is the assumption that

$$\left. \frac{\partial \mathbf{p}_\beta[k]}{\partial \beta} \right|_{\beta=0} \neq 0 \text{ for some } k. \tag{13.32}$$

Constraining ourselves to some right neighborhood of 0, $\beta \in [0, \beta_0)$, we could expand $\mathbf{p}_\beta[i]$ using Taylor expansion on this interval and then repeat similar arguments as in the proof of the SRL under assumption (13.13).

13.2.2   Experimental verification of the SRL

In the previous section, under certain assumptions on the cover source we learned that the secure payload of imperfect stegosystems grows only as the square root of the number of usable elements of the cover. In this section, this law is experimentally verified on the example of the F5 algorithm that embeds messages in JPEG images (Section 7.3.2). For this experiment, the matrix embedding in F5 was disabled to obtain a linear relationship between relative payload and the change rate.

As in the rest of this book, for JPEG images, the relative payload α is measured with respect to all non-zero DCT coefficients in the cover JPEG file. To verify the SRL, care needs to be taken when preparing the image sets for the experiment. Basically, what is needed is a sequence of image sets, $\mathcal{S}_i$, $i = 1, \ldots, L$, such that all images from one fixed set contain approximately the same number of non-zero DCT coefficients, n_i, with n_i forming a strictly increasing sequence. Each set needs to contain sufficiently many images for training and testing a blind steganalyzer to evaluate the statistical detectability. Moreover, the images should all have the same quality factor to avoid biasing the experimental results. Ideally, the sets $\mathcal{S}_i$ should contain images from different sources. When attempting to comply with this requirement, however, an infeasibly high number of images would be needed. A more economical approach would be to use one large database of raw, never-compressed images, $\mathcal{D}$, and create the individual image sets by cropping, scaling, and compressing the images from $\mathcal{D}$. Downscaling would, however, likely introduce an unwanted bias because downsampled images contain more high-frequency content, which would produce JPEG images with a different profile of non-zero DCT coefficients across the DCT modes. While cropping avoids this problem, it may produce images with singular content, such as portions of sky or grass. To avoid creating such pathological images, the cropping was carried out so that the ratio of non-zero coefficients to the number of pixels in the image was kept approximately the same as for the original-size image.

The images used in the experiments below were all created from a database of 6000 never-compressed raw images stored as 8-bit grayscales. From this database, a total of $L = 15$ image sets that contained images with $n = 20 \times 10^3, 40 \times 10^3, \ldots, 300 \times 10^3$ non-zero DCT coefficients were prepared by cropping. The cover images from all image sets were singly compressed JPEGs with quality factor 80.

The statistical detectability was evaluated using the targeted blind steganalyzer containing 274 features described in Section 12.4 implemented using SVMs with Gaussian kernel. Each image set, $\mathcal{S}_i$, was split into two disjoint subsets, one of which was used for training and the other for testing the steganalyzer. The training set contained 3500 images and the corresponding 3500 stego images, while the testing set consisted of 2500 cover and the corresponding 2500 stego images. The ability of the steganalyzer to detect the embedding on the testing set was evaluated using $1 - P_{\mathrm{E}}$, where $P_{\mathrm{E}} = \frac{1}{2}(P_{\mathrm{FA}} + P_{\mathrm{MD}})$ is the minimal average classification error under equal prior probabilities of cover and stego images defined in equation (10.15). For an undetectable steganography, $P_{\mathrm{E}} \approx 0.5$, while $P_{\mathrm{E}} \approx 1$ corresponds to perfect detection. Finally, the training and testing of the steganalyzer was repeated 100 times with different splitting of $\mathcal{S}_i$ into the training and testing subsets and the average P_{E} reported as the result.

First, the same fixed payload was embedded across all image sets. Intuitively, we expect the detectability to be the largest for the image set with the smallest images, $\mathcal{S}_1$, while the smallest detectability, close to 0.5, is expected for the set with the largest images, $\mathcal{S}_{15}$. Figure 13.1 confirms this expectation. Next, the payload embedded in images from set $\mathcal{S}_i$ was made linearly proportional to the number of non-zero DCT coefficients, n_i. As can be seen from Figure 13.2, the statistical detectability increases with increasing n_i. Finally, the detectability stays approximately constant when the payload is made proportional to $\sqrt{n_i}$ (Figure 13.3). The spread around the data points in all three figures corresponds to a 90% confidence interval.

Figure 13.4 shows the result of the fourth experiment, where for each image set, $\mathcal{S}_i$, the largest absolute payload, $M(n_i)$, for which the steganalyzer obtained

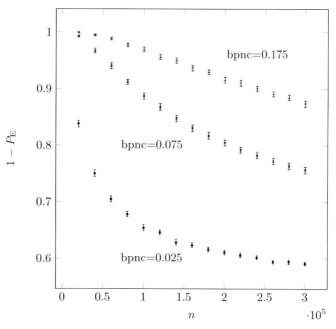

Figure 13.1 Detectability $1 - P_{\mathrm{E}}$ of embedding in JPEG images with increasing number of non-zero DCT coefficients, n. Fixed payload.

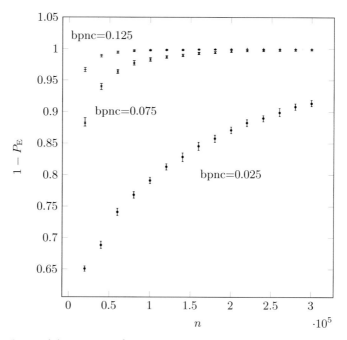

Figure 13.2 Detectability $1 - P_{\mathrm{E}}$ of embedding in JPEG images with increasing number of non-zero DCT coefficients, n. Payload proportional to n.

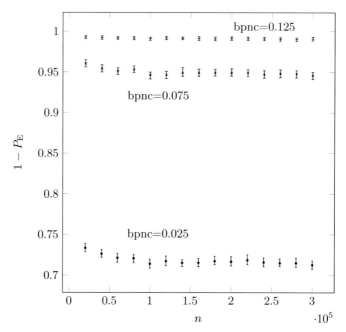

Figure 13.3 Detectability $1 - P_\mathrm{E}$ of embedding in JPEG images with increasing number of non-zero DCT coefficients, n. Payload proportional to $\sqrt{n}$.

a fixed value of P_E was iteratively found. A linear fit through the experimental data displayed in a log–log plot confirms the SRL.

The SRL of imperfect steganography was verified for other forms of steganography and other measures of security in [140].

Summary

- Informally, the steganographic capacity is the size of the maximal payload that can be undetectably embedded in the sense that the KL divergence between the set of cover and stego images is zero.
- The steganographic capacity can be rigorously defined in the limit when the number of elements forming the cover object grows to infinity. It is the largest communication rate taken across all stegosystem satisfying the requirement of perfect security, possibly with constraints on the embedding and/or channel distortion.
- The capacity can be computed analytically for sufficiently simple cover sources. It is rather difficult to estimate the capacity for real digital media due to its complexity.
- The secure payload is the largest payload that can be embedded using a given steganographic scheme with a fixed statistical detectability expressed by the KL divergence between cover and stego objects.

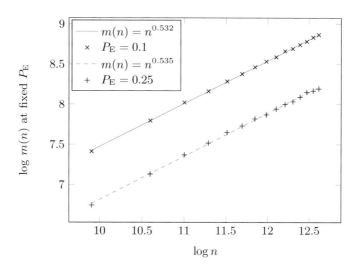

Figure 13.4 The largest payload $m(n_i)$ that produces the same value of P_E for the ith image set containing images with n_i non-zero DCT coefficients. The straight lines are the corresponding linear fits. The slope of the lines is 0.53 and 0.54, which is in good agreement with the SRL.

- The secure payload can be estimated experimentally by estimating it from a large database of images using blind steganalyzers. The disadvantage is that this estimate depends on the database of images and the choice of blind steganalyzers. On the other hand, it provides useful information to the steganographer, given the current state of the art of steganalysis.
- The secure payload of imperfect steganographic systems increases with the square root of the number of cover elements. This result is known as the SRL of imperfect steganography and holds for virtually all known steganographic techniques for real digital media.
- The SRL has been confirmed experimentally using blind steganalyzers.

Exercises

   13.1 [Berry–Esséen theorem] A version of the Berry–Esséen theorem says that for a sequence of iid random variables $x[i]$, $i = 1, \ldots, n$, whose distribution satisfies $E[x] = 0$, $E[x^2] = \sigma^2 > 0$, and $E[|x|^3] = \rho < \infty$, the cumulative density function $F_n(x)$ of the random variable

$$\sqrt{n}\frac{\bar{x}}{\sigma} \tag{13.33}$$

satisfies

$$|F_n(x) - \Phi(x)| < \frac{\rho C}{\sigma^3 \sqrt{n}}, \quad \text{for all } x. \tag{13.34}$$

The symbol $\bar{x}$ stands for the sample mean computed from n realizations, $\Phi(x)$ is the cdf of a normal random variable, and C is a positive constant. The theorem essentially says that the cdf of the scaled sample mean uniformly converges to the cdf of its limiting random variable and the rate of convergence is $1/\sqrt{n}$.

Use this theorem to carry out the arguments in the proof of the SRL of imperfect steganography without the Gaussian assumption (13.17). Note that the same bound holds true for the corresponding tail probability functions $1 - F_n(x)$ and $Q(x) = 1 - \Phi(x)$.

A   Statistics

A.1   Descriptive statistics

The purpose of this appendix is to provide the reader with some basic concepts from statistics needed throughout the main text. A good introductory text on statistics for signal-processing applications is [59]. A discrete-valued random variable $\times$ reaching values from a finite alphabet $\mathcal{A}$ with probabilities $p(x)$, $x \in \mathcal{A}$, $0 \leq p(x) \leq 1$, $\sum_x p(x) = 1$, is described by its probability mass function (pmf) $p(x)$. The probability that $\times$ reaches a value in $\mathcal{S} \subset \mathcal{A}$ is $\Pr\{\times \in \mathcal{S}\} = \sum_{x \in \mathcal{S}} p(x)$.

Example A.1: [Fair die] $\times \in \{1, 2, 3, 4, 5, 6\}$, $p(i) = \frac{1}{6}$. The probability that $\times$ is even is $\Pr\{\times \in \{2, 4, 6\}\} = 3 \times \frac{1}{6} = \frac{1}{2}$.

A continuous random variable accepting values in $\mathbb{R}$ is characterized by a probability density function (pdf) $f(x)$, which is a non-negative and Lebesgue-integrable function satisfying $\int_{\mathbb{R}} f(x)\mathrm{d}x = 1$. For any Borel set $\mathcal{B} \subset \mathbb{R}$, $\Pr\{\times \in \mathcal{B}\} = \int_{\mathcal{B}} f(x)\mathrm{d}x$. Informally, a set that can be written as a countable union or intersection of open (closed) intervals in $\mathbb{R}$ is called a Borel set. In particular, $\Pr\{\times \in [a, b]\} = \int_a^b f(x)\mathrm{d}x$.

For a real-valued random variable, we define the cumulative distribution function (cdf)

$$F(x) = \Pr\{\times \leq x\} = \int_{-\infty}^{x} f(t)\mathrm{d}t. \tag{A.1}$$

The cdf has the following properties.

- $F'(x) = \mathrm{d}F(x)/\mathrm{d}x = f(x)$ by the fundamental theorem of calculus.
- $F(x)$ is non-decreasing, $F(x) \leq F(y)$, for $x \leq y$.
- $F(x)$ is right-continuous, $\lim_{x \to x_0^+} F(x) = F(x_0)$.
- $\lim_{x \to -\infty} F(x) = 0$.
- $\lim_{x \to \infty} F(x) = 1$.

The complementary cdf is defined as

$$1 - F(x) = \Pr\{x > x\} = \int_x^\infty f(t)\mathrm{d}t. \tag{A.2}$$

Example A.2: [Uniform distribution] A random variable uniformly distributed on interval $[a, b]$ is denoted $U(a, b)$. Its pdf and cdf are

$$f(x) = \begin{cases} \frac{1}{b-a} & x \in [a, b] \\ 0 & \text{otherwise}, \end{cases} \tag{A.3}$$

$$F(x) = \begin{cases} 0 & x < a \\ \frac{x-a}{b-a} & a \le x \le b \\ 1 & x > b. \end{cases} \tag{A.4}$$

A.1.1   Measures of central tendency and spread

For a real-valued random variable x, we define its mean value (or expected value) and variance as

$$\bar{x} = E[x] = \int_{\mathbb{R}} x f(x)\mathrm{d}x, \tag{A.5}$$

$$\mathrm{Var}[x] = E[(x - \bar{x})^2] \tag{A.6}$$

$$= E[x^2 - 2\bar{x}x + \bar{x}^2] = E[x^2] - 2\bar{x}^2 + \bar{x}^2 = E[x^2] - \bar{x}^2 \tag{A.7}$$

assuming the integrals $\int_{\mathbb{R}} |x|^k f(x)\mathrm{d}x$ exist ($k = 1$ for the mean and $k = 2$ for the variance). For a discrete random variable, $E[x] = \sum_x x p(x)$, $\mathrm{Var}[x] = \sum_x (x - \bar{x})^2 p(x)$. The mean value is commonly denoted with symbol μ, while the standard deviation defined as the square root of variance, $\sqrt{\mathrm{Var}[x]}$, is often denoted with the Greek letter σ.

Note that there are probability distributions for which the mean or the variance does not exist. An example of a pdf with undefined mean is the Cauchy distribution, which has "thick tails"

$$f(x) = \frac{1}{\pi(1 + x^2)}. \tag{A.8}$$

This distribution arises as the pdf of the ratio of two Gaussian random variables with zero mean. The integral $\int_{-\infty}^\infty |x|/[\pi(1 + x^2)]\mathrm{d}x$ is divergent because

$$\int_{-\infty}^\infty \frac{|x|}{\pi(1 + x^2)}\mathrm{d}x = \frac{2}{\pi}\int_0^\infty \frac{x}{1 + x^2}\mathrm{d}x \ge \frac{2}{\pi}\int_1^\infty \frac{1}{1 + x}\mathrm{d}x \tag{A.9}$$

and the last integral is divergent.

An example of a distribution for which the variance is undefined is Student's t-distribution (Section A.8) with the tail index $\nu \leq 2$.

The issue of undefined mean or variance is not a pure academic construct but rather a real phenomenon. Thick-tail distributions are often encountered in steganalysis and signal processing. For example, in this book the error of quantitative steganalyzers (Section 10.3.2) is well modeled using Student's t-distribution. For thick-tail distributions, the spread of the random variable may not be well described using variance because the sample variance may not converge to a finite number with increasing number of samples or may converge very slowly.

Robust alternatives to the mean and variance are the median and Median Absolute Deviation (MAD). For a finite statistical sample, the median is obtained by ordering the samples from the lowest to the largest value and choosing the middle value. When the number of samples is even, one usually takes the mean of the two middle values. The median is sometimes denoted as $\mu_{1/2}(\mathsf{x})$. It satisfies

$$\Pr\{\mathsf{x} \leq \mu_{1/2}(\mathsf{x})\} \geq \frac{1}{2} \ \text{ and } \ \Pr\{\mathsf{x} \geq \mu_{1/2}(\mathsf{x})\} \geq \frac{1}{2}. \tag{A.10}$$

When more than one value $\mu_{1/2}(\mathsf{x})$ satisfies this equation, we again take the mean of all such values. For example, for the fair die from Example A.1, any value in the interval $(3, 4)$ satisfies (A.10). The median is thus defined as $\mu_{1/2} = \frac{3+4}{2} = 3.5$, which coincides with the mean. In general, the median of any probability distribution that is symmetrical about its mean must be equal to the mean. Moreover, for a random variable with continuous pdf,

$$\int_{-\infty}^{\mu_{1/2}(\mathsf{x})} f(x)\mathrm{d}x = \frac{1}{2}. \tag{A.11}$$

The MAD is defined as

$$\mathrm{MAD} = \mu_{1/2}(|\mathsf{x} - \mu_{1/2}(\mathsf{x})|). \tag{A.12}$$

The sample MAD for a random variable that represents an error term is called the Median Absolute Error (MAE).

Another common robust measure of spread is the Inter-Quartile Range (IQR). For a random variable x, it is defined as the interval $[q_1, q_3]$, where q_i is the ith quartile determined from $\Pr\{\mathsf{x} < q_i\} = i/4$, $i = 1, \ldots, 4$.

The next proposition gives a formula for the expected value of a random variable obtained as a transformation of another random variable x.

Proposition A.3. *For any piecewise continuous function g*

$$E[g(\mathsf{x})] = \int_{\mathbb{R}} g(x)f(x)\mathrm{d}x. \tag{A.13}$$

This statement is more intuitive for a discrete variable, where the transformed random variable $g(\mathsf{x})$ reaches values $g(x)$ with probabilities $p(x)$ and thus $E[g(\mathsf{x})] = \sum_x g(x)p(x)$. To see this statement for a simpler case of a continuous variable, consider g strictly monotonic and differentiable. Then, assuming for example that g is increasing, we have for the cdf of $\mathsf{y} = g(\mathsf{x})$

$$F_{\mathsf{y}}(y) = \Pr\{\mathsf{y} \le y\} = \Pr\{g(\mathsf{x}) \le y\} = \Pr\{\mathsf{x} \le g^{-1}(y)\} = \int\limits_{-\infty}^{g^{-1}(y)} f(x)\mathrm{d}x. \quad (A.14)$$

Thus, the pdf of y is obtained by differentiating its cdf,

$$f_{\mathsf{y}}(y) = F_{\mathsf{y}}'(y) = f\left(g^{-1}(y)\right) \frac{\mathrm{d}g^{-1}(y)}{\mathrm{d}y}. \quad (A.15)$$

We can now calculate its mean value:

$$E[\mathsf{y}] = \int\limits_{y_1}^{y_2} y f\left(g^{-1}(y)\right) \frac{\mathrm{d}g^{-1}(y)}{\mathrm{d}y} \mathrm{d}y = \{x = g^{-1}(y)\} = \int\limits_{\mathbb{R}} g(x) f(x)\mathrm{d}x. \quad (A.16)$$

In the derivation above, we denoted the image of $\mathbb{R}$ under g as $(y_1, y_2) = g(-\infty, \infty)$.

A special case of Proposition A.3 states that for any $a, b \in \mathbb{R}$

$$E[a\mathsf{x} + b] = aE[\mathsf{x}] + b \quad (A.17)$$

due to linearity of the integral. In this special case, the pdf of $\mathsf{y} = a\mathsf{x} + b$ for $a > 0$ is

$$f_{\mathsf{y}}(y) = f_{\mathsf{x}}\left(\frac{y - b}{a}\right) \frac{1}{a}. \quad (A.18)$$

A.1.2   Construction of PRNGs using compounding

Proposition A.3 can be conveniently used for constructing pseudo-random number generators (PRNGs) with arbitrary pdf from a uniform random number generator on $(0, 1)$. In order to obtain a PRNG producing numbers according to cdf $F(x)$, note that for $\mathsf{u} \sim U(0, 1)$

$$\Pr\{F^{-1}(\mathsf{u}) \le x\} = \Pr\{\mathsf{u} \le F(x)\} = F(x), \quad (A.19)$$

which implies that the cdf of $F^{-1}(\mathsf{u})$ is $F(x)$.

As an example, let us imagine that we wish to construct a PRNG that generates numbers following the Cauchy distribution

$$f(x) = \frac{1}{\pi(1 + x^2)}. \quad (A.20)$$

First, we obtain the cdf

$$F(x) = \int_{-\infty}^{x} \frac{\mathrm{d}t}{\pi(1+t^2)} = \frac{1}{2} + \frac{1}{\pi}\arctan(x). \tag{A.21}$$

The inverse function to $y = F(x)$ is $x = \tan\left(\pi\left(y - \frac{1}{2}\right)\right)$. Finally, if $u \sim U(0,1)$, the variable $\tan\left(\pi\left(u - \frac{1}{2}\right)\right)$ follows the Cauchy distribution.

A.2 Moment-generating function

Statistical moments are often useful for description of the shape of a probability distribution. For a positive integer k, the kth moment of x is defined as

$$\mu[k] = E[\mathsf{x}^k] \tag{A.22}$$

and the kth central moment is

$$\mu_{\mathrm{c}}[k] = E[(\mathsf{x} - \bar{\mathsf{x}})^k]. \tag{A.23}$$

The moments do not exist if the integral $E[|\mathsf{x}|^k] = \int_{\mathbb{R}} |x|^k f(x)\mathrm{d}x$ does not converge. The first moment is the mean and the second moment $\mu[2] = E[\mathsf{x}^2] = \bar{\mathsf{x}}^2 + \mathrm{Var}[\mathsf{x}] = \mu^2 + \sigma^2$. Of course, the first and second central moments are $\mu_{\mathrm{c}}[1] = 0$ and $\mu_{\mathrm{c}}[2] = \mathrm{Var}[\mathsf{x}]$.

Central moments normalized by the corresponding power of the standard deviation, $\mu_{\mathrm{c}}[k]/\sigma^k$, are called normalized moments. The third normalized central moment, $\mu_{\mathrm{c}}[3]/\sigma^3$, is called skewness and it measures the asymmetry of the distribution. The fourth normalized central moment, $\mu_{\mathrm{c}}[4]/\sigma^4$, is called kurtosis and is often used to evaluate the Gaussianity of a random variable (the kurtosis of a standard Gaussian variable is 3).

Moments can be conveniently computed using the moment-generating function defined as

$$M_{\mathsf{x}}(t) = E[e^{t\mathsf{x}}] = \int_{\mathbb{R}} e^{tx} f(x)\mathrm{d}x. \tag{A.24}$$

This is a well-defined transform because, if $M_{\mathsf{x}}(t) = M_{\mathsf{y}}(t)$, then $\mathsf{x} = \mathsf{y}$ in the sense that they have the same distribution almost everywhere with respect to the pdf $f(x)$.

The moment-generating function can be used to calculate moments of random variables in the following manner:

$$\frac{\mathrm{d}M_{\mathsf{x}}(t)}{\mathrm{d}t} = \frac{\mathrm{d}}{\mathrm{d}t}\int_{\mathbb{R}} e^{tx} f(x)\mathrm{d}x = \int_{\mathbb{R}} x e^{tx} f(x)\mathrm{d}x. \tag{A.25}$$

By differentiating k times

$$\frac{\mathrm{d}^k M_\mathsf{x}(t)}{\mathrm{d}t^k} = \int_{\mathbb{R}} x^k e^{tx} f(x)\mathrm{d}x. \tag{A.26}$$

Thus,

$$\left.\frac{\mathrm{d}^k M_\mathsf{x}(t)}{\mathrm{d}t^k}\right|_{t=0} = \int_{\mathbb{R}} x^k f(x)\mathrm{d}x = E[\mathsf{x}^k] = \boldsymbol{\mu}[k]. \tag{A.27}$$

Example A.4: [Gaussian moment-generating function] Let x be a Gaussian random variable with mean μ and variance σ^2, $\mathsf{x} \sim N(\mu, \sigma^2)$ (see the definition in Section A.4),

$$M_\mathsf{x}(t) = \int_{\mathbb{R}} e^{tx}\frac{1}{\sqrt{2\pi}\sigma}e^{-\frac{(x-\mu)^2}{2\sigma^2}}\mathrm{d}x = \frac{1}{\sqrt{2\pi}\sigma}\int_{\mathbb{R}} e^{-\frac{x^2 - 2\mu x + \mu^2 - 2\sigma^2 tx}{2\sigma^2}}\mathrm{d}x. \tag{A.28}$$

Completing the exponent to the form $(x-A)^2 + B$, we obtain

$$x^2 - 2\mu x + \mu^2 - 2\sigma^2 tx = x^2 - 2x(\mu + t\sigma^2) + (\mu + t\sigma^2)^2 + \mu^2 - (\mu + t\sigma^2)^2 \tag{A.29}$$

$$= \left(x - (\mu + t\sigma^2)\right)^2 - 2\mu t\sigma^2 - t^2\sigma^4. \tag{A.30}$$

Thus, we can write

$$M_\mathsf{x}(t) = \frac{1}{\sqrt{2\pi}\sigma}\int_{\mathbb{R}} e^{-\frac{(x-(\mu+t\sigma^2))^2}{2\sigma^2}}e^{\mu t + \frac{t^2\sigma^2}{2}}\mathrm{d}x = e^{\mu t + \frac{t^2\sigma^2}{2}}, \tag{A.31}$$

which is the moment-generating function of a Gaussian variable $N(\mu, \sigma^2)$.

From the moment-generating function, we can obtain the first four moments of a Gaussian random variable simply by differentiating $M_\mathsf{x}(t)$ and evaluating the derivatives at 0. Writing higher-order derivatives as roman numerals,

$$M_\mathsf{x}'(t) = (\mu + t\sigma^2)M_\mathsf{x}(t) = \{\text{at } t = 0\} = \mu, \tag{A.32}$$

$$M_\mathsf{x}^{II}(t) = \left(\sigma^2 + (\mu + t\sigma^2)^2\right)M_\mathsf{x}(t) = \{\text{at } t = 0\} = \mu^2 + \sigma^2, \tag{A.33}$$

$$M_\mathsf{x}^{III}(t) = \left(2\sigma^2(\mu + t\sigma^2) + \left(\sigma^2 + (\mu + t\sigma^2)^2\right)(\mu + t\sigma^2)\right)M_\mathsf{x}(t), \tag{A.34}$$

$$= \{\text{at } t = 0\} = \mu^3 + 3\mu\sigma^2, \tag{A.35}$$

$$M_\mathsf{x}^{IV}(t) = \{\text{at } t = 0\} = 3\sigma^4 + 6\mu^2\sigma^2 + \mu^4. \tag{A.36}$$

The first four central moments ($\mu = 0$) are $0, \sigma^2, 0, 3\sigma^4$. In general, for an even pdf, all odd moments are 0.

A.3 Jointly distributed random variables

Let x and y be two real-valued random variables. Their joint probability density function $f(x,y) \geq 0$ defines the probability that a joint event (x, y) will fall into a Borel subset $\mathcal{B} \subset \mathbb{R}^2$,

$$\Pr\{(x, y) \in \mathcal{B}\} = \int_{\mathcal{B}} f(x, y) \mathrm{d}x \mathrm{d}y. \tag{A.37}$$

As with the one-dimensional counterpart, $\int_{\mathbb{R}^2} f(x, y)\mathrm{d}x\mathrm{d}y = 1$.

The marginal probabilities for each random variable are

$$f_x(x) = \int_{\mathbb{R}} f(x, y)\mathrm{d}y, \ f_y(y) = \int_{\mathbb{R}} f(x, y)\mathrm{d}x. \tag{A.38}$$

We say that x and y are independent if

$$f(x, y) = f_x(x)f_y(y) \quad \text{for all } x, y. \tag{A.39}$$

These concepts generalize to more than two variables in a straightforward manner.

An analogy of Proposition A.3 for jointly distributed random variables is the following useful statement.

Proposition A.5. *Consider random variables* $x_1, \ldots, x_n$ *with joint pdf* $f(x_1, \ldots, x_n)$. *Then for any piecewise continuous function* $g : \mathbb{R}^n \to \mathbb{R}$

$$E[g(x_1, \ldots, x_n)] = \int_{-\infty}^{\infty} \cdots \int_{-\infty}^{\infty} g(x_1, \ldots, x_n)f(x_1, \ldots, x_n)\mathrm{d}x_1 \ldots \mathrm{d}x_n. \tag{A.40}$$

For any two random variables x and y, their covariance

$$\mathrm{Cov}[x, y] = E[(x - \bar{x})(y - \bar{y})] = E[xy] - \bar{x}\bar{y} - \bar{y}\bar{x} + \bar{x}\bar{y} = E[xy] - \bar{x}\bar{y} \tag{A.41}$$

measures the extent of linear dependence between them. If $\mathrm{Cov}[x, y] = 0$, we say that the variables are uncorrelated. Thus, $E[xy] = E[x]E[y]$ for two uncorrelated random variables. Note that $\mathrm{Cov}[x, x] = \mathrm{Var}[x]$.

If x, y are independent, then they are uncorrelated because by (A.41) and Proposition A.5

$$E[xy] = \iint_{\mathbb{R}^2} xyf(x, y)\mathrm{d}x\mathrm{d}y = \iint_{\mathbb{R}^2} xyf_x(x)f_y(y)\mathrm{d}x\mathrm{d}y \tag{A.42}$$

$$= \int_{\mathbb{R}} xf_x(x)\mathrm{d}x \int_{\mathbb{R}} yf_y(y)\mathrm{d}y = E[x]E[y]. \tag{A.43}$$

Example A.7 shows that the converse is not generally true. Two uncorrelated variables may be dependent.

Linear dependence between two random variables is evaluated using the correlation coefficient

$$\rho(\mathsf{x}, \mathsf{y}) = \frac{\mathrm{Cov}[\mathsf{x}, \mathsf{y}]}{\sqrt{\mathrm{Var}[\mathsf{x}]\mathrm{Var}[\mathsf{y}]}}. \tag{A.44}$$

The correlation coefficient is always between -1 and 1,

$$|\rho(\mathsf{x}, \mathsf{y})| \leq 1, \tag{A.45}$$

which is the Cauchy–Schwartz inequality in the vector space of zero-mean random variables with inner product defined as the covariance (see Section D.10).

Using Proposition A.5, we can write for any random variables $\mathsf{x}_1, \ldots, \mathsf{x}_n$ and a constant vector $\mathbf{a}$

$$E\left[\sum_{i=1}^{n} \mathbf{a}[i]\mathsf{x}_i\right] = \sum_{i=1}^{n} \mathbf{a}[i]E[\mathsf{x}_i]. \tag{A.46}$$

Often, we will also need to know the variance of a sum of random variables $\mathsf{y} = \sum_{i=1}^{n} \mathbf{a}[i]\mathsf{x}_i$. The variance of y is

$$\mathrm{Var}[\mathsf{y}] = E[\mathsf{y}^2] - (E[\mathsf{y}])^2 = E\left[\left(\sum_{i=1}^{n} \mathbf{a}[i]\mathsf{x}_i\right)^2\right] - \left(\sum_{i=1}^{n} \mathbf{a}[i]E[\mathsf{x}_i]\right)^2 \tag{A.47}$$

$$= E\left[\sum_{i=1}^{n}\sum_{j=1}^{n} \mathbf{a}[i]\mathbf{a}[j]\mathsf{x}_i\mathsf{x}_j\right] - \sum_{i=1}^{n}\sum_{j=1}^{n} \mathbf{a}[i]\mathbf{a}[j]E[\mathsf{x}_i]E[\mathsf{x}_j] \tag{A.48}$$

$$= \sum_{i=1}^{n}\sum_{j=1}^{n} \mathbf{a}[i]\mathbf{a}[j]\left(E[\mathsf{x}_i\mathsf{x}_j] - E[\mathsf{x}_i]E[\mathsf{x}_j]\right) \tag{A.49}$$

$$= \sum_{i=1}^{n}\sum_{j=1}^{n} \mathbf{a}[i]\mathbf{a}[j]\mathrm{Cov}[\mathsf{x}_i, \mathsf{x}_j] = \mathbf{a}\mathbf{C}\mathbf{a}', \tag{A.50}$$

where $\mathbf{C}[i, j] = \mathrm{Cov}[\mathsf{x}_i, \mathsf{x}_j] = E[\mathsf{x}_i\mathsf{x}_j] - E[\mathsf{x}_i]E[\mathsf{x}_j]$ is the covariance between random variables x_i and x_j forming the covariance matrix $\mathbf{C}$.

If x_i are pairwise uncorrelated, i.e., $\mathrm{Cov}[\mathsf{x}_i, \mathsf{x}_j] = 0$ for all $i \neq j$, then

$$\mathrm{Var}\left[\sum_{i=1}^{n} \mathbf{a}[i]\mathsf{x}_i\right] = \sum_{i=1}^{n} \mathbf{a}[i]^2\mathrm{Var}[\mathsf{x}_i]. \tag{A.51}$$

Often, it is necessary to determine the distribution of a sum of two random variables. Let x and y be two real-valued random variables with joint pdf $f(x, y)$.

Then, the cdf of their sum is

$$F_{x+y}(z) = \Pr\{x + y \le z\} = \iint\limits_{x+y\le z} f(x, y)\mathrm{d}x\mathrm{d}y \tag{A.52}$$

$$= \int\limits_{-\infty}^{\infty} \mathrm{d}x \int\limits_{-\infty}^{z-x} f(x, y)\mathrm{d}y = \{u = x + y\} = \int\limits_{-\infty}^{\infty} \mathrm{d}x \int\limits_{-\infty}^{z} f(x, u - x)\mathrm{d}u. \tag{A.53}$$

By differentiating

$$f_{x+y}(z) = \int\limits_{-\infty}^{\infty} \mathrm{d}x \frac{\mathrm{d}}{\mathrm{d}z} \int\limits_{-\infty}^{z} f(x, u - x)\mathrm{d}u = \int\limits_{-\infty}^{\infty} f(x, z - x)\mathrm{d}x. \tag{A.54}$$

If x and y are independent, then $f(x, y) = f_x(x)f_y(y)$ and we have

$$f_{x+y}(z) = \int\limits_{-\infty}^{\infty} f_x(x)f_y(z - x)\mathrm{d}x = f_x \star f_y(z). \tag{A.55}$$

In other words, the pdf of the sum of two independent random variables is the convolution of their pdfs.

We now show that the sum of two independent Gaussian variables is again Gaussian, with mean equal to the sum of means and variance equal to the sum of variances. To this end, we use the following lemma.

Lemma A.6. *For two independent random variables* x, y, *the moment-generating function of their sum is the product of moment-generating functions for* x *and* y

$$M_{x+y}(t) = E\left[e^{t(x+y)}\right] = \iint e^{t(x+y)} f(x, y)\mathrm{d}x\mathrm{d}y \tag{A.56}$$

$$= \int e^{tx} f_x(x)\mathrm{d}x \int e^{ty} f_y(y)\mathrm{d}y = M_x(t)M_y(t). \tag{A.57}$$

For $x \sim N(\mu_1, \sigma_1^2)$ and $y \sim N(\mu_2, \sigma_2^2)$, we have

$$M_{x+y}(t) = e^{\mu_1 + \frac{1}{2}\sigma_1^2 t^2} e^{\mu_2 + \frac{1}{2}\sigma_2^2 t^2} = e^{(\mu_1 + \mu_2) + \frac{1}{2}(\sigma_1^2 + \sigma_2^2)t^2}, \tag{A.58}$$

which is the pdf of the Gaussian random variable $N(\mu_1 + \mu_2, \sigma_1^2 + \sigma_2^2)$. In general, a linear combination of independent Gaussian variables is again a Gaussian random variable, with mean equal to the same linear combination of means and variance equal to a linear combination of variances, where the coefficients in the linear combination are squared.

A.4   Gaussian random variable

The Gaussian distribution is, without any doubts, the most famous and most important distribution for a signal-processing engineer. The pdf for a Gaussian (also called normal) random variable $N(\mu, \sigma^2)$ with mean μ and variance σ^2 is

$$f(x) = \frac{1}{\sqrt{2\pi}\sigma} e^{-\frac{(x-\mu)^2}{2\sigma^2}}. \qquad (A.59)$$

By numerical evaluation, we obtain the following probabilities of outliers:

$$\Pr\{|x - \mu| \le \sigma\} \doteq 68.27\%, \qquad (A.60)$$
$$\Pr\{|x - \mu| \le 2\sigma\} \doteq 95.45\%, \qquad (A.61)$$
$$\Pr\{|x - \mu| \le 3\sigma\} \doteq 99.73\%. \qquad (A.62)$$

The cdf for a standard Gaussian random variable $x \sim N(0,1)$ will be denoted $\Phi(x) = \int_{-\infty}^{x}(1/\sqrt{2\pi})e^{-\frac{t^2}{2}}\,dt$. We also define the complementary cumulative distribution function (the tail probability) as the probability that x reaches a value larger than x,

$$Q(x) = \Pr\{x > x\} = 1 - \Phi(x) = \int_{x}^{\infty} \frac{1}{\sqrt{2\pi}} e^{-\frac{t^2}{2}}\,dt. \qquad (A.63)$$

This tail probability can be expressed using the error function

$$\mathrm{Erf}(x) = \frac{2}{\sqrt{\pi}} \int_{0}^{x} e^{-t^2}\,dt, \qquad (A.64)$$

which is available in Matlab as `erf.m`. Here is the connection between $\mathrm{Erf}(x)$ and $Q(x)$:

$$1 - Q(x) = \int_{-\infty}^{x} \frac{1}{\sqrt{2\pi}} e^{-\frac{t^2}{2}}\,dt = \{\text{substitution } u = t/\sqrt{2}\} = \int_{-\infty}^{x/\sqrt{2}} \frac{1}{\sqrt{\pi}} e^{-u^2}\,du$$
$$ \qquad (A.65)$$

$$= \frac{1}{\sqrt{\pi}} \left(\int_{-\infty}^{0} e^{-u^2}\,du + \int_{0}^{x/\sqrt{2}} e^{-u^2}\,du \right) = \frac{1}{2} + \frac{1}{2}\mathrm{Erf}\left(\frac{x}{\sqrt{2}}\right). \qquad (A.66)$$

Thus,

$$Q(x) = \frac{1}{2}\left(1 - \mathrm{Erf}\left(\frac{x}{\sqrt{2}}\right)\right). \qquad (A.67)$$

Note that for $x \sim N(\mu, \sigma^2)$,

$$\Pr\{x \geq x\} = \int_x^\infty \frac{1}{\sqrt{2\pi}\sigma} e^{-\frac{(t-\mu)^2}{2\sigma^2}} \, dt = \{u = (t-\mu)/\sigma\} \qquad (A.68)$$

$$= \int_{\frac{x-\mu}{\sigma}}^\infty \frac{1}{\sqrt{2\pi}} e^{-\frac{u^2}{2}} \, du = Q\left(\frac{x-\mu}{\sigma}\right). \qquad (A.69)$$

The tail probability $Q(x)$ cannot be expressed in a closed form. However, for large x there exists an approximate asymptotic form

$$Q(x) \approx \frac{e^{-x^2/2}}{\sqrt{2\pi}x}, \qquad (A.70)$$

which should be understood in the sense that $f(x) \approx g(x)$ for $x \to \infty$ if $\lim_{x\to\infty} f(x)/g(x) = 1$. This asymptotic expression can be easily obtained by integrating by parts ($\int uv' = uv - \int u'v$),

$$Q(x) = \int_x^\infty \frac{t}{\sqrt{2\pi}t} e^{-\frac{t^2}{2}} \, dt = \{u = 1/(\sqrt{2\pi}t), \ v' = te^{-t^2/2}\} \qquad (A.71)$$

$$= \left[-\frac{e^{-t^2/2}}{\sqrt{2\pi}t}\right]_x^\infty + \int_x^\infty \frac{e^{-t^2/2}}{\sqrt{2\pi}t^2} \, dt = \frac{e^{-x^2/2}}{\sqrt{2\pi}x} - R(x), \qquad (A.72)$$

where

$$R(x) = \int_x^\infty \frac{e^{-t^2/2}}{\sqrt{2\pi}t^2} \, dt \leq \frac{1}{x^2} \int_x^\infty \frac{e^{-t^2/2}}{\sqrt{2\pi}} \, dt = \frac{Q(x)}{x^2}. \qquad (A.73)$$

Thus,

$$\frac{e^{-x^2/2}/(\sqrt{2\pi}x)}{Q(x)} = 1 + \frac{R(x)}{Q(x)} = 1 + O\left(\frac{1}{x^2}\right), \qquad (A.74)$$

which proves (A.70).

A.5   Multivariate Gaussian distribution

A multivariate Gaussian distribution, also called jointly Gaussian, is obtained as a linear transformation of independent standard normal variables. Let $y_1, \ldots, y_n$ be all mutually independent and $N(0, 1)$ and let $x_1, \ldots, x_n$ be random variables obtained as linear transformations of $y_1, \ldots, y_n$,

$$x_i = \mathbf{T}[i, 1]y_1 + \mathbf{T}[i, 2]y_2 + \cdots + \mathbf{T}[i, n]y_n + \boldsymbol{\mu}[i] \text{ for each } i = 1, \ldots, n, \quad (A.75)$$

or in matrix notation

$$\mathbf{x} = \mathbf{T}\mathbf{y} + \boldsymbol{\mu}, \qquad (A.76)$$

where $\mathbf{y} = (y_1, \ldots, y_n)'$ and $\boldsymbol{\mu} = (\boldsymbol{\mu}[1], \ldots, \boldsymbol{\mu}[n])'$ is a vector of constants. We also assume that the $n \times n$ matrix $\mathbf{T}$ is regular, so that $\mathbf{T}^{-1}$ exists (we denote $\mathbf{T} = \mathbf{D}^{-1}$). The vector $\mathbf{x} = (x_1, \ldots, x_n)'$ follows the multivariate Gaussian distribution. We now derive the pdf of this vector random variable.

Because y_i are independent, the joint pdf of $\mathbf{y}$ is the product of pdfs of each y_i

$$f(\mathbf{y}) = \prod_{i=1}^{n} \frac{1}{\sqrt{2\pi}} e^{-\frac{1}{2}\mathbf{y}[i]^2} = \frac{1}{\sqrt{(2\pi)^n}} e^{-\frac{1}{2}(\mathbf{y}[1]^2 + \cdots + \mathbf{y}[n]^2)} = \frac{1}{\sqrt{(2\pi)^n}} e^{-\frac{1}{2}\mathbf{y}'\mathbf{y}}, \quad (A.77)$$

because $\mathbf{y}$ is a column vector.

We now derive the pdf for the transformed variables x_i by substituting

$$\mathbf{y} = \mathbf{T}^{-1}(\mathbf{x} - \boldsymbol{\mu}) = \mathbf{D}(\mathbf{x} - \boldsymbol{\mu}) \qquad (A.78)$$

into (A.77). First, note that $\mathbf{y}'\mathbf{y} = (\mathbf{D}(\mathbf{x} - \boldsymbol{\mu}))' \mathbf{D}(\mathbf{x} - \boldsymbol{\mu}) = (\mathbf{x} - \boldsymbol{\mu})'\mathbf{D}'\mathbf{D}(\mathbf{x} - \boldsymbol{\mu}) = (\mathbf{x} - \boldsymbol{\mu})'\mathbf{C}^{-1}(\mathbf{x} - \boldsymbol{\mu})$, where we denoted

$$\mathbf{C}^{-1} = \mathbf{D}'\mathbf{D}. \qquad (A.79)$$

We remark that $\mathbf{C}^{-1}$ is a positive-definite matrix because, for any $\mathbf{z} \in \mathbb{R}^n$, $\mathbf{z}'\mathbf{C}^{-1}\mathbf{z} = \mathbf{z}'\mathbf{D}'\mathbf{D}\mathbf{z} = \|\mathbf{D}\mathbf{z}\|^2 \geq 0$ and equality occurs if and only if $\mathbf{z} = 0$ because $\mathbf{D}$ is regular. To obtain the full pdf expressed in variables $\mathbf{x}[i]$, we need to multiply the pdf by the Jacobian of the linear transform. This is because

$$\Pr\{\mathbf{y} \in \mathcal{B} \subset \mathbb{R}^n\} = \int_{\mathcal{B}} f(\mathbf{y}) d\mathbf{y} = \{\text{substitution } \mathbf{y} = h(\mathbf{x})\} = \int_{\mathcal{B}} f(h(\mathbf{x})) \left| \frac{\partial \mathbf{y}}{\partial \mathbf{x}} \right| d\mathbf{x}. \qquad (A.80)$$

Thus, the pdf expressed in terms of $\mathbf{x}$ is $f(h(\mathbf{x})) |\partial\mathbf{y}/\partial\mathbf{x}|$. From (A.78), $\partial\mathbf{y}[i]/\partial\mathbf{x}[j] = \mathbf{D}[i,j]$ and we have for the Jacobian

$$\left| \frac{\partial \mathbf{y}}{\partial \mathbf{x}} \right| = |\mathbf{D}| = \sqrt{|\mathbf{D}| \cdot |\mathbf{D}'|} = \sqrt{|\mathbf{C}^{-1}|} = \frac{1}{\sqrt{|\mathbf{C}|}}. \qquad (A.81)$$

The pdf of $\mathbf{x}$ is thus

$$f(\mathbf{x}) = \frac{1}{\sqrt{(2\pi)^n |\mathbf{C}|}} e^{-\frac{1}{2}(\mathbf{x}-\boldsymbol{\mu})'\mathbf{C}^{-1}(\mathbf{x}-\boldsymbol{\mu})}. \qquad (A.82)$$

Notice that from (A.75) the mean values $E[x_i] = \boldsymbol{\mu}[i]$ (or in vector form, $E[\mathbf{x}] = \boldsymbol{\mu}$) and the covariances

$$E[(x_i - \boldsymbol{\mu}[i])(x_j - \boldsymbol{\mu}[j])] = E\left[\sum_{k=1}^{n} \mathbf{D}^{-1}[i,k]y_k \times \sum_{k=1}^{n} \mathbf{D}^{-1}[j,k]y_k \right] \qquad (A.83)$$

$$= E\left[\sum_{k=1}^{n} \mathbf{D}^{-1}[i,k]\mathbf{D}^{-1}[j,k]y_k^2 \right] \qquad (A.84)$$

$$= \sum_{k=1}^{n} \mathbf{D}^{-1}[i,k]\mathbf{D}^{-1}[j,k] \qquad (A.85)$$

$$= \mathbf{D}^{-1}(\mathbf{D}^{-1})'[i,j] = (\mathbf{D}'\mathbf{D})^{-1}[i,j] = \mathbf{C}[i,j]. \qquad (A.86)$$

In the derivations above, we used the fact that $E[y_i y_j] = \delta(i - j)$, because y_i are iid, and $(\mathbf{D}')^{-1} = (\mathbf{D}^{-1})'$ for any regular matrix $\mathbf{D}$. Thus, the matrix $\mathbf{C}$ is the covariance matrix of variables $x_1, \ldots, x_n$.

Conversely, if a vector of random variables $\mathbf{x}$ follows the distribution (A.82) for a symmetric positive-definite matrix $\mathbf{C}$, it is jointly Gaussian because any symmetric positive-definite matrix $\mathbf{C}$ can be written as $\mathbf{C}^{-1} = \mathbf{D}'\mathbf{D}$ for some regular matrix $\mathbf{D}$ (this is called Choleski decomposition). By transforming $\mathbf{x}$ using the linear transform $\mathbf{y} = \mathbf{D}(\mathbf{x} - \boldsymbol{\mu})$, we can make all $y_i \sim N(0,1)$ and independent, simply by reversing the steps above. This will also prove that $E[\mathbf{x}] = \boldsymbol{\mu}$ and $E[(x_i - \boldsymbol{\mu}[i])(x_j - \boldsymbol{\mu}[j])] = \mathbf{C}[i,j]$, the covariance matrix.

Note that jointly Gaussian random variables that are uncorrelated must be also independent. This is a rare case when uncorrelatedness does imply independence. This is because, if x_i and x_j are uncorrelated for every $i \neq j$, their covariance matrix $\mathbf{C} = \text{diag}(\boldsymbol{\sigma}[1]^2, \ldots, \boldsymbol{\sigma}[n]^2)$ is diagonal, $|\mathbf{C}| = \boldsymbol{\sigma}[1] \times \cdots \times \boldsymbol{\sigma}[n]$, and the pdf (A.82) can be factorized,

$$f(\mathbf{x}) = \frac{1}{\sqrt{(2\pi)^n |\mathbf{C}|}} e^{-\frac{1}{2}(\mathbf{x}-\boldsymbol{\mu})'\mathbf{C}^{-1}(\mathbf{x}-\boldsymbol{\mu})} = \prod_{i=1}^{n} \frac{1}{\sqrt{2\pi\boldsymbol{\sigma}[i]^2}} e^{-\frac{(\mathbf{x}[i]-\boldsymbol{\mu}[i])^2}{2\boldsymbol{\sigma}[i]^2}}, \quad (A.87)$$

which means that the pdf is the product of its marginals, establishing thus their independence.

Note that uncorrelated Gaussians that are not jointly distributed do not have to be independent. Consider this simple example.

Example A.7: [Uncorrelated but dependent] Let $\xi \sim N(0,1)$ and s be uniformly distributed on $\{-1,1\}$, independent of ξ (s is a random variable that equally likely attains values -1 and 1). Then, $E[\xi] = E[s\xi] = 0$ and ξ and $s\xi$ are two uncorrelated variables because $E[\xi \cdot s\xi] = 0$, and thus $E[\xi \cdot s\xi] - E[\xi]E[s\xi] = 0$, because the pdf of $s\xi^2$ is symmetrical about 0 (the symmetry of s is responsible for this). The variables ξ and $s\xi$ are, however, not independent because the knowledge of ξ determines $s\xi$ up to its sign. In fact, the marginal distributions of each variable are standard Gaussian $N(0,1)$ (a Gaussian $N(0,1)$ with randomized sign is again $N(0,1)$). Thus, if ξ and $s\xi$ were independent, their joint pdf would have to be a product of two Gaussian pdfs, which is clearly not the case because the joint pdf $f(x,y) = 0$ whenever $|x| \neq |y|$ (because we always have $|\xi| = |s\xi|$).

A.6   Asymptotic laws

There exist many important asymptotic results in statistics when a sequence of random variables converges in some sense to another random variable giving us the option in practice to replace a variable with a rather complicated distribution with a much simpler distribution.

The weak law of large numbers states that the sample mean of a sequence of iid random variables $x_1, \ldots, x_n$ with finite mean μ converges in probability to the mean

$$\bar{x}_n = \frac{1}{n}(x_1 + \cdots + x_n) \xrightarrow{P} \mu \qquad (A.88)$$

in the sense that for any $\epsilon > 0$,

$$\Pr\{|\bar{x}_n - \mu| > \epsilon\} \to 0, \qquad (A.89)$$

as $n \to \infty$.

The Central Limit Theorem (CLT) is one of the most fundamental theorems in probability. It states that given a sequence of iid random variables $x_1, \ldots, x_n$ with finite mean value μ and variance σ^2, the distribution of the average $(x_1 + \cdots + x_n)/n$ approaches a Gaussian with mean μ and variance σ^2/n independently of the distribution of x_i.

More precisely, denoting the partial sum $s_n = x_1 + \cdots + x_n$, the following variable converges in distribution to $N(0,1)$:

$$\frac{s_n - n\mu}{\sigma\sqrt{n}} \xrightarrow{D} N(0,1). \qquad (A.90)$$

We say that sequence z_n converges in distribution to y if $\lim_{n \to \infty} F_{z_n}(x) = F_y(x)$ for all x where the cdf F_y is continuous.

A.7   Bernoulli and binomial distributions

In this section, we introduce two discrete distributions that appear in the book.

A Bernoulli random variable $B(p)$ is a random variable x with range $\{0,1\}$ with

$$\Pr\{x = 1\} = p, \qquad (A.91)$$
$$\Pr\{x = 0\} = 1 - p. \qquad (A.92)$$

The mean and variance of a Bernoulli random variable are

$$E[x] = (1 - p) \times 0 + p \times 1 = p, \qquad (A.93)$$
$$\mathrm{Var}[x] = (1 - p) \times (0 - p)^2 + p \times (1 - p)^2 = p(1 - p). \qquad (A.94)$$

Assume we have an experiment with two possible outcomes "yes" or "no," where "yes" occurs with probability p and "no" with probability $1 - p$. If the experiment is repeated n times, the number of occurrences of "yes" is a random variable y

with range $y \in \{0, 1, \ldots, n\}$ that follows the binomial distribution $Bi(n, p)$

$$\Pr\{y = k\} = \binom{n}{k} p^k (1-p)^{n-k}, \quad k = 0, 1, \ldots, n, \qquad (A.95)$$

$$E[y] = np, \qquad (A.96)$$

$$\text{Var}[y] = np(1-p). \qquad (A.97)$$

Given n independent Bernoulli random variables $B(p)$, $x_1, \ldots, x_n$, $\sum_{i=1}^{n} x_i \sim Bi(n, p)$. Thus, by the CLT, $Bi(n, p)$ converges in distribution to a Gaussian in the sense defined in Section A.6.

A.8   Generalized Gaussian, generalized Cauchy, Student's t-distributions

The generalized Gaussian distribution

$$f(x; \alpha, \beta, \mu) = \frac{\beta}{2\alpha \Gamma\left(\frac{1}{\beta}\right)} e^{-\left|\frac{x-\mu}{\alpha}\right|^{\beta}} \qquad (A.98)$$

is a reasonably good model of DCT coefficients in a JPEG file and in general a good fit for any high-pass-filtered natural image. The model depends on three parameters – the mean μ, the shape parameter $\beta > 0$, and the width parameter $\alpha > 0$.

The influence of the parameter β on the distribution shape is shown in Figure A.1. For $\beta > 1$, the distribution is continuously differentiable at zero, while for $0 < \beta \leq 1$, the distribution has a "spike" at zero (it is not differentiable there). The smaller β, the spikier the distribution looks. For $\beta = 1$, the distribution is called Laplacian and for $\beta = 2$, we obtain the Gaussian distribution. Values of $\beta > 2$ lead to distributions with increasingly flatter maximum at zero. For $\beta \to \infty$, the distribution converges to a uniform distribution on $(-\alpha, \alpha)$.

A simple method for fitting the generalized Gaussian model through sample data $x[i]$ is the method of moments [172]. It works by first calculating the sample mean and the first two absolute central sample moments,

$$\hat{\mu}_c[1] = \frac{1}{n} \sum_{i=1}^{n} |x[i] - \hat{\mu}|, \qquad (A.99)$$

$$\hat{\mu}_c[2] = \frac{1}{n} \sum_{i=1}^{n} |x[i] - \hat{\mu}|^2, \qquad (A.100)$$

where

$$\hat{\mu} = \frac{1}{n} \sum_{i=1}^{n} x[i].$$

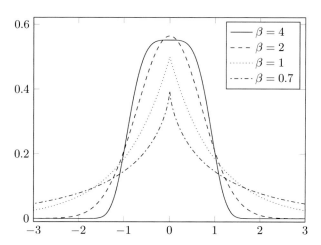

Figure A.1 Examples of the generalized Gaussian distribution for different values of the shape parameter β and $\alpha = 1$, $\mu = 0$.

Estimates of the generalized Gaussian parameters are then obtained as

$$\hat{\beta} = G^{-1}\left(\frac{\hat{\boldsymbol{\mu}}_c[1]^2}{\hat{\boldsymbol{\mu}}_c[2]}\right), \tag{A.101}$$

$$\hat{\alpha} = \hat{\boldsymbol{\mu}}_c[1]\frac{\Gamma\left(\frac{1}{\hat{\beta}}\right)}{\Gamma\left(\frac{2}{\hat{\beta}}\right)}, \tag{A.102}$$

where

$$G(x) = \frac{\Gamma^2\left(\frac{2}{x}\right)}{\Gamma\left(\frac{1}{x}\right)\Gamma\left(\frac{3}{x}\right)}, \tag{A.103}$$

and $\Gamma(x)$ is the Gamma function

$$\Gamma(x) = \int_0^\infty t^{x-1}e^{-t}\mathrm{d}t. \tag{A.104}$$

Note that the method fails to estimate the parameters if $\hat{\boldsymbol{\mu}}_c[1]^2/\hat{\boldsymbol{\mu}}_c[2] > 3/4$ because the range of G is $\left[0, \frac{3}{4}\right]$. The Gamma function is implemented in Matlab as gamma.m. The inverse function $G^{-1}(x)$ must be implemented numerically (e.g., through a binary search).

Another common model for the distribution of DCT coefficients is the Cauchy distribution, which arises from the ratio of two normally distributed random variables. Its generalized version has the following pdf:

$$f(x; p, s, \mu) = \frac{p-1}{2s}\left(1 + \frac{|x-\mu|}{s}\right)^{-p} \tag{A.105}$$

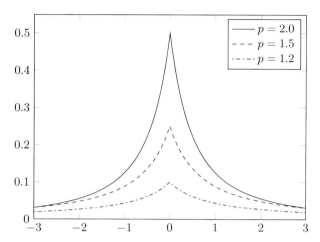

Figure A.2 Generalized Cauchy distribution for different values of the shape parameter p and $s = 1$, $\mu = 0$.

with three parameters – the mean μ, the width parameter $s > 0$, and the shape parameter $p > 1$. The generalized Cauchy distribution has thick tails. It becomes spikier with larger p and flatter with smaller p (see Figure A.2).

Student's t-distribution finds applications in modeling the output from quantitative steganalyzers (see Chapter 10). It arises in the following problem. Let $x_1, \ldots, x_n$ be iid Gaussian variables $N(\mu, \sigma^2)$ and let $\bar{x}_n = (x_1 + \cdots + x_n)/n$, $\hat{\sigma}_n^2 = [1/(n-1)] \sum_{i=1}^{n} (x_i - \bar{x})^2$ be the sample mean and sample variance. While the normalized mean is Gaussian distributed if μ and σ^2 are known,

$$\frac{\bar{x}_n - \mu}{\sigma/\sqrt{n}} \sim N(0, 1), \tag{A.106}$$

replacing the standard deviation σ with its sample form leads to a more complicated random variable,

$$\frac{\bar{x} - \mu}{\hat{\sigma}_n/\sqrt{n}}, \tag{A.107}$$

which follows Student's t-distribution with $\nu = n - 1$ degrees of freedom:

$$f(x) = \frac{\Gamma\left(\frac{\nu+1}{2}\right)}{\sqrt{\nu\pi}\,\Gamma\left(\frac{\nu}{2}\right)} \left(1 + \frac{(x-\mu)^2}{\nu}\right)^{-\frac{\nu+1}{2}}. \tag{A.108}$$

The distribution is symmetric about its mean, which is μ for $\nu > 1$ and is undefined otherwise. The parameter ν is called the tail index. When $\nu = 1$, the distribution is Cauchy. As $\nu \to \infty$, the distribution approaches the Gaussian distribution. The variance is $\nu/(\nu - 2)$ for $\nu > 2$ and undefined otherwise. In general, only moments strictly smaller than ν are defined. The tail probability of Student's distribution satisfies $1 - F(x) \approx x^{-\nu}$, where $F(x)$ is the cumulative density. Maximum likelihood estimators (see Section D.7 and [229]) can be used in practice to fit Student's distribution to data.

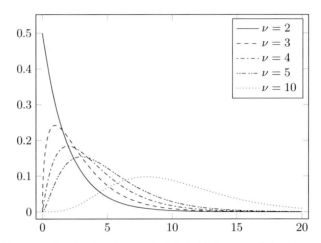

Figure A.3 Chi-square distribution for $\nu = 2, 3, 4, 5, 10$ degrees of freedom.

A.9   Chi-square distribution

The chi-square distribution often arises in many engineering applications. In this book, we will encounter it in Section 5.1.1 dealing with the histogram attack. There, it is used in Pearson's chi-square test to determine whether a discrete random variable follows a known distribution. A chi-square-distributed random variable with ν degrees of freedom arises as a sum of squares of ν standard normal Gaussian variables

$$\mathsf{x} = \sum_{i=1}^{\nu} \xi_i^2, \quad \xi_i \sim N(0,1). \tag{A.109}$$

The probability distribution of x is denoted with χ_ν^2 and its pdf is

$$f(x) = \frac{e^{-\frac{x}{2}} x^{\frac{\nu}{2}-1}}{2^{\frac{\nu}{2}} \Gamma\left(\frac{\nu}{2}\right)} \quad \text{for } x \geq 0 \tag{A.110}$$

and $f(x) = 0$ otherwise. The mean and variance of x are

$$E[\mathsf{x}] = \nu, \tag{A.111}$$

$$\text{Var}[\mathsf{x}] = 2\nu. \tag{A.112}$$

Examples of chi-square probability distributions for several values of ν are shown in Figure A.3.

A.10   Log–log empirical cdf plot

For a given scalar random variable, x, the log–log empirical cdf plot shows the tail probability $\Pr\{\mathsf{x} > x\}$ (or complementary cdf) as a function of x in a log–log plot. A separate plot is usually made for the right and left tails when the

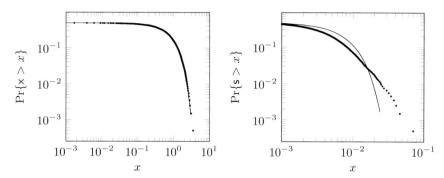

Figure A.4 The log–log empirical cdf plot for a Gaussian, x, and a t-distributed random variable, s, and the corresponding Gaussian fits (thin line).

distribution is non-symmetrical. The purpose of the plot is to analyze the tails of the distribution, which are important for error estimation in signal detection (Appendix D) and quantitative steganalysis (Section 10.3.2).

Figure A.4 shows the log–log empirical cdf plot obtained from 10,000 samples of a Gaussian random variable $N(0,1)$ and for the same number of samples of a t-distributed random variable with 3 degrees of freedom. The thin line is the plot of the corresponding Gaussian fit. Note that the t-distribution has thicker tails (the curve lies above the Gaussian fit in the log–log plot). Also, observe that the plot for the t-distributed variable approaches a straight line for large x because the tail probability $\Pr\{x > x\} = 1 - F(x) \approx x^{-\nu}$, which produces a straight line in the log–log plot.

B   Information theory

The purpose of this text is to introduce selected basic concepts of information theory necessary to understand the material in this book. An excellent textbook on information theory is [50].

B.1   Entropy, conditional entropy, mutual information

Let x be a discrete random variable attaining values in alphabet $\mathcal{A}$ with probability mass function $p(x) = \Pr\{x = x\}$. The entropy of x is defined as the expected value of $-\log p(x)$

$$H(x) = -\sum_{x \in \mathcal{A}} p(x) \log p(x). \tag{B.1}$$

If the base of the log is 2, we measure H in bits; if the base is the Euler number, $e = 2.71828...$, we speak of "nats." Since $0 \le p(x) \le 1$, $-\log p(x) \ge 0$, and thus $H(x) \ge 0$. Note that entropy does not depend on the values x, only on their probabilities $p(x)$.

Entropy measures the uncertainty of the outcome of x. It is the average number of bits communicated by one realization of x. Also, $H(x)$ is the minimum average number of bits needed to describe x.

In Chapter 6, we will encounter a less common notion of minimal entropy defined as

$$H_{\min}(x) = \min_{x \in \mathcal{A}} -\log p(x). \tag{B.2}$$

The equivalent of entropy for real-valued random variables with probability density function $f(x)$ is the differential entropy

$$h(x) = -\int_{\mathbb{R}} f(x) \log f(x) \mathrm{d}x. \tag{B.3}$$

The lower-case letter is used to stress the fact that, although entropy and differential entropy share many properties, they also behave very differently. For example, $h(x)$ can be negative because we can certainly have $f(x) > 1$ on its support $(h(u) = -\log 2$ for $u \sim U(0, 1/2))$. Moreover, h may not be preserved under one-to-one mapping (c.f., Proposition B.3).

Example B.1: [Entropy of Bernoulli random variable $B(p)$]

$$H(p) = -p \log p - (1-p) \log (1-p).$$

This function is called the binary entropy function and is displayed in Figure 8.2.

Example B.2: [Differential entropy of Gaussian variable] $\xi \sim N(\mu, \sigma^2)$

$$h(\xi) = \int_{\mathbb{R}} \frac{1}{\sqrt{2\pi\sigma^2}} e^{-\frac{(x-\mu)^2}{2\sigma^2}} \left(\frac{1}{2} \log 2\pi\sigma^2 + \frac{(x-\mu)^2}{2\sigma^2} \right) dx \qquad (B.4)$$

$$= \frac{1}{2} \log 2\pi\sigma^2 \int_{\mathbb{R}} \frac{1}{\sqrt{2\pi}} e^{-\frac{y^2}{2}} dy + \frac{1}{2} \int_{\mathbb{R}} \frac{y^2}{\sqrt{2\pi}} e^{-\frac{y^2}{2}} dy \qquad (B.5)$$

$$= \frac{1}{2} \left(1 + \log 2\pi\sigma^2 \right). \qquad (B.6)$$

Proposition B.3. [Processing cannot increase entropy] *Let f be a map from $\mathcal{A}$ to $\mathcal{B}$, $f : \mathcal{A} \to \mathcal{B}$. If f is injective,[1] then the entropy of the random variable $f(\mathsf{x})$ on $\mathcal{B}$ is $H(f(\mathsf{x})) = H(\mathsf{x})$. In general, $H(f(\mathsf{x})) \le H(\mathsf{x})$.*

Proof. If f is injective, then f is one-to-one from $\mathcal{A}$ to $f(\mathcal{A}) \subset \mathcal{B}$ and the equality is clear because we are just "renaming" the elements. To prove the inequality, take any $b \in f(\mathcal{A})$ and denote $f^{-1}(b) = \{x \in \mathcal{A} | f(x) = b\}$, the set of all elements from $\mathcal{A}$ that are mapped to b under f. Note that when f is injective, $f^{-1}(b)$ contains only one element.

Let us denote the random variable $f(\mathsf{x})$ as x' and its pmf on $\mathcal{B}$ as p'. The probability of observing b for x' is the probability of observing any $x \in f^{-1}(b)$, or $p'(b) = \sum_{x \in f^{-1}(b)} p(x)$. We will show that

$$-p'(b) \log p'(b) \le - \sum_{x \in f^{-1}(b)} p(x) \log p(x). \qquad (B.7)$$

The inequality $H(f(\mathsf{x})) \le H(\mathsf{x})$ will then be proved by summing (B.7) over all $b \in f(\mathcal{A})$ because

$$H(f(\mathsf{x})) = - \sum_{b \in f(\mathcal{A})} p'(b) \log p'(b) \le \sum_{b \in f(\mathcal{A})} \sum_{x \in f^{-1}(b)} -p(x) \log p(x) \qquad (B.8)$$

$$= - \sum_{x \in \mathcal{A}} p(x) \log p(x) = H(\mathsf{x}). \qquad (B.9)$$

---

[1] f is injective if $x \ne y \Rightarrow f(x) \ne f(y)$.

Because $p(x)/p'(b)$ is itself a pmf on $f^{-1}(b)$ (it sums to 1), from the non-negativity of entropy,

$$0 \le - \sum_{x \in f^{-1}(b)} \frac{p(x)}{p'(b)} \log \frac{p(x)}{p'(b)} = \frac{1}{p'(b)} \sum_{x \in f^{-1}(b)} -p(x) \log \frac{p(x)}{p'(b)} \qquad \text{(B.10)}$$

$$= \frac{1}{p'(b)} \left\{ \sum_{x \in f^{-1}(b)} -p(x) \log p(x) + \sum_{x \in f^{-1}(b)} p(x) \log p'(b) \right\} \qquad \text{(B.11)}$$

$$= \frac{1}{p'(b)} \left\{ \sum_{x \in f^{-1}(b)} -p(x) \log p(x) + p'(b) \log p'(b) \right\}, \qquad \text{(B.12)}$$

which proves (B.7) and the whole proposition. $\qquad\square$

The proposition simply states that by transforming a random variable, one cannot make its output more uncertain. This is intuitively clear because either f maps elements of $\mathcal{A}$ to different elements, which does not change entropy, or it maps several elements to one (not a one-to-one map), which should decrease uncertainty.

Entropy is the measure of uncertainty. The conditional entropy of x is the uncertainty of x given the outcome of another random variable y and it can be expressed using the conditional probability $p(x|y = y)$:

$$H(\mathsf{x}|\mathsf{y} = y) = - \sum_{x \in \mathcal{A}_x} p(x|y) \log p(x|y). \qquad \text{(B.13)}$$

Since, potentially, the alphabets for x and y may be different, we use the subscript to denote this fact, $x \in \mathcal{A}_x$, $y \in \mathcal{A}_y$. Using (B.13), we define the conditional entropy

$$H(\mathsf{x}|\mathsf{y}) = - \sum_{x \in \mathcal{A}_x} \sum_{y \in \mathcal{A}_y} p(x|y) \log p(x|y) p(y) \qquad \text{(B.14)}$$

$$= - \sum_{x \in \mathcal{A}_x} \sum_{y \in \mathcal{A}_y} p(x, y) \log p(x|y). \qquad \text{(B.15)}$$

Here, $p(x, y) = \Pr\{\mathsf{x} = x, \mathsf{y} = y\}$ is the joint probability.

The joint entropy of two random variables is defined as the entropy of the vector random variable (x, y):

$$H(\mathsf{x}, \mathsf{y}) = - \sum_{x \in \mathcal{A}_x} \sum_{y \in \mathcal{A}_y} p(x, y) \log p(x, y) \qquad \text{(B.16)}$$

$$= - \sum_{x \in \mathcal{A}_x} \sum_{y \in \mathcal{A}_y} p(x, y) \log p(x|y) - \sum_{x \in \mathcal{A}_x} \sum_{y \in \mathcal{A}_y} p(x, y) \log p(y) \qquad \text{(B.17)}$$

$$= H(\mathsf{x}|\mathsf{y}) - \sum_{y \in \mathcal{A}_y} p(y) \log p(y) = H(\mathsf{x}|\mathsf{y}) + H(\mathsf{y}). \qquad \text{(B.18)}$$

Thus, the joint entropy is the sum of the entropy of y and the conditional entropy $H(x|y)$,

$$H(x, y) = H(x|y) + H(y). \tag{B.19}$$

The mutual information $I(x; y)$ measures how much information about x is conveyed by y,

$$I(x; y) = H(x) - H(x|y). \tag{B.20}$$

If $H(x|y) = H(x)$, or $I(x; y) = 0$, the uncertainty of x is unaffected by y and thus knowledge of y does not give us any information about x. On the other hand, if $H(x|y) = 0$, x is completely determined by y and thus it delivers all information about x. In other words, the mutual information $I(x; y) = H(x)$.

Mutual information is symmetrical,

$$I(x; y) = I(y; x), \tag{B.21}$$

because, expressing $H(x|y)$ using (B.19),

$$I(x; y) = H(x) - H(x|y) = H(x) + H(y) - H(x, y), \tag{B.22}$$

which is obviously symmetrical in both random variables.

There exists a fundamental relationship between mutual information and KL divergence (introduced in the next section):

$$I(x; y) = - \sum_{x \in A_x} p(x) \log p(x) + \sum_{x \in A_x} \sum_{y \in A_y} p(x, y) \log p(x|y) \tag{B.23}$$

$$= - \sum_{x \in A_x} \sum_{y \in A_y} p(x, y) \log p(x) + \sum_{x \in A_x} \sum_{y \in A_y} p(x, y) \log p(x|y) \tag{B.24}$$

$$= \sum_{x \in A_x} \sum_{y \in A_y} p(x, y) \log \frac{p(x|y)}{p(x)} = \sum_{x \in A_x} \sum_{y \in A_y} p(x, y) \log \frac{p(x, y)}{p(x)p(y)} \tag{B.25}$$

$$= D_{\mathrm{KL}} \left(p(x, y) \| p(x)p(y) \right). \tag{B.26}$$

From the non-negativity of KL divergence (Proposition B.5), we have $I(x; y) \geq 0$ and $I(x; y) = 0$ if and only if $p(x, y) = p(x)p(y)$ for all x, y, which means that x and y are independent.

Moreover, the non-negativity of mutual information in connection with (B.20) implies

$$H(x) \geq H(x|y), \tag{B.27}$$

which is known as the fact that conditioning reduces entropy.

B.2   Kullback–Leibler divergence

The Kullback–Leibler divergence is also called the KL distance or relative entropy. This fundamental concept is defined for two probability mass functions p

and q on $\mathcal{A}$ as

$$D_{\mathrm{KL}}(p||q) = \sum_{x \in \mathcal{A}} p(x) \log \frac{p(x)}{q(x)}. \qquad (B.28)$$

For real-valued random variables with distributions $p(x)$ and $q(x)$

$$D_{\mathrm{KL}}(p||q) = \int p(x) \log \frac{p(x)}{q(x)} dx. \qquad (B.29)$$

The convention is that $0 \log 0 = 0$ because $\lim_{x \to 0} x \log x = 0$.

One can view the KL divergence as a measure of difference between two probability mass functions p, q. Only when $p = q$ is the KL divergence equal to zero. The more different p and q are, the larger their KL divergence is.

Example B.4: [KL divergence on a set of two elements] For $\mathcal{A} = \{0, 1\}$ the KL divergence (B.28) becomes

$$D_{\mathrm{KL}}(p||q) = p(0) \log \frac{p(0)}{q(0)} + (1 - p(0)) \log \frac{1 - p(0)}{1 - q(0)}. \qquad (B.30)$$

Proposition B.5. [Non-negativity of KL divergence] *For any two distributions, p and q on $\mathcal{A}$,*

$$D_{\mathrm{KL}}(p||q) \geq 0 \qquad (B.31)$$

and equality holds if and only if $p(x) = q(x)$ for all $x \in \mathcal{A}$.

Proof. For the proof, it will be convenient to use the natural logarithm in the definition of the KL divergence. We will use the log inequality

$$\log t \leq t - 1 \qquad (B.32)$$

which holds for all $t > 0$ with equality if and only if $t = 1$. This can be seen by noting that $\log t$ is a concave function (because $(\log t)'' = -t^{-2} < 0$ for $t > 0$) and $y = t - 1$ is its tangent at $t = 1$.

Substituting $t = q(x)/p(x)$ into (B.32), we obtain

$$\log \frac{q(x)}{p(x)} \leq \frac{q(x)}{p(x)} - 1. \qquad (B.33)$$

After multiplying by $-p(x)$ (do not forget to flip the inequality sign because $-p(x) \leq 0$)

$$p(x) \log \frac{p(x)}{q(x)} \geq p(x) - q(x). \qquad (B.34)$$

Summing over all $x \in \mathcal{A}$, we obtain the required inequality

$$D_{\mathrm{KL}}(p||q) = \sum_{x \in \mathcal{A}} p(x) \log \frac{p(x)}{q(x)} \geq \sum_{x \in \mathcal{A}} p(x) - \sum_{x \in \mathcal{A}} q(x) = 1 - 1 = 0. \qquad (\mathrm{B.35})$$

The equality can hold if and only if we had equality *for each* x, or $q(x)/p(x) = 1$ for all x, which means that the distributions are identical. $\qquad\qquad \square$

The KL divergence is not a metric because it is in general not symmetrical, $D_{\mathrm{KL}}(p||q) \neq D_{\mathrm{KL}}(q||p)$. It does not satisfy the triangle inequality $D_{\mathrm{KL}}(p||q) + D_{\mathrm{KL}}(q||r) \geq D_{\mathrm{KL}}(p||r)$ either. It is, nevertheless, useful to think of it as a distance. To get a better feeling for it, consider the following special case. Choose $q = u(x) = 1/|\mathcal{A}|$, the uniform distribution on $\mathcal{A}$. Then,

$$D_{\mathrm{KL}}(p||u) = \sum_{x \in \mathcal{A}} p(x) \log \frac{p(x)}{u(x)} = -H(\mathsf{x}) + \sum_{x \in \mathcal{A}} p(x) \log |\mathcal{A}| \qquad (\mathrm{B.36})$$

or

$$D_{\mathrm{KL}}(p||u) = \log |\mathcal{A}| - H(\mathsf{x}). \qquad (\mathrm{B.37})$$

From here, we can draw some interesting conclusions. First, because the KL divergence is always non-negative, we see that we must always have $H(\mathsf{x}) \leq \log |\mathcal{A}|$, the entropy is bounded by the logarithm of the number of elements in $\mathcal{A}$. The difference between this maximal value and the entropy $H(\mathsf{x})$ is just the KL divergence between the pmf of x and the uniformly distributed random variable (which always has the maximal entropy, as can be easily verified).

The KL divergence comes up very frequently in information theory and it can be argued that it is a more fundamental concept than the entropy itself. For example, let us assume that we wish to compress a random variable x with pmf p using Huffman code but we know only an estimate of p that we denote $\hat{p}$. If we construct a Huffman code based on our imprecise pmf $\hat{p}$, we will on average need $D_{\mathrm{KL}}(p||\hat{p})$ more bits for compression of x due to the fact that we know its pmf only approximately. The KL divergence has also a fundamental relationship to hypothesis testing (see Appendix D). Moreover, in Section B.1 we showed that the mutual information $I(\mathsf{x}; \mathsf{y})$ between two random variables x and y can be written as the KL divergence between their joint probability mass function $p(x, y)$ and the product distribution of their marginals $p(x)q(y)$ (this would be the joint pmf if x and y were independent): $I(\mathsf{x}; \mathsf{y}) = D_{\mathrm{KL}}(p(x, y)||p(x)q(y))$.

In order to understand the information-theoretic definition of steganographic security, we will need the following proposition, which is an equivalent of Proposition B.3 for KL divergence.

Proposition B.6. [Processing cannot increase KL divergence] *Let* x *and* y *be two random variables on* $\mathcal{A}$ *with pmfs* p *and* q. *Let* $f : \mathcal{A} \to \mathcal{B}$ *be a map from* $\mathcal{A}$ *to some set* $\mathcal{B}$. *Denoting the pmfs of the transformed random variables* $f(\mathsf{x})$

and $f(\mathsf{y})$ with p' and q', their KL divergence cannot increase,

$$D_{\mathrm{KL}}(p||q) \geq D_{\mathrm{KL}}(p'||q'). \tag{B.38}$$

Proof. We will need the following log-sum inequality, which holds for any non-negative $r_1, \ldots, r_k$ and positive $s_1, \ldots, s_k$:

$$\sum_{i=1}^{k} r_i \log \frac{r_i}{s_i} \geq \sum_{i=1}^{k} r_i \log \frac{\sum_{j=1}^{k} r_j}{\sum_{j=1}^{k} s_j}, \tag{B.39}$$

which can be proved again using the log inequality (B.32), in which we substitute

$$t = \frac{s_i \sum r_j}{r_i \sum s_j} : \tag{B.40}$$

$$\log \frac{s_i}{r_i} + \log \frac{\sum r_j}{\sum s_j} \leq \frac{s_i \sum r_j}{r_i \sum s_j} - 1. \tag{B.41}$$

We now multiply (B.41) by r_i and sum over i,

$$\sum_{i=1}^{k} r_i \log \frac{s_i}{r_i} + \sum_{i=1}^{k} r_i \log \frac{\sum r_j}{\sum s_j} \leq \sum_{i=1}^{k} s_i \frac{\sum r_j}{\sum s_j} - \sum_{i=1}^{k} r_i = 0, \tag{B.42}$$

which is the log-sum inequality.

To prove the proposition, we again take an arbitrary $b \in f(\mathcal{A})$ and note that $p'(b) = \sum_{x \in f^{-1}(b)} p(x)$ and $q'(b) = \sum_{x \in f^{-1}(b)} q(x)$, which correspond to the probabilities of observing b for $f(\mathsf{x})$ and $f(\mathsf{y})$, respectively. Let $x_1, \ldots, x_k$ be all the elements of $f^{-1}(b)$. We now use the log-sum inequality for $r_i = p(x_i)$ and $s_i = q(x_i)$, noting that $\sum r_i = p'(b)$ and $\sum s_i = q'(b)$ to obtain

$$\sum_{x \in f^{-1}(b)} p(x) \log \frac{p(x)}{q(x)} = \sum_{i=1}^{k} p(x_i) \log \frac{p(x_i)}{q(x_i)} \geq \sum_{i=1}^{k} p(x_i) \log \frac{\sum_{j=1}^{k} p(x_j)}{\sum_{j=1}^{k} q(x_j)} \tag{B.43}$$

$$= p'(b) \log \frac{p'(b)}{q'(b)}. \tag{B.44}$$

Taking similar steps as in the proof of Proposition B.3, we obtain

$$D_{\mathrm{KL}}(p||q) = \sum_{x \in \mathcal{A}} p(x) \log \frac{p(x)}{q(x)} = \sum_{b \in f(\mathcal{A})} \sum_{x \in f^{-1}(b)} p(x) \log \frac{p(x)}{q(x)} \tag{B.45}$$

$$\geq \sum_{b \in f(\mathcal{A})} p'(b) \log \frac{p'(b)}{q'(b)} = D_{\mathrm{KL}}(p'||q'). \tag{B.46}$$

$\square$

Proposition B.7. [KL divergence is locally quadratic] *Let $p(x; \beta)$ be a family of probability mass functions parametrized by a scalar parameter β. Then*

$$D_{\mathrm{KL}}\left(p(x;0)||p(x;\beta)\right) = \frac{\beta^2}{2} I(0) + O(\beta^3), \tag{B.47}$$

where $I(\beta)$ is the Fisher information

$$I(\beta) = \sum_x \frac{1}{p(x;\beta)} \left(\frac{\partial p}{\partial \beta}(x;\beta) \right)^2 = \sum_x p(x;\beta) \left(\frac{\partial \log p}{\partial \beta}(x;\beta) \right)^2. \qquad (B.48)$$

Proof. We expand $\log p(x;\beta)$ using Taylor series in the parameter:

$$\log p(x;\beta) = \log \left(p(x;0) + \beta \frac{\partial p}{\partial \beta}(x;0) + \frac{\beta^2}{2} \frac{\partial^2 p}{\partial \beta^2}(x;0) + O(\beta^3) \right) \qquad (B.49)$$

$$= \log \left(p(x;0) \left[1 + \frac{\beta}{p(x;0)} \frac{\partial p}{\partial \beta}(x;0) + \frac{\beta^2}{2p(x;0)} \frac{\partial^2 p}{\partial \beta^2}(x;0) + O(\beta^3) \right] \right) \qquad (B.50)$$

$$= \log p(x;0) + \frac{\beta}{p(x;0)} \frac{\partial p}{\partial \beta}(x;0) + \frac{\beta^2}{2p(x;0)} \frac{\partial^2 p}{\partial \beta^2}(x;0)$$

$$- \frac{\beta^2}{2} \left(\frac{1}{p(x;0)} \frac{\partial p}{\partial \beta}(x;0) \right)^2 + O(\beta^3), \qquad (B.51)$$

which will hold if $p(x;\beta)$ is sufficiently smooth in β for all x. Thus,

$$p(x;0) \left(\log p(x;0) - \log p(x;\beta) \right) = -\beta \frac{\partial p}{\partial \beta}(x;0) - \frac{\beta^2}{2} \frac{\partial^2 p}{\partial \beta^2}(x;0)$$

$$+ \frac{\beta^2}{2p(x;0)} \left(\frac{\partial p}{\partial \beta}(x;0) \right)^2 + O(\beta^3). \qquad (B.52)$$

We can now write for the KL divergence

$$D_{\mathrm{KL}} \left(p(x;0) \| p(x;\beta) \right) = \sum_x p(x;0) \left(\log p(x;0) - \log p(x;\beta) \right) \qquad (B.53)$$

$$= \frac{\beta^2}{2} \sum_x \frac{1}{p(x;0)} \left(\frac{\partial p}{\partial \beta}(x;0) \right)^2 + O(\beta^3) \qquad (B.54)$$

because

$$\sum_x \frac{\partial p}{\partial \beta}(x;0) = \frac{\partial}{\partial \beta} \sum_x p(x;0) = \frac{\partial}{\partial \beta} 1 = 0, \qquad (B.55)$$

$$\sum_x \frac{\partial^2 p}{\partial \beta^2}(x;0) = \frac{\partial^2}{\partial \beta^2} \sum_x p(x;0) = \frac{\partial^2}{\partial \beta^2} 1 = 0. \qquad (B.56)$$

$\square$

Proposition B.8. [Additivity of KL divergence for independent variables] *Let $\mathsf{x}_1, \ldots, \mathsf{x}_n$ and $\mathsf{y}_1, \ldots, \mathsf{y}_n$ be independent random variables with distributions $p_1(x), \ldots, p_n(x)$ and $q_1(x), \ldots, q_n(x)$, respectively. Considering the vector random variables $\mathbf{x} = (\mathsf{x}_1, \ldots, \mathsf{x}_n)$ and $\mathbf{y} = (\mathsf{y}_1, \ldots, \mathsf{y}_n)$ with distributions $p_{\mathbf{x}}$ and $q_{\mathbf{y}}$,*

$$D_{\mathrm{KL}}(p_{\mathbf{x}} \| p_{\mathbf{y}}) = \sum_{i=1}^{n} D_{\mathrm{KL}}(p_i \| q_i). \qquad (B.57)$$

Proof. From the definition of the KL divergence

$$D_{\mathrm{KL}}(p_\mathbf{x}||p_\mathbf{y}) = \sum_{x_1,\ldots,x_n} p_1(x_1),\ldots,p_n(x_n) \log \prod_{i=1}^{n} \frac{p_i(x_i)}{q_i(x_i)} \tag{B.58}$$

$$= \sum_{i=1}^{n} \sum_{x_1,\ldots,x_n} p_1(x_1),\ldots,p_n(x_n) \log \frac{p_i(x_i)}{q_i(x_i)} \tag{B.59}$$

$$= \sum_{i=1}^{n} \sum_{x_i} p_i(x_i) \log \frac{p_i(x_i)}{q_i(x_i)} = \sum_{i=1}^{n} D_{\mathrm{KL}}(p_i||q_i), \tag{B.60}$$

because on the second line we can sum over all x_j, $j \neq i$, and use the fact that $\sum_{x_j} p(x_j) = 1$. □

B.3   Lossless compression

A lossless compression scheme, C, is a mapping that assigns a bit string $\mathbf{c}_i$ consisting of $l(a_i)$ bits (a codeword) to each symbol $a_i \in \mathcal{A}$. Given a sequence of independent realizations of x, $a_{i_1}, a_{i_2}, \ldots$, the compression maps it into a concatenation of bit strings $\mathbf{c}_{i_1} \& \mathbf{c}_{i_2} \& \ldots$. Thus, on average each symbol is encoded using $l(C) = \sum_{j=1}^{|\mathcal{A}|} p(a_j) l(a_j)$ bits. The goal of lossless compression is to compress a sequence of n independent realizations of x into a bit string that is as short as possible. The best possible (perfect) lossless compression scheme will do this task with $nH(\mathbf{x})$ bits. A practical way to achieve this is to divide the sequence of symbols into groups of k symbols and work with these blocks rather than individual symbols. The vector random variable y consisting of k independent realizations of x attains values from $\mathcal{A} \times \cdots \times \mathcal{A}$, which is an alphabet of $|\mathcal{A}|^k$ k-ary symbols, with probabilities $\Pr\{\mathbf{y} = (a_{i_1},\ldots,a_{i_k})\} = \prod_{j=1}^{k} p(a_{i_j})$. It is possible to design lossless compression schemes C_k that with increasing k asymptotically reach the best possible performance and compress n symbols from $\mathcal{A}$ using approximately $nH(\mathbf{x})$ bits. For example, Huffman codes can be shown to be asymptotically optimal in this sense.

The codeword lengths in an asymptotically perfect lossless compression scheme must satisfy

$$l(a_j) \approx -\log p(a_j). \tag{B.61}$$

This is because the average length of a codeword is

$$l(C) = \sum_{j} p(a_j) l(a_j) = -\sum_{j} p(a_j) \log p(a_j) = H(\mathbf{x}) \tag{B.62}$$

and this is the minimal average codeword length possible.

B.3.1   Prefix-free compression scheme

We say that a compression scheme is prefix-free if no codeword is a prefix of any other codeword. A compression scheme with codewords $\{0, 10, 110, 111\}$ is prefix-free, while $\{0, 1, 10, 11\}$ is not because $'1'$ is a prefix of both $'10'$ and $'11'$. There exist asymptotically perfect prefix-free compression schemes (e.g., the Huffman codes).

Every perfect prefix-free scheme C has the following interesting property. Given any sequence of bits $\mathbf{b}[i], i = 1, 2, \ldots$, there must exist $i_1 \geq 1$ such that the bit string $(\mathbf{b}[1], \ldots, \mathbf{b}[i_1])$ is a codeword. To see this, let k be the length of the longest codeword. If $(\mathbf{b}[1], \ldots, \mathbf{b}[i_1])$ is a codeword for some $i_1 \leq k$, we are done. In the other case, we obtain a contradiction in the following manner. We know that $(\mathbf{b}[1], \ldots, \mathbf{b}[k-1])$ is not a codeword and we show that it is not a prefix of any codeword either. The only codeword with this prefix would have to be $(\mathbf{b}[1], \ldots, \mathbf{b}[k-1], 1 - \mathbf{b}[k])$, in which case we could remove the last bit from the codeword and shorten the average code length. Therefore, $(\mathbf{b}[1], \ldots, \mathbf{b}[k-1])$ is not a prefix of any codeword and we can shorten the code by replacing one of the codewords with length k with $(\mathbf{b}[1], \ldots, \mathbf{b}[k-1])$.

Thus, any bit sequence $\mathbf{b}[i]$ can be divided into disjoint bit strings of codewords $\mathbf{b} = \mathbf{b}[1], \ldots, \mathbf{b}[i_1], \mathbf{b}[i_1 + 1], \ldots, \mathbf{b}[i_2], \ldots$. If the bit sequence is random, each codeword, $\mathbf{c}_j$, will appear with probability $p(a_j)$. This is because the probability that the first codeword occurring in the bit sequence is $\mathbf{c}_j$ is the same as the probability of generating $\mathbf{c}_j$ randomly (because the compression scheme is prefix-free). Because the length of $\mathbf{c}_j$ is $l(a_j)$, for a perfect compression scheme, this probability is $2^{-l(a_j)} = 2^{\log p(a_j)} = p(a_j)$.

Example B.9: [Biased tetrahedron] The tetrahedron t has four sides, $\mathcal{A} = \{1, 2, 3, 4\}$. The probabilities that the tetrahedron falls on side 1, 2, 3, or 4 when thrown are $p(1) = \frac{1}{2}$, $p(2) = \frac{1}{4}$, $p(3) = \frac{1}{8}$, $p(4) = \frac{1}{8}$. The entropy $H(\mathsf{t}) = -\frac{1}{2}\log_2\frac{1}{2} - \frac{1}{4}\log_2\frac{1}{4} - 2 \times \left(\frac{1}{8}\log_2\frac{1}{8}\right) = 1 + \frac{3}{4}$. In this special case, there exists a simple perfect lossless encoding scheme that encodes each toss with $H(\mathsf{t})$ bits. Assign the codewords in the following manner

$$1 \to \,'0' \tag{B.63}$$

$$2 \to \,'10' \tag{B.64}$$

$$3 \to \,'110' \tag{B.65}$$

$$4 \to \,'111'. \tag{B.66}$$

It is clear that this compression scheme is prefix-free. Assume we keep tossing the tetrahedron while registering the results. We know that we will need at least $n \times H(\mathsf{t})$ bits to describe the tosses. The average number of bits needed to describe the realization of one toss *with our encoding* is $\frac{1}{2} \times 1 + \frac{1}{4} \times 2 + \frac{1}{8} \times 3 + \frac{1}{8} \times 3 = 1.75$. This is also the best encoding that we can have because $H(\mathsf{t}) = 1.75$. Note that in this special case the codeword lengths do satisfy $l(a_j) = -\log_2 p(a_j)$ for each j.

C Linear codes

This appendix contains the basics of linear covering codes needed to explain the material in Chapter 8 on matrix embedding and Chapter 9 on non-shared selection channels. Coding theory is the appropriate mathematical discipline to formulate and solve problems associated with minimal-embedding-impact steganography introduced in Chapter 5. An excellent text on finite fields and coding theory is [248].

We first start with the concept of a finite field and then introduce linear codes while focusing on selected material relevant to the topics covered in this book.

C.1 Finite fields

Many steganographic schemes work by first mapping the numerical values of pixels/coefficients onto a finite field where embedding tasks can be formulated and solved in terms of linear codes. This enables us to import powerful tools from a well-developed discipline and in the end obtain more secure steganographic schemes.

A field is an alphabet $\mathcal{A}$ with two operations of addition, $'+'$, and multiplication, $'\cdot'$, that are both associative, commutative, and distributive. Also, in a field there must exist a zero element 0, $a + 0 = a, \forall a \in \mathcal{A}$, and a unit element 1, $a \cdot 1 = a, \forall a \in \mathcal{A}$. Moreover, all elements must have an additive inverse: $\forall a \in \mathcal{A}, \exists b \in \mathcal{A}, a + b = 0$; and all non-zero elements must have a multiplicative inverse: $\forall a \in \mathcal{A}, a \neq 0, \exists b \in \mathcal{A}, a \cdot b = 1$.

Example C.1: The alphabet $\mathcal{A}_2 = \{0, 1\}$ equipped with modulo 2 arithmetic is a finite field. The operations of addition and multiplication satisfy $0 + 1 = 1 + 0 = 1, 0 + 0 = 1 + 1 = 0$, and $1 \cdot 0 = 0 \cdot 1 = 0 \cdot 0 = 0, 1 \cdot 1 = 1$. The zero element is 0; the additive and multiplicative inverse of 1 is 1.

Example C.2: For q a prime number, the alphabet $\mathcal{A}_q = \{0, 1, 2, \ldots, q-1\}$ with arithmetic modulo q is a finite field. The required properties of addition and multiplication are obviously satisfied. The existence of a multiplicative inverse follows from the little Fermat's theorem ($a^{q-1} = 1 \bmod q$ for all positive integers a).

When q is not prime, $\mathcal{A}_q$ with modulo q arithmetic is not a field because the factors of q would not have a multiplicative inverse. For example, for $q = 4$, there is no multiplicative inverse for 2.

A finite field with q elements exists if and only if $q = p^m$ for some positive integer m and p prime. The field is formed by polynomials of degree $m - 1$ modulo an irreducible polynomial.

All finite fields with q elements are isomorphic and are called Galois fields $\mathbb{F}_q$. By isomorphism, we mean a one-to-one mapping, $\Psi : \mathbb{F}_q \leftrightarrow \mathbb{G}_q$, that preserves all operations, e.g., $\forall a, b \in \mathbb{F}_q$, $\Psi(ab) = \Psi(a)\Psi(b)$ and $\Psi(a + b) = \Psi(a) + \Psi(b)$.

C.2   Linear codes

A q-ary code $\mathcal{C}$ of length n is any subset of the Cartesian product $\mathbb{F}_q^n = \mathbb{F}_q \times \cdots \times \mathbb{F}_q$ (n times) and its elements are called codewords. The code $\mathcal{C}$ is linear if $\mathbb{F}_q$ is a finite field and $\mathcal{C}$ is a vector subspace of $\mathbb{F}_q^n$. A vector subspace is a subset closed with respect to addition and multiplication by an element from the finite field: $\mathcal{C}$ is a vector subspace if $\forall \mathbf{x}, \mathbf{y} \in \mathcal{C}$ and $\forall a \in \mathbb{F}_q$, $\mathbf{x} + \mathbf{y} \in \mathcal{C}$ and $a\mathbf{x} \in \mathcal{C}$. We say that $\mathcal{C}$ is closed under linear combination of its elements.

The Hamming distance $d_{\mathrm{H}}(\mathbf{x}, \mathbf{y})$ between $\mathbf{x}, \mathbf{y} \in \mathcal{C}$ is defined as the number of elements in which $\mathbf{x}, \mathbf{y}$ differ. The Hamming weight $w(\mathbf{x})$ of $\mathbf{x}$ is the number of non-zero elements in $\mathbf{x}$ or $w(\mathbf{x}) = d_{\mathrm{H}}(\mathbf{x}, \mathbf{0})$.

Because linear code is a vector subspace of $\mathbb{F}_q^n$, it has a basis consisting of $k \leq n$ linearly independent vectors. Let us write the basis vectors as rows of a $k \times n$ matrix $\mathbf{G}$. All codewords can thus be obtained as linear combinations of rows of $\mathbf{G}$. We call $\mathbf{G}$ the generator matrix of the code and k is the code dimension. We also say that $\mathcal{C}$ is an $[n, k]$ code. Note that there are exactly q^k codewords in such a code.

The ball of radius r with center at $\mathbf{x}$ is the set of all $\mathbf{y} \in \mathbb{F}_q^n$ whose distances from $\mathbf{x}$ are less than or equal to r,

$$\mathcal{B}(\mathbf{x}, r) = \{\mathbf{y} \in \mathbb{F}_q^n \,|\, d_{\mathrm{H}}(\mathbf{x}, \mathbf{y}) \leq r\}. \tag{C.1}$$

The ball volume is the cardinality of $\mathcal{B}(\mathbf{x}, r)$,

$$V_q(r, n) \triangleq |\mathcal{B}(\mathbf{x}, r)| = 1 + \binom{n}{1}(q - 1) + \binom{n}{2}(q - 1)^2 + \cdots + \binom{n}{r}(q - 1)^r, \tag{C.2}$$

because there are $\binom{n}{i}$ possible places for i changes in n symbols and each change can attain $q - 1$ different values.

The minimal distance of a code is determined by the two closest codewords. For linear codes, it is also the smallest Hamming weight among all non-zero codewords,

$$d = \min_{\mathbf{x},\mathbf{y}\in\mathcal{C},\mathbf{x}\neq\mathbf{y}} d_H(\mathbf{x},\mathbf{y}) = \min_{\mathbf{x},\mathbf{y}\in\mathcal{C},\mathbf{x}\neq\mathbf{y}} w(\mathbf{x}-\mathbf{y}) = \min_{\mathbf{c}\in\mathcal{C},\, \mathbf{c}\neq\mathbf{0}} w(\mathbf{c}). \tag{C.3}$$

The distance to code is the distance to the closest codeword,

$$d_H(\mathbf{x},\mathcal{C}) = \min_{\mathbf{c}\in\mathcal{C}} d_H(\mathbf{x},\mathbf{c}). \tag{C.4}$$

The average distance to code is the expected value of $d_H(\mathbf{x},\mathcal{C})$ over randomly uniformly distributed $\mathbf{x}\in\mathbb{F}_q^n$,

$$R_a = \frac{1}{q^n}\sum_{\mathbf{x}\in\mathbb{F}_q^n} d_H(\mathbf{x},\mathcal{C}). \tag{C.5}$$

The covering radius of a code is determined by the most distant word from the code,

$$R = \max_{\mathbf{x}\in\mathbb{F}_q^n} d_H(\mathbf{x},\mathcal{C}). \tag{C.6}$$

Note that

$$R \geq R_a, \tag{C.7}$$

$$\bigcup_{\mathbf{x}\in\mathcal{C}} \mathcal{B}(\mathbf{x},R) = \mathbb{F}_q^n. \tag{C.8}$$

Example C.3: Consider a binary code given by the generator matrix

$$\mathbf{G} = \begin{pmatrix} 1\,0\,1\,1\,1 \\ 0\,1\,1\,0\,1 \\ 1\,1\,0\,0\,0 \end{pmatrix}. \tag{C.9}$$

The code length is $n = 5$ and its dimension is $k = 3$ (the number of rows in $\mathbf{G}$). The codewords are elements of $\mathbb{F}_2^5$ (they are five-tuples of bits). We know that there must be $2^3 = 8$ codewords. Three of them already appear as rows of $\mathbf{G}$. The fourth one is the all-zero codeword, $(0,0,0,0,0)$, which is always an element of every linear code. The remaining four codewords are obtained by adding the rows of $\mathbf{G}$: $(1,1,0,1,0)$ is the sum of the first two rows, $(1,0,1,0,1)$ is the sum of the last two rows, and $(0,1,1,1,1)$ is the sum of the first and third row. The last codeword is the sum of all three rows, $(0,0,0,1,0)$. Thus, a complete list of all codewords is

$$
C = \begin{Bmatrix}
0\,0\,0\,0\,0 \\
1\,0\,1\,1\,1 \\
0\,1\,1\,0\,1 \\
1\,1\,0\,0\,0 \\
1\,1\,0\,1\,0 \\
1\,0\,1\,0\,1 \\
0\,1\,1\,1\,1 \\
0\,0\,0\,1\,0
\end{Bmatrix}. \tag{C.10}
$$

This code has many other generator matrices formed by any triple of linearly independent codewords. The minimal distance of C is 1 because one of the codewords has Hamming weight 1. The covering radius of C is $R = 1$ because no other word in $\mathbb{F}_2^5$ is farther away from C than 1 (the reader is encouraged to verify this statement by listing the distance to C for all 32 words in $\mathbb{F}_2^5$). In Section C.2.4, we will learn a better and faster method for how to determine the covering radius for such small codes. Because the distance to C is 0 for all 8 codewords and 1 for the remaining $32 - 8 = 24$ words, the average distance to code $R_a = \frac{24}{32} = \frac{2}{3}$.

The ball of radius 1 centered at codeword $\mathbf{c} = (1,0,1,1,1)$ is the set of six words

$$
\mathcal{B}(\mathbf{c}, 1) = \begin{Bmatrix}
1\,0\,1\,1\,1 \\
0\,0\,1\,1\,1 \\
1\,1\,1\,1\,1 \\
1\,0\,0\,1\,1 \\
1\,0\,1\,0\,1 \\
1\,0\,1\,1\,0
\end{Bmatrix}. \tag{C.11}
$$

C.2.1   Isomorphism of codes

We say that a code C is isomorphic to D if there exists a one-to-one map $\Psi : C \leftrightarrow D$ that preserves the distance between codewords, $d_H(\mathbf{c}_1, \mathbf{c}_2) = d_H(\Psi(\mathbf{c}_1), \Psi(\mathbf{c}_2))$ for all $\mathbf{c}_1, \mathbf{c}_2 \in C$. In other words, two isomorphic codes are geometrically identical in the sense that they are rotations or mirror images of each other.

Two generator matrices correspond to two isomorphic codes if we can transform the generator matrix of one code to the other using the following operations:

- Swap two rows (linear combinations of rows stay the same)
- Add a multiple of a row to another row (linear combinations of rows stay the same)
- Multiply a row by a non-zero scalar (linear combinations of rows stay the same)
- Multiply a column by a non-zero scalar (the distance between every pair of codewords does not change)

• Swap two columns (symmetry)

Using these operations, every linear code can be mapped to an isomorphic code with generator matrix $\mathbf{G} = [\mathbf{I}_k; \mathbf{A}]$, where $\mathbf{I}_k$ is the $k \times k$ unity matrix and $\mathbf{A}$ is a $k \times (n - k)$ matrix. This form of the generator matrix is called the systematic form.

The reader is encouraged to think why adding two columns may not lead to an isomorphic code (come up with an example where adding two rows changes the minimal distance).

C.2.2   Orthogonality and dual codes

We define the dot product between two words

$$\mathbf{x} \cdot \mathbf{y} = x_1 y_1 + \cdots + x_n y_n, \tag{C.12}$$

all operations in the corresponding finite field. Note that when $\mathbf{x} \in \mathbb{F}_2^n$ contains an even number of ones, $\mathbf{x} \cdot \mathbf{x} = 0$, which is true for the standard dot product in Euclidean space only when $\mathbf{x} = \mathbf{0}$.

The orthogonal complement to $\mathcal{C}$ is the set of all vectors orthogonal to every codeword:

$$\mathcal{C}^{\perp} = \{\mathbf{x} | \mathbf{x} \cdot \mathbf{c} = 0, \forall \mathbf{c} \in \mathcal{C}\}. \tag{C.13}$$

We can find $\mathcal{C}^{\perp}$ by solving the system of linear equations $\mathbf{Gx} = \mathbf{0}$. From linear algebra, the set of all solutions of this equation is a vector subspace of dimension $n - k$. Thus, $\mathcal{C}^{\perp}$ is an $[n, n - k]$ code called the dual code of $\mathcal{C}$. The dimension of $\mathcal{C}^{\perp}$ is the codimension of $\mathcal{C}$. Note that the dual of the dual is again $\mathcal{C}$.

To find the generator matrix of $\mathcal{C}^{\perp}$, we assume that $\mathbf{G} = [\mathbf{I}_k, \mathbf{A}]$ is in the systematic form. The basis of the dual code is formed by solutions to the equations

$$0 = \mathbf{Gx} = [\mathbf{I}_k, \mathbf{A}]\mathbf{x} = \begin{pmatrix} \mathbf{x}[1] + \mathbf{A}[1, .] \begin{pmatrix} \mathbf{x}[k+1] \\ \cdots \\ \mathbf{x}[n] \end{pmatrix} \\ \cdots \\ \mathbf{x}[k] + \mathbf{A}[k, .] \begin{pmatrix} \mathbf{x}[k+1] \\ \cdots \\ \mathbf{x}[n] \end{pmatrix} \end{pmatrix}, \tag{C.14}$$

where we wrote $\mathbf{A}[j, .]$ for the jth row of $\mathbf{A}$. The equation above implies

$$\mathbf{x}[1] = -\mathbf{A}[1, .] \begin{pmatrix} \mathbf{x}[k+1] \\ \cdots \\ \mathbf{x}[n] \end{pmatrix}, \ldots, \quad \mathbf{x}[k] = -\mathbf{A}[k, .] \begin{pmatrix} \mathbf{x}[k+1] \\ \cdots \\ \mathbf{x}[n] \end{pmatrix}. \tag{C.15}$$

We can choose $\mathbf{x}[k + 1], \ldots, \mathbf{x}[n]$ arbitrarily and always find $\mathbf{x}[1], \ldots, \mathbf{x}[k]$ so that all k equations hold. Choose

$$\begin{pmatrix} \mathbf{x}[k+1] \\ \cdots \\ \mathbf{x}[n] \end{pmatrix} \in \left\{ \begin{pmatrix} 1 \\ 0 \\ \cdots \\ 0 \end{pmatrix}, \begin{pmatrix} 0 \\ 1 \\ \cdots \\ 0 \end{pmatrix}, \ldots, \begin{pmatrix} 0 \\ 0 \\ \cdots \\ 1 \end{pmatrix} \right\}. \tag{C.16}$$

By writing the solutions as rows of a matrix, we obtain the generator matrix, $\mathbf{H}$, of the dual code

$$\mathbf{H} = \begin{pmatrix} \mathbf{A}[1,1] & \mathbf{A}[2,1] & \cdots & \mathbf{A}[k,1] & 1 & 0 & \cdots & 0 \\ \mathbf{A}[1,2] & \mathbf{A}[2,2] & \cdots & \mathbf{A}[k,2] & 0 & 1 & \cdots & 0 \\ \cdots & \cdots & \cdots & \cdots & & 0 & 0 & \cdots & 0 \\ \mathbf{A}[1,k] & \mathbf{A}[2,k] & \cdots & \mathbf{A}[k,k] & 0 & 0 & \cdots & 1 \end{pmatrix} = [\mathbf{A}', \mathbf{I}_{n-k}], \tag{C.17}$$

which is called the parity-check matrix of $\mathcal{C}$. The parity-check matrix is an equivalent description of the code; $\mathbf{G}$ is $k \times n$, $\mathbf{H}$ is $(n - k) \times n$. The codewords are defined implicitly through $\mathbf{H}$ because the rows of $\mathbf{H}$ are orthogonal to rows of $\mathbf{G}$, $\mathbf{Hc} = \mathbf{0}$ for all codewords $\mathbf{c} \in \mathcal{C}$.

The parity-check matrix can be used to find the minimal distance of a code (at least for small codes). The minimal distance d is the smallest number of columns in $\mathbf{H}$ whose sum is $\mathbf{0}$ because $\mathbf{Hc} = \mathbf{0}$ can be written as a linear combination of columns of $\mathbf{H}$: $\mathbf{Hc} = \mathbf{c}[1]\mathbf{H}[.,1] + \cdots + \mathbf{c}[n]\mathbf{H}[.,n] = \mathbf{0}$. (Recall that $d = \min_{\mathbf{c} \in \mathcal{C}, \mathbf{x} \neq \mathbf{0}} w(\mathbf{c})$.)

Example C.4: Consider the code from Example C.3. We will first find the systematic form of the generator matrix and then the parity-check matrix. Using the operations that lead to isomorphic codes, we can write

$$\mathbf{G} = \begin{pmatrix} 1 & 0 & 1 & 1 & 1 \\ 0 & 1 & 1 & 0 & 1 \\ 1 & 1 & 0 & 0 & 0 \end{pmatrix} \sim \begin{pmatrix} 1 & 0 & 1 & 1 & 1 \\ 0 & 1 & 1 & 0 & 1 \\ 0 & 1 & 1 & 1 & 1 \end{pmatrix} \sim \begin{pmatrix} 1 & 0 & 1 & 1 & 1 \\ 0 & 1 & 1 & 0 & 1 \\ 0 & 0 & 0 & 1 & 0 \end{pmatrix} \tag{C.18}$$

$$\sim \begin{pmatrix} 1 & 0 & 1 & 1 & 1 \\ 0 & 1 & 0 & 1 & 1 \\ 0 & 0 & 1 & 0 & 0 \end{pmatrix} \sim \begin{pmatrix} 1 & 0 & 0 & 1 & 1 \\ 0 & 1 & 0 & 1 & 1 \\ 0 & 0 & 1 & 0 & 0 \end{pmatrix} = [\mathbf{I}_3; \mathbf{A}]. \tag{C.19}$$

The first operation involved adding the first row to the third row to manufacture zeros in the first column (besides the first one). In the second operation, we added the second row to the third one to obtain zeros in the second column. Then, we swapped the third and fourth columns to obtain an upper-diagonal matrix. The last operation turned the matrix to the desired systematic form by adding the third row to the first row.

Thus, the parity-check matrix is obtained as

$$\mathbf{H} = [\mathbf{A}'; \mathbf{I}_{5-3}] = \begin{pmatrix} 1 & 1 & 0 & 1 & 0 \\ 1 & 1 & 0 & 0 & 1 \end{pmatrix}. \tag{C.20}$$

The reader is encouraged to verify that the rows of $\mathbf{H}$ are orthogonal to all rows of $\mathbf{G}$ in systematic form.

C.2.3   Perfect codes

Proposition C.5. [The sphere-covering bound] *For any $[n, k]$ code*

$$V_q(R, n) \geq q^{n-k}. \tag{C.21}$$

Proof. Because the union of balls with radius equal to R, the covering radius, is the whole space, we must have $\bigcup_{\mathbf{x}\in\mathcal{C}} \mathcal{B}(\mathbf{x}, R) = \mathbb{F}_q^n$. Now, if all balls were disjoint, we would have $\left|\bigcup_{\mathbf{x}\in\mathcal{C}} \mathcal{B}(\mathbf{x}, R)\right| = |\mathbb{F}_q^n| = q^n = V_q(R, n) \times q^k$, which is equality in the sphere-covering bound, because all balls have the same volume $V_q(R, n)$ and there are q^k of them. If the balls are not disjoint, we will count some words more than once and thus obtain inequality, $q^n = \left|\bigcup_{\mathbf{x}\in\mathcal{C}} \mathcal{B}(\mathbf{x}, R)\right| \leq V_q(R, n) \times q^k$. $\quad\square$

An $[n, k]$ code is called perfect if the minimal distance d is odd and the following equality holds:

$$V_q\left(\frac{d-1}{2}, n\right) = q^{n-k}, \tag{C.22}$$

In other words, the balls with radius $\lfloor\frac{d-1}{2}\rfloor = \frac{d-1}{2}$ centered at codewords cover the whole space without overlap.

Proposition C.6. [Perfect codes] *The only perfect codes are*

- The repetition code $[n, 1]$, $d = \frac{n-1}{2}$ for n odd,
- Hamming codes,
- Binary and ternary Golay codes.

The q-ary Hamming code $\left[\frac{q^p-1}{q-1}, \frac{q^p-1}{q-1} - p\right]$ is a linear code of length $n = \frac{q^p-1}{q-1}$ and codimension p. The code parity-check matrix has dimensions $p \times \frac{q^p-1}{q-1}$ and contains as its columns all different non-zero p-tuples of symbols from $\mathbb{F}_q$ up to multiplication by a scalar. The length is $n = \frac{q^p-1}{q-1}$ because there are $q^p - 1$ non-zero p-tuples of q-ary symbols and each such tuple appears in $q - 1$ versions (there are $q - 1$ non-zero multiples).

Because we need to add at least three columns of $\mathbf{H}$ to obtain a zero vector, e.g., $(1, 0, 0, \ldots 0)'$, $(0, 1, 0, \ldots, 0)'$, and $(1, 1, 0, \ldots, 0)'$, the minimal distance $d = 3$.

The Hamming code is perfect because

$$V_q(1, n) = 1 + n(q - 1) = q^p. \tag{C.23}$$

This means that balls of radius 1 centered at codewords cover the whole space without overlap. This also implies that the covering radius $R = 1$. Because there are q^p words in every ball out of which there is only one codeword, the average distance to code is $R_{\mathrm{a}} = \frac{q^p - 1}{q^p} = 1 - q^{-p}$.

Description of the Golay codes and the proof of Proposition C.6 can be found, e.g., in [248].

C.2.4   Cosets of linear codes

For any $\mathbf{x} \in \mathbb{F}_q^n$, we define the concept of a syndrome $\mathbf{H}\mathbf{x} = \mathbf{s} \in \mathbb{F}_q^{n-k}$. The set of all $\mathbf{x} \in \mathbb{F}_q^n$ with the same syndrome, $\mathbf{s} \in \mathbb{F}_q^{n-k}$, is a coset $\mathcal{C}(\mathbf{s})$:

$$\mathcal{C}(\mathbf{s}) = \{\mathbf{x} \in \mathbb{F}_q^n | \mathbf{H}\mathbf{x} = \mathbf{s}\}. \tag{C.24}$$

Note that $\mathcal{C}(\mathbf{0}) = \mathcal{C}$, $\mathcal{C}(\mathbf{s}_1) \cap \mathcal{C}(\mathbf{s}_2) = \emptyset$ for $\mathbf{s}_1 \neq \mathbf{s}_2$. The whole space $\mathbb{F}_q^n$ can be thus decomposed into q^{n-k} cosets, each coset containing q^k elements,

$$\bigcup_{\mathbf{s} \in \mathbb{F}_q^{n-k}} \mathcal{C}(\mathbf{s}) = \mathbb{F}_q^n. \tag{C.25}$$

Also, from linear algebra $\mathcal{C}(\mathbf{s}) = \tilde{\mathbf{x}} + \mathcal{C}$, where $\tilde{\mathbf{x}}$ is an arbitrary coset member (arbitrary solution $\mathbf{H}\tilde{\mathbf{x}} = \mathbf{s}$).

Coset leader $\mathbf{e}(\mathbf{s})$ is a member of the coset $\mathcal{C}(\mathbf{s})$ with the smallest Hamming weight. The Hamming weight of every coset leader satisfies

$$w(\mathbf{e}(\mathbf{s})) \leq R \tag{C.26}$$

and this bound is tight. This is because for any $\mathbf{x} \in \mathcal{C}(\mathbf{s})$

$$R = \max_{\mathbf{z} \in \mathbb{F}_q^n} d_{\mathrm{H}}(\mathbf{z}, \mathcal{C}) \geq d_{\mathrm{H}}(\mathbf{x}, \mathcal{C}) = \min_{\mathbf{c} \in \mathcal{C}} w(\mathbf{x} - \mathbf{c}) = w(\mathbf{e}(\mathbf{s})) \tag{C.27}$$

because as $\mathbf{c}$ goes through $\mathcal{C}$, $\mathbf{x} - \mathbf{c}$ goes through all members of the coset $\mathcal{C}(\mathbf{s})$.

This result also implies that any syndrome, $\mathbf{s} \in \mathbb{F}_q^{n-k}$, can be obtained by adding at most R columns of $\mathbf{H}$. This is because $\mathcal{C}(\mathbf{s}) = \{\mathbf{x} | \mathbf{H}\mathbf{x} = \mathbf{s}\}$ and the weight of a coset leader is the smallest number of columns that need to be summed to obtain the coset syndrome $\mathbf{s}$. Thus, one method to determine the covering radius of a linear code is to first form its parity-check matrix and then find the smallest number of columns of $\mathbf{H}$ that can generate any syndrome.

Example C.7: Using the code from Example C.3 again, this last result can immediately give us the covering radius of the code from the parity-check matrix

$$\mathbf{H} = \begin{pmatrix} 1 & 1 & 0 & 1 & 0 \\ 1 & 1 & 0 & 0 & 1 \end{pmatrix}. \tag{C.28}$$

Because the columns of $\mathbf{H}$ cover all four binary vectors, each syndrome can be written as a linear combination of one column and thus $R = 1$. Because $k = 3$ and $n = 5$, there are $2^{n-k} = 4$ cosets, each containing $2^k = 8$ words. The coset corresponding to syndrome $\mathbf{s}_{00} = (0, 0)'$ is the code $\mathcal{C}$ and the coset leader is the all-zero codeword. The coset leader of the coset $\mathcal{C}(\mathbf{s}_{01})$, $\mathbf{s}_{01} = (0,1)'$, is $\mathbf{e}(\mathbf{s}_{01}) = (0,0,0,0,1)'$ because this is the word with the smallest Hamming weight for which $\mathbf{He} = (0,1)'$. The coset leader of the coset $\mathcal{C}(\mathbf{s}_{10})$, $\mathbf{s}_{10} = (1,0)'$, is $\mathbf{e}(\mathbf{s}_{10}) = (0,0,0,1,0)'$ because this is the word with the smallest Hamming weight for which $\mathbf{He} = (1,0)'$. Finally, the coset leader of the coset $\mathcal{C}(\mathbf{s}_{11})$, $\mathbf{s}_{11} = (1,1)'$, is $\mathbf{e}(\mathbf{s}_{11}) = (0,1,0,0,0)'$ because this is a word with the smallest Hamming weight for which $\mathbf{He} = (1,1)'$. Note that the coset leader for this coset is not unique because $(1,0,0,0,0)'$ is also a coset leader.

The space $\mathbb{F}_2^5 = \mathcal{C} \cup \{\mathbf{s}_{01} + \mathcal{C}\} \cup \{\mathbf{s}_{10} + \mathcal{C}\} \cup \{\mathbf{s}_{11} + \mathcal{C}\}$ can be written as a disjoint union of four cosets.

Example C.8: Consider a linear code with parity-check matrix

$$\mathbf{H} = \begin{pmatrix} 1\,0\,0\,1\,1 \\ 0\,1\,0\,1\,0 \\ 0\,0\,1\,0\,1 \end{pmatrix} \tag{C.29}$$

already in the systematic form. The covering radius of this code is $R = 2$ because, for some syndromes, we need to add two columns to obtain the syndrome. For example, the syndrome $\mathbf{s} = (1,1,1)'$ is the sum of the second and fifth columns, $\mathbf{s} = \mathbf{H}[.,2] + \mathbf{H}[.,5]$. This tells us that a coset leader of $\mathcal{C}(\mathbf{s})$ is $\mathbf{e}(\mathbf{s}) = (0,1,0,0,1)'$. Note that because also $\mathbf{s} = \mathbf{H}[.,3] + \mathbf{H}[.,4]$, another coset leader is $\mathbf{e}(\mathbf{s}) = (0,0,1,1,0)'$. The coset $\mathcal{C}(1,0,1)$, on the other hand, has a unique coset leader $\mathbf{e}(1,0,1) = (0,0,0,0,1)'$.

Example C.9: Consider a linear code with parity-check matrix

$$\mathbf{H} = \begin{pmatrix} 1\,1\,0\,0\,0\,0 \\ 0\,0\,1\,1\,0\,0 \\ 0\,0\,0\,0\,1\,1 \end{pmatrix}. \tag{C.30}$$

The code covering radius is $R = 3$ because for example the syndrome $(1,1,1)'$ is obtained by adding no fewer than three columns of $\mathbf{H}$. And three is the number of columns that can generate every syndrome because the first, third, and fifth columns of $\mathbf{H}$ form a basis of the vector space of all triples of bits.

D Signal detection and estimation

In this appendix, we explain some elementary facts from statistical signal detection and estimation. The material is especially relevant to Chapter 6 on definition of steganographic security and Chapters 10–12 that deal with steganalysis because the problem of detection of secret messages can be formulated within the framework of statistical hypothesis testing. The reader is referred to [126] and [125] for an in-depth treatment of signal detection and estimation for engineers.

We note that the material in this appendix applies to discrete random variables represented with probability mass functions on simply replacing the integrals with summations.

D.1 Simple hypothesis testing

Assume that we carried out a measurement represented as a vector of scalar values $\mathbf{x}[i]$, $i = 1, \ldots, n$, and we desire to know whether repetitive measurements follow distribution p_0 or p_1 defined on $\mathbb{R}^n$,

$$\mathrm{H}_0 : \quad \mathbf{x} \sim p_0, \tag{D.1}$$
$$\mathrm{H}_1 : \quad \mathbf{x} \sim p_1. \tag{D.2}$$

In (D.1)–(D.2), we resorted to a common notational abuse of denoting the measurements and the random variable using the same letter. In signal detection, we speak of the null hypothesis H_0 as the "signal-absent" or noise-only hypothesis, while H_1 is the signal-present hypothesis. In steganalysis, usually (but not always), H_0 is the hypothesis that $\mathbf{x}$ is a cover image, while H_1 means that a secret message is present in $\mathbf{x}$.

In hypothesis testing, we want to make the best possible decision that optimizes some fundamental criterion selected by the user. Every decision-making process is an algorithm that assigns an index of the hypothesis (e.g., 0 or 1) to every possible vector of measurements $\mathbf{x} \in \mathbb{R}^n$. We call such an algorithm a detector. Mathematically, it is a map $F : \mathbb{R}^n \to \{0, 1\}$. The sets

$$\mathcal{R}_0 = \{\mathbf{x} \in \mathbb{R}^n | F(\mathbf{x}) = 0\}, \tag{D.3}$$
$$\mathcal{R}_1 = \{\mathbf{x} \in \mathbb{R}^n | F(\mathbf{x}) = 1\} \tag{D.4}$$

form a disjoint partition $\mathbb{R}^n = \mathcal{R}_0 \cup \mathcal{R}_1$ and they fully describe the detector F. The set $\mathcal{R}_1$ is called the critical region because the detector decides H_1 if and only if $\mathbf{x} \in \mathcal{R}_1$.

The detector will make two types of error: false alarms and missed detections.[1] The probability of a false alarm, P_{FA}, is the probability that a random variable distributed according to p_0 is detected as stego (the detector decides "1"), while the probability of missed detection, P_{MD}, is the probability that a random variable distributed according to p_1 is incorrectly detected as cover:

$$P_{FA} = \Pr\{F(\mathbf{x}) = 1 | \mathbf{x} \sim p_0\} = \int_{\mathcal{R}_1} p_0(\mathbf{x}) \mathrm{d}\mathbf{x}, \tag{D.5}$$

$$P_{MD} = \Pr\{F(\mathbf{x}) = 0 | \mathbf{x} \sim p_1\} = \int_{\mathcal{R}_0} p_1(\mathbf{x}) \mathrm{d}\mathbf{x} = 1 - \int_{\mathcal{R}_1} p_1(\mathbf{x}) \mathrm{d}\mathbf{x}. \tag{D.6}$$

The two most frequently used criteria for constructing optimal detectors are:

- **[Neyman–Pearson]** Impose a bound on the probability of false alarms, $P_{FA} \leq \epsilon_{FA}$, and maximize the probability of detection, $P_D(\epsilon_{FA}) = 1 - P_{MD}(\epsilon_{FA})$, or, equivalently, minimize the probability of missed detection. The optimization process here needs to find among all possible subsets of $\mathbb{R}^n$ the critical region $\mathcal{R}_1$ that maximizes the detection probability

$$P_D = \Pr\{\mathbf{x} \in \mathcal{R}_1 | \mathbf{x} \sim p_1\} = \int_{\mathcal{R}_1} p_1(\mathbf{x}) \mathrm{d}\mathbf{x} \tag{D.7}$$

  subject to the condition

$$\int_{\mathcal{R}_1} p_0(\mathbf{x}) \mathrm{d}\mathbf{x} \leq \epsilon_{FA}. \tag{D.8}$$

- **[Bayesian]** Assign positive costs to each error, $C_{10} > 0$ (cost of false alarm), $C_{01} > 0$ (cost of missed detection), and non-positive "costs" or gains to each correct decision $C_{00} \leq 0$ (H_0 correctly detected as H_0), $C_{11} \leq 0$ (H_1 correctly detected as H_1), and prior probabilities $P(H_0)$ and $P(H_1)$ that $\mathbf{x} \sim p_0$ and $\mathbf{x} \sim p_1$, and minimize the total cost

$$\sum_{i,j=0}^{1} C_{ij} \Pr\{\mathbf{x} \in \mathcal{R}_i | \mathbf{x} \sim p_j\} P(H_j). \tag{D.9}$$

Theorem D.1. [Likelihood-ratio test] *The optimal detector for both scenarios above is the Likelihood-Ratio Test (LRT):*

$$Decide\ H_1\ if\ and\ only\ if\ L(\mathbf{x}) = \frac{p_1(\mathbf{x})}{p_0(\mathbf{x})} > \gamma, \tag{D.10}$$

---

[1] In statistics, these are called Type I and Type II errors.

where $L(\mathbf{x})$ is called the likelihood ratio, and the threshold γ is a solution to the following equation, for the Neyman–Pearson scenario,

$$\int\limits_{L(\mathbf{x})>\gamma} p_0(\mathbf{x})d\mathbf{x} = \epsilon_{\mathrm{FA}}, \tag{D.11}$$

and γ can be computed from costs and prior probabilities,

$$\gamma = \frac{C_{10} - C_{00}}{C_{01} - C_{11}} \frac{P(\mathrm{H}_0)}{P(\mathrm{H}_1)}, \tag{D.12}$$

for the Bayesian scenario.

Proof. In the Neyman–Pearson scenario, we need to determine the critical region $\mathcal{R}_1$ so that $P_{\mathrm{D}} = \int_{\mathcal{R}_1} p_1(\mathbf{x})d\mathbf{x}$ is maximal subject to the constraint $P_{\mathrm{FA}} = \int_{\mathcal{R}_1} p_0(\mathbf{x})d\mathbf{x} \leq \epsilon_{\mathrm{FA}}$. We can assume equality in the constraint because one can always enlarge $\mathcal{R}_1$ to obtain equality $P_{\mathrm{FA}} = \epsilon_{\mathrm{FA}}$ and thus further increase P_{D} (or, at worst, make it the same). Thus, using the method of Lagrangian multipliers, we maximize the functional

$$G(\mathcal{R}_1) = P_{\mathrm{D}} + \lambda(P_{\mathrm{FA}} - \epsilon_{\mathrm{FA}}) \tag{D.13}$$

$$= \int_{\mathcal{R}_1} p_1(\mathbf{x})d\mathbf{x} + \lambda \left(\int_{\mathcal{R}_1} p_0(\mathbf{x})d\mathbf{x} - \epsilon_{\mathrm{FA}} \right) \tag{D.14}$$

$$= \int_{\mathcal{R}_1} (p_1(\mathbf{x}) + \lambda p_0(\mathbf{x}))d\mathbf{x} - \lambda \epsilon_{\mathrm{FA}}. \tag{D.15}$$

The last expression will be maximized if $\mathbf{x} \in \mathcal{R}_1$ whenever the integrand is positive, or, equivalently,

$$\frac{p_1(\mathbf{x})}{p_0(\mathbf{x})} > -\lambda. \tag{D.16}$$

The Lagrange multiplier λ must be non-positive because otherwise all $\mathbf{x} \in \mathcal{R}_1$, which would lead to $P_{\mathrm{FA}} = 1$.

In order to control the false-positive probability, we determine the constant $\gamma = -\lambda$ so that

$$\int\limits_{\frac{p_1(\mathbf{x})}{p_0(\mathbf{x})}>\gamma} p_0(\mathbf{x})d\mathbf{x} = \epsilon_{\mathrm{FA}}, \tag{D.17}$$

which gives one equation for one unknown scalar γ.

The proof for the Bayesian case is even simpler. Here, we wish to find the critical region $\mathcal{R}_1$ to minimize the cost $\sum_{i,j=0}^{1} C_{ij}\Pr\{\mathbf{x} \in \mathcal{R}_i | \mathbf{x} \sim p_j\} P(\mathrm{H}_j)$, which we express as

$$C_{00}\left(1 - \int_{\mathcal{R}_1} p_0(\mathbf{x})d\mathbf{x}\right)P(H_0) + C_{10}\int_{\mathcal{R}_1} p_0(\mathbf{x})d\mathbf{x}P(H_0)$$

$$+ C_{11}\int_{\mathcal{R}_1} p_1(\mathbf{x})d\mathbf{x}P(H_1) + C_{01}\left(1 - \int_{\mathcal{R}_1} p_1(\mathbf{x})d\mathbf{x}\right)P(H_1)$$

$$= \int_{\mathcal{R}_1} P(H_0)(C_{10} - C_{00})p_0(\mathbf{x}) + P(H_1)(C_{11} - C_{01})p_1(\mathbf{x})d\mathbf{x}$$

$$+ C_{00} + C_{01}. \tag{D.18}$$

The cost will be minimized if $\mathbf{x} \in \mathcal{R}_1$ whenever the integrand is negative, or

$$P(H_0)(C_{10} - C_{00})p_0(\mathbf{x}) + P(H_1)(C_{11} - C_{01})p_1(\mathbf{x}) < 0, \tag{D.19}$$

which can be rewritten as

$$\frac{p_1(\mathbf{x})}{p_0(\mathbf{x})} > \frac{C_{10} - C_{00}}{C_{01} - C_{11}}\frac{P(H_0)}{P(H_1)}, \tag{D.20}$$

which concludes the proof of the LRT. $\qquad\qquad\qquad\qquad\qquad\square$

In steganography, useful detectors must have low probability of a false alarm. This is because images detected as potentially containing secret messages are likely to be subjected to further forensic analysis (see Section 10.8) to determine the steganographic program, the stego key, and eventually to extract the secret message. This may require running expensive and time-consuming dictionary attacks for the stego key and further cryptanalysis if the message is encrypted. Thus, it is more valuable to have a detector with very low P_{FA} even though its probability of missed detection may be quite high (e.g., $P_{MD} > 0.5$ or higher). Because typical steganographic communication is repetitive, a detector with $P_{MD} = 0.5$ is still quite useful as long as its false-alarm probability is small.

Thus, the hypothesis-testing problem in steganalysis is almost exclusively formulated using the Neyman–Pearson setting. We repeat that the goal here is to construct a detector with the highest detection probability $P_D(\epsilon_{FA}) = 1 - P_{MD}(\epsilon_{FA})$ while imposing a bound on the probability of false alarms, $P_{FA} \leq \epsilon_{FA}$. Even though it is possible to associate cost with both types of error, Bayesian detectors are typically not used in steganalysis, because the prior probabilities of encountering a cover or stego image, $P(H_0)$ and $P(H_1)$, are rarely available.

When the measurements $\mathbf{x}[i]$, $i = 1, \ldots, n$, are independent and identically distributed, it is intuitively clear that with increasing n, we should be able to build increasingly more accurate detectors. The performance of an optimal Neyman–Pearson detector in the limit $n \to \infty$ is captured by the Chernoff–Stein lemma, which gives the error exponent for the probability of missed detection given a bound on the probability of false alarms.

Lemma D.2. [The Chernoff–Stein lemma] *Let* $\mathbf{x}[i]$, $i = 1, \ldots, n$, *be a sequence of iid realizations of random variable* $\mathbf{x}$. *Given the simple binary hypothesis-testing problem* $H_0 : \mathbf{x}[i] \sim p_0$, $H_1 : \mathbf{x}[i] \sim p_1$ *with* $D_{\mathrm{KL}}(p_0 \| p_1) < \infty$ *and an upper bound on the probability of false alarms,* $P_{\mathrm{FA}} \leq \epsilon_{\mathrm{FA}}$, *the probability of missed detection of the optimal Neyman–Pearson detector approaches* 0 *exponentially fast,*

$$\lim_{n \to \infty} \frac{1}{n} \log P_{\mathrm{MD}}(\epsilon_{\mathrm{FA}}) = -D_{\mathrm{KL}}(p_0 \| p_1). \tag{D.21}$$

Proof. See Section 12.8 in [50]. □

D.1.1 Receiver operating characteristic

The performance of every detector is described using its Receiver-Operating-Characteristic (ROC) curve, which expresses the trade-off between the probability of false alarms and missed detections. In this book, it is defined as the probability of detection as a function of P_{FA}, $P_{\mathrm{D}}(P_{\mathrm{FA}})$. The ROC of the optimal detector (D.10) must be a concave curve. To see this, let $(u, P_{\mathrm{D}}(u))$ and $(v, P_{\mathrm{D}}(v))$ be two points on the ROC curve, $u < v$, and F_u and F_v the corresponding optimal detectors for $P_{\mathrm{FA}} = u$ and $P_{\mathrm{FA}} = v$. For each $x \in (u, v)$, we can construct a detector

$$F_x(\mathbf{x}) = \begin{cases} F_u(\mathbf{x}) & \text{with probability } (v - x)/(v - u) \\ F_v(\mathbf{x}) & \text{with probability } (x - u)/(v - u). \end{cases} \tag{D.22}$$

The probability of detection, P_{D}, for this detector is

$$P_{\mathrm{D}}^*(x) = P_{\mathrm{D}}(u) \frac{v - x}{v - u} + P_{\mathrm{D}}(v) \frac{x - u}{v - u}, \tag{D.23}$$

which defines a line segment connecting the points $(u, P_{\mathrm{D}}(u))$ and $(v, P_{\mathrm{D}}(v))$. Because the best detector must have $P_{\mathrm{D}}(x) \geq P_{\mathrm{D}}^*(x)$, it means that the ROC curve lies above this line segment and is thus concave.

A perfect detector with $P_{\mathrm{D}}(x) = 1$ for all x can be obtained only when the supports of p_0 and p_1 are disjoint. The opposite case is when $p_0 = p_1$, in which case, no detector can be built and we have $P_{\mathrm{D}}(x) = x$, which corresponds to a randomly-guessing detector.

D.1.2 Detection of signals corrupted by white Gaussian noise

In this section, we give a simple example to illustrate the apparatus introduced above. Consider the situation when one needs to decide about the presence or absence of a known signal $\mathbf{w}[i]$, $i = 1, \ldots, n$, in additive white Gaussian noise with known variance σ^2. Formally, given an iid sequence of variables $\boldsymbol{\xi}[i] \sim N(0, \sigma^2)$, the hypothesis-testing problem is

$$H_0: \quad \mathbf{x}[i] = \boldsymbol{\xi}[i], \tag{D.24}$$
$$H_1: \quad \mathbf{x}[i] = \mathbf{w}[i] + \boldsymbol{\xi}[i], \tag{D.25}$$

where H_0 is the noise-only hypothesis, while the alternative hypothesis states that the signal of interest is present.

The LRT for this problem is

$$L(\mathbf{x}) = \frac{p_1(\mathbf{x})}{p_0(\mathbf{x})} = \frac{\prod_{i=1}^{n} \frac{1}{\sqrt{2\pi}\sigma} e^{-\frac{(\mathbf{x}[i]-\mathbf{w}[i])^2}{2\sigma^2}}}{\prod_{i=1}^{n} \frac{1}{\sqrt{2\pi}\sigma} e^{-\frac{\mathbf{x}[i]^2}{2\sigma^2}}} \tag{D.26}$$

$$= e^{\frac{1}{\sigma^2}\left(\sum_{i=1}^{n} \mathbf{x}[i]\mathbf{w}[i] - \frac{1}{2}\sum_{i=1}^{n}\mathbf{w}[i]^2\right)} > \gamma. \tag{D.27}$$

An equivalent LRT is obtained by taking the natural logarithm of this inequality,

$$\log L(\mathbf{x}) = \frac{1}{\sigma^2}\left(\sum_{i=1}^{n} \mathbf{x}[i]\mathbf{w}[i] - \frac{1}{2}\sum_{i=1}^{n}\mathbf{w}[i]^2\right) > \log\gamma, \tag{D.28}$$

or

$$T(\mathbf{x}) = \sum_{i=1}^{n} \mathbf{x}[i]\mathbf{w}[i] > \sigma^2 \log\gamma + \frac{1}{2}\sum_{i=1}^{n}\mathbf{w}[i]^2 = \gamma'. \tag{D.29}$$

The result we have just obtained is known as the thesis that the optimal detector for a signal corrupted by additive white Gaussian noise is the correlation. We now determine the threshold γ' for the Neyman–Pearson test and characterize the detector performance. The test statistic $T(\mathbf{x})$ is Gaussian under both hypotheses because it is a linear combination of iid Gaussians (see Lemma A.6). The reader can easily verify that

$$T(\mathbf{x}) \sim \begin{cases} N(\mu_0, \nu^2) & \text{under } H_0 \\ N(\mu_1, \nu^2) & \text{under } H_1, \end{cases} \tag{D.30}$$

where $\mu_0 = 0$, $\mu_1 = \mathcal{E}$, and $\nu^2 = \sigma^2\mathcal{E}$ with $\mathcal{E} = \sum_{i=1}^{n} \mathbf{w}[i]^2$, the energy of the known signal. We now face a situation when the test statistic is a mean-shifted Gauss–Gauss (Gaussians with the same variances and different means). Whenever this situation occurs, the detector's performance is completely described using the so-called deflection coefficient

$$d^2 = \frac{(\mu_1 - \mu_0)^2}{\nu^2}. \tag{D.31}$$

To see this, let us compute P_{FA} and P_{D},

$$P_{\text{FA}} = \Pr\left\{T(\mathbf{x}) > \gamma' | T(\mathbf{x}) \sim N(\mu_0, \nu^2)\right\} = Q\left(\frac{\gamma' - \mu_0}{\nu}\right), \tag{D.32}$$

$$P_{\text{D}} = \Pr\left\{T(\mathbf{x}) > \gamma' | T(\mathbf{x}) \sim N(\mu_1, \nu^2)\right\} = Q\left(\frac{\gamma' - \mu_1}{\nu}\right), \tag{D.33}$$

where $Q(x) = \int_x^\infty (1/\sqrt{2\pi}) e^{-\frac{t^2}{2}} dt$ is the right-tail probability of a Gaussian variable $N(0,1)$ introduced in Section A.4.

From the first equation, we can determine the decision threshold for the Neyman–Pearson test, $\gamma' = \mu_0 + \nu Q^{-1}(P_{\text{FA}})$ and substitute it into the equation for P_{D},

$$P_{\text{D}}(P_{\text{FA}}) = Q\left(\frac{\mu_0 - \mu_1 + \nu Q^{-1}(P_{\text{FA}})}{\nu}\right) \tag{D.34}$$

$$= Q\left(Q^{-1}(P_{\text{FA}}) - \frac{\mu_1 - \mu_0}{\nu}\right) \tag{D.35}$$

$$= Q\left(Q^{-1}(P_{\text{FA}}) - \sqrt{d^2}\right), \tag{D.36}$$

where d^2 is the deflection coefficient defined in (D.31). The equation (D.36) is the mathematical form of the ROC. The reader is encouraged to use the form of $Q(x)$ to see that the ROC curve goes through the origin $(0,0)$ and the point $(1,1)$ and, also, $P_{\text{D}}'(0) = \infty$, $P_{\text{D}}'(1) = 0$, and $P_{\text{D}}''(P_{\text{FA}}) < 0$ for $0 \leq P_{\text{FA}} \leq 1$ (the ROC is concave).

D.2 Hypothesis testing and Fisher information

The basic goal of steganalysis is construction of a detector distinguishing between two distributions, p_0 and p_β, where $\beta > 0$ is a parameter (change rate). In the limit of $\beta \to 0$ (the so-called small-payload limit), the performance of the optimal likelihood-ratio detector is completely described by the leading term in Taylor expansion of the KL divergence $D_{\text{KL}}(p_0 \| p_\beta)$. The coefficient in the leading term is proportional to the Fisher information, which makes it useful, for example for comparing security of steganographic schemes. We now provide detailed technical explanation of these claims.

Let $\mathbf{x}[i]$, $i = 1, \ldots, n$, be n iid realizations of some scalar random variable x. We desire to determine whether the individual samples follow p_0 or p_β:

$$\text{H}_0: \quad \mathbf{x}[i] \sim p_0, \tag{D.37}$$

$$\text{H}_1: \quad \mathbf{x}[i] \sim p_\beta, \quad \beta > 0, \tag{D.38}$$

where β is a known parameter and $p_\beta(x)$ is three times continuously differentiable on some right neighborhood of $\beta = 0$ for all x. The likelihood-ratio test in logarithmic form is (D.10)

$$L_\beta(\mathbf{x}) = \sum_{i=1}^{n} \log \frac{p_\beta(\mathbf{x}[i])}{p_0(\mathbf{x}[i])}. \tag{D.39}$$

Assuming the random variable $\log(p_\beta(\mathbf{x})/p_0(\mathbf{x}))$ has finite mean and variance, by the central limit theorem, $(1/n)L_\beta(\mathbf{x})$ converges to a Gaussian distribution whose mean and variance we now determine.

The expected value of $(1/n)L_\beta(\mathbf{x})$ under hypothesis H_0 is $-D_{\mathrm{KL}}(p_0\|p_\beta)$ because

$$E_{p_0}\left[\frac{1}{n}L_\beta(\mathbf{x})\right] = E_{p_0}\left[\log \frac{p_\beta(\mathbf{x})}{p_0(\mathbf{x})}\right] = -D_{\mathrm{KL}}(p_0\|p_\beta). \tag{D.40}$$

It can be expanded using Proposition B.7 as

$$-D_{\mathrm{KL}}(p_0\|p_\beta) = -\frac{\beta^2}{2}I(0) + O(\beta^3), \tag{D.41}$$

where $I(\beta)$ is the Fisher information of one observation,

$$I(\beta) = E_{p_\beta}\left[\left(\frac{\partial \log p_\beta(x)}{\partial \beta}\right)^2\right] = \sum_x \frac{1}{p_\beta(x)}\left(\frac{\partial p_\beta(x)}{\partial \beta}\right)^2, \tag{D.42}$$

where the second equality holds for the discrete case (in the continuous case, the sum is replaced with an integral). See Section D.6 for more details about Fisher information.

The expected value of $(1/n)L_\beta(\mathbf{x})$ under H_1 is

$$E_{p_\beta}\left[\frac{1}{n}L_\beta(\mathbf{x})\right] = D_{\mathrm{KL}}(p_\beta\|p_0). \tag{D.43}$$

Because the leading term of $p_\beta(x)$ in β is $p_0(x)$, the leading terms of $D_{\mathrm{KL}}(p_\beta\|p_0)$ and $D_{\mathrm{KL}}(p_0\|p_\beta)$ are the same up to the sign,

$$E_{p_\beta}\left[\frac{1}{n}L_\beta(\mathbf{x})\right] = \frac{\beta^2}{2}I(0) + O(\beta^3). \tag{D.44}$$

To compute the variance of $(1/n)L_\beta$ under either hypothesis, notice that

$$\mathrm{Var}\left[\log \frac{p_\beta(\mathbf{x})}{p_0(\mathbf{x})}\right] = E\left[\left(\log \frac{p_\beta(\mathbf{x})}{p_0(\mathbf{x})}\right)^2\right] - \left(E\left[\log \frac{p_\beta(\mathbf{x})}{p_0(\mathbf{x})}\right]\right)^2. \tag{D.45}$$

We already know that the second term is $O(\beta^4)$ under both hypotheses. For the first expectation, we expand

$$\log \frac{p_\beta(x)}{p_0(x)} = \log\left(1 + \frac{\beta}{p_0}\frac{\partial p_\beta(x)}{\partial \beta}\Big|_{\beta=0} + O(\beta^2)\right) \tag{D.46}$$

$$= \frac{\beta}{p_0}\frac{\partial p_\beta(x)}{\partial \beta}\Big|_{\beta=0} + O(\beta^2), \tag{D.47}$$

and thus both $p_0(x)\left(\log(p_\beta(x)/p_0(x))\right)^2$ and $p_\beta(x)\left(\log(p_\beta(x)/p_0(x))\right)^2$ have the same leading term,

$$\frac{\beta^2}{p_0(x)} \left(\frac{\partial p_\beta(x)}{\partial \beta} \bigg|_{\beta=0} \right)^2. \tag{D.48}$$

Therefore, under both hypotheses, the leading term of the variance is

$$\operatorname{Var}\left[\frac{1}{n} \sum_{i=1}^{n} \log \frac{p_\beta(\mathbf{x}[i])}{p_0(\mathbf{x}[i])} \right] = \frac{1}{n} E\left[\left(\log \frac{p_\beta(\mathbf{x})}{p_0(\mathbf{x})} \right)^2 \right] \tag{D.49}$$

$$\doteq \frac{1}{n} \sum_x \frac{\beta^2}{p_0(x)} \left(\frac{\partial p_\beta(x)}{\partial \beta} \bigg|_{\beta=0} \right)^2 \tag{D.50}$$

$$= \frac{1}{n} \beta^2 I(0). \tag{D.51}$$

Finally, we can conclude that for small β the likelihood-ratio test is the mean shifted Gauss–Gauss problem

$$\frac{1}{n} L_\beta(\mathbf{x}) \sim \begin{cases} N\left(-\beta^2 I(0)/2, \ \beta^2 I(0)/n \right) & \text{under } H_0 \\ N\left(\beta^2 I(0)/2, \ \beta^2 I(0)/n \right) & \text{under } H_1 \end{cases} \tag{D.52}$$

and its performance is thus completely described using the deflection coefficient (accurate up to the order of β^2), which is in turn proportional to the Fisher information

$$d^2 = \frac{\left(\beta^2 I(0)/2 + \beta^2 I(0)/2 \right)^2}{\beta^2 I(0)/n} = n\beta^2 I(0). \tag{D.53}$$

Using the so-called J-divergence (sometimes called symmetric KL divergence)

$$J(p_0 \| p_\beta) = D_{\mathrm{KL}}(p_0 \| p_\beta) + D_{\mathrm{KL}}(p_\beta \| p_0) \tag{D.54}$$

the deflection coefficient can also be written as

$$d^2 = n \frac{\left(J(p_0 \| p_\beta) \right)^2}{\beta^2 I(0)} = \frac{4n \left(D_{\mathrm{KL}}(p_0 \| p_\beta) \right)^2}{\beta^2 I(0)}. \tag{D.55}$$

D.3   Composite hypothesis testing

When one of the probability distributions in the hypothesis test is not known or depends on unknown parameters, we obtain the so-called composite hypothesis-testing problem, which is substantially more complicated than the simple test. In general, no optimal detectors exist in this case and one has to resort to sub-optimal detectors by accepting simplifying or additional assumptions.

In this book, we are primarily interested in the case when the distribution under the alternative hypothesis depends on an unknown scalar parameter β, such as the change rate (relative number of embedding changes due to message hiding). In this case, we know that $\beta \geq 0$, obtaining thus the one-sided hypothesis test

$$\mathrm{H}_0: \quad \mathbf{x} \sim p_0, \tag{D.56}$$

$$\mathrm{H}_1: \quad \mathbf{x} \sim p_\beta, \quad \beta > 0, \tag{D.57}$$

where $\mathbf{x}$ is again a vector of measurements (e.g., the histogram). Examples of specific instances of this problem appear in Chapter 10.

If the threshold in the LRT can be set so that the detector has the highest probability of detection P_D for any value of the unknown parameter β, we speak of a Universally Most Powerful (UMP) detector. Because UMP detectors rarely exist, other approaches are often explored. One possibility is to constrain ourselves to small values of the parameter β. In steganalysis, this is the case of main interest because small payloads are harder to detect than large ones. If β is the only unknown parameter, one can derive the Locally Most Powerful (LMP) detector that will have a constant false-alarm rate for small β and thus one will be able to set a decision threshold in Neyman–Pearson setting. To see this, we expand $\log p_\beta$ using Taylor expansion around $\beta = 0$ and write the log-likelihood ratio test as

$$L_\beta(\mathbf{x}) = \log p_\beta(\mathbf{x}) - \log p_0(\mathbf{x}) = \beta \frac{\partial \log p_\beta(\mathbf{x})}{\partial \beta}\bigg|_{\beta=0} + O(\beta^2) > \log \gamma \tag{D.58}$$

or, for the leading term,

$$\frac{\partial \log p_\beta(\mathbf{x})}{\partial \beta}\bigg|_{\beta=0} > \frac{\log \gamma}{\beta}. \tag{D.59}$$

Thus, for small β, the test statistic is

$$T(\mathbf{x}) = \frac{\partial \log p_\beta(\mathbf{x})}{\partial \beta}\bigg|_{\beta=0}. \tag{D.60}$$

Under H_0, its expected value

$$E_{p_0}\left[\frac{\partial \log p_\beta(\mathbf{x})}{\partial \beta}\bigg|_{\beta=0}\right] = 0 \tag{D.61}$$

by the regularity condition (D.89) and the variance

$$\mathrm{Var}\left[\frac{\partial \log p_\beta(\mathbf{x})}{\partial \beta}\bigg|_{\beta=0}\right] = E_{p_0}\left[\left(\frac{\partial \log p_\beta(\mathbf{x})}{\partial \beta}\bigg|_{\beta=0}\right)^2\right], \tag{D.62}$$

which is the Fisher information for the vector of observations $\mathbf{x}$.

Additionally, if $\mathbf{x}[i]$ are iid realizations of some scalar variable, we can invoke the central limit theorem and state that $T(\mathbf{x})$ is Gaussian,

$$T(\mathbf{x}) = \sum_{i=1}^{n} \frac{\partial \log p_\beta(\mathbf{x}[i])}{\partial \beta}\bigg|_{\beta=0} \sim N\left(0, nI(0)\right), \tag{D.63}$$

where $I(0)$ is the Fisher information of one observation $\mathbf{x}[i]$ (D.42). We also denoted the marginal distribution of $p_\beta(\mathbf{x})$ constrained to the ith component with the same symbol. Notice that the distribution of the test statistic under H_0 does not depend on the unknown parameter β. This allows us to compute the threshold γ for $T(\mathbf{x})$ for a given bound on $P_{\mathrm{FA}} < \epsilon_{\mathrm{FA}}$ from

$$\Pr\{T(\mathbf{x}) > \gamma | H_0\} = \int_{T(\mathbf{x}) > \gamma} p_0(\mathbf{x})\mathrm{d}\mathbf{x} = \epsilon_{\mathrm{FA}} \tag{D.64}$$

as in simple hypothesis testing. On the other hand, the probability of missed detection P_{D} will depend on β and thus will generally be unknown.

We now briefly mention two more alternative approaches to the composite hypothesis test. A popular choice is to use the Generalized Likelihood-Ratio Test (GLRT) or impose an a priori probability distribution on the unknown parameter and use a Bayesian approach.

The GLRT detector has the form

$$\text{Decide } H_1 \text{ if and only if } L(\mathbf{x}) = \frac{p_1(\mathbf{x}; \hat{\beta}_1)}{p_0(\mathbf{x}; \hat{\beta}_0)} > \gamma, \tag{D.65}$$

where $\hat{\beta}_0$ and $\hat{\beta}_1$ are maximum-likelihood estimates of β under the corresponding hypotheses

$$\hat{\beta}_0 = \arg\max_{\beta} p_0(\mathbf{x}; \beta), \tag{D.66}$$

$$\hat{\beta}_1 = \arg\max_{\beta} p_1(\mathbf{x}; \beta). \tag{D.67}$$

In a Bayesian detector, a pdf is imposed on the unknown parameter, $p(\beta)$, at which point the composite hypothesis-testing problem is converted to a simple one and the detector takes the form of (D.10),

$$\text{Decide } H_1 \text{ when } \frac{\int p_1(\mathbf{x}|\beta)p(\beta)\mathrm{d}\beta}{\int p_0(\mathbf{x}|\beta)p(\beta)\mathrm{d}\beta} > \gamma. \tag{D.68}$$

D.4   Chi-square test

Pearson's chi-square test is a popular test for the composite hypothesis-testing problem of whether or not a given iid signal $\mathbf{x}[i]$, $i = 1, \ldots, n$, follows a known distribution, p,

$$H_0 : \quad \mathbf{x}[i] \sim p, \tag{D.69}$$

$$H_1 : \quad \mathbf{x}[i] \nsim p. \tag{D.70}$$

The test is directly applicable only to discrete random variables but can be used for continuous variables after binning. The binning has typically little influence on the result if the pdf is "reasonable." For example, for a unimodal pdf, the bin width can be chosen as $\hat{\sigma}/2$, where $\hat{\sigma}$ is the sample standard deviation of the data.

Suppose we have d bins (also called categories). Let $\mathbf{o}[k]$ be the observed number of data samples in the kth bin, $k = 1, \ldots, d$, and $\mathbf{e}[i]$ the expected occupancy of the ith bin if the data followed the known distribution p. Under the null hypothesis, the test statistic

$$S = \sum_{k=1}^{d} \frac{(\mathbf{o}[k] - \mathbf{e}[k])^2}{\mathbf{e}[k]} \tag{D.71}$$

is a random variable with distribution approaching the chi-square distribution χ_{d-1}^2 with $d - 1$ degrees of freedom (Section A.9). This approximation is valid only if the expected occupancy of each bin $\mathbf{e}[k] > 4$. If this is not the case, the bins should be merged to satisfy the minimum-occupancy condition.

The detector is

$$\text{Reject H}_0 \text{ (decide H}_1\text{) when } S > \gamma, \tag{D.72}$$

where γ is determined from the bound on the probability of the Type I error (deciding H$_1$ when H$_0$ is correct), $\epsilon_{\text{FA}} = \Pr\{\chi_{d-1}^2 > \gamma\}$, given by the complementary cumulative distribution function for the chi-square variable with $d - 1$ degrees of freedom,

$$\epsilon_{\text{FA}} = \Pr\{\chi_{d-1}^2 > \gamma\} = 1 - \frac{1}{2^{\frac{d-1}{2}} \Gamma\left(\frac{d-1}{2}\right)} \int_0^{\gamma} e^{-\frac{t}{2}} t^{\frac{d-1}{2}-1} dt. \tag{D.73}$$

In Matlab, this function can be evaluated as `1-chi2cdf`$(\gamma, d-1)$.

Example D.3: [Fair-die test] Throw a die $n = 1000$ times and calculate $\mathbf{o}[k] =$ number of throws with outcome k, $k = 1, \ldots, 6$. For a fair die, $\mathbf{e}[k] = n/6$ for all $k = 1, \ldots, 6$ (we have $k = 6$ bins). Because $\mathbf{e}[k] = \frac{1000}{6} > 4$, the statistic

$$S = \sum_{k=1}^{6} \frac{(\mathbf{o}[k] - n/6)^2}{n/6} \tag{D.74}$$

is approximately chi-square distributed with $d - 1 = 5$ degrees of freedom. Let us assume that we threw the die 1000 times and obtained the following observed frequencies $\mathbf{o}[k] = (183, 162, 162, 155, 168, 170)$. Then, $S = 2.765$. For $\epsilon_{\text{FA}} = 0.05$ and 5 degrees of freedom, we have $\gamma = 11.07$. In Matlab, $\gamma = $ `chi2inv`$(1 - \epsilon_{\text{FA}}, d - 1)$. Because $S = 2.765 < 11.07$, we accept the null hypothesis (the die is fair).

D.5   Estimation theory

Assume we have a data set $\mathbf{x}[i]$, $i = 1, \ldots, n$, that depends on some unknown scalar parameter θ. Our task is to estimate the parameter from the data using some function $\hat{\theta}(\mathbf{x})$, which we will call an estimator of θ. We encounter this situation in quantitative steganalysis when the unknown parameter is the change rate due to steganographic embedding and $\mathbf{x}$ is the vector of elements from the stego object.

An estimator is thus a mapping from $\mathbb{R}^n \rightarrow \mathbb{R}$ that assigns an estimate, $\hat{\theta}(\mathbf{x})$, of the parameter to each data set $\mathbf{x}$. If we were to repeatedly collect more data sets, each time the value of the estimate would be different, following some distribution $p(\mathbf{x}; \theta)$. Thus, the estimator itself is a random variable and we can speak of its mean value and variance.

We say that the estimator $\hat{\theta}$ is unbiased if

$$E[\hat{\theta}] = \theta \text{ for all } \theta \tag{D.75}$$

or, using the pdf of our measurements,

$$\int_{\mathbb{R}^n} \hat{\theta}(\mathbf{x}) p(\mathbf{x}; \theta) \mathrm{d}\mathbf{x} = \theta. \tag{D.76}$$

In other words, an unbiased estimator on average yields the correct answer.

Example D.4: [Unknown DC level in AWGN] Let

$$\mathbf{x}[i] = A + \boldsymbol{\xi}[i], \ i = 1, \ldots, n \tag{D.77}$$

be a DC signal (constant signal) corrupted by Additive White Gaussian Noise (AWGN), meaning that $\boldsymbol{\xi}[i] \sim N(0, \sigma^2)$ is iid. The variance of the noise, σ^2, is not necessarily known. Our task is to estimate the parameter A. One possibility would be to set the estimate equal to the first sample

$$\hat{A} = \mathbf{x}[1]. \tag{D.78}$$

This is an unbiased estimator because $E[\mathbf{x}[1]] = A$. Its variance is $\mathrm{Var}[\hat{A}] = \mathrm{Var}[\mathbf{x}[1]] = \sigma^2$. But we are not using the observations very efficiently because we are ignoring all the other measurements. Thus, perhaps a better estimator would be the arithmetic average

$$\hat{A} = \frac{1}{n} \sum_{i=1}^{n} \mathbf{x}[i]. \tag{D.79}$$

Indeed, its variance is

$$\mathrm{Var}[\frac{1}{n} \sum_{i=1}^{n} \mathbf{x}[i]] = \frac{1}{n^2} \sum_{i=1}^{n} \mathrm{Var}[\mathbf{x}[i]] = \frac{1}{n^2} n\sigma^2 = \frac{\sigma^2}{n}, \tag{D.80}$$

which is n times smaller than for the previous estimator. In fact, this is the best unbiased estimator of A that we can hope for, in some well-defined sense.

We just posed an interesting question of comparing different estimators. Can we somehow decide which estimator is better and then select the best one among them? One seemingly natural criterion would be to rank estimators by their Mean-Square Error (MSE)

$$\text{MSE}(\hat{\theta}) = E[(\hat{\theta} - \theta)^2]. \tag{D.81}$$

However, best estimators according to this criterion may not be constructable because the best estimator may depend on the unknown parameter that we are estimating. To see this, imagine a slightly different estimator of A from the previous example:

$$\hat{A} = \frac{a}{n} \sum_{i=1}^{n} \mathbf{x}[i], \tag{D.82}$$

where $a > 0$ is a constant. The MSE of any estimator can be written as the sum of the estimator variance and the square of its bias, $b(\hat{\theta}) = E[\hat{\theta}] - \theta$, because

$$\text{MSE}(\hat{\theta}) = E[(\hat{\theta} - E[\hat{\theta}] + E[\hat{\theta}] - \theta)^2] \tag{D.83}$$

$$= E[(\hat{\theta} - E[\hat{\theta}])^2] + E[(E[\hat{\theta}] - \theta)^2] - 2E[(\hat{\theta} - E[\hat{\theta}])(E[\hat{\theta}] - \theta)] \tag{D.84}$$

$$= \text{Var}[\hat{\theta}] + b^2(\hat{\theta}) - 2 \times 0 \times b(\hat{\theta}) = \text{Var}[\hat{\theta}] + b^2(\hat{\theta}). \tag{D.85}$$

The last equality follows from the fact that $E[\hat{\theta} - E[\hat{\theta}]] = 0$ and that $E[\hat{\theta}] - \theta$ is a number and not a random variable. Thus, the MSE of estimator (D.82) is

$$\text{MSE}(\hat{A}) = \text{Var}[\hat{A}] + b^2(\hat{A}) = \frac{a^2 \sigma^2}{n} + (a-1)^2 A^2. \tag{D.86}$$

By differentiating $\text{MSE}(\hat{A})$ with respect to a, setting to zero, and solving for a, we obtain the value of a that minimizes the MSE:

$$\frac{\text{d}}{\text{d}a}\text{MSE}(\hat{A}) = \frac{2a\sigma^2}{n} + 2(a-1)A^2 = 0, \tag{D.87}$$

$$a = \frac{A^2}{A^2 + \sigma^2/n}. \tag{D.88}$$

Because the second derivative is positive, we have indeed found a minimum of the MSE. We are witnessing here an interesting trade-off. It is better to allow the estimator to be biased if its variance becomes lower. The optimal value of a, however, depends on the unknown parameter A. Thus, this best estimator (in the sense of the smallest MSE) cannot be constructed.

D.6   Cramer–Rao lower bound

Because of the problem with realizability, we will next only consider unbiased estimators. Among them, we select as the best the one with the smallest variance. Such estimators are called Minimum-Variance Unbiased (MVU).

Even MVU estimators may not always exist because the minimum variance for one fixed estimator may not be minimum for all θ. Again, we do not know when to switch between estimators if there is an MVU for $\theta \in (-\infty, \theta_0)$ and another MVU for $\theta \in (\theta_0, \infty)$. There are three possible approaches that we can choose in practice [125]. We can first obtain the Cramer–Rao Lower Bound (CRLB) on the estimator variance and show that our estimator's variance is close to the theoretical bound, or we can apply the Rao–Blackwell–Lehmann–Sheffe theorem, or we can restrict our estimator to the class of linear estimators. In this appendix, we explain the CRLB.

Before formulating the CRLB, we derive a few useful facts and introduce some terminology.

We say that a pdf $p(\mathbf{x}; \theta)$ satisfies the regularity condition if

$$E\left[\frac{\partial \log p(\mathbf{x}; \theta)}{\partial \theta}\right] = 0 \text{ for all } \theta, \tag{D.89}$$

where the expected value is taken with respect to the pdf $p(\mathbf{x}; \theta)$. As in Section D.1, we adopt the same convention and use $\mathbf{x}$ to denote a random variable. The regularity condition means that

$$\int p(\mathbf{x}; \theta) \frac{\partial \log p(\mathbf{x}; \theta)}{\partial \theta} d\mathbf{x} = \int \frac{\partial p(\mathbf{x}; \theta)}{\partial \theta} d\mathbf{x} = \frac{\partial}{\partial \theta} \int p(\mathbf{x}; \theta) d\mathbf{x} = \frac{\partial 1}{\partial \theta} = 0 \tag{D.90}$$

if we can exchange the partial derivative and the integral. It turns out that for most real-life cases this exchange will be justified (there will be an integrable majorant for the pdf for some range of the parameter). Thus, the regularity condition is, in fact, not a strong condition and is practically always satisfied.

By differentiating the regularity condition partially with respect to θ, we establish one more useful fact,

$$E\left[\frac{\partial^2 \log p(\mathbf{x}; \theta)}{\partial \theta^2}\right] = -E\left[\left(\frac{\partial \log p(\mathbf{x}; \theta)}{\partial \theta}\right)^2\right], \tag{D.91}$$

which follows from

$$\frac{\partial}{\partial \theta} \int p(\mathbf{x};\theta) \frac{\partial \log p(\mathbf{x};\theta)}{\partial \theta} \mathrm{d}\mathbf{x} = 0, \tag{D.92}$$

$$\int p(\mathbf{x};\theta) \frac{\partial^2 \log p(\mathbf{x};\theta)}{\partial \theta^2} \mathrm{d}\mathbf{x} = -\int \frac{\partial p(\mathbf{x};\theta)}{\partial \theta} \frac{\partial \log p(\mathbf{x};\theta)}{\partial \theta} \mathrm{d}\mathbf{x}, \tag{D.93}$$

$$\int p(\mathbf{x};\theta) \frac{\partial^2 \log p(\mathbf{x};\theta)}{\partial \theta^2} \mathrm{d}\mathbf{x} = -\int p(\mathbf{x};\theta) \frac{\partial \log p(\mathbf{x};\theta)}{\partial \theta} \frac{\partial \log p(\mathbf{x};\theta)}{\partial \theta} \mathrm{d}\mathbf{x}, \tag{D.94}$$

$$E\left[\frac{\partial^2 \log p(\mathbf{x};\theta)}{\partial \theta^2} \right] = -E\left[\left(\frac{\partial \log p(\mathbf{x};\theta)}{\partial \theta} \right)^2 \right]. \tag{D.95}$$

Theorem D.5. [Cramer–Rao lower bound] *Assume the pdf of measurements $p(\mathbf{x};\theta)$ satisfies the regularity condition (D.89). Then, the variance of any unbiased estimator $\hat{\theta}$ must satisfy*

$$\mathrm{Var}[\hat{\theta}] \geq \frac{1}{-E\left[\frac{\partial^2 \log p(\mathbf{x};\theta)}{\partial \theta^2} \right]}. \tag{D.96}$$

Moreover, an unbiased estimator that attains the bound exists for all θ if and only if

$$\frac{\partial \log p(\mathbf{x};\theta)}{\partial \theta} = I(\theta)\,(g(\mathbf{x}) - \theta) \tag{D.97}$$

for some functions g and I. The MVU estimator is

$$\hat{\theta} = g(\mathbf{x}) \tag{D.98}$$

and its variance is

$$\mathrm{Var}[\hat{\theta}] = \frac{1}{I(\theta)}. \tag{D.99}$$

Proof. The proof of this theorem is a direct consequence of the Cauchy–Schwartz inequality (see Section D.10.1). Let $g(\mathbf{x})$ be the function that realizes an unbiased estimator. Because it is unbiased, we have

$$\int g(\mathbf{x}) p(\mathbf{x};\theta) \mathrm{d}\mathbf{x} = \theta. \tag{D.100}$$

Differentiating this with respect to θ, we obtain

$$\int \frac{\partial p(\mathbf{x};\theta)}{\partial \theta} g(\mathbf{x}) \mathrm{d}\mathbf{x} = 1, \tag{D.101}$$

$$\int p(\mathbf{x};\theta) \frac{\partial \log p(\mathbf{x};\theta)}{\partial \theta} g(\mathbf{x}) \mathrm{d}\mathbf{x} = 1, \tag{D.102}$$

$$\int p(\mathbf{x};\theta) \frac{\partial \log p(\mathbf{x};\theta)}{\partial \theta} (g(\mathbf{x}) - \theta) \mathrm{d}\mathbf{x} = 1. \tag{D.103}$$

The last equality follows from the fact that

$$\int p(\mathbf{x};\theta)\frac{\partial\log p(\mathbf{x};\theta)}{\partial\theta}\theta\mathrm{d}\mathbf{x} = \theta E\left[\frac{\partial\log p(\mathbf{x};\theta)}{\partial\theta}\right] = 0 \qquad \text{(D.104)}$$

due to the regularity condition. We now apply the Cauchy–Schwartz inequality in the space of quadratically integrable functions on the support of the pdf $p(\mathbf{x};\theta)$ with the inner product $\langle f,h\rangle = \int w(\mathbf{x})f(\mathbf{x})h(\mathbf{x})\mathrm{d}\mathbf{x}$, where the weight function w is the pdf itself. The inequality says that $\langle f,h\rangle^2 \leq \|f\|^2\,\|h\|^2$ or $\left(\int w(\mathbf{x})f(\mathbf{x})h(\mathbf{x})\mathrm{d}\mathbf{x}\right)^2 \leq \int w(\mathbf{x})(f(\mathbf{x}))^2\mathrm{d}\mathbf{x} \times \int w(\mathbf{x})(h(\mathbf{x}))^2\mathrm{d}\mathbf{x}$. In our case, $w(\mathbf{x}) = p(\mathbf{x};\theta)$, $f(\mathbf{x}) = \partial\log p(\mathbf{x};\theta)/\partial\theta$, $h(\mathbf{x}) = g(\mathbf{x}) - \theta$. Thus,

$$1 \leq \int p(\mathbf{x};\theta)\left(\frac{\partial\log p(\mathbf{x};\theta)}{\partial\theta}\right)^2\mathrm{d}\mathbf{x} \times \int p(\mathbf{x};\theta)(g(\mathbf{x})-\theta)^2\mathrm{d}\mathbf{x}, \qquad \text{(D.105)}$$

$$1 \leq E\left[\left(\frac{\partial\log p(\mathbf{x};\theta)}{\partial\theta}\right)^2\right] \times \mathrm{Var}[\hat{\theta}], \qquad \text{(D.106)}$$

which proves the bound (together with equality (D.91)).

The equality in the Cauchy–Schwartz inequality occurs if and only if the functions f and h are collinear or one is a multiple of the other. Since the independent variable for the functions is $\mathbf{x}$, the multiplicative constant could depend on θ. The equality thus occurs if and only if

$$\frac{\partial\log p(\mathbf{x};\theta)}{\partial\theta} = I(\theta)(g(\mathbf{x})-\theta), \qquad \text{(D.107)}$$

which proves half of the statement in the CRLB.

We now need to prove that if (D.107) holds, then $g(\mathbf{x})$ is an MVU estimator and its variance is $1/I(\theta)$. Taking the expected value of (D.107) with respect to $p(\mathbf{x};\theta)$, the left-hand side is the regularity condition and thus equal to 0. The right-hand side is $I(\theta)$ times the estimator bias, $I(\theta)E[g(\mathbf{x})-\theta]$. Thus, $\hat{\theta} = g(\mathbf{x})$ is an unbiased estimator. To show the variance, we first calculate the analytic form of $I(\theta)$. We differentiate (D.107) with respect to θ and take the expected value again:

$$\frac{\partial^2\log p(\mathbf{x};\theta)}{\partial\theta^2} = I'(\theta)(g(\mathbf{x})-\theta) - I(\theta), \qquad \text{(D.108)}$$

$$E\left[\frac{\partial^2\log p(\mathbf{x};\theta)}{\partial\theta^2}\right] = E[I'(\theta)(g(\mathbf{x})-\theta)] - I(\theta) = -I(\theta). \qquad \text{(D.109)}$$

Thus,

$$I(\theta) = -E\left[\frac{\partial^2\log p(\mathbf{x};\theta)}{\partial\theta^2}\right] = E\left[\left(\frac{\partial\log p(\mathbf{x};\theta)}{\partial\theta}\right)^2\right]. \qquad \text{(D.110)}$$

To calculate the variance of $\hat{\theta} = g(\mathbf{x})$, we rewrite (D.107), square, and take the expected value of both sides:

$$\frac{1}{I(\theta)}\frac{\partial \log p(\mathbf{x};\theta)}{\partial \theta} = g(\mathbf{x}) - \theta, \tag{D.111}$$

$$E\left[\left(\frac{1}{I(\theta)}\frac{\partial \log p(\mathbf{x};\theta)}{\partial \theta}\right)^2\right] = E[(g(\mathbf{x}) - \theta)^2] = \mathrm{Var}[\hat{\theta}], \tag{D.112}$$

$$\frac{1}{I(\theta)^2}E\left[\left(\frac{\partial \log p(\mathbf{x};\theta)}{\partial \theta}\right)^2\right] = \mathrm{Var}[\hat{\theta}], \tag{D.113}$$

$$\frac{1}{I(\theta)} = \mathrm{Var}[\hat{\theta}]. \tag{D.114}$$

The last equality follows from the analytic expression for $I(\theta)$. $\square$

The quantity $I(\theta) = E[(\partial \log p(\mathbf{x};\theta)/\partial \theta)^2]$ is called Fisher information. It is a measure of how fast the pdf changes with θ at $\mathbf{x}$. Intuitively, the faster the pdf changes, the more accurately we can estimate θ from the measurements. The larger it is, the smaller the bound on the unbiased estimator variance, $\mathrm{Var}[\hat{\theta}] \geq 1/I(\theta)$.

The Fisher information is always non-negative and it is additive when the observations are independent. This is because for independent observations

$$p(\mathbf{x};\theta) = \prod_{i=1}^{n} p(\mathbf{x}[i];\theta) \tag{D.115}$$

and we have for the Fisher information

$$I(\theta) = -E\left[\frac{\partial^2 \sum_{i=1}^{n}\log p(\mathbf{x}[i];\theta)}{\partial \theta^2}\right] = -\sum_{i=1}^{n} E\left[\frac{\partial^2 \log p(\mathbf{x}[i];\theta)}{\partial \theta^2}\right] = \sum_{i=1}^{n} I_i(\theta), \tag{D.116}$$

where $I_i(\theta)$ is the Fisher information of each observation.

Example D.6: We can use the CRLB to prove that the sample mean is the MVU estimator for DC level in AWGN (Example D.4). The pdf of each observation is a Gaussian $N(A,\sigma^2)$. Since all n observations are independent random variables, their joint pdf is multiplication of their marginal pdfs and we have

$$p(\mathbf{x};A) = \prod_{i=1}^{n}\frac{1}{\sqrt{2\pi}\sigma}e^{-\frac{(\mathbf{x}[i]-A)^2}{2\sigma^2}} = (2\pi\sigma^2)^{-\frac{n}{2}}e^{-\frac{1}{2\sigma^2}\sum_{i=1}^{n}(\mathbf{x}[i]-A)^2}. \tag{D.117}$$

Thus,

$$\frac{\partial \log p(\mathbf{x}; A)}{\partial A} = -\frac{1}{2\sigma^2} \sum_{i=1}^{n} -2(\mathbf{x}[i] - A)$$

$$= \frac{1}{\sigma^2} \left(\sum_{i=1}^{n} \mathbf{x}[i] - nA \right) = \frac{n}{\sigma^2}(\overline{\mathbf{x}} - A).$$

From here, the CRLB tells us that the sample mean $g(\mathbf{x}) = \overline{\mathbf{x}}$ is the MVU estimator with variance σ^2/n.

Example D.7: Consider the situation in Example D.4,

$$\mathbf{x}[i] = A + \boldsymbol{\xi}[i], \ i = 1, \ldots, n, \ \boldsymbol{\xi}[i] \sim N(0, \sigma^2), \tag{D.118}$$

but now let us say we know the DC level A and we wish to estimate the noise variance σ^2 instead. We start by writing down the pdf of the n observations, with σ^2 being the unknown parameter,

$$p(\mathbf{x}; \sigma^2) = \prod_{i=1}^{n} \frac{1}{\sqrt{2\pi}\sigma} e^{-\frac{(\mathbf{x}[i]-A)^2}{2\sigma^2}} = (2\pi\sigma^2)^{-\frac{n}{2}} e^{-\frac{1}{2\sigma^2}\sum_{i=1}^{n}(\mathbf{x}[i]-A)^2}. \tag{D.119}$$

We have

$$\frac{\partial \log p(\mathbf{x}; \sigma^2)}{\partial \sigma^2} = \frac{\partial}{\partial \sigma^2} \left(-\frac{n}{2}\log(2\pi\sigma^2) - \frac{1}{2\sigma^2}\sum_{i=1}^{n}(\mathbf{x}[i] - A)^2 \right) \tag{D.120}$$

$$= -\frac{n}{2\sigma^2} + \frac{1}{2\sigma^4}\sum_{i=1}^{n}(\mathbf{x}[i] - A)^2 \tag{D.121}$$

$$= \frac{n}{2\sigma^4} \left(\frac{1}{n}\sum_{i=1}^{n}(\mathbf{x}[i] - A)^2 - \sigma^2 \right). \tag{D.122}$$

Thus, the sample variance $\hat{\sigma}^2 = (1/n)\sum_{i=1}^{n}(\mathbf{x}[i] - A)^2$ is an MVU estimator and its variance is $2\sigma^4/n$. Note that while the MVU estimator of A does not need knowledge of the noise variance σ^2, the MVU estimator of variance needs knowledge of A.

If both A and σ^2 are unknown, it is tempting to write an estimator for the variance as

$$\hat{\sigma}^2 = \frac{1}{n}\sum_{i=1}^{n}(\mathbf{x}[i] - \hat{A})^2 = \frac{1}{n}\sum_{i=1}^{n}(\mathbf{x}[i] - \overline{\mathbf{x}})^2. \tag{D.123}$$

However, this plug-in estimator is biased. In fact, plug-in estimators are generally biased unless they are linear in the parameter. The reader is challenged to verify that

$$\hat{\sigma}^2 = \frac{1}{n-1} \sum_{i=1}^{n} (\mathbf{x}[i] - \overline{\mathbf{x}})^2 \tag{D.124}$$

is an MVU estimate of variance.

D.7   Maximum-likelihood and maximum a posteriori estimation

Even though the CRLB can be used for derivation of MVU estimators, it is rarely used in this way except for some simple cases. In practice, alternative approaches that typically give good estimators are used. The two most common principles are maximum-likelihood and maximum a posteriori estimation.

In Maximum-Likelihood Estimation (MLE), the parameters are estimated from measurements $\mathbf{x}$ using the following optimization problem:

$$\hat{\theta} = \arg \max_{\theta} p(\mathbf{x}|\theta). \tag{D.125}$$

When the measurements are independent,

$$p(\mathbf{x}|\theta) = \prod_{i=1}^{n} p(\mathbf{x}[i]|\theta). \tag{D.126}$$

Because maximizing the likelihood in (D.125) is the same as maximizing its logarithm, we obtain

$$\hat{\theta} = \arg \max_{\theta} \sum_{i=1}^{n} \log p(\mathbf{x}[i]|\theta) \tag{D.127}$$

and the parameter can be obtained by solving the following algebraic equation:

$$\sum_{i=1}^{n} \frac{\partial \log p(\mathbf{x}[i]|\theta)}{\partial \theta} = 0. \tag{D.128}$$

When θ is a vector parameter, (D.128) generalizes to a system of k algebraic equations for k unknowns $\boldsymbol{\theta}[k]$,

$$\sum_{i=1}^{n} \frac{\partial \log p(\mathbf{x}[i]|\boldsymbol{\theta})}{\partial \boldsymbol{\theta}[k]} = 0 \text{ for all } k. \tag{D.129}$$

Maximum-likelihood estimation is often used to fit a parametric distribution to experimental data. We illustrate this particular application on the example of estimating the parameters of the generalized Cauchy distribution.[2]

---

[2] This estimation procedure is used in the implementation of Model-Based Steganography described in Section 7.1.2.

Example D.8: [MLE for generalized Cauchy distribution] The generalized Cauchy distribution with zero mean has the form (Section A.8)

$$f(x; p, s, 0) = \frac{p-1}{2s}\left(1 + \frac{|x|}{s}\right)^{-p} \tag{D.130}$$

with parameters $\boldsymbol{\theta} = (s, p)$. The partial derivatives are

$$\frac{\partial \log f(\mathbf{x}[i]|\boldsymbol{\theta})}{\partial s} = -\frac{1}{s} + \frac{p|x|}{s^2 + s|x|}, \tag{D.131}$$

$$\frac{\partial \log f(\mathbf{x}[i]|\boldsymbol{\theta})}{\partial p} = \frac{1}{p-1} - \log\left(1 + \frac{|x|}{s}\right). \tag{D.132}$$

Substituting them into (D.129), we obtain two equations for two unknowns,

$$\sum_{i=1}^{n} \frac{1}{1 + \frac{s}{|\mathbf{x}[i]|}} = \frac{n}{p}, \tag{D.133}$$

$$\sum_{i=1}^{n} \log\left(1 + \frac{|\mathbf{x}[i]|}{s}\right) = \frac{n}{p-1}, \tag{D.134}$$

which can be solved for p and s. In this particular case, we can substitute for n/p from the first equation into the second one and thus deal with only one algebraic equation for s.

When we have prior information about the distribution of the parameters, we can utilize this knowledge to obtain a better estimate. This approach to parameter estimation is called Maximum A Posteriori (MAP) estimation,

$$\hat{\theta} = \arg\max_{\theta} p(\theta|\mathbf{x}) = \arg\max_{\theta} p(\mathbf{x}|\theta)p(\theta), \tag{D.135}$$

where $p(\theta)$ is the a priori distribution of the parameter. Note that when $p(\theta)$ is uniform, ML and MAP estimates coincide. The reader is referred to [125] for more information about these two important estimation methods.

D.8   Least-square estimation

In practice, we often need to quantitatively describe a relationship between a scalar quantity y and a vector $\mathbf{x} \in \mathbb{R}^d$. In steganalysis, y may stand for the change rate and $\mathbf{x}$ for a stego image histogram or some other (vector) feature derived from the stego image. In Least-Square Estimation (LSE), one selects a parametric model for this relationship in the form

$$y = f(\mathbf{x}; \boldsymbol{\theta}) + \eta, \tag{D.136}$$

where $\boldsymbol{\theta}$ is an unknown vector parameter, η is the modeling noise, and f maps to $\mathbb{R}$. The unknown parameter will be determined from l tuples $(\mathbf{y}[i], \mathbf{x}_i)$, $i = 1, \ldots, l$, usually obtained experimentally, by minimizing the scalar functional

$$J(\boldsymbol{\theta}) = \sum_{i=1}^{l} (f(\mathbf{x}_i; \boldsymbol{\theta}) - \mathbf{y}[i])^2 \qquad (\text{D.137})$$

with respect to $\boldsymbol{\theta}$. The minimization leads to a set of k algebraic equations for k unknowns $\boldsymbol{\theta}[1], \ldots, \boldsymbol{\theta}[k]$:

$$\frac{\partial J(\boldsymbol{\theta})}{\partial \boldsymbol{\theta}[k]} = 2 \sum_{i=1}^{l} (f(\mathbf{x}_i; \boldsymbol{\theta}) - \mathbf{y}[i]) \frac{\partial f(\mathbf{x}_i; \boldsymbol{\theta})}{\partial \boldsymbol{\theta}[k]} = 0. \qquad (\text{D.138})$$

In general, this problem needs to be solved numerically, e.g., using the Newton–Raphson method.

When the model is linear, a closed-form solution to the optimization can be obtained. Considering the vectors as column vectors, $f(\mathbf{x}; \boldsymbol{\theta}) = \mathbf{x}'\boldsymbol{\theta}$, $\boldsymbol{\theta} \in \mathbb{R}^d$, and (D.136) can be written in a compact matrix form as

$$\mathbf{y} = \mathbf{H}\boldsymbol{\theta} + \boldsymbol{\eta}, \qquad (\text{D.139})$$

where the ith row of the $l \times d$ matrix $\mathbf{H}$ is $\mathbf{x}_i'$. We will further assume that $\mathbf{H}$ is of full rank.

Writing the functional $J(\boldsymbol{\theta})$ as

$$J(\boldsymbol{\theta}) = (\mathbf{y} - \mathbf{H}\boldsymbol{\theta})'(\mathbf{y} - \mathbf{H}\boldsymbol{\theta}) = \mathbf{y}'\mathbf{y} - \mathbf{y}'\mathbf{H}\boldsymbol{\theta} - (\mathbf{H}\boldsymbol{\theta})'\mathbf{y} + (\mathbf{H}\boldsymbol{\theta})'\mathbf{H}\boldsymbol{\theta} \qquad (\text{D.140})$$

$$= \mathbf{y}'\mathbf{y} - 2\mathbf{y}'\mathbf{H}\boldsymbol{\theta} + \boldsymbol{\theta}'\mathbf{H}'\mathbf{H}\boldsymbol{\theta}, \qquad (\text{D.141})$$

we can easily obtain the result of differentiating as

$$\frac{\partial J(\boldsymbol{\theta})}{\partial \boldsymbol{\theta}} = -2\mathbf{H}'\mathbf{y} + 2\mathbf{H}'\mathbf{H}\boldsymbol{\theta}. \qquad (\text{D.142})$$

The least-square estimate of the parameter, $\hat{\boldsymbol{\theta}}$, is obtained by setting the gradient to zero and solving for $\boldsymbol{\theta}$,

$$\hat{\boldsymbol{\theta}} = (\mathbf{H}'\mathbf{H})^{-1}\mathbf{H}'\mathbf{y}. \qquad (\text{D.143})$$

The LSE is usually applied when the properties of the modeling noise are not known. We note that LSE becomes MLE when the modeling noise η is Gaussian $N(0, \sigma^2)$. This is because MLE maximizes

$$\log p(\mathbf{y}|\boldsymbol{\theta}) = \frac{n}{2} \log(2\pi\sigma^2) - \frac{1}{2\sigma^2} \sum_{i=1}^{l} (\mathbf{y}[i] - \mathbf{H}[i, .]\boldsymbol{\theta})^2, \qquad (\text{D.144})$$

which is equivalent to minimizing $J(\boldsymbol{\theta}) = (\mathbf{y} - \mathbf{H}\boldsymbol{\theta})'(\mathbf{y} - \mathbf{H}\boldsymbol{\theta})$, the task of LSE.

D.9   Wiener filter

An important tool in steganalysis is a denoising filter. The task of denoising belongs to signal estimation (we are estimating the noise-free image). Here, we describe in detail one of the simplest filters, the Wiener filter. It is an adaptive linear filter that removes additive white Gaussian noise of a given (known) variance from a signal so that the filtered signal is closest to the original (non-noisy) signal in the least-square sense. We assume we have n samples of a noisy signal $\mathbf{x}[i]$

$$\mathbf{x}[i] = \mathbf{S}[i] + \boldsymbol{\xi}[i], \ i = 1, \ldots, n, \tag{D.145}$$

which is a superposition of the original noise-free signal $\mathbf{S}[i]$, the true scene, and Gaussian noise $\boldsymbol{\xi}[i]$. The following assumptions are made about the original signal and the noise:

1. $\mathbf{S}[i]$ are zero-mean jointly Gaussian random variables with a known covariance matrix $\mathbf{C}[i, j] = \text{Cov}\,[\mathbf{S}[i], \mathbf{S}[j]] = E\,[\mathbf{S}[i]\mathbf{S}[j]]$.
2. The noise is a sequence of Gaussian random variables with a known covariance matrix. Here, we will assume that the covariance matrix is diagonal with variance σ^2 on its diagonal (in other words, the sequence of random variables $\boldsymbol{\xi}[i] \sim N(0, \sigma^2)$ is iid). We also assume that $\mathbf{S}$ and $\boldsymbol{\xi}$ are independent of each other.

Our task is to estimate $\mathbf{S}$ from $\mathbf{x}$ in the least-square sense so that the expected value of the error

$$E\left[\left(\mathbf{S}[i] - \hat{\mathbf{S}}[i]\right)^2\right] \tag{D.146}$$

is minimum for each i. We seek the estimate as a linear function of the (noisy) observations

$$\hat{\mathbf{S}}[i] = \sum_{j=1}^{n} \mathbf{W}[i, j]\mathbf{x}[j]. \tag{D.147}$$

The requirement of minimal mean-square error means that we need to find the matrix $\mathbf{W}$ so that the expected value of the error for the ith sample, $\mathbf{e}[i]$, is minimal,

$$\mathbf{e}[i](\mathbf{W}[i, 1], \ldots, \mathbf{W}[i, n]) = E\left[\left(\mathbf{S}[i] - \sum_{j=1}^{n} \mathbf{W}[i, j]\mathbf{x}[j]\right)^2\right]. \tag{D.148}$$

We do so by differentiating $\mathbf{e}[i]$ with respect to its arguments and solving for the local minimum,

$$\frac{\partial \mathbf{e}[i]}{\partial \mathbf{W}[i, j]} = E\left[-2\left(\mathbf{S}[i] - \sum_{k=1}^{n} \mathbf{W}[i, k]\mathbf{x}[k]\right)\mathbf{x}[j]\right] = 0 \text{ for all } i, j. \tag{D.149}$$

Rewriting further and substituting for $\mathbf{x}[j] = \mathbf{S}[j] + \boldsymbol{\xi}[j]$, we obtain for all i and j

$$E\left[\mathbf{S}[i]\left(\mathbf{S}[j] + \boldsymbol{\xi}[j]\right)\right] = E\left[\sum_{k=1}^{n} \mathbf{W}[i,k]\left(\mathbf{S}[k] + \boldsymbol{\xi}[k]\right)\left(\mathbf{S}[j] + \boldsymbol{\xi}[j]\right)\right]. \qquad (D.150)$$

The right-hand side can be written as

$$\sum_{k=1}^{n} \mathbf{W}[i,k]\left(E\left[\mathbf{S}[k]\mathbf{S}[j]\right] + E\left[\mathbf{S}[k]\boldsymbol{\xi}[j]\right] + E\left[\mathbf{S}[j]\boldsymbol{\xi}[k]\right] + E\left[\boldsymbol{\xi}[k]\boldsymbol{\xi}[j]\right]\right). \quad (D.151)$$

Using the fact that $\boldsymbol{\xi}[j]$ and $\boldsymbol{\xi}[k]$ are independent for $j \neq k$ and the fact that $\mathbf{S}$ and $\boldsymbol{\xi}$ are also independent signals, we have $E\left[\mathbf{S}[k]\boldsymbol{\xi}[j]\right] = 0$ and $E\left[\boldsymbol{\xi}[k]\boldsymbol{\xi}[j]\right] = \delta(k-j)\sigma^2$, where δ is the Kronecker delta. Thus, we can rewrite (D.150) in matrix notation as

$$\mathbf{C} = \mathbf{W}\left(\mathbf{C} + \sigma^2\mathbf{I}\right) \qquad (D.152)$$

and obtain the general form of the Wiener filter $\mathbf{W}$ as

$$\mathbf{W} = \mathbf{C}\left(\mathbf{C} + \sigma^2\mathbf{I}\right)^{-1}. \qquad (D.153)$$

It should be clear that we have indeed found a *minimum* because the second partial derivative with respect to $\mathbf{W}[i,j]$ is $\partial^2\mathbf{e}[i]/\partial\mathbf{W}^2[i,j] = E\left[2\mathbf{x}^2[j]\right] > 0$.

In the special case when the covariance matrix $\mathbf{C}$ is diagonal,[3] we have $\mathbf{C} = \mathrm{diag}(\boldsymbol{\sigma}^2[1],\ldots,\boldsymbol{\sigma}^2[n])$ and the inverse $(\mathbf{C} + \sigma^2\mathbf{I})^{-1}$ as well as the matrix $\mathbf{W}$ are also diagonal, $\mathbf{W} - \mathrm{diag}(\mathbf{w}[1],\ldots,\mathbf{w}[n])$,

$$\mathbf{w}[i] = \frac{\boldsymbol{\sigma}^2[i]}{\boldsymbol{\sigma}^2[i] + \sigma^2}. \qquad (D.154)$$

The denoised signal is

$$\hat{\mathbf{S}}[i] = \frac{\boldsymbol{\sigma}^2[i]}{\boldsymbol{\sigma}^2[i] + \sigma^2}\mathbf{x}[i]. \qquad (D.155)$$

D.9.1   Practical implementation for images

Let $\mathbf{x}[i,j]$, $i = 1,\ldots,M$, $j = 1,\ldots,N$, be the pixel values of a grayscale $M \times N$ image. The values $\mathbf{x}[i,j]$ are a superposition of the noise-free image $\mathbf{S}[i,j]$ and an iid Gaussian component $\boldsymbol{\xi}[i,j] \sim N(0,\sigma^2)$,

$$\mathbf{x}[i,j] = \mathbf{S}[i,j] + \boldsymbol{\xi}[i,j]. \qquad (D.156)$$

We wish to filter out the noise and obtain an approximation $\hat{\mathbf{S}}[i,j]$ to $\mathbf{S}$. It would not be reasonable to assume that $\mathbf{S}[i,j]$ is a zero-mean Gaussian. We can,

---

[3] $\mathbf{S}[i]$ is a non-stationary sequence of independent Gaussian variables.

however, make that assumption about $\mathbf{S}[i,j] - \hat{\boldsymbol{\mu}}[i,j]$, where

$$\hat{\boldsymbol{\mu}}[i,j] = \frac{1}{|\mathcal{N}[i,j]|} \sum_{(k,l) \in \mathcal{N}[i,j]} \mathbf{x}[k,l] \qquad (D.157)$$

is the average grayscale value in the neighborhood $\mathcal{N}[i,j]$ of pixel (i,j) containing $|\mathcal{N}[i,j]|$ pixels (e.g., we can take the 3×3 neighborhood). Note that

$$\mathrm{Var}\left[\mathbf{x}[i,j]\right] = \mathrm{Var}\left[\mathbf{S}[i,j]\right] + \sigma^2, \qquad (D.158)$$

where $\mathrm{Var}\left[\mathbf{x}[i,j]\right]$ can be approximated[4] by the sample variance in the neighborhood $\mathcal{N}[i,j]$,

$$\mathrm{Var}\left[\mathbf{x}[i,j]\right] \approx \hat{\boldsymbol{\sigma}}^2[i,j] = \frac{1}{|\mathcal{N}[i,j]|} \sum_{(k,l) \in \mathcal{N}[i,j]} \mathbf{x}^2[k,l] - \hat{\boldsymbol{\mu}}^2[i,j]. \qquad (D.159)$$

Now, consider the signal $\mathbf{x}[i,j] - \hat{\boldsymbol{\mu}}[i,j] = \mathbf{S}[i,j] - \hat{\boldsymbol{\mu}}[i,j] + \boldsymbol{\xi}[i,j]$ as the equivalent of the noisy signal $\mathbf{x}[i]$ in the derivation of the Wiener filter, and $\mathbf{S}[i,j] - \hat{\boldsymbol{\mu}}[i,j]$ the equivalent of the noise-free signal $\mathbf{S}[i]$. From (D.158), the variance of the noise-free signal $\mathrm{Var}\left[\mathbf{S}[i,j] - \hat{\boldsymbol{\mu}}[i,j]\right] = \mathrm{Var}\left[\mathbf{S}[i,j]\right] = \mathrm{Var}\left[\mathbf{x}[i,j]\right] - \sigma^2 \approx \hat{\boldsymbol{\sigma}}^2[i,j] - \sigma^2$. Thus, we obtain the Wiener filter for denoising images in the form (D.155)

$$\hat{\mathbf{S}}[i,j] = \hat{\boldsymbol{\mu}}[i,j] + \frac{\hat{\boldsymbol{\sigma}}^2[i,j] - \sigma^2}{\hat{\boldsymbol{\sigma}}^2[i,j]} \left(\mathbf{x}[i,j] - \hat{\boldsymbol{\mu}}[i,j]\right). \qquad (D.160)$$

We summarize that the Wiener filter is an adaptive linear filter that is optimal in the sense that it minimizes the MSE between the denoised image and the noise-free image under the assumption that the zero-meaned pixel values form a sequence of independent zero-mean Gaussian variables (not necessarily with equal variances – we allow non-stationary signals) and the image is corrupted with white Gaussian noise of known variance σ^2. This is exactly the implementation of `wiener2.m` in Matlab. The command `wiener2(X,[N N],`σ^2`)` returns a denoised version of X, where N is the size of the square neighborhood $\mathcal{N}$. If the parameters are not specified by the user, Matlab determines the default value of σ^2 as the average variance in 3×3 neighborhoods over the whole image.

D.10   Vector spaces with inner product

Some results in this appendix and in Appendix A use the Cauchy–Schwartz inequality in abstract spaces. To help the reader understand these results on a deeper level, we introduce in this section the concept of an abstract vector space and formulate and prove the Cauchy–Schwartz inequality.

---

[4] We are really making a tacit assumption that the pixel values are locally stationary.

A vector space $\mathcal{V}$ is a set of objects that can be added and multiplied by a scalar from some field $\mathcal{T}$. The addition of elements from $\mathcal{V}$ is associative and commutative and there exists a zero vector (so that $x + 0 = x$ for all $x \in \mathcal{V}$). Every element $x \in \mathcal{V}$ also has an inverse element, $-x \in \mathcal{V}$, so that $x + (-x) = 0$. The multiplication by a scalar, $\alpha \in \mathcal{T}$, is distributive in the sense that for all $\alpha, \beta \in \mathcal{T}$, $x, y \in \mathcal{V}$, $\alpha(x + y) = \alpha x + \alpha y$ and $(\alpha + \beta)x = \alpha x + \beta x$, and associative, $\alpha(\beta x) = (\alpha\beta)x$. Also, for the identity element in $\mathcal{T}$, $1 \cdot x = x$ for all $x \in \mathcal{V}$.

An inner product in vector space is a mapping $\langle x, y \rangle : \mathcal{V} \times \mathcal{V} \to \mathbb{R}$ with the following properties.

1. $\langle x, y \rangle = \langle y, x \rangle$ for all $x, y \in \mathcal{V}$ (symmetry).[5]
2. $\langle \alpha x + y, z \rangle = \alpha \langle x, z \rangle + \langle y, z \rangle$ for all $x, y, z \in \mathcal{V}$, $\alpha \in \mathcal{T}$ (linearity).
3. $\langle x, x \rangle \geq 0$ for all $x \in \mathcal{V}$. $\sqrt{\langle x, x \rangle} = \|x\|$ is called the norm of x.
4. $\langle x, x \rangle = 0$ if and only if $x = 0$ (non-degeneracy).

We now provide several examples of important vector spaces with inner product.

Example D.9: [Euclidean space] $\mathcal{V} = \mathbb{R}^n, \mathcal{T} = \mathbb{R}$.

$$\langle \mathbf{x}, \mathbf{y} \rangle = \langle (\mathbf{x}[1], \ldots, \mathbf{x}[n])(\mathbf{y}[1], \ldots, \mathbf{y}[n]) \rangle = \sum_{i=1}^{n} \mathbf{x}[i]\mathbf{y}[i] \qquad (D.161)$$

is the usual dot product in Euclidean space. In Euclidean spaces, it is customary to denote the dot product with a dot, $\mathbf{x} \cdot \mathbf{y}$.

Example D.10: [L_2 space] $\mathcal{V} = L_2(a, b)$ is the space of all quadratically integrable functions on $[a, b]$ with $\mathcal{T} = \mathbb{R}$. To be absolutely precise here, we consider two functions f and g from this space as identical if they differ on a set of Lebesgue measure 0. Thus, formally speaking, the elements of this vector space are classes of equivalence of functions. We endow this space with an inner product

$$\langle f, g \rangle = \int_a^b f(t)g(t)\mathrm{d}t. \qquad (D.162)$$

It is easy to verify that the four defining properties of an inner product are satisfied. The fourth property is satisfied because if $\int f^2(t)dt = 0$, then $f(t) = 0$ almost everywhere (in the sense of being zero up to a set of Lebesgue measure 0).

---

[5] For complex inner products that map to $\mathbb{C}$, the set of all complex numbers, the symmetry is replaced with conjugate symmetry $\langle x, y \rangle = \overline{\langle y, x \rangle}$.

In this book, we will also encounter spaces with a slightly different inner product defined as

$$\langle f, g \rangle = \int_a^b w(t) f(t) g(t) dt, \tag{D.163}$$

where $w(t) > 0$ is a weight function on $[a, b]$.

Example D.11: [Vector space of random variables] Let $\mathcal{V}$ be the space of all real-valued random variables with zero mean and the inner product defined as

$$\langle \mathsf{x}, \mathsf{y} \rangle = \text{Cov}[\mathsf{x}, \mathsf{y}]. \tag{D.164}$$

It is easy to verify that this is, indeed, a correctly defined inner product. The fourth property is satisfied because $\langle \mathsf{x}, \mathsf{x} \rangle = \text{Var}[\mathsf{x}] \geq 0$. If the variance of a real-valued random variable is zero and its mean is zero, it means that its pdf is zero almost everywhere and thus $\mathsf{x} \equiv 0$.

D.10.1 Cauchy–Schwartz inequality

The Cauchy–Schwartz inequality holds in any vector space $\mathcal{V}$ with inner product. For any $x, y \in \mathcal{V}$

$$|\langle x, y \rangle| \leq \|x\| \|y\|, \tag{D.165}$$

where the equality occurs if and only if the vectors x and y are collinear (one is a scalar multiple of the other).

First, note that if $y = 0$, the inequality is satisfied. Let λ be an arbitrary number. Then,

$$0 \leq \langle x - \lambda y, \ x - \lambda y \rangle = \|x\|^2 + \lambda^2 \|y\|^2 - 2\lambda \langle x, y \rangle. \tag{D.166}$$

By selecting a specific value of $\lambda = \langle x, y \rangle / \|y\|^2$, we obtain the Cauchy–Schwartz inequality

$$0 \leq \|x\|^2 + \frac{\langle x, y \rangle^2}{\|y\|^4} \|y\|^2 - 2 \frac{\langle x, y \rangle}{\|y\|^2} \langle x, y \rangle = \|x\|^2 - \frac{|\langle x, y \rangle|^2}{\|y\|^2}. \tag{D.167}$$

Note that equality in the Cauchy–Schwartz inequality occurs if and only if there exists λ such that $x - \lambda y = 0$, or in other words when x and y are collinear (one is a multiple of the other).

E   Support vector machines

In this appendix, we describe the basic principles of classification using Support Vector Machines (SVM). SVMs recently gained popularity because they offer performance comparable to or better than other machine-learning tools and are relatively easy to use. This material is especially relevant to Chapter 12 on blind steganalysis, where SVMs are used to construct blind steganalyzers. For a more detailed tutorial on SVMs the reader is referred to [33, 54].

In the first, rather theoretic section, we explain the main principles of SVMs on linearly separable problems. The methodology is then extended to problems where linear separation is not possible. Finally, we describe the most general kernelized SVMs. The second part of the appendix deals with practical implementation issues when applying SVMs to real problems.

We start by defining the binary classification problem.

E.1   Binary classification

Let $\mathcal{X}$ be an arbitrary non-empty input space and $\mathcal{Y}$ be the label set $\mathcal{Y} = \{-1, +1\}$. For example, in blind steganalysis the points $\mathbf{x}_i \in \mathcal{X}$ are feature vectors extracted from images and the binary label stands for either cover or stego image. In this appendix, the input space is $\mathcal{X} = \mathbb{R}^n$. Let us suppose that a set of l training examples $\mathbf{x}_i$ is available together with their associated labels $\mathbf{y}[i]$: $(\mathbf{x}_1, \mathbf{y}[1]), \ldots, (\mathbf{x}_l, \mathbf{y}[l]) \in \mathcal{X} \times \mathcal{Y}$. We will also assume that the pairs $(\mathbf{x}_i, \mathbf{y}[i])$ are realizations of a random variable described by a joint probability measure $P(\mathbf{x}, y)$ on $\mathcal{X} \times \mathcal{Y}$. Our goal is to use the training examples to build a decision function $f : \mathcal{X} \to \mathcal{Y}$ that assigns a label to each $\mathbf{x} \in \mathcal{X}$ while making as few errors as possible. The function f is a binary classifier because it classifies every input $\mathbf{x}$ into two classes.

The ability of f to classify will be evaluated using the risk functional

$$R(f) = \int_{\mathcal{X} \times \mathcal{Y}} u(-yf(\mathbf{x}))\mathrm{d}P(\mathbf{x}, y), \qquad (\text{E.1})$$

where u is the step function

$$u(z) = \begin{cases} 1, & z > 0 \\ 0, & z \leq 0. \end{cases} \qquad \text{(E.2)}$$

Clearly, $0 \leq R(f) \leq 1$, and $R(f) = 0$ when $f(\mathbf{x})$ correctly assigns the labels to all $\mathbf{x} \in \mathcal{X}$ up to a set of measure zero (with respect to P). We also stress at this point that unless we know the measure $P(\mathbf{x}, y)$, we cannot guarantee that any *estimated* decision function is optimal (that it minimizes the risk functional R). Unfortunately, in most practical applications the measure will not be known. Note that if $P(\mathbf{x}, y)$ were known, we could apply the apparatus of classical detection theory and derive optimal detectors (classifiers) using the Neyman–Pearson or Bayesian approach as explained in Appendix D.

E.2   Linear support vector machines

E.2.1   Linearly separable training set

We say that the training set is linearly separable if there exists a hyperplane with normal vector $\mathbf{w} \in \mathbb{R}^n$ and $b \in \mathbb{R}$,

$$f(\mathbf{x}) = \text{sign}(\mathbf{w} \cdot \mathbf{x} - b), \qquad \text{(E.3)}$$

such that the empirical risk (error) on the training set

$$R_{\text{emp}}(f) = \frac{1}{l} \sum_{i=1}^{l} u\left(-\mathbf{y}[i]f(\mathbf{x}_i)\right) = 0. \qquad \text{(E.4)}$$

The function classifies the point $\mathbf{x} \in \mathbb{R}^n$ according to on which side of the hyperplane $\mathbf{w} \cdot \mathbf{x} - b$ the point $\mathbf{x}$ lies. Because the decision function is fully described by the separating hyperplane, we will use the terms decision function, hyperplane, or classifier interchangeably depending on the context.

If the training set is linearly separable, there may exist infinitely many decision functions $f(\mathbf{x}) = \text{sign}(\mathbf{w} \cdot \mathbf{x} - b)$ perfectly classifying the training set with $R_{\text{emp}}(f) = 0$. To lower the chance of making an incorrect decision on $\mathbf{x}$ not contained in the training set, we select the separating hyperplane with maximum distance from positive and negative training examples. This hyperplane, which we denote f^*, is uniquely defined. It can be found by solving the following optimization problem:

$$[\mathbf{w}^*, b^*] = \arg \max_{\mathbf{w} \in \mathbb{R}^n, b \in \mathbb{R}} \min_{\mathbf{x}, i} \{\|\mathbf{x} - \mathbf{x}_i\| \,|\, \mathbf{x} \in \mathbb{R}^n, \ \mathbf{w} \cdot \mathbf{x} - b = 0, \ i \in \{1, \ldots, l\}\}$$
$$\text{(E.5)}$$

subject to

$$\mathbf{y}[i]\left(\mathbf{w} \cdot \mathbf{x}_i - b\right) > 0, \quad \text{for all } i \in \{1, \ldots, l\}. \qquad \text{(E.6)}$$

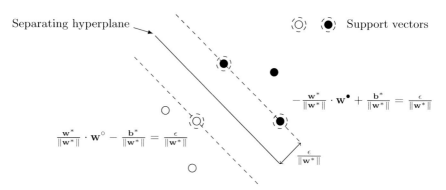

Figure E.1 Example of a linearly separable training set in $\mathcal{X} = \mathbb{R}^2$. The separating hyperplane is defined by the support vectors. Notice that other examples do not affect the solution of the optimization problem. Using the notation from the text, $\mathbf{w}^* \cdot \mathbf{x} - b^* = 0$ for all examples lying on the separating hyperplane, while for the support vectors $(\mathbf{w}^*/\epsilon) \cdot \mathbf{x}^\circ - b^*/\epsilon = +1$ and $(\mathbf{w}^*/\epsilon) \cdot \mathbf{x}^\bullet - b^*/\epsilon = -1$, depending on which side of the hyperplane the support vector lies. The distance between the support vectors and the separating hyperplane is $|\mathbf{w}^* \cdot \mathbf{x}^\bullet/\|\mathbf{w}^*\| - b^*/\|\mathbf{w}^*\|| = \epsilon/\|\mathbf{w}^*\|$.

This optimization problem is quite difficult to solve in practice. It can, however, be reformulated as convex quadratic programming,

$$[\mathbf{w}^*, b^*] = \arg \min_{\mathbf{w} \in \mathbb{R}^n, b \in \mathbb{R}} \frac{1}{2} \|\mathbf{w}\|^2 \tag{E.7}$$

subject to

$$\mathbf{y}[i] (\mathbf{w} \cdot \mathbf{x}_i - b) \geq 1, \quad \text{for all } i \in \{1, \dots, l\}. \tag{E.8}$$

And this problem can be solved using standard quadratic-programming tools [27].

We now show that problems (E.5) and (E.7) are indeed equivalent. Let us assume that $[\mathbf{w}^*, b^*]$ is the solution to (E.5). Denoting by $(\mathbf{x}^\circ, y^\circ)$ and $(\mathbf{x}^\bullet, y^\bullet)$ the examples from the positive and negative classes that are closest to the hyperplane, we must have

$$\min_{\mathbf{x} \in \mathbb{R}^n} \{\|\mathbf{x} - \mathbf{x}^\circ\| | \mathbf{w}^* \cdot \mathbf{x} - b^* = 0\} = \min_{\mathbf{x} \in \mathbb{R}^n} \{\|\mathbf{x} - \mathbf{x}^\bullet\| | \mathbf{w}^* \cdot \mathbf{x} - b^* = 0\}. \tag{E.9}$$

In other words the distances from the separating hyperplane to the closest points from each class must be equal (see Figure E.1). If they were not, we could move the hyperplane away from the closer class and thus decrease the minimum in (E.5), which would contradict the optimality of the solution $[\mathbf{w}^*, b^*]$. The closest examples $\mathbf{x}^\circ, \mathbf{x}^\bullet$ cannot lie on the separating hyperplane because we have strict inequality in (E.6). Keeping in mind that $\mathbf{x}^\circ$ is from the positive class, there has to exist $\epsilon > 0$ so that

$$\mathbf{w}^* \cdot \mathbf{x}^\circ - b^* = +\epsilon, \tag{E.10}$$

$$\mathbf{w}^* \cdot \mathbf{x}^\bullet - b^* = -\epsilon. \tag{E.11}$$

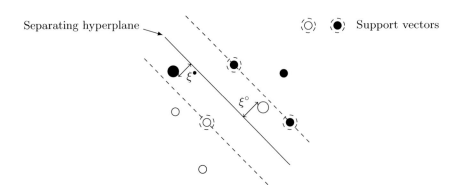

Separating hyperplane ——→   ⦶ ● Support vectors

Figure E.2 Example of a training set on $\mathcal{X} = \mathbb{R}^2$ that cannot be linearly separated. The separating hyperplane is again defined by support vectors. Incorrectly classified examples are displayed as large circles with their slack variables $\xi^\bullet, \xi^\circ > 0$.

By normalizing the last pair of equations by ϵ, they can be rewritten as

$$\frac{\mathbf{w}^*}{\epsilon} \cdot \mathbf{x}^\circ - \frac{b^*}{\epsilon} = +1, \tag{E.12}$$

$$\frac{\mathbf{w}^*}{\epsilon} \cdot \mathbf{x}^\bullet - \frac{b^*}{\epsilon} = -1. \tag{E.13}$$

Because $\mathbf{x}^\circ, \mathbf{x}^\bullet$ are the closest examples, it is obvious that the hyperplane $[\mathbf{w}^*/\epsilon, b^*/\epsilon]$ satisfies the conditions (E.8). We now calculate the margin between the classes, which is defined as the sum of distances of the points $\mathbf{x}^\circ, \mathbf{x}^\bullet$ from the separating hyperplane $[\mathbf{w}^*/\epsilon, b^*/\epsilon]$. From (E.12)–(E.13), we obtain $(\mathbf{w}^*/\epsilon) \cdot (\mathbf{x}^\circ - \mathbf{x}^\bullet) = 2$, and, after normalizing by $\|\mathbf{w}^*/\epsilon\|$, $(\mathbf{w}^*/\|\mathbf{w}^*\|) \cdot (\mathbf{x}^\circ - \mathbf{x}^\bullet) = 2\epsilon/\|\mathbf{w}^*\|$. Realizing that the distance from any $\mathbf{x}$ to the hyperplane $[\mathbf{w}, b]$ is equal to $(\mathbf{w}/\|\mathbf{w}\|) \cdot \mathbf{x} - b$, we see that the margin is equal to $2\epsilon/\|\mathbf{w}^*\|$. Thus, maximizing the margin $2\epsilon/\|\mathbf{w}\|$ in (E.5) is the same as minimizing $\|\mathbf{w}\|^2/2$ in (E.7), which finishes the proof of the equivalence of problems (E.5) and (E.7).

We note that our choice of the maximum margin hyperplane f^* does not guarantee that f^* is optimal in terms of minimizing the overall risk $R(f)$. There might exist f such that $R_{\mathrm{emp}}(f) > R_{\mathrm{emp}}(f^*)$ and yet $R(f) < R(f^*)$.

E.2.2   Non-separable training set

Now, we consider the case when the training data $(\mathbf{x}_1, \mathbf{y}[1]), \ldots, (\mathbf{x}_l, \mathbf{y}[l])$ cannot be linearly separated by a hyperplane without error. In this case, we would like to find a linear classifier f' that minimizes the number of errors $\sum_{i=1}^{l} u(-\mathbf{y}[i]f'(\mathbf{x}_i))$ on the training set. If we exclude all examples incorrectly classified by f' from the training set, $(\mathbf{x}_1, \mathbf{y}[1]), \ldots, (\mathbf{x}_l, \mathbf{y}[l])$, the training set will become linearly separable and we can find the maximum margin classifier f^*, as we did in the previous section. The classifier f^* has the following important properties. First, its empirical risk on the training set is minimal. Second, it has maximum distance from correctly classified training examples.

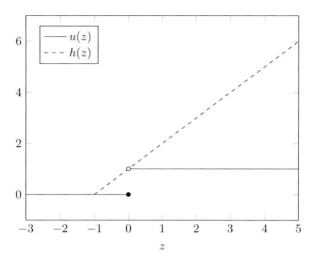

Figure E.3 Comparison of the step function $u(z)$ and its convex majorant hinge loss $h(z) = \max\{0, 1 + z\}$.

The classifier f^* with maximum margin and minimal loss $R(f^*)$ is found by solving the following optimization problem:

$$[\mathbf{w}^*, b^*] = \arg \min_{\mathbf{w}, b, \boldsymbol{\xi}} \frac{1}{2} \|\mathbf{w}\|^2 + C \cdot \sum_{i=1}^{l} u(\boldsymbol{\xi}[i]) \qquad (E.14)$$

subject to constraints

$$\mathbf{y}[i]\left((\mathbf{w} \cdot \mathbf{x}_i) - b\right) \geq 1 - \boldsymbol{\xi}[i], \quad \text{for all } i \in \{1, \dots, l\}, \qquad (E.15)$$

$$\boldsymbol{\xi}[i] \geq 0, \quad \text{for all } i \in \{1, \dots, l\}, \qquad (E.16)$$

for some suitably chosen value of the penalization constant C. The "slack" variables $\boldsymbol{\xi}[i]$ in (E.14) measure the distance of incorrectly classified examples $\mathbf{x}_i$ from the separating hyperplane. Of course, if $\mathbf{x}_i$ is classified correctly, $\boldsymbol{\xi}[i]$ is zero and thus $u(\boldsymbol{\xi}[i]) = 0$.

Unfortunately, the optimization problem (E.14) is NP-complete. The complexity can be significantly reduced by replacing the step function $u(z)$ with the hinge loss function $h(z) = \max\{0, 1 + z\}$. Because $h(z)$ is convex and $h(z) \geq u(z)$ for $z \geq 0$, it transforms (E.14) to a convex quadratic-programming problem

$$[\mathbf{w}^*, b^*] = \arg \min_{\mathbf{w}, b, \boldsymbol{\xi}} \frac{1}{2} \|\mathbf{w}\|^2 + C \cdot \sum_{i=1}^{l} \boldsymbol{\xi}[i] \qquad (E.17)$$

subject to constraints (E.15)–(E.16). Notice that the optimization problem (E.17) minimizes the *overall distance* of incorrectly classified training examples from the hyperplane instead of their number. Support vector machines that classify by solving (E.17) are called soft-margin SVMs (C-SVMs).

Even though the optimization problem (E.17) can be easily solved, in the context of SVMs, it is usually solved in its dual form. The reason for this will become clear once we move to kernelized SVMs in Section E.3. The constrained optimization problem (E.17) can be approached in a standard manner using Lagrange multipliers. Since the constraints are in the form of inequalities, the multipliers must be non-negative (for equality constraints, there is no limit on the multipliers' values). The Lagrangian is

$$L(\mathbf{w}, b, \boldsymbol{\xi}, \boldsymbol{\alpha}, \mathbf{r}) = \frac{1}{2}\mathbf{w}'\mathbf{w} + C\sum_{i=1}^{l}\boldsymbol{\xi}[i] \tag{E.18}$$

$$- \sum_{i=1}^{l}\boldsymbol{\alpha}[i]\left\{\mathbf{y}[i]\left((\mathbf{w}\cdot\mathbf{x}_i) - b\right) - 1 + \boldsymbol{\xi}[i]\right\} - \sum_{i=1}^{l}\mathbf{r}[i]\boldsymbol{\xi}[i] \tag{E.19}$$

with Lagrange multipliers $\boldsymbol{\alpha} = (\alpha[1], \dots, \alpha[l])$ and $\mathbf{r} = (\mathbf{r}[1], \dots, \mathbf{r}[l])$, $\boldsymbol{\alpha} \geq 0$, $\mathbf{r} \geq 0$. A standard result from constraint optimization says that the solution of the problem (E.17) is in the saddle point of the Lagrangian $L(\mathbf{w}, b, \boldsymbol{\xi}, \boldsymbol{\alpha}, \mathbf{r})$ – minimum with respect to variables $(\mathbf{w}, b, \boldsymbol{\xi})$ and maximum with respect to the Lagrange multipliers $(\boldsymbol{\alpha}, \mathbf{r})$.

The conditions for the minimum of $L(\mathbf{w}, b, \boldsymbol{\xi}, \boldsymbol{\alpha}, \mathbf{r})$ at the extrema point (labeled again with superscript $*$) are

$$\left.\frac{\partial L}{\partial \mathbf{w}}\right|_{\mathbf{w}=\mathbf{w}^*} = \mathbf{w}^* - \sum_{i=1}^{l}\alpha[i]\mathbf{y}[i]\mathbf{x}_i = 0, \tag{E.20}$$

$$\left.\frac{\partial L}{\partial b}\right|_{b=b^*} = \sum_{i}^{l}\alpha[i]\mathbf{y}[i] = 0, \tag{E.21}$$

$$\left.\frac{\partial L}{\partial \boldsymbol{\xi}}\right|_{\boldsymbol{\xi}[i]=\boldsymbol{\xi}[i]^*} = C - \alpha[i] - \mathbf{r}[i] = 0, \quad \text{for all } i \in \{1, \dots, l\}. \tag{E.22}$$

After substituting (E.20)–(E.22) into the Lagrangian (E.19), we obtain the formulation of the dual problem

$$\max_{\boldsymbol{\alpha}, \mathbf{r}} L(\boldsymbol{\alpha}, \mathbf{r}) = \sum_{i=1}^{l}\alpha[i] - \frac{1}{2}\sum_{i,j=1}^{l}\alpha[i]\alpha[j]\mathbf{y}[i]\mathbf{y}[j](\mathbf{x}_i\cdot\mathbf{x}_j) \tag{E.23}$$

subject to constraints

$$\sum_{i=1}^{l}\alpha[i]\mathbf{y}[i] = 0, \tag{E.24}$$

$$C \geq \alpha[i] \geq 0, \quad \text{for all } i \in \{1, \dots, l\}. \tag{E.25}$$

Note that the formulation of the dual problem does not contain the Lagrange multipliers $\mathbf{r}[i]$.

The main advantage of solving the dual problem (E.23) over the primal problem (E.17) is that the complexity (measured by the number of free variables) of the dual problem depends on the number of training examples, while the complexity of the primal problem depends on the dimension of the input space $\mathcal{X}$.

After we introduce kernelized SVMs in Section E.3, we will see that the dimension of the primal problem can be much larger (even infinite) than the number of training examples.

Denoting again the solutions of the dual problem (E.23) with superscript $*$, we need to recover the solution of the primal problem (E.17), which is the pair $[\mathbf{w}^*, b^*]$, from $\boldsymbol{\alpha}^* = (\boldsymbol{\alpha}^*[1], \ldots, \boldsymbol{\alpha}^*[l])$. From (E.20), we can easily obtain the hyperplane normal $\mathbf{w}^*$ as

$$\mathbf{w}^* = \sum_{i=1}^{l} \boldsymbol{\alpha}^*[i]\mathbf{y}[i]\mathbf{x}_i. \tag{E.26}$$

The computation of the threshold b^* is more involved and here we include only its most frequently used form without proof,

$$b^* = \frac{1}{|\mathcal{J}|} \sum_{j\in\mathcal{J}}(\mathbf{w}^* \cdot \mathbf{x}_j) - \mathbf{y}[j], \tag{E.27}$$

where $\mathcal{J} = \{i \in \{1, \ldots, l\}|0 < \boldsymbol{\alpha}^*[i] < C\}$. Equation (E.27) can be obtained from the so-called Karush–Kuhn–Tucker conditions for the primal problem [33, 240].

Solving either the primal or dual optimization problem is commonly called training of SVMs. Technically, any optimization library that includes a routine for quadratic programming can be used. Most general-purpose libraries, however, are usually able to solve only small-scale problems. Therefore, we highly recommend using algorithms developed specifically for SVMs, such as LibSVM, `http://www.csie.ntu.edu.tw/~cjlin/libsvm/`.

E.3   Kernelized support vector machines

The linear SVMs described in the previous section can implement only linear decision functions f, which is rarely sufficient for real-world problems. In this section, we extend SVMs to non-linear decision functions. The extension is surprisingly simple.

The main idea is to map the input space $\mathcal{X}$, which is the space where the observed data lives, to a different space $\mathcal{H}$, using a non-linear data-driven mapping $\phi : \mathcal{X} \to \mathcal{H}$, and then find the separating hyperplane in $\mathcal{H}$ (for linear SVMs described above, $\mathcal{X} = \mathcal{H}$). The non-linearity introduced through the mapping ϕ allows implementation of a non-linear decision boundary in the input space $\mathcal{X}$ as a linear decision boundary in $\mathcal{H}$.

While the input space $\mathcal{X}$ is usually given by the nature of the application (it is the space of features extracted from images), the space $\mathcal{H}$ can be freely chosen as long as it satisfies the following two conditions.

1. $\mathcal{H}$ must be a Hilbert space (a complete space endowed with an inner product $\langle \cdot, \cdot \rangle_{\mathcal{H}}$).

2. There must exist a positive-definite[1] function $k : \mathcal{X} \times \mathcal{X} \to \mathbb{R}$ called a kernel, so that $\forall \mathbf{x}, \mathbf{x}' \in \mathcal{X}$, $k(\mathbf{x}, \mathbf{x}') = \langle \phi(\mathbf{x}), \phi(\mathbf{x}') \rangle_{\mathcal{H}}$. The kernel function $k : \mathcal{X} \times \mathcal{X} \to \mathbb{R}$ can be understood as a similarity measure on the input space $\mathcal{X}$.

These conditions ensure that the maximum margin hyperplane exists in $\mathcal{H}$ and that it can be found by solving the dual problem (E.23).

The space $\mathcal{H}$ is a function space obtained by completing the set of all functions on $\mathcal{X}$ that are linear combinations

$$f(x) = \sum_i \mathbf{a}[i] k(\mathbf{x}_i, x) \tag{E.28}$$

whose inner product is defined as

$$\langle f(x), g(x) \rangle_{\mathcal{H}} = \sum_{i,j} \mathbf{a}[i] \mathbf{b}[j] k(\mathbf{x}_i, \mathbf{x}_j), \tag{E.29}$$

where $g(x) = \sum_j \mathbf{b}[j] k(\mathbf{x}_j, x)$. Note that the Hilbert space is driven by the data $\mathbf{x}_i$. The positive definiteness of the kernel guarantees that (E.29) is an inner product.

One of the most popular kernel functions is the Gaussian kernel

$$k(\mathbf{x}, \mathbf{x}') = \exp\left(-\gamma \|\mathbf{x} - \mathbf{x}'\|^2\right), \tag{E.30}$$

where $\gamma > 0$ is a parameter controlling the width of the kernel and $\|\mathbf{x}\|$ is the Euclidean norm of $\mathbf{x}$. The Hilbert space $\mathcal{H}$ induced by the Gaussian kernel has infinite dimension. Other popular kernels are the linear kernel $k(\mathbf{x}, \mathbf{x}') = \mathbf{x} \cdot \mathbf{x}'$, the polynomial kernel of degree d, defined as $k(\mathbf{x}, \mathbf{x}') = (r + \gamma \mathbf{x} \cdot \mathbf{x}')^d$, and the sigmoid kernel $k(\mathbf{x}, \mathbf{x}') = \tanh(r + \gamma \mathbf{x} \cdot \mathbf{x}')$, both with two parameters γ, r.

Non-linear SVMs are implemented in the same way as in Section E.2.2, with the exception that now all operations should be carried out in the Hilbert space $\mathcal{H}$. Because the dimensionality of the feature space $\mathcal{H}$ can be infinite, we now must use the dual optimization problem (E.23) because its dimensionality is determined by the cardinality of the training set. Fortunately, because $\mathbf{x}_i$ always appears in the inner product, we can simply replace the inner product with its kernel-based expression $k(\mathbf{x}, \mathbf{x}') = \langle \phi(\mathbf{x}), \phi(\mathbf{x}') \rangle_{\mathcal{H}}$. This substitution, called the "kernel trick," is possible in all algorithms where the calculation with data appears exclusively in the form of inner products.

---

[1] $k : \mathcal{X} \times \mathcal{X} \to \mathbb{R}$ is positive definite if and only if $\forall n \geq 1$, $\forall \mathbf{x}_1, \ldots, \mathbf{x}_n \in \mathcal{X}$, $\forall \mathbf{c}[1], \ldots, \mathbf{c}[n] \in \mathbb{R}$, $\sum_{i,j=1}^{n} \mathbf{c}[i] \mathbf{c}[j] k(\mathbf{x}_i, \mathbf{x}_j) \geq 0$.

The dual optimization problem (E.23) can thus be rewritten as follows:

$$\max_{\boldsymbol{\alpha},\mathbf{r}} L(\boldsymbol{\alpha},\mathbf{r}) = \sum_{i=1}^{l} \boldsymbol{\alpha}[i] - \frac{1}{2}\sum_{i,j=1}^{l} \boldsymbol{\alpha}[i]\boldsymbol{\alpha}[j]\mathbf{y}[i]\mathbf{y}[j]\,\langle\phi(\mathbf{x}_i),\phi(\mathbf{x}_j)\rangle_{\mathcal{H}} \qquad \text{(E.31)}$$

$$= \sum_{i=1}^{l} \boldsymbol{\alpha}[i] - \frac{1}{2}\sum_{i,j=1}^{l} \boldsymbol{\alpha}[i]\boldsymbol{\alpha}[j]\mathbf{y}[i]\mathbf{y}[j]k\,(\mathbf{x}_i,\mathbf{x}_j), \qquad \text{(E.32)}$$

with constraints

$$\sum_{i=1}^{l} \boldsymbol{\alpha}[i]\mathbf{y}[i] = 0, \qquad \text{(E.33)}$$

$$C \geq \boldsymbol{\alpha}[i] \geq 0, \forall i \in \{1,\dots,l\}. \qquad \text{(E.34)}$$

As the constraints do not contain any vector manipulation, they stay the same. In the equation $\mathbf{w}^* = \sum_{i=1}^{l} \boldsymbol{\alpha}^*[i]\mathbf{y}[i]\phi(\mathbf{x}_i)$ of the optimal hyperplane $\mathbf{w}^*$, all manipulations are carried out in $\mathcal{H}$ and we cannot convert it to the input space $\mathcal{X}$. Fortunately, we do not need to know $\mathbf{w}^*$ explicitly because the decision function (E.3) can be rewritten as

$$f(\mathbf{x}) = \text{sign}\left(\langle\mathbf{w}^*,\phi(\mathbf{x})\rangle_{\mathcal{H}} - b^*\right) \qquad \text{(E.35)}$$

$$= \text{sign}\left(\sum_{j=1}^{l} \boldsymbol{\alpha}^*[j]\mathbf{y}[j]\,\langle\phi(\mathbf{x}_j),\phi(\mathbf{x})\rangle_{\mathcal{H}} - b^*\right) \qquad \text{(E.36)}$$

$$= \sum_{j=1}^{l} \boldsymbol{\alpha}^*[j]\mathbf{y}[j]k\,(\mathbf{x}_j,\mathbf{x}) - b^*. \qquad \text{(E.37)}$$

By the same mechanism, equation (E.27) for the threshold b^* becomes

$$b^* = \frac{1}{|\mathcal{J}|}\sum_{j\in\mathcal{J}}(\mathbf{w}^* \cdot \phi(\mathbf{x}_j)) - \mathbf{y}[j] = \frac{1}{|\mathcal{J}|}\sum_{j\in\mathcal{J}}\sum_{i=1}^{l}\boldsymbol{\alpha}^*[i]\mathbf{y}[i]k\,(\mathbf{x}_i,\mathbf{x}_j) - \mathbf{y}[j], \quad \text{(E.38)}$$

with $\mathcal{J} = \{i \in \{1,\dots,l\}|0 < \boldsymbol{\alpha}^*[i] < C\}$ as before.

Soft-margin SVMs were proved to converge to an optimal classifier minimizing the risk functional $R(f)$ as the number of training examples increases [223].

E.4   Weighted support vector machines

In steganalysis, it is important to control the false-positive rate of the steganalyzer. To allow SVMs to adjust the false alarms, weighted SVMs use different penalization coefficients for false positives and missed detections.

Denoting $\mathcal{I}^-, \mathcal{I}^+$ the indices of negative and positive examples with $\mathbf{y}[i] = -1$ and $\mathbf{y}[i] = 1$, the primal problem of weighted SVMs accepts the form

$$\min_{\mathbf{w}, b, \boldsymbol{\xi}} \frac{1}{2} \|\mathbf{w}\|_{\mathcal{H}}^2 + C^- \sum_{i \in \mathcal{I}^-} \boldsymbol{\xi}[i] + C^+ \sum_{i \in \mathcal{I}^+} \boldsymbol{\xi}[i] \tag{E.39}$$

subject to constraints

$$\mathbf{y}[i] \left(\langle \mathbf{w}, \mathbf{x}_i \rangle_{\mathcal{H}} - b \right) \geq 1 - \boldsymbol{\xi}[i], \quad \text{for all } i \in \{1, \ldots, l\}, \tag{E.40}$$

$$\boldsymbol{\xi}[i] \geq 0, \quad \text{for all } i \in \{1, \ldots, l\}. \tag{E.41}$$

We denoted by $\| \cdot \|_{\mathcal{H}}$ the norm in the space $\mathcal{H}$. If we compare the original primal problem with equal costs of both detection errors (E.17) with the new formulation (E.39), we can see how differently the costs are expressed. By adjusting the penalization costs C^+ and C^-, we can now put more importance on one or the other error type.

Following the same steps as in Section E.2.2, the dual form of (E.39) can be derived:

$$\max_{\boldsymbol{\alpha}, \mathbf{r}} L(\boldsymbol{\alpha}, \mathbf{r}) = \sum_{i=1}^{l} \boldsymbol{\alpha}[i] - \frac{1}{2} \sum_{i,j=1}^{l} \boldsymbol{\alpha}[i] \boldsymbol{\alpha}[j] \mathbf{y}[i] \mathbf{y}[j] k(\mathbf{x}_i, \mathbf{x}_j), \tag{E.42}$$

subject to constraints

$$\sum_{i=1}^{l} \boldsymbol{\alpha}[i] \mathbf{y}[i] = 0, \tag{E.43}$$

$$C^+ \geq \boldsymbol{\alpha}[i] \geq 0, \quad \text{for all } i \in \mathcal{I}^+, \tag{E.44}$$

$$C^- \geq \boldsymbol{\alpha}[i] \geq 0, \quad \text{for all } i \in \mathcal{I}^-. \tag{E.45}$$

The dual problem of weighted SVMs (E.42) is again a convex quadratic-programming problem, which is almost identical to (E.31), with the only exception that now the Lagrange multipliers $\boldsymbol{\alpha}$ are bounded by different constants depending on what training example they correspond to (e.g., whether the training example is a cover or a stego image).

The equation for b^* becomes

$$b^* = \frac{1}{|\mathcal{J}^-| + |\mathcal{J}^+|} \sum_{j \in \mathcal{J}^- \cup \mathcal{J}^+} \sum_{i=1}^{l} \boldsymbol{\alpha}^*[i] \mathbf{y}[i] k(\mathbf{x}_i, \mathbf{x}_j) - \mathbf{y}[j], \tag{E.46}$$

where $\mathcal{J}^- = \{i \in \mathcal{I}^- | 0 < \boldsymbol{\alpha}^*[i] < C^-\}$ and $\mathcal{J}^+ = \{i \in \mathcal{I}^+ | 0 < \boldsymbol{\alpha}^*[i] < C^+\}$.

The decision function of a weighted SVM remains unchanged,

$$f(\mathbf{x}) = \text{sign} \left(\sum_{j=1}^{l} \boldsymbol{\alpha}^*[j] \mathbf{y}[j] k(\mathbf{x}_j, \mathbf{x}) - b^* \right). \tag{E.47}$$

E.5   Implementation of support vector machines

The kernel type and its parameters as well as the penalization parameter(s) have a great impact on the accuracy of the classifier. Unfortunately, there is no universal methodology regarding how to best select them. We provide some guidelines that often give good results.

E.5.1   Scaling

First, the input data needs to be preprocessed. Assuming $\mathcal{X} = \mathbb{R}^n$, which is the case for steganalysis, the input data is scaled so that all elements of vectors $\mathbf{x}_i$ from the training set are in the range $[-1, +1]$. This scaling is very important as it ensures that features with large numeric values do not dominate features with small values. It also increases the numerical stability of the learning algorithm.

E.5.2   Kernel selection

The next step is the selection of a proper kernel. Unless we have some side-information about the problem we are facing, the Gaussian kernel $k(\mathbf{x}, \mathbf{x}') = \exp(-\gamma\|\mathbf{x} - \mathbf{x}'\|^2)$ is typically a good first choice. This kernel is flexible enough to solve many problems, yet it has only one free parameter, in comparison with polynomial or sigmoid kernels, which depend on more free parameters. Moreover, for all values of $\gamma > 0$, the Gaussian kernel is positive definite.

E.5.3   Determining parameters

Before training, we need to determine the kernel parameters (the width γ if we use the Gaussian kernel) and the penalization parameter(s) (C^+, C^-). A common way to find (C^+, C^-, γ) is to carry out an exhaustive search on predefined points from a grid $\mathcal{G}$. At each point $(C^+, C^-, \gamma) \in \mathcal{G}$, we train the SVM and estimate its performance on unknown data. The estimated probabilities of false positives and missed detection of an SVM trained with parameters (C^+, C^-, γ) will be denoted as $\hat{P}_{\mathrm{FA}}(C^+, C^-, \gamma)$ and $\hat{P}_{\mathrm{MD}}(C^+, C^-, \gamma)$, respectively. The parameters (C^+, C^-, γ) are then selected as a point from the grid $\mathcal{G}$ using a Bayesian or Neyman–Pearson approach (see below).

The requirement of estimating the performance on unknown data is very important. It is easy to find (C^+, C^-, γ) so that the error on the training set is zero, but this classifier will most likely exhibit a high error rate on the unknown data (this problem is called overtraining or overfitting).

A popular way of estimating $\hat{P}_{\mathrm{FA}}(C^+, C^-, \gamma)$ and $\hat{P}_{\mathrm{MD}}(C^+, C^-, \gamma)$ on unknown data is k-fold cross-validation. The available training examples are divided into k subsets of approximately equal size. Then, the union of $k-1$ subsets is used as a training set to train the SVM, while the remaining subset is used to estimate the error on "unknown" examples. This is repeated k times, each time with different subsets. k-fold cross-validation usually gives estimates of error close to the error we can expect on truly unknown data.

When the costs of both errors are known and equal to w^+ and w^- for an error on positive and negative classes, we can use weighted SVMs with the Bayesian approach and minimize the total cost

$$w^- p^- P_{\text{FA}} + w^+ p^+ P_{\text{MD}}, \tag{E.48}$$

where p^- and p^+ are a priori probabilities of the negative and positive classes. In this case, the search for the parameters can be described as

$$(C^+, C^-, \gamma) = \arg\min w^- p^- \hat{P}_{\text{FA}}(C^+, C^-, \gamma) + w^+ p^+ \hat{P}_{\text{MD}}(C^+, C^-, \gamma), \tag{E.49}$$

where the arg min is taken over all $(C^+, C^-, \gamma) \in \mathcal{G}$,

$$\mathcal{G} = \left\{ (w^- p^- 2^i, w^+ p^+ 2^i, 2^k) | i \in \{-3, \ldots, 9\}, \ k \in \{-5, \ldots, 3\} \right\}. \tag{E.50}$$

Note that even though the grid $\mathcal{G}$ has 3 dimensions, its effective dimension is 2 because the ratio $C^+/C^- = w^+ p^+/(w^- p^-)$ must stay constant.

In a Neyman–Pearson setting, we impose an upper bound on the probability of false alarms, $P_{\text{FA}} \leq \epsilon_{\text{FA}} < 1$, and minimize the probability of missed detection P_{MD}. The search for (C^+, C^-, γ) becomes

$$(C^+, C^-, \gamma) = \arg\min \hat{P}_{\text{MD}}(C^+, C^-, \gamma), \tag{E.51}$$

where the arg min is carried over all $(C^+, C^-, \gamma) \in \mathcal{G}$ satisfying $\hat{P}_{\text{FA}}(C^+, C^-, \gamma) \leq \epsilon_{\text{FA}}$,

$$\mathcal{G} = \left\{ (2^i, 2^j, 2^k) | i, j \in \{-3, \ldots, 9\}, \ k \in \{-5, \ldots, 3\} \right\}. \tag{E.52}$$

In this case, the grid $\mathcal{G}$ has the effective dimension 3, which makes the search computationally expensive. Frequently, suboptimal search [44] is used to alleviate the computational complexity.

E.5.4   Final training

After selecting the kernel and determining its parameters including the penalization parameter C, we use the whole training set to train the final SVM. This process, which involves solving a quadratic-programming problem, will determine the vector $(\boldsymbol{\alpha}^*[1], \ldots, \boldsymbol{\alpha}^*[l])$ and b^*. The decision function of the final SVM is

$$f(\mathbf{x}) = \text{sign}\left(\sum_{j=1}^{l} \boldsymbol{\alpha}^*[j] \mathbf{y}[j] k(\mathbf{x}_j, \mathbf{x}) - b^* \right). \tag{E.53}$$

E.5.5   Evaluating classification performance

The probability of detection as a function of the probability of false alarms, $P_D(P_{FA})$, is called the receiver operating characteristic (see Section D.1.1) and it is often used to visualize the performance of detectors, including those implemented as SVM classifiers. Since the SVM depends on several parameters, we essentially have a family of parametrized classifiers. The usual method to draw an ROC curve for a specific SVM is to train with fixed values of the penalization parameters, C^-, C^+, and the kernel width γ, and change the threshold b^*. The ROC curve is then obtained in a parametric form

$$P_{FA} = P_{FA}(b^*), \qquad (E.54)$$
$$P_D = P_D(b^*), \ b^* \in \mathbb{R}. \qquad (E.55)$$

It is clear that by changing the threshold b^*, we can achieve all values of $P_{FA} \in [0, 1]$. While this method to draw the ROC curve is simple and fast, it does not really reflect the performance of the SVM. For example, if the penalization parameters and the kernel width were determined under a Neyman–Pearson setting for a given bound on false alarms, ϵ_{FA}, the ROC obtained this way is suboptimal in the sense that the probability of detection $P_D(\epsilon'_{FA})$, for $\epsilon'_{FA} \neq \epsilon_{FA}$, may be lower than what we would obtain if we trained for ϵ'_{FA}. We could alternatively draw the ROC by varying the penalization parameters as well as the kernel parameter. However, such an approach is often computationally very expensive and thus almost never used in practice. An interesting approach with reduced computational cost for fixed kernel while varying the penalization parameters C^+ and C^- was proposed in [13].

Notation and symbols

The mathematical notation in this book was chosen to be compact and visually distinct to make it easier for the reader to interpret formulas. For this reason, we adhere to short names of variables, sets, and functions rather than long descriptive names similar to variable naming in programming as the latter would lead to long and rather unsightly mathematical formulas. The price paid for the short variable names is an occasional reuse of some symbols across the text. However, the author strongly believes that the meaning of the symbols should always be clear and unambiguous from the context. The most frequently occurring key symbols, such as the relative message length α, change rate β, cover object $\mathbf{x}$, or stego object $\mathbf{y}$, are used exclusively and never reused. Whenever possible, variable names are chosen mnemotechnically by the first letter of the concept they stand for, such as $\mathbf{h}$ for histogram, $\mathcal{A}$ for alphabet, etc. In some cases, however, if there exists a widely accepted notation, rather than coining a new symbol, we accept the established notation. A good example is the parity-check matrix, which is almost exclusively denoted as $\mathbf{H}$ in the coding literature.

Everywhere in this book, vectors and matrices of predefined dimensions are printed in boldface with indices following the symbol in square brackets. Thus, the ijth element of the parity-check matrix $\mathbf{H}$ is $\mathbf{H}[i,j]$ and the jth element of the histogram is $\mathbf{h}[j]$. This notation generalizes to higher-dimensional arrays. The transpose of a matrix is denoted with a prime: $\mathbf{H}'$ is the transpose of $\mathbf{H}$. Sequences of vectors or matrices will be indexed with a subscript. Thus, $\mathbf{f}_k$ will be a sequence of vectors $\mathbf{f}$. If it is meaningful and useful, we may sometimes arrange a sequence of vectors into a matrix and address individual vector elements using two indices, e.g., $\mathbf{f}[i,k]$. Often, quantities may depend on parameters, such as the relative payload α. The histogram $\mathbf{h}_\alpha[j]$ stands for the histogram of the stego image embedded with relative payload α. If the vector or matrix quantities depend on other parameters or indices, we may put them as superscripts. An example would be the histogram of the DCT mode (k,l) from a JPEG stego image, $\mathbf{h}_\alpha^{(k,l)}[j]$. Strings of symbols will also be typeset as vectors. The length of string $\mathbf{s}$ is $|\mathbf{s}|$.

Random variables are typeset in sans serif font x, y. A vector/matrix random variable will then be boldface sans serif, $\mathbf{x}, \mathbf{y}$. When a random variable x follows probability distribution f, we will write $\mathsf{x} \sim f$. By a common abuse of notation, we will sometimes write $\mathbf{x} \sim f$, meaning that $\mathbf{x}$, whose realization is vector $\mathbf{x}$, follows the distribution f. When the probability distribution is clear from the context or unspecified, we will write $\Pr\{\cdot\}$, where the dot stands for an expression involving a random variable. For example, $\Pr\{\mathsf{x} \leq x\}$ stands for the probability that random variable x attains the value x or smaller. The expected value, variance, and covariance will be denoted as $E[\mathsf{x}]$, $\mathrm{Var}[\mathsf{x}]$, and $\mathrm{Cov}[\mathsf{x}, \mathsf{y}]$ when it is clear from the context with respect to what probability distribution we are taking the expected value or variance. Sometimes, we may stress the probability distribution by writing $E_{\mathsf{x} \sim f}[\mathsf{y}(x)]$ or simply $E_{\mathsf{x}}[\mathsf{y}(x)]$, which is the expected value of y when x follows the distribution f of variable x. The sample mean will be denoted with a bar, $\bar{\mathsf{x}}$.

Calligraphic font will be used for sets or regions, $\mathcal{X}, \mathcal{Y}$, with $|\mathcal{X}|$ denoting the cardinality of $\mathcal{X}$.

Estimated quantities will be denoted with a hat, meaning that $\hat{\theta}$ is an estimator of θ. A tilde is often used for quantities subjected to distortion, e.g., $\tilde{\mathbf{y}}$ is a noisy version of $\mathbf{y}$.

In this book, we use Landau's big O symbol $f = O(g)$ meaning that $|f(x)| < Ag(x)$ for all x and a constant $A > 0$. The asymptotic equality $f(x) \approx g(x)$ holds for two functions if and only if $\lim_{x \to \infty} f(x)/g(x) = 1$. We will also use the symbol $\approx$ with scalars, in which case it stands for approximate equality (e.g., $\alpha \approx 1$ or $k \approx m$).

Among other, less common symbols, we point out the eXclusive OR operation on bits, which we denote with $\oplus$. Concatenation of strings is denoted as $\&$. We reserve $\star$ for convolution, and $\lfloor x \rfloor$, $\lceil x \rceil$ for rounding down and up, respectively. The function $\mathrm{round}(x)$ is rounding to the closest integer, while $\mathrm{trunc}(x)$ is the operation of truncation to a finite dynamic range (e.g., to bring the real values of pixels back to the $0, \ldots, 255$ range of integers).

Below, we provide definitions for the most common symbols used throughout the text.

α	relative payload
$A(\tilde{\mathbf{y}}\|\mathbf{y})$	probability of receiving $\tilde{\mathbf{y}}$ when sending $\mathbf{y}$ through a noisy channel
$\mathbf{A}[., i]$	ith column of matrix $\mathbf{A}$
$\mathbf{A}[j, .]$	jth row of matrix $\mathbf{A}$
$b(\hat{\theta})$	bias of estimator $\hat{\theta}$
$\mathcal{B}$	bin
$\mathcal{B}(\mathbf{x}, R)$	ball of radius R centered at $\mathbf{x}$

β	change rate
$B(p)$	Bernoulli random variable when probability of 0 is p
$Bi(n,p)$	binomial random variable with n trials of probability p
B_γ	measure of blockiness with $\gamma = 1, 2$
χ_d^2	chi-square distribution with d degrees of freedom
$\mathbf{c}[j] = (\mathbf{r}[j], \mathbf{g}[j], \mathbf{b}[j])$	an RGB color
$\mathcal{C}$	code
$\mathcal{C}(\mathbf{s})$	coset corresponding to syndrome $\mathbf{s}$
$\mathcal{C}(\mathbf{x}_{\mathrm{inv}})$	set of all covers with invariant component $\mathbf{x}_{\mathrm{inv}}$
$\mathcal{C}_i, \mathcal{E}_i, \mathcal{O}_i$	trace sets used in structural steganalysis
C_{steg}	steganographic capacity
$\mathbf{C}$	co-occurrence matrix, covariance matrix
$\delta(x)$	Kronecker delta, $$\delta(x) = \begin{cases} 1 & \text{when } x = 0 \\ 0 & \text{when } x \neq 0. \end{cases}$$
$\mathbf{D}$	matrix of quantized DCT coefficients, also a matrix used in wet paper codes
$\mathbf{d}$	matrix of unquantized DCT coefficients
$D_{\mathrm{KL}}(P\|Q)$	Kullback–Leibler (KL) divergence between probability distributions P and Q
$d_{\mathrm{H}}(\mathbf{x}, \mathbf{y})$	Hamming distance between two words
$d_\gamma(\mathbf{x}, \mathbf{y})$	distortion measure $$d_\gamma(\mathbf{x}, \mathbf{y}) = \sum_{i=1}^{n} (\mathbf{x}[i] - \mathbf{y}[i])^\gamma$$

$d_\rho(\mathbf{x}, \mathbf{y})$	measure of embedding impact with profile $\boldsymbol{\rho}[i]$

$$d_\rho(\mathbf{x}, \mathbf{y}) = \sum_{i=1}^{n} \boldsymbol{\rho}[i] \left(1 - \delta(\mathbf{x}[i] - \mathbf{y}[i])\right)$$

$d_{\mathrm{RGB}}(\mathbf{c}, \mathbf{c}')$	distance between two RGB colors
d^2	deflection coefficient
DCT	discrete cosine transform
$\mathrm{diag}(\mathbf{x})$	diagonal matrix with diagonal $\mathbf{x}$
$e, \underline{e}$	embedding efficiency, lower embedding efficiency
$\mathbf{e}(\mathbf{s})$	coset leader of coset corresponding to syndrome $\mathbf{s}$
$\mathbf{e}[i]$	quantization error at element i, noise residual
$\mathrm{Erf}(x)$	error function
$\mathbf{f}$	feature vector
$\mathbb{F}_2$	finite field containing two elements
$\mathbb{F}_q$	finite field containing q elements
$\Phi(x)$	cumulative distribution function of standard normal variable, also a mapping used in targeted steganalysis
$\mathbf{G}$	generator matrix
$\mathbf{g}^{(r)}$	dual histogram for value r
$\Gamma(x)$	gamma function
γ	gamma factor in gamma correction, parameter in d_γ, generic threshold
$h(\mathbf{x})$	hash (message digest) function
$\mathbf{h}$	histogram
$\mathbf{h}^{(k,l)}$	histogram of the (k, l)th DCT mode (in a JPEG file)

$H(\mathsf{x})$	entropy of random variable x
$H_{\min}(\mathsf{x})$	minimal entropy of random variable x
$H(\mathsf{x}\vert\mathsf{y})$	conditional entropy of x given y
$H(x),\ H^{-1}(x)$	binary entropy function, its inverse
$H_q(x),\ H_q^{-1}(x)$	q-ary entropy function, its inverse
$\mathbf{H}$	parity-check matrix
$\mathsf{H}_0,\ \mathsf{H}_1$	null and alternative hypothesis
$\mathcal{H}_p$	binary Hamming code with codimension p
$\mathbf{I}_k$	$k \times k$ identity matrix
$I(\beta)$	Fisher information w.r.t. β
$I(\mathsf{x};\mathsf{y})$	mutual information between random variables x and y
IDCT	inverse DCT transform
$J_1,\ J_2$	JPEG images
$\mathbf{k} \in \mathcal{K}$	secret stego key from the set of all stego keys
$k(x,y)$	kernel (used in support vector machines)
Λ	lattice
$\log$	logarithm to the base $e = 2.718281828045\ldots$
$L(\mathbf{x})$	likelihood ratio for measurements $\mathbf{x}$
$L_2(a,b)$	vector space of all quadratically integrable functions on interval $[a,b]$
$\mu,\ \boldsymbol{\mu}[k],\ \boldsymbol{\mu}_{\mathrm{c}}[k]$	mean, kth moment, kth central moment
$\mathbf{m} \in \mathcal{M}$	message from the set of all messages

M_x	moment-generating function of random variable x
$\mathrm{MSE}(\hat\theta)$	mean-square error of estimator $\hat\theta$
$N(\mu,\sigma^2)$	Gaussian random variable with mean μ and variance σ^2
n_{01}, n_{AC}, n_{c}	number of DCT coefficients different from 0 and 1, number of AC DCT coefficients, number of bits to represent a color
$O_{h,b}$	cover oracle drawing the next b cover bits based on the history of h cover bits
$\pi(x)$	symbol-assignment function
$\mathcal{P}$	set of pixel pairs
Ψ	isomorphism
$\mathbf{p}[i]$	probability mass function
$\mathbf{p}_\mathrm{v}[i]$	p-value (in histogram attack)
P_c, p_c, P_s, p_s	distributions of cover and stego images
Path	random path through image
P_E	minimum average probability of error $(P_{\mathrm{MD}}+P_{\mathrm{FA}})/2$
P_D, P_{MD}, P_{FA}	probability of detection, missed detection, and false alarm
$P_\mathrm{D}^{-1}\left(\frac{1}{2}\right)$	false-alarm probability at 50% detection
$Q(x)$	the complementary cumulative distribution function of a standard normal variable
$\mathbf{Q}[i,j]$	quantization matrix
Q_Λ	quantizer to lattice Λ
$Q(\mathbf{y},u\vert\mathbf{x})$	covert channel (matrix of conditional probabilities)
q_f	quality factor
q	size of a q-ary alphabet

$\boldsymbol{\rho}[i]$	embedding impact at pixel i
$\boldsymbol{\rho}(\mathsf{x}, \mathsf{y})$	correlation coefficient for random variables x and y
$\mathbf{r_H, r_V, r_D}$	horizontal, vertical, and diagonal noise residuals
R	covering radius
R_{a}	average distance to code
$R(f)$	risk functional
$\mathcal{R}_0$	region of acceptance of null hypothesis
$\mathcal{R}_1$	critical region (acceptance of alternative hypothesis)
$\mathbf{s}$	syndrome
$\mathcal{S}$	set of changeable (dry) pixels in writing on wet paper
$S_{\mathbf{f}}(\mathbf{x})$	sparsity measure defined by the feature set $\mathbf{f}$
θ	unknown parameter to be estimated
$\vartheta(\mathbf{x}, \mathbf{y})$	number of embedding changes
$t(\mathbf{B})$	texture measure of block $\mathbf{B}$
σ^2	variance
V	variation
$V_q(n, r)$	volume of a ball with radius r in an n-dimensional space
$U(a, b)$	uniform distribution on the interval $[a, b]$
$\mathcal{V}$	Voronoi region, also an abstract vector space
$\mathbf{W}$	noise residual
W	Wiener filter

$w(\mathbf{x})$	Hamming weight of $\mathbf{x} \in \{0,1\}^n$
$\mathbf{w}_\theta$	weighted stego image
$\mathbf{x} \in \mathcal{C}$	cover image from the set of all possible covers $\mathcal{C}$
$\mathbf{x}_{\mathrm{inv}}, \mathbf{x}_{\mathrm{emb}}$	random variables describing the cover component that is invariant with respect to embedding and the one used for embedding (in model-based steganography)
$\mathcal{X}, \mathcal{Y}, \mathcal{Z}, \mathcal{V}, \mathcal{W}$	primary sets used in Sample Pairs Analysis
$\mathbf{y}$	stego image
Y, C_r, C_b	luminance and two chrominance signals used in JPEG
$\mathbf{Z}(\mathbf{c}_1, \mathbf{c}_2)$	color cut for colors $\mathbf{c}_1$, $\mathbf{c}_2$ (in Pairs Analysis)
$\mathbb{Z}, \mathbb{R}$	the set of all integers, real numbers
$\mathbb{Z}_M$	finite cyclic group of order M
$\{0,1\}^\star$	bit strings of arbitrary length
$[a,b], (a,b)$	closed, open interval

Operations and relations

The following operations are used throughout the text:

$\rightarrow$	mapping
$\xrightarrow{D}$	convergence in distribution
$\xrightarrow{P}$	convergence in probability
$\triangleq$	symbol used in definitions
$\sim$	$x \sim f$, random variable x follows distribution f
$\approx$	approximate equality for scalars, asymptotic equality for functions

$\&$	concatenation of strings
$\oplus$	XOR (eXclusive OR)
$|\mathcal{A}|$	cardinality of set $\mathcal{A}$
$|\mathbf{s}|$	length of string $\mathbf{s}$
$\langle \mathbf{x}, \mathbf{y} \rangle$	inner (dot) product between vectors in abstract vector spaces
$\|\mathbf{x}\| = \sqrt{\langle \mathbf{x}, \mathbf{x} \rangle}$	norm of vector $\mathbf{x}$
$\mathbf{x} \cdot \mathbf{y}$	dot product between vectors in Euclidean space
$\lceil x \rceil$	rounding to the closest integer larger than or equal to x
$\lfloor x \rfloor$	rounding down to the closest integer smaller than or equal to x
Emb, Ext	embedding and extraction algorithms (mappings)
PEmb, PExt	probabilistic embedding and extraction algorithms (mappings)
$\mathrm{LSBflip}(x)$	operation that flips the LSB of integer x
$\mathrm{LSB}(x)$	the least significant bit of integer x
$\mathrm{LSB}_{\mathrm{F5}}(x)$	the least significant bit of integer x as redefined in the F5 algorithm
$\mathrm{round}(x)$	rounding x to the closest integer
$\mathrm{sign}(x)$	sign of x
$\mathrm{trunc}(x)$	truncating x to a finite dynamic range. For an 8-bit range,

$$\mathrm{trunc}(x) = \begin{cases} 255 & x > 255 \\ x & x \in [0, 255] \\ 0 & x < 0 \end{cases}$$

Glossary

± 1 **embedding.** A *steganographic method* in which message bits are embedded by randomly modifying *cover elements* by at most 1 to match the *LSB* with the message bit (also called *LSB matching*).

–F5. A variation of the *steganographic algorithm F5* in which the *embedding operation* of decreasing the absolute value of the *DCT* coefficient is replaced with increasing its absolute value.

ABG. Anti-Blooming Gate built into an imaging sensor to bleed off overflow from saturated *pixels*.

AC coefficient. *DCT* coefficient other than the *DC term* (from alternating current).

Achievable rate. *Relative payload* that can be undetectably embedded in a *cover source* using a given *steganographic method*.

Acrostics. *Linguistic steganography.*

A/D. Analog/Digital.

Adaptive
  – **selection channel.** A *selection channel* determined by the *cover* content.
  – **steganography.** *Steganography* in which the data-hiding process is dependent on the *cover* content.

Additive color model. Model of color in which color is created by adding several base colors, such as red, green, and blue.

Adjacency histogram. Histogram of adjacent *cover elements*.

Advantage. Success rate of a *warden* (judge) in detecting steganographic content.

AES. Advanced Encryption Standard. A *symmetric cryptosystem*.

ALE. Amplitude of Local Extrema. A *steganalytic method* for detection of ± 1 *embedding*.

Alice. A fictitious character from the *prisoners' problem* who secretly communicates with *Bob*.

Alphabet. A set of symbols used to construct *codes*.

Amplifier noise. Image noise component due to signal amplification.

APS. Active Pixel Sensor.

Arithmetic coder. A lossless compression algorithm.

Attack. An algorithm whose goal is to detect the usage of *steganography*.

AUC/AUR. Two acronyms for Area Under *ROC* Curve.

Average distance to code. The average *Hamming distance* from a randomly (uniformly) selected *word* to a *code*.

Ball (of radius r and center $\mathbf{x}$). A set of points at distance at most r from $\mathbf{x}$.

Basis pattern. One of the 64 patterns forming the *discrete cosine transform* (DCT).

Batch steganography. Steganographic communication by splitting *payload* into multiple *cover objects*.

Bayer pattern. A popular arrangement of red, green, and blue *color filters* allowing a sensor to register color.

BCH. A class of parametrized error-correcting codes invented by Bose, Ray-Chaudhuri, and Hocquenghem.

Benchmarking. A procedure aimed to compare and evaluate systems.

Bernoulli random variable. A random variable $B(p)$ reaching values in $\{0, 1\}$ with probabilities p and $1 - p$.

Between-image error. A component of the estimation error of *quantitative steganalyzers* specific to each image.

Bias. A systematic *estimator* error, $b(\theta) = E[\hat{\theta}] - \theta$, where θ is a parameter that is being estimated.

Binary
  – **arithmetic.** Arithmetic in *finite field* $\{0, 1\}$.
  – **classification.** A *classification* to two classes.
  – **entropy function.** *Entropy* of a *Bernoulli random variable* $B(x)$, $H(x) = -x \log x - (1 - x) \log(1 - x)$.

Blind steganalysis. An approach to *steganalysis* whose aim is to detect an arbitrary *steganographic method*.

Blockiness. Sum of discontinuities at boundaries of 8×8 blocks in a decompressed *JPEG* file.

Blooming. An imaging artifact caused by charge overflow from saturated *pixels*.

BMP. Bitmap raster image format.

Bob. A fictitious character from the *prisoners' problem* who secretly communicates with *Alice*.

Borel set. A subset of real numbers that can be written as a countable union or intersection of intervals.

bpnc. Unit of *relative payload* for *JPEG* images (bits per non-zero *DCT* coefficient).

bpp. Unit of *relative payload* for spatial formats (bits per pixel).

C-SVM. *Soft-margin SVM.*

Cachin's definition of steganographic security. An information-theoretic definition of *steganographic security*.

Calibration. Process of estimating the *cover image* from the *stego image*.

Cardan's grille. A *linguistic-steganography* method in which a secret message is extracted by applying a mask over text.

Cauchy distribution. A thick-tail probability distribution used to model the distribution of *DCT* coefficients in a *JPEG* file.

Cauchy–Schwartz inequality. For any two vectors from an abstract vector space $\mathbf{x}, \mathbf{y} \in \mathcal{V}$, $|\langle \mathbf{x}, \mathbf{y} \rangle| \leq \sqrt{\|\mathbf{x}\| \|\mathbf{y}\|}$.

CCD. Charge-Coupled Device (type of *imaging sensor*).

cdf. Cumulative distribution function.

Central moment. Statistical moment of a zero-meaned random variable.

CFA. Color Filter Array. An array of filters bonded to *pixels* of an *imaging sensor* used to acquire color.

Change rate. The relative number of *embedding* changes with respect to the number of all possible changes a *steganographic method* can introduce. Usually denoted β.

Changeable pixels/elements. *Cover elements* that are allowed to be modified during steganographic *embedding*.

Channel distortion. Signal *distortion* in a communication channel.

Charge transfer. Process of moving charge from a *CCD sensor*.

Charge-transfer efficiency. Percentage of signal preserved during *charge transfer*.

Chebyshev sum-inequality. For any non-decreasing sequence $\mathbf{a}[i]$, non-increasing sequence $\mathbf{b}[i]$, and non-negative sequence $\mathbf{p}[i]$

$$\sum_{i=0}^{N-1}\mathbf{p}[i]\sum_{i=0}^{N-1}\mathbf{p}[i]\mathbf{a}[i]\mathbf{b}[i] \leq \sum_{i=0}^{N-1}\mathbf{p}[i]\mathbf{a}[i]\sum_{i=0}^{N-1}\mathbf{p}[i]\mathbf{b}[i].$$

Chernoff–Stein lemma. A statement that binds *KL divergence* between distributions and the probability of missed detection in *Neyman–Pearson hypothesis testing*.

Chi-square
  – **distribution.** Probability distribution of a sum of squares of *iid* zero-mean *Gaussian variables*.
  – **test.** A statistical test used to determine whether or not discrete data follows a known distribution.

Chrominance. Component of signal communicating color.

Ciphertext. Stream of symbols obtained by encrypting *plaintext*.

Classifier. An algorithm capable of distinguishing objects from several classes.

CLT. Central Limit Theorem.

CMOS. Complementary Metal–Oxide–Semiconductor (type of *imaging sensor*).

CMY. Cyan, Magenta, Yellow, subtractive *color model*.

CMYK. Cyan, Magenta, Yellow, blacK *subtractive color model*.

Code. A set of strings of symbols from an *alphabet*.
  – **length.** The number of *alphabet* symbols in each *codeword*.
  – **dimension.** The dimension of a linear *code* considered as a vector subspace.
  – **codimension.** Dimension of the *orthogonal complement* of a *code*. For a *linear code* with length n and *dimension* k, the codimension is $n - k$.

Codeword. An element of a *code*.

COG. *Center Of Gravity*. The mass center of data points assuming all data points have the same mass.

Color
  – **channel.** A component of color signal covering one color.
  – **correction.** A transformation of color registered by the *sensor*.
  – **cube.** The set of all *RGB* colors expressible as triples of integers $(R, G, B) \in \{0, \ldots, 255\}^3$.
  – **cut.** A binary sequence obtained by scanning an image by rows and registering 0 and 1 for two fixed colors.
  – **filter array.** See *CFA*.
  – **interpolation.** See *demosaicking*.
  – **model.** See *color representation*.

- **quantization.** Process of decreasing the number of unique colors in an image. Part of conversion to *palette format.*
- **representation.** Also called *color model.* A method for representing color in a computer using a vector of numerical values.
- **transformation.** Process of converting one *color representation* to another.

Complementary cumulative distribution function. The tail probability of a random variable, $1 - F(x)$, where $F(x)$ is the cumulative distribution function.

Complete feature set. A set of numerical *features* that completely characterizes natural images.

Composite hypothesis testing. A hypothesis-testing problem in which some distributions are not known or depend on unknown parameters.

Compression ratio. Ratio between the original size of a file and its compressed version.

Conditional entropy. *Entropy* of a random variable conditional on another random variable.

Co-occurrence matrix. Histogram of pairs of cover elements along a certain direction (e.g., horizontal pairs, vertical, diagonal, etc.).

Correlation. Optimal detector of known signals corrupted by additive white *Gaussian* noise.
- **coefficient.** A scalar value expressing linear dependence between two random variables.

Coset. The set of all words with the same *syndrome.*
- **leader.** A member of a *coset* with the smallest *Hamming weight.*

Cover
- **element.** An individual entity (e.g., *pixel, DCT coefficient*) from which the *cover object* is constituted.
- **modification.** A *steganographic method* in which the *cover object* is modified to embed the secret message.
- **object.** The original unmodified object before embedding a secret message in it.
- **selection.** A *steganographic method* in which the *cover object* is selected from a database so that the *cover* communicates the required message.
- **source.** Random variable defining the process of selecting objects as *cover objects* for *steganographic communication.*
- **synthesis.** A *steganographic method* in which the *cover object* is synthesized to communicate the required message.

Covering radius. The smallest radius of a *ball* needed to cover all possible *words* with balls centered at all *codewords.*

Covert communication/channel. A non-obvious information exchange usually obtained by hiding messages in other objects. Another term for *steganography*.

Cramer–Rao lower bound. A bound on variance of an *estimator*.

Critical region. A set of measurements for which a *detector* decides the alternative hypothesis (signal present).

Cross-validation. A method for determining parameters of a *support vector machine*.

CRT. Cathode-Ray Tube. An old monitor construction that uses a glass tube.

CRW. Canon raw image format.

Cryptographically secure *PRNG*. A pseudo-random number generator satisfying the property that it be computationally intractable to derive its seed from a sequence of pseudo-random numbers generated by it.

Cryptosystem
- **asymmetric (public key).** An encryption system in which the process of encrypting (controlled by the encryption or public key) is different from decryption (controlled by the decryption key). The system is conceived so that knowledge of the encryption key does not allow deriving the decryption key or decrypt because the task of finding the decryption key leads to an intractable problem. An example of a public-key cryptosystem is *RSA*.
- **symmetric (private key).** An encryption scheme controlled by a secret key that needs to be shared between the communicating parties before starting the communication. Examples are *DES* and *AES*.

CTE. Charge-Transfer Efficiency.

Dark current. *Sensor* response when it is not lit by light.

Data masking. A *covert communication* method in which data is given statistical properties of typical *cover objects*.

DC term. The *DCT* coefficient corresponding to *spatial frequency* $(0,0)$ (from direct current).

DCT. Discrete Cosine Transform.
- **coefficient.** A coefficient obtained via *DCT*.
- **mode.** *DCT* coefficients corresponding to a particular *spatial frequency*.

Dead pixel. A defective *pixel* always registering no (low) signal.

Decision function. Another word for *detector* used in *support vector machines*.

Decompression. Process of obtaining the original data from its compressed form.

Defective
- **memory.** A memory in which some cells are defective or permanently stuck to zero or one.
- **pixel.** A *pixel* whose response to light does not comply with design specifications.

Deflection coefficient. A numerical quantity measuring the performance of a *detector*.

Demosaicking. Process of interpolating colors from a signal acquired by an imaging *sensor* equipped with a *color filter array*.

Denoising. Process of removing noise from a signal.

DES. Data-Encryption Standard. A *symmetric cryptosystem*.

Detection probability. Probability of correctly detecting a *stego image* as stego.

Detector. An algorithm that detects presence of a signal.

DFT. Discrete Fourier Transform.

Digital image acquisition. Process of acquiring an image through an *imaging sensor*.

Discrete cosine transformation. An orthogonal transformation used in *JPEG* format.

Distance to code. Distance between a given *word* and the closest *codeword* from the *code*.

Distortion. Signal degradation typically due to noise or processing.

Distortion-limited embedder. An *embedding algorithm* whose *embedding distortion* is bounded.

Dithering. Process of spreading quantization error to neighboring *pixels* to prevent creating bands during *color quantization*.

DNG. Digital NeGative file format.

Double JPEG compression. Image that was *JPEG* compressed twice, each time with a different *quantization matrix*.

Dry elements. See *changeable elements*.

Dual
- **code.** The *orthogonal complement* of a *linear code*.
- **histogram.** A *feature* used in *blind steganalysis* of *JPEG* images.

DWT. Discrete Wavelet Transform.

Embedding
- **algorithm/mapping.** Algorithm that embeds a secret message in a *cover object*.
- **capacity.** Maximal number of bits that can be embedded using a given *steganographic method*.
- **distortion.** Distortion imposed onto the *cover object* due to embedding a secret message.
- **efficiency.** The expected number of bits communicated per unit expected *embedding distortion*.
- **impact.** Increase in statistical detectability of embedding changes.
- **operation.** Procedure that modifies individual *cover elements* during embedding.
- **path.** A path along which message bits are embedded.
- **while dithering.** A *steganographic method* for *palette* images.

Empirical risk. *Risk function* evaluated from data samples.

Entropy. Uncertainty (information content) of a random variable.
- **encoder/compressor.** A lossless compression algorithm capable of compressing a stream of symbols close to its *entropy*.
- **decoder/decompressor.** Algorithm inverting the action of an *entropy encoder*.

Erasure channel. A communication channel in which some symbols may get replaced by an erasure symbol.

Error function. $\mathrm{Erf}(x) = (2/\sqrt{\pi}) \int_0^x e^{-t^2} \mathrm{d}t$.

Estimator. An algorithm that computes an estimate of an unknown parameter.

Eve. A fictitious character (the *warden*) from the *prisoners' problem*.

Extraction algorithm/mapping. Algorithm that extracts secret message from the *stego object*.

EzStego. A *steganographic program* for *palette* images.

F5. A *steganographic algorithm* for the *JPEG* format.

False alarm. An error when a *cover object* is mistakenly identified as *stego*.
- **probability.** Probability of encountering a *false alarm*.

Feasible covert channel. A *covert channel* that complies with the bound on the *embedding distortion* and *perfect security*.

Feature. A numerical quantity extracted from an object.

Finite field. A finite *alphabet* of symbols that can be added or multiplied so that the operations have similar properties as operations of addition and multiplication on real numbers.

Fisher information. A fundamental quantity expressing the amount of information that measurements (that depend on an unknown parameter) give about the parameter value.

Floyd–Steinberg dithering. Algorithm for *dithering* by diffusing *color quantization* error to neighboring *pixels*.

Forensic steganalysis. Effort directed towards extracting the message from a *stego object*. It may include various tasks, such as estimating attributes of the message (e.g., its length) and determining the *steganographic method* used for embedding the message and/or the *stego* and encryption keys.

Galois field. See *finite field*.

Gamma
   – **correction.** Non-linear transform of light intensity expressed as power law.
   – **function.** $\Gamma(x) = \int_0^\infty t^{x-1} e^{-t} \mathrm{d}t$.

Gaussian
   – **kernel.** *Kernel* used in *SVMs*,

$$k(\mathbf{x}, \mathbf{x}') = e^{-\gamma \|\mathbf{x} - \mathbf{x}'\|^2}.$$

   – **variable.** A random variable with mean μ and variance σ^2 with the following probability distribution:

$$\frac{1}{\sqrt{2\pi}\sigma} e^{-\frac{(x-\mu)^2}{2\sigma^2}}.$$

Generalized Gaussian distribution. A unimodal probability distribution which can be thought of as a generalized form of the Gaussian distribution.

Generator matrix. A matrix whose rows form the basis of a *linear code*.

GIF. Graphic Interchange File format for images.

Gifshuffle. A *steganographic program* for *palette* images.

GLRT. Generalized Likelihood-Ratio Test.

Golay code. One of the *perfect codes*.

Hamming
   – **code.** A *perfect linear code* with *parity-check matrix* containing all non-zero *words* of fixed length as columns (all columns are different up to a multiplication by an element from the *finite field*).
   – **distance.** The number of elements in which two *words* differ.
   – **weight.** The number of non-zero symbols in a *word*.

Hash. A bit string extracted from an object (digest).
   – **function** (one-way function). A function that assigns a bit string of fixed length to any bit string on its input. A cryptographic hash $f(x)$ must

satisfy the property that it be computationally intractable to invert f (i.e., find y such that $f(y) = x$) and to find collisions (i.e., x and y such that $f(x) = f(y)$).

HCF. Histogram Characteristic Function. Fourier transform of the *histogram*.

Hilbert kernel density estimator. A specific *sparsity measure*.

Hilbert space. A complete linear vector space with inner product.

Hinge loss function. A convex loss function used in definition of *risk function* in *soft-margin support vector machines*.

Histogram
- **attack.** A statistical attack on *LSB embedding* based on detecting artifacts in the *histogram* of the *stego image*.
- **characteristic function.** Amplitude of the Fourier transform of a *histogram*.

Homogeneous bit pair. A pair of bits 00 or 11.

Hot pixel. A *defective pixel* with very high response independent of the incident light.

iid. Independent and identically distributed.

IIF. Type of an image format.

Imaging sensor. A semiconductor device used to acquire digital images in cameras.

Inter-block feature. A *feature* obtained by quantifying a relationship between *DCT* coefficients from different blocks.

Intra-block feature. A *feature* obtained by quantifying a relationship between *DCT* coefficients from the same block.

IQR. Inter-Quartile Range. A robust measure of spread corresponding to the range defined by the first and third quartile.

Isolation. Distance between a *palette* color and another, different, closest *palette* color.

Isomorphic code. A geometrically identical *code*.

J-divergence. A symmetrized *KL divergence* $D_{\mathrm{KL}}(f\|g) + D_{\mathrm{KL}}(g\|f)$.

JPEG. Joint Photographic Expert Group image format.

JPEG2000. A modern version of *JPEG* based on discrete wavelet transform.

JPEG80. Image database *RAW* compressed with 80% quality *JPEG*.

JPEG compatibility steganalysis. A *steganalysis* method that detects presence of *embedding* changes by verifying the compatibility of each 8×8 *pixel* block with *JPEG* compression.

JP Hide&Seek. A *steganographic program* for *JPEG* images.

Jsteg. A *steganographic program* for *JPEG* images.

Kernel. A similarity function used to define kernelized *SVMs*.

Kronecker delta. Function defined as $\delta(0) = 1$ and $\delta(x) = 0$ otherwise.

Kullback–Leibler divergence (distance). A fundamental concept from information theory. It is a measure of distance between two random variables. Also called *relative entropy*.

Kurtosis. The fourth normalized *central moment*.

Large payload. *Payload* whose length is close to *embedding capacity*.

Lattice. A discrete set of points in space, usually exhibiting a regular structure.

Law of rare events. Probabilistic description of random phenomena that occur independently and uniformly in time (see *Poisson distribution*).

LCD. Liquid-Crystal Display.

LibSVM. A popular *support-vector-machine* library.

Likelihood-ratio test. An optimal *detector* for a simple hypothesis-testing problem.

Linear
- **CCD.** A *CCD sensor* formed by a one-dimensional row of *photodetectors*.
- **code.** A set of *codewords* forming a vector space.
- **SVM.** A *support vector machine* realized with a *separating hyperplane* directly in the feature space.

LMP. Locally Most Powerful *detector*.

Log inequality. $\log x \leq x - 1$ valid for all $x > 0$.

Log-sum inequality. For any non-negative numbers $r_1, \ldots, r_k$, and positive $s_1, \ldots, s_k$

$$\sum_{i=1}^{k} r_i \log \frac{r_i}{s_i} \geq \sum_{i=1}^{k} r_i \log \frac{\sum_{j=1}^{k} r_j}{\sum_{j=1}^{k} s_j}.$$

Lossless compression. Compression of data that does not incur any loss of information.

Lower embedding efficiency. The ratio between the *payload* length and the largest number of *embedding* changes that may be needed to embed that *payload.*

LSB. Least Significant Bit. The last bit in a big-endian representation of an integer.
   – **embedding.** A *steganographic method* in which message bits are embedded in a sequence of natural numbers by replacing their least significant bits with message bits.
   – **matching.** See ± 1 *embedding.*
   – **pair.** A pair of values differing only in their *LSBs.*
   – **plane.** Array of *LSBs.*

LSE. Least-Square Estimation. A general *estimation* method in which data model parameters are determined by minimizing the squared error between the model and observations.

Luminance. Color component carrying information about light intensity.

LT
   – **code.** Sparse *linear code* designed for the *erasure channel.*
   – **process.** An algorithm based on *LT codes* used for realizing *non-shared selection channels.*

Machine learning. The field of artificial intelligence.

Macroblock. A block of 16×16 *pixels* used in *JPEG format.*

MAD. Median Absolute Deviation. A robust measure of spread.

MAE. Median Absolute Error. A robust measure of error spread.

MAP. Maximum A Posteriori Estimation. A general *estimation* method for determining an unknown parameter of a distribution. The parameter is determined by maximizing the probability of the parameter given the measurements.

Margin. The distance between the *support vector* closest to the *separating hyperplane.*

Markov
   – **chain.** A sequence of random variables x_i in which each variable depends only on the previous variable ($\Pr\{x_i|x_{i-1}\} = \Pr\{x_i|x_{i-1}, \ldots, x_1\}$). For discrete variables, a Markov chain is completely described using the transition probability matrix $\mathbf{A}[k, l] = \Pr\{x_i = l|x_{i-1} = k\}$.
   – **features.** *Features* for *blind steganalysis* of *JPEG* images obtained by quantifying the relationship between neighboring *DCT* coefficients of the same block.

– **random field.** A two-dimensional generalization of a *Markov chain*. An array of random variables in which each variable depends only on the variables from its neighborhood.

Matrix embedding. A *syndrome coding* method for decreasing the *embedding distortion*.

– **theorem.** A general methodology for constructing *matrix-embedding* methods from *linear codes*.

Matrix LT process. An algorithm based on *LT codes*. It permutes columns and rows of a matrix and brings it to an upper-diagonal form.

Maximum a posteriori estimator. An *estimator* of unknown parameter θ from data $\mathbf{x}$ that maximizes $\Pr\{\theta|\mathbf{x}\}$, $\hat{\theta} = \arg\max_\theta \Pr\{\theta|\mathbf{x}\} = \arg\max_\theta \Pr\{\mathbf{x}|\theta\}P(\theta)$.

Maximum-likelihood estimator. An *estimator* of unknown parameter θ from data $\mathbf{x}$ that maximizes $\Pr\{\mathbf{x}|\theta\}$, $\hat{\theta} = \arg\max_\theta \Pr\{\mathbf{x}|\theta\}$.

Maximum mean discrepancy. A two-sample statistic used to classify between two categories.

MB1 (MB2). *Model-Based Steganography* with (without) deblocking.

Median absolute error. The median of absolute values of error realizations.

Meet-in-the-middle. A brute-force algorithm for finding *coset leaders*.

Message digest function. A function that returns a digest (*hash*) of a given object.

Message source. Random variable defining the statistical properties of communicated messages, such as their length.

MEX. Matlab EXecutable file.

Microdot. *Stego object* downsized to such minuscule dimensions that it resembles a speck of dirt.

Microlens. A miniature lens bonded to each *pixel* to help direct light to the *photodetector*.

Mimic function. A mimic function changes a file to match its statistical properties to those of another file.

Minimal

– **distance.** *Hamming distance* between two closest *codewords* from a *code*.

– **entropy.** An information-theoretic concept characterizing the uncertainty of a random variable.

– **embedding impact.** A general principle for constructing *steganographic schemes*.

Missed detection. An error made when a *stego object* is mistakenly identified as a *cover object*.
– **probability.** The probability of failing to recognize a *stego image* as containing a secret message.

MLE. Maximum-Likelihood Estimation. A general estimation method for determining an unknown parameter of a distribution. The parameter is determined by maximizing the probability of observing the measurements conditioned on the unknown parameter.

MMx. A *steganographic algorithm* for the *JPEG* format.

Model-based steganography. A *steganographic system* designed to preserve a model of *cover source*.

MPEG. A video format based on *JPEG*.

MSE. Mean-Square Error.

Multi-classification. *Classification* into more than two classes.

Mutual information (between two random variables). Information about a random variable conveyed by another random variable.

MVU estimator. Minimum-Variance Unbiased *estimator*.

Negligible function. A function of x that falls to zero faster than any power of x.

NEF. Nikon raw image format.

Neyman–Pearson hypothesis testing. A principle for constructing *detectors* in which one imposes an upper bound on the probability of a *false alarm* and minimizes the probability of *missed detection*.

Noise
– **moments.** Statistical moments of the *noise residual*.
– **reduction.** Process of suppressing noise in a signal.
– **residual.** The result of subtracting the signal and its *denoised* version.

Non-shared selection channel. A *selection channel* that is not shared between the sender and the recipient.

Normalized histogram. A sample probability mass function.

NRCS. Natural Resources Conservation Service image database.

nsF5. An improved version of the *F5* algorithm with removed *shrinkage*.

OC-NM. One-Class Neighbor Machine, a *classifier* that recognizes a single class.

OG. An abbreviation for *OutGuess*.

One-time pad. A simple *symmetric cryptosystem* in which message bits are XORed with a random stream of bits shared between *Alice* and *Bob*.

One-way function. A mapping h for which it is computationally infeasible to find x for given $h(x)$.

Optimal parity embedding. A *steganographic algorithm* for *palette* images.

Oracle. A device that can be queried to produce a cover object from a given source.

Orthogonal complement. The set of all vectors orthogonal to the object.

OutGuess. A *steganographic program* for hiding messages in *JPEG* files.

***p*-value.** The tail probability of a *chi-square distribution*.

Pairs Analysis. A *quantitative steganalysis* method for *palette* or grayscale images.

Palette. A look-up table of all colors that may occur in a palette image.
 – **color.** A color stored in a *palette*.

Parity. A bit assigned to a *cover element*.
 – **check matrix.** A *generator matrix* of the *dual code*.
 – **function.** A *symbol-assignment function* that assigns bits.

Passive warden. A *warden* who does not interfere with communication.

Payload. Length of the secret message.

pdf. Probability density function (defined for continuous variables).

Perfect
 – **code.** A *code* that saturates the *sphere-covering inequality*.
 – **security.** Impossibility to distinguish between *cover* and *stego objects*.

Perturbed quantization. A steganographic *embedding* principle in which *embedding* occurs during processing the *cover object* while limiting the changes to those elements of the *cover object* that experience the largest *quantization error*.

Photodetector. Another term for *pixel*.

Photodiode. Another term for *pixel*.

Photoelectric effect. A phenomenon during which a photon creates an electron–hole pair.

Photonic noise. See *shot noise*.

Pixel. An element of an *imaging sensor* capable of registering incident light intensity.

– **well.** A portion of *pixel* collecting free electrons released due to the *photo-electric effect* at image acquisition.

Plaintext. Message to be encrypted.

pmf. Probability mass function (defined for discrete variables).

PNG. Portable Network Graphics image format.

Poisson distribution. A mathematical form of the *law of rare events*.

PQe. A version of the *perturbed quantization* algorithm for *JPEG* images.

PQt. A version of the *perturbed quantization* algorithm for *JPEG* images.

Precover. A hypothetical *cover* used to justify a priori distribution of some *cover-image* statistics.

Primary set. A set of pixel pairs with specified relationship (used to construct *Sample Pairs Analysis* attack on *LSB* embedding).

Prisoners' problem. A fictitious scenario involving *Alice*, *Bob*, and *warden Eve*, used to demonstrate the problem of *steganography* and *steganalysis*.

PRNG. Pseudo-Random Number Generator.

PRNU. Pixel Photo-Response Non-Uniformity. A *systematic imperfection* due to varying response to light of individual *pixels*.

Probabilistic algorithm. An algorithm whose mechanism involves randomness.

Pseudo-random selection channel. A *selection channel* determined using a *PRNG*.

PSNR. Peak Signal-to-Noise Ratio.

Public-key
– **encryption.** *Asymmetric cryptosystem* (realized using the so-called public key) and decryption, realized with the private key).
– **steganography.** *Steganographic channel* with public access to the secret message, which is encrypted using a private key.

q-**ary.** Related to an *alphabet* consisting of q symbols.

QIM. Quantization Index Modulation (a robust *watermarking* technique).

Quality factor. A scalar value that controls the quality of a *JPEG file*.

Quantitative steganalysis. *Steganalysis* whose objective is to estimate the embedded *payload* or the *change rate* (or the number of *embedding* changes).

Quantization. Process of rounding real-valued quantities to represent them with bits.
– **error.** Distortion induced by *quantization*.

   – **noise.** Distortion introduced by *quantization.*

   – **step.** During *quantization,* values are rounded to multiples of the quantization step.

   – **matrix (table).** A matrix of *quantization steps* used in *JPEG* compression.

Quantum efficiency. The probability that a photodetector absorbs a photon.

Rainbow coloring. Assignment of colors to *lattice* points so that the colors of all neighboring *lattice* points are different.

Random

   – **binning.** A methodology for creating *random codes.*

   – **code.** A *code* generated randomly.

Raster format. Image representation by storing *pixel* values as one or more matrices of the same dimensions as the image.

RAW. A database of never-compressed natural images taken by digital cameras.

Raw sensor output. Unprocessed signal obtained by reading out the charge of individual *pixels.*

Readout noise. Noise introduced during the process of extracting the charge at each *pixel.*

Regularity condition. We say that a pdf $p(\mathbf{x}; \theta)$ satisfies the regularity condition if

$$E_{p(\mathbf{x};\theta)} \left[\frac{\partial \log p(\mathbf{x}; \theta)}{\partial \theta} \right] = 0 \text{ for all } \theta.$$

This condition is satisfied for most well-behaved distributions.

Rejection sampling. A *steganographic method* in which the *cover* or its elements are selected by randomly drawing them from their corresponding sets till a correct message is embedded.

Relative

   – **distortion.** *Distortion* measured per *cover element.*

   – **embedding capacity.** *Embedding capacity* per *cover element.*

   – **entropy.** See *KL divergence.*

   – **message length/payload.** Ratio between the *payload* and *embedding capacity.* Usually denoted α. Also called relative payload.

Repetition code. A *code* whose *codewords* are strings of repeating symbols.

RGB. Red, Green, and Blue channels in a color signal.

Risk functional. Performance measure of *SVMs.*

Robustness. Ability to withstand *distortion.*

ROC curve. Receiver Operating Characteristic of a detector showing the *probability of detection* versus *probability of false alarm*.

RS analysis. A *quantitative steganalysis* method for detection of *LSB embedding* in images.

RSA. (Rivest Shamir Adleman) The first and the most popular public-key cryptosystem created from the fact that it is computationally hard to factorize a composite number into the product of primes.

RSD. Robust Soliton Distribution.

S-Tools. A *steganographic program*.

Sample Pairs Analysis. A *targeted steganalysis* method for *LSB embedding* in spatial domain.

SCAN. An image database containing scans of natural images.

SDCS. Sum and Difference Covering Set.

Secure payload. The number of bits that can be communicated at a given security level using a specific imperfect *steganographic scheme*. The secure payload grows only with the square root of the number of pixels in the *cover*.

Selection channel. A portion of the *cover object* used for embedding the secret message.

Sensor. See *imaging sensor*.
 – **noise.** The complex of noise sources introduced during image acquisition.
 – **resolution.** The number of *pixels* on the *sensor*.

Separating hyperplane. Hyperplane separating two classes.

Sequential selection channel. A *selection channel* defined by some simple scan of the image, e.g., by rows or columns.

Shot noise. Noise in signal acquired by an *imaging sensor* due to quantum properties of light.

Shrinkage. Shrinkage occurs when a non-zero *DCT* coefficient is modified to zero after message embedding.

Side-information. Information available to *Alice* but not *Bob*.

Skewness. The third normalized *central moment*.

SLR. Single-Lens Reflex camera.

SNR. Signal-to-Noise Ratio.

Soft-margin SVM. An *SVM* that allows misclassifications by penalizing them.

SPA. Sample Pairs Analysis, a *quantitative steganalysis* of *LSB embedding* in the spatial domain.

Space-filling curve (planar). A curve that goes through every point in a unit square.

Sparse codes. *Codes* with *codewords* of small *Hamming weight*.

Sparsity measure. A function evaluating sparsity of data points at a given point.

Spatial frequency. A pair of integers characterizing a *DCT mode*.

Sphere-covering bound. An inequality that binds the number of *codewords* in a *code*, its length, and its *covering radius*.

Spread-spectrum (data hiding). A method for information hiding in which a bit is spread into a longer random sequence that is superimposed on the *cover* object. Usually used to achieve *robustness* with respect to channel *distortion*.

Square-root law. Thesis claiming that the *steganographic capacity* of *imperfect stegosystems* grows only as the square root of the number of *cover elements*.

Standard quantization matrix. A family of *quantization matrices* parametrized by quality factor as specified in the *JPEG* format standard.

Statistical
  – **detectability.** A measure of how detectable *embedding* changes are using standard methods of hypothesis testing.
  – **restoration.** A general principle for constructing *steganographic methods* that preserve selected statistics of the *cover object*.

Steganalysis. The counterpart of *steganography*. Effort directed towards discovering the presence of a secret message.

Steganalysis-aware steganography. *Steganography* constructed to avoid known *steganalysis attacks*.

Steganalyzer. A specific implementation of some *steganalysis* attack.

Steganographic
  – **capacity.** Maximum number of message bits that can be embedded in a *cover object* without introducing statistically detectable artifacts.
  – **channel.** A *covert* communication system consisting of the *cover source*, *stego key source*, and *message source*, and the physical communication channel used to exchange messages.
  – **communication.** See *steganographic channel*.
  – **file system.** A tool to thwart "rubber-hose attacks" that allows the user to plausibly deny that encrypted files reside on the disk.
  – **scheme.** See *steganographic channel*.
  – **security.** Impossibility to construct an attack on a *steganographic scheme*.

Steganography. The art of communicating messages in a *covert* manner.

Steghide. A *steganographic program* that embeds messages in *JPEG* files while preserving the *histogram* of *DCT* coefficients.

Stego
- **key.** A secret shared between the sender and the recipient that drives the *embedding* and *extraction algorithms*.
- **key source.** Random variable defining the process of selecting the *stego key* in *steganographic communication*.
- **noise.** The difference between *cover* and *stego objects*.
- **object.** The result of embedding a message in the *cover object*.

Stegosystem. Another term for *steganographic channel*.

Stirling's formula. An approximate formula for factorial.

Stochastic modulation. A *steganographic scheme* that embeds messages in raster images by adding noise with predefined probability density function.

Structural steganalysis. A generalized extensible formulation of *SPA* aimed at detection of *LSB embedding*.

Student's *t*-distribution. A statistical distribution obtained by normalizing zero-mean *Gaussian* samples by their sample variance.

Subtractive color model. *Color model* in which colors are created by removing base colors.

Support vector. A data point determining the *separating hyperplane* in *SVMs*.

Support vector machine (SVM). A tool for machine learning.
- **kernelized.** An SVM implemented using a *kernel*.
- **soft-margin.** An SVM capable of discriminating between non-separable data samples.
- **weighted.** An SVM whose decision errors are controlled by penalization parameters.

SVM. *Support vector machine*.

Symbol-assignment function. A mapping that assigns an *alphabet* symbol (or an element of a *finite field*) to every *cover element* (every *pixel* or *DCT* coefficient). Symbol-assignment functions enable using the apparatus of coding to improve *embedding efficiency* (*matrix embedding*) and to communicate using *non-shared selection channels*.

Syndrome. A bit stream obtained by multiplying a *word* and the *parity-check matrix*.
- **coding.** A method for communicating a secret message as a *syndrome* of a *linear code*.

System attack. An attack on a *steganographic scheme* that utilizes implementation errors or other auxiliary information rather than the statistical distribution of *cover-object elements*.

Systematic form. A *linear code* is in systematic form if its *generator matrix* is $[\mathbf{I}; \mathbf{A}]$, where $\mathbf{I}$ is the identity matrix.

Systematic imperfection. Distortion of image signal that is not random in nature and repeats in all images.

Tail inequality. An upper bound on partial sum of binomial coefficients

$$\sum_{i=0}^{\beta n} \binom{n}{i} \leq 2^{nH(\beta)}.$$

Targeted steganalysis. An approach to *steganalysis* designed to detect a specific *steganographic method*.

TDI. Time-Delay Integration imaging.

Ternary
  – **alphabet.** An *alphabet* consisting of three symbols.
  – **code.** A *code* built over a ternary *alphabet*.

TIFF. Tagged-Image File Format.

Timestamped bits. Bits tagged with a scalar value (time).

Trace set. A set of *pixel* pairs used in *structural steganalysis*.

Transform image format. Format that represents an image in a transform domain.

Tristimulus theory. A theory of color vision based on the fact that the human retina contains three types of cones sensitive to red, green, and blue colors.

True-color image. An image in which each *pixel* is represented using 24 bits (8 bits per *color channel*).

UMP. Universally Most Powerful Detector.

Unbiased estimator. An *estimator* whose *bias* is zero, $E[\hat{\theta}] = \theta$ for all θ.

Undetectability. The inability to prove the existence of a secret message in a *stego object*.

Variation. A *feature* used in *blind steganalysis* of *JPEG* images.

Voronoi cell. A region defined for a *lattice*.

WAM. *Blind steganalyzer* constructed to detect spatial-domain *steganography* using Wavelet Absolute Moments of noise residuals.

Warden. The subject or a computer program that monitors the traffic between *Alice* and *Bob* in the *prisoners' problem*.
– **active.** A *warden* that slightly distorts the communication so that it still conveys the same overt meaning. The goal is to prevent *steganography*.
– **malicious.** A *warden* who tries to trick the communicating parties by impersonating them or by other means that are based on the warden's knowledge of the steganographic protocol.
– **passive.** A *warden* that passively observes the communication.

Watermarking. Data-hiding application in which the hidden data supplements the *cover object*.

Wavelet transform. A signal transform with basis functions that are localized simultaneously both in the spatial and in the frequency domain.

Weak Law of large numbers. The law that states that the sample mean converges to the mean in probability.

Weighted SVM. A *support vector machine* in which misclassifications are penalized on the basis of weights assigned to false alarms and missed detections.

Wet paper codes. *Codes* that enable communication using *non-shared selection channels*.

Wet pixels. *Pixels* that are not to be changed during *steganographic embedding*.

White balance. *Color transformation* adjusting the gain of each *color channel*.

Wiener filter. An adaptive linear *denoising* filter.

Within-image error. A component of the estimation error of *quantitative steganalyzers* that depends on the placement of *embedding* changes in the *stego image*.

WMF. Windows Meta File image format.

Word. A string of symbols from an *alphabet*.

WS. Weighted stego method for detection of *LSB embedding*.

XOR. Exclusive or.

YUV. *Luminance* and two *chrominance* signals in a color signal.

Zig-zag scan. Scanning order of quantized *DCT* coefficients used during *JPEG* compression.

ZZW construction. A general method for constructing *matrix-embedding* methods from existing methods.

References

[1] Fingerprints for car parts. *The Economist*, December 8, 2005.

[2] G. F. Amelio, M. F. Tompsett, and G. E. Smith. Experimental verification of the charge coupled device concept. *Bell Systems Technical Journal*, 49:593–600, 1970.

[3] R. Anderson. Stretching the limits of steganography. In R. J. Anderson, editor, *Information Hiding, 1st International Workshop*, volume 1174 of Lecture Notes in Computer Science, pages 39–48, Cambridge, May 30–June 1, 1996. Springer-Verlag, Berlin.

[4] R. J. Anderson, R. Needham, and A. Shamir. The steganographic file system. In D. Aucsmith, editor, *Information Hiding, 2nd International Workshop*, volume 1525 of Lecture Notes in Computer Science, pages 73–82, Portland, OR, April 14–17, 1998. Springer-Verlag, Berlin.

[5] R. J. Anderson and F. A. P. Petitcolas. On the limits of steganography. *IEEE Journal of Selected Areas in Communication*, 16(4):474–481, 1998.

[6] N. Aoki. A band extension technique for G.711 speech using steganography. *IEICE Transactions on Communications*, E89-B(6):1896–1898, 2006.

[7] N. Aoki. Potential of value-added speech communications by using steganography. In B.-Y. Liao, editor, *Proceedings of the 3rd International Conference on Intelligent Information Hiding and Multimedia Signal Processing*, volume 2, pages 251–254, Kaohsiung, Taiwan, November 26–28, 2007.

[8] A. Aspect, J. Dalibard, and G. Roger. Experimental test of Bell's inequalities using time-varying analyzers. *Physical Review Letters*, 49:1804, 1982.

[9] I. Avcibas. Audio steganalysis with content-independent distortion measures. *IEEE Signal Processing Letters*, 13:92–95, February 2006.

[10] I. Avcibas, M. Kharrazi, N. D. Memon, and B. Sankur. Image steganalysis with binary similarity measures. *EURASIP Journal on Applied Signal Processing*, 17:2749–2757, 2005.

[11] I. Avcibas, N. D. Memon, and B. Sankur. Steganalysis using image quality metrics. In E. J. Delp and P. W. Wong, editors, *Proceedings SPIE, Electronic Imaging, Security and Watermarking of Multimedia Contents III*, volume 4314, pages 523–531, San Jose, CA, January 22–25, 2001.

[12] I. Avcibas, N. D. Memon, and B. Sankur. Image steganalysis with binary similarity measures. In *Proceedings IEEE, International Conference on Image Processing, ICIP 2002*, volume 3, pages 645–648, Rochester, NY, September 22–25, 2002.

[13] F. Bach, D. Heckerman, and E. Horvitz. On the path to an ideal ROC curve: Considering cost asymmetry in learning classifiers. In R. G. Cowell and Z. Ghahramani, editors, *Proceedings of the 10th International Workshop on Artificial Intelligence and Statistics (AISTATS)*, pages 9–16. Society for Artificial Intelligence and Statistics, 2005. Available

electronically at http://www.gatsby.ucl.ac.uk/aistats/.

[14] M. Backes and C. Cachin. Public-key steganography with active attacks. In J. Kilian, editor, *2nd Theory of Cryptography Conference*, volume 3378 of Lecture Notes in Computer Science, pages 210–226, Cambridge, MA, February 10–12, 2005. Springer-Verlag, Heidelberg.

[15] F. Bacon. *Of the Advancement and Proficiencie of Learning or the Partitions of Sciences*, volume VI. Leon Lichfield, Oxford, for R. Young and E. Forest, 1640.

[16] M. Barni and F. Bartolini. *Watermarking Systems Engineering: Enabling Digital Assets Security and Other Applications*, volume 21 of *Signal Processing and Communications*. Boca Raton, FL: CRC Press, 2004.

[17] R. J. Barron, B. Chen, and G. W. Wornell. The duality between information embedding and source coding with side information and some applications. *IEEE Transactions on Information Theory*, 49(5):1159–1180, 2003.

[18] R. Bergmair. Towards linguistic steganography: A systematic investigation of approaches, systems, and issues. Final year thesis, April 2004. University of Derby, http://bergmair.cjb.net/pub/towlingsteg-rep-inoff-a4.ps.gz.

[19] J. Bierbrauer. *Introduction to Coding Theory*. London: Chapman & Hall/CRC, 2004.

[20] J. Bierbrauer and J. Fridrich. Constructing good covering codes for applications in steganography. *LNCS Transactions on Data Hiding and Multimedia Security*, 4920:1–22, 2008.

[21] R. Böhme. Weighted stego-image steganalysis for JPEG covers. In K. Solanki, K. Sullivan, and U. Madhow, editors, *Information Hiding, 10th International Workshop*, volume 5284 of Lecture Notes in Computer Science, pages 178–194, Santa Barbara, CA, June 19–21, 2007. Springer-Verlag, New York.

[22] R. Böhme. *Improved Statistical Steganalysis Using Models of Heterogeneous Cover Signals*. PhD thesis, Faculty of Computer Science, Technische Universität Dresden, 2008.

[23] R. Böhme and A. D. Ker. A two-factor error model for quantitative steganalysis. In E. J. Delp and P. W. Wong, editors, *Proceedings SPIE, Electronic Imaging, Security, Steganography, and Watermarking of Multimedia Contents VIII*, volume 6072, pages 59–74, San Jose, CA, January 16–19, 2006.

[24] R. Böhme and A. Westfeld. Feature-based encoder classification of compressed audio streams. *ACM Multimedia System Journal*, 11(2):108–120, 2005.

[25] I. A. Bolshakov. Method of linguistic steganography based on collocationally-verified synonymy. In J. Fridrich, editor, *Information Hiding, 6th International Workshop*, volume 3200 of Lecture Notes in Computer Science, pages 180–191, Toronto, May 23–25, 2005. Springer-Verlag, Berlin.

[26] K. Borders and A. Prakash. Web tap: Detecting covert web traffic. In V. Atluri, B. Pfitzmann, and P. Drew McDaniel, editors, *Proceedings 11th ACM Conference on Computer and Communications Security (CCS)*, pages 110–120, Washington, DC, October 25–29, 2004.

[27] S. Boyd and L. Vandenberghe. *Convex Optimization*. Cambridge: Cambridge University Press, 2004.

[28] W. S. Boyle and G. E. Smith. Charge coupled semiconductor devices. *Bell Systems Technical Journal*, 49:587–593, 1970.

[29] J. Brassil, S. Low, N. F. Maxemchuk, and L. O'Gorman. Hiding information in document images. In *Proceedings of the Conference on Information Sciences and Systems, CISS*, pages 482–489, Johns Hopkins University, Baltimore, MD, March 22–24, 1995.

[30] R. P. Brent, S. Gao, and A. G. B. Lauder. Random Krylov spaces over finite fields. *SIAM Journal of Discrete Mathematics*, 16(2):276–287, 2003.

[31] D. Brewster. *Microscope*, volume XIV. Encyclopaedia Britannica or the Dictionary of Arts, Sciences, and General Literature, Edinburgh, IX – Application of photography to the microscope, 8th edition, 1857.

[32] C. W. Brown and B. J. Shepherd. *Graphics File Formats*. Greenwich, CT: Manning Publications Co., 1995.

[33] C. J. C. Burges. A tutorial on support vector machines for pattern recognition. *Data Mining and Knowledge Discovery*, 2(2):121–167, 1998.

[34] S. Cabuk, C. E. Brodley, and C. Shields. IP covert timing channels: Design and detection. In V. Atluri, B. Pfitzmann, and P. Drew McDaniel, editors, *Proceedings 11th ACM Conference on Computer and Communications Security (CCS)*, pages 178–187, Washington, DC, October 25–29, 2004.

[35] C. Cachin. An information-theoretic model for steganography. In D. Aucsmith, editor, *Information Hiding, 2nd International Workshop*, volume 1525 of Lecture Notes in Computer Science, pages 306–318, Portland, OR, April 14–17, 1998. Springer-Verlag, New York.

[36] G. Cancelli and M. Barni. MPSteg-color: A new steganographic technique for color images. In T. Furon, F. Cayre, G. Doërr, and P. Bas, editors, *Information Hiding, 9th International Workshop*, volume 4567 of Lecture Notes in Computer Science, pages 1–15, Saint Malo, June 11–13, 2007. Springer-Verlag, Berlin.

[37] G. Cancelli, G. Doërr, I. J. Cox, and M. Barni. A comparative study of ± 1 steganalyzers. In *Proceedings IEEE International Workshop on Multimedia Signal Processing*, pages 791–796, Cairns, Australia, October 8–10, 2008.

[38] G. Cancelli, G. Doërr, I. J. Cox, and M. Barni. Detection of ± 1 LSB steganography based on the amplitude of histogram local extrema. In *Proceedings IEEE, International Conference on Image Processing, ICIP 2008*, pages 1288–1291, San Diego, CA, October 12–15, 2008.

[39] R. Chandramouli and N. D. Memon. A distributed detection framework for steganalysis. In J. Dittmann, K. Nahrstedt, and P. Wohlmacher, editors, *Proceedings of the 3rd ACM Multimedia & Security Workshop*, pages 123–126, Los Angeles, CA, November 4, 2000.

[40] R. Chandramouli and N. D. Memon. Analysis of LSB based image steganography techniques. In *Proceedings IEEE International Conference on Image Processing, ICIP 2001*, Thessaloniki, October 7–10, 2001. CD ROM version.

[41] M. Chapman, G. I. Davida, and M. Rennhard. A practical and effective approach to large-scale automated linguistic steganography. In *Proceedings of the 4th International Conference on Information Security*, volume 2200 of Lecture Notes in Computer Science, pages 156–165, Malaga, October 1–3, 2001. Springer-Verlag, Berlin.

[42] B. Chen and G. Wornell. Quantization index modulation: A class of provably good methods for digital watermarking and information embedding. *IEEE Transactions on Information Theory*, 47(4):1423–1443, 2001.

[43] M. Chen, J. Fridrich, and M. Goljan. Determining image origin and integrity using sensor noise. *IEEE Transactions on Information Forensics and Security*, 1(1):74–90, March 2008.

[44] H. G. Chew, R. E. Bogner, and C. C. Lim. Dual-ν support vector machine with error rate and training size biasing. In *Proceedings IEEE, International Conference on*

Acoustics, Speech, and Signal Processing, volume 2, pages 1269–1272, Salt Lake City, UT, May 7–11, 2001.

[45] C. T. Clelland, V. Risca, and C. Bancroft. Hiding messages in DNA microdots. *Nature*, 399:533–534, June 10, 1999.

[46] G. D. Cohen, I. Honkala, S. Litsyn, and A. Lobstein. *Covering Codes*, volume 54. Amsterdam: Elsevier, North-Holland Mathematical Library, 1997.

[47] E. Cole. *Hiding in Plain Sight: Steganography and the Art of Covert Communication.* New York: Wiley Publishing Inc., 2003.

[48] S. Coll and S. B. Glasser. Terrorists turn to web as base of operations. *Washington Post*, page A01, August 7, 2005.

[49] P. Comesana and F. Pérez-Gonzáles. On the capacity of stegosystems. In J. Dittmann and J. Fridrich, editors, *Proceedings of the 9th ACM Multimedia & Security Workshop*, pages 3–14, Dallas, TX, September 20–21, 2007.

[50] T. M. Cover and J. A. Thomas. *Elements of Information Theory.* New York: John Wiley & Sons, Inc., 1991.

[51] I. J. Cox, M. L. Miller, J. A. Bloom, J. Fridrich, and T. Kalker. *Digital Watermarking and Steganography.* Morgan Kaufman Publishers Inc., San Francisco, CA, 2007.

[52] R. Crandall. Some notes on steganography. *Steganography Mailing List*, available from http://os.inf.tu-dresden.de/~westfeld/crandall.pdf, 1998.

[53] S. Craver. On public-key steganography in the presence of an active warden. In D. Aucsmith, editor, *Information Hiding, 2nd International Workshop*, volume 1525 of Lecture Notes in Computer Science, pages 355–368, Portland, OR, April 14–17, 1998. Springer-Verlag, New York.

[54] N. Cristianini and J. Shawe-Taylor. *Support Vector Machines and Other Kernel-Based Learning Methods.* Cambridge: Cambridge University Press, 2000.

[55] O. Dabeer, K. Sullivan, U. Madhow, S. Chandrasekaran, and B. S. Manjunath. Detection of hiding in the least significant bit. *IEEE Transactions on Signal Processing*, 52:3046–3058, 2004.

[56] N. Dedic, G. Itkis, L. Reyzin, and S. Russell. Upper and lower bounds on black-box steganography. In J. Kilian, editor, *Theory of Cryptography*, volume 3378 of Lecture Notes in Computer Science, pages 227–244, Cambridge, MA, February 10–12, 2005. Springer-Verlag, London.

[57] Y. Desmedt. Subliminal-free authentication and signature. In C. G. Günther, editor, *Advances in Cryptology – EUROCRYPT '88, Workshop on the Theory and Application of Cryptographic Techniques*, volume 330 of Lecture Notes in Computer Science, pages 22–33, Davos, May 25–27, 1988. Springer-Verlag, Berlin.

[58] J. Dittmann, T. Vogel, and R. Hillert. Design and evaluation of steganography for voice-over-IP. In *Proceedings IEEE International Symposium on Circuits and Systems (ISCAS)*, Kos, May 21–24, 2006.

[59] E. R. Dougherty. *Random Processes for Image and Signal Processing*, volume Monograph PM44. Washington, DC: SPIE Press, International Society for Optical Engineering, 1998.

[60] I. Dumer. *Handbook of Coding Theory: Volume II*, chapter 23, Concatenated Codes and Their Multilevel Generalizations, pages 1911–1988. Elsevier Science, Amsterdam, 1998.

[61] S. Dumitrescu and X. Wu. LSB steganalysis based on higher-order statistics. In A. M. Eskicioglu, J. Fridrich, and J. Dittmann, editors, *Proceedings of the 7th ACM Multi-*

media & Security Workshop, pages 25–32, New York, NY, August 1–2, 2005.

[62] S. Dumitrescu, X. Wu, and N. D. Memon. On steganalysis of random LSB embedding in continuous-tone images. In *Proceedings IEEE, International Conference on Image Processing, ICIP 2002*, pages 324–339, Rochester, NY, September 22–25, 2002.

[63] S. Dumitrescu, X. Wu, and Z. Wang. Detection of LSB steganography via Sample Pairs Analysis. In F. A. P. Petitcolas, editor, *Information Hiding, 5th International Workshop*, volume 2578 of Lecture Notes in Computer Science, pages 355–372, Noordwijkerhout, October 7–9, 2002. Springer-Verlag, New York.

[64] J. Eggers, R. Bäuml, and B. Girod. A communications approach to steganography. In E. J. Delp and P. W. Wong, editors, *Proceedings SPIE, Electronic Imaging, Security and Watermarking of Multimedia Contents IV*, volume 4675, pages 26–37, San Jose, CA, January 21–24, 2002.

[65] T. Ernst. Schwarzweisse Magie. Der Schlüssel zum dritten Buch der *Steganographia* des Trithemius. *Daphnis*, 25(1), 1996.

[66] J. M. Ettinger. Steganalysis and game equilibria. In D. Aucsmith, editor, *Information Hiding, 2nd International Workshop*, volume 1525 of Lecture Notes in Computer Science, pages 319–328, Portland, OR, April 14–17, 1998. Springer-Verlag, New York.

[67] H. Farid and L. Siwei. Detecting hidden messages using higher-order statistics and support vector machines. In F. A. P. Petitcolas, editor, *Information Hiding, 5th International Workshop*, volume 2578 of Lecture Notes in Computer Science, pages 340–354, Noordwijkerhout, October 7–9, 2002. Springer-Verlag, New York.

[68] T. Filler and J. Fridrich. Binary quantization using belief propagation over factor graphs of LDGM codes. In *45th Annual Allerton Conference on Communication, Control, and Computing*, Allerton, IL, September 26–28, 2007.

[69] T. Filler and J. Fridrich. Complete characterization of perfectly secure stegosystems with mutually independent embedding. In *Proceedings IEEE, International Conference on Acoustics, Speech, and Signal Processing*, Taipei, April 19–24, 2009.

[70] T. Filler and J. Fridrich. Fisher information determines capacity of ϵ-secure steganography. In S. Katzenbeisser and A.-R. Sadeghi, editors, *Information Hiding, 11th International Workshop*, volume 5806 of Lecture Notes in Computer Science, pages 31–47, Darmstadt, June 7–10, 2009. Springer-Verlag, New York.

[71] T. Filler, A. D. Ker, and J. Fridrich. The Square Root Law of steganographic capacity for Markov covers. In N. D. Memon, E. J. Delp, P. W. Wong, and J. Dittmann, editors, *Proceedings SPIE, Electronic Imaging, Security and Forensics of Multimedia XI*, volume 7254, pages 08 1–08 11, San Jose, CA, January 18–21, 2009.

[72] G. Fisk, M. Fisk, C. Papadopoulos, and J. Neil. Eliminating steganography in Internet traffic with active wardens. In F. A. P. Petitcolas, editor, *Information Hiding, 5th International Workshop*, volume 2578 of Lecture Notes in Computer Science, pages 18–35, Noordwijkerhout, October 7–9, 2003. Springer-Verlag, New York.

[73] M. A. Fitch and R. E. Jamison. Minimum sum covers of small cyclic groups. *Congressus Numerantium*, 147:65–81, 2000.

[74] J. D. Foley, A. van Dam, S. K. Feiner, J. F. Hughes, and R. L. Phillips. *Introduction to Computer Graphics*. New York: Addison-Wesley, 1997.

[75] E. Franz. Steganography preserving statistical properties. In F. A. P. Petitcolas, editor, *Information Hiding, 5th International Workshop*, volume 2578 of Lecture Notes in Computer Science, pages 278–294, Noordwijkerhout, October 7–9, 2002. Springer-Verlag, New York.

[76]   E. Franz. Embedding considering dependencies between pixels. In E. J. Delp and P. W. Wong, editors, *Proceedings SPIE, Electronic Imaging, Security, Forensics, Steganography, and Watermarking of Multimedia Contents X*, volume 6819, pages D 1–D 12, San Jose, CA, January 27–31, 2008.

[77]   E. Franz and A. Schneidewind. Pre-processing for adding noise steganography. In M. Barni, J. Herrera, S. Katzenbeisser, and F. Pérez-González, editors, *Information Hiding, 7th International Workshop*, volume 3727 of Lecture Notes in Computer Science, pages 189–203, Barcelona, June 6–8, 2005. Springer-Verlag, Berlin.

[78]   J. Fridrich. Feature-based steganalysis for JPEG images and its implications for future design of steganographic schemes. In J. Fridrich, editor, *Information Hiding, 6th International Workshop*, volume 3200 of Lecture Notes in Computer Science, pages 67–81, Toronto, May 23–25, 2004. Springer-Verlag, New York.

[79]   J. Fridrich. Asymptotic behavior of the ZZW embedding construction. *IEEE Transactions on Information Forensics and Security*, 4(1):151–153, March 2009.

[80]   J. Fridrich and R. Du. Secure steganographic methods for palette images. In A. Pfitzmann, editor, *Information Hiding, 3rd International Workshop*, volume 1768 of Lecture Notes in Computer Science, pages 47–60, Dresden, September 29–October 1, 1999. Springer-Verlag, New York.

[81]   J. Fridrich, P. Lisoněk, and D. Soukal. On steganographic embedding efficiency. In J. L. Camenisch, C. S. Collberg, N. F. Johnson, and P. Sallee, editors, *Information Hiding, 8th International Workshop*, volume 4437 of Lecture Notes in Computer Science, pages 282–296, Alexandria, VA, July 10–12, 2006. Springer-Verlag, New York.

[82]   J. Fridrich and M. Goljan. Digital image steganography using stochastic modulation. In E. J. Delp and P. W. Wong, editors, *Proceedings SPIE, Electronic Imaging, Security and Watermarking of Multimedia Contents V*, volume 5020, pages 191–202, Santa Clara, CA, January 21–24, 2003.

[83]   J. Fridrich and M. Goljan. On estimation of secret message length in LSB steganography in spatial domain. In E. J. Delp and P. W. Wong, editors, *Proceedings SPIE, Electronic Imaging, Security, Steganography, and Watermarking of Multimedia Contents VI*, volume 5306, pages 23–34, San Jose, CA, January 19–22, 2004.

[84]   J. Fridrich, M. Goljan, and R. Du. Detecting LSB steganography in color and gray-scale images. *IEEE Multimedia, Special Issue on Security*, 8(4):22–28, October–December 2001.

[85]   J. Fridrich, M. Goljan, and R. Du. Steganalysis based on JPEG compatibility. In A. G. Tescher, editor, *Special Session on Theoretical and Practical Issues in Digital Watermarking and Data Hiding, SPIE Multimedia Systems and Applications IV*, volume 4518, pages 275–280, Denver, CO, August 20–24, 2001.

[86]   J. Fridrich, M. Goljan, and D. Hogea. Steganalysis of JPEG images: Breaking the F5 algorithm. In *Information Hiding, 5th International Workshop*, volume 2578 of Lecture Notes in Computer Science, pages 310–323, Noordwijkerhout, October 7–9, 2002. Springer-Verlag, New York.

[87]   J. Fridrich, M. Goljan, D. Hogea, and D. Soukal. Quantitative steganalysis of digital images: Estimating the secret message length. *ACM Multimedia Systems Journal*, 9(3):288–302, 2003.

[88]   J. Fridrich, M. Goljan, and D. Soukal. Higher-order statistical steganalysis of palette images. In E. J. Delp and P. W. Wong, editors, *Proceedings SPIE, Electronic Imaging, Security and Watermarking of Multimedia Contents V*, pages 178–190, Santa Clara,

CA, January 21–24, 2003.

[89] J. Fridrich, M. Goljan, and D. Soukal. Perturbed quantization steganography using wet paper codes. In J. Dittmann and J. Fridrich, editors, *Proceedings of the 6th ACM Multimedia & Security Workshop*, pages 4–15, Magdeburg, September 20–21, 2004.

[90] J. Fridrich, M. Goljan, and D. Soukal. Searching for the stego key. In E. J. Delp and P. W. Wong, editors, *Proceedings SPIE, Electronic Imaging, Security, Steganography, and Watermarking of Multimedia Contents VI*, volume 5306, pages 70–82, San Jose, CA, January 19–22, 2004.

[91] J. Fridrich, M. Goljan, and D. Soukal. Efficient wet paper codes. In M. Barni, J. Herrera, S. Katzenbeisser, and F. Pérez-González, editors, *Information Hiding, 7th International Workshop*, Lecture Notes in Computer Science, pages 204–218, Barcelona, June 6–8, 2005. Springer-Verlag, Berlin.

[92] J. Fridrich, M. Goljan, and D. Soukal. Perturbed quantization steganography. *ACM Multimedia System Journal*, 11(2):98–107, 2005.

[93] J. Fridrich, M. Goljan, and D. Soukal. Steganography via codes for memory with defective cells. In *43rd Annual Allerton Conference on Communication, Control, and Computing*, Allerton, IL, September 28–30, 2005.

[94] J. Fridrich, M. Goljan, D. Soukal, and T. Holotyak. Forensic steganalysis: Determining the stego key in spatial domain steganography. In E. J. Delp and P. W. Wong, editors, *Proceedings SPIE, Electronic Imaging, Security, Steganography, and Watermarking of Multimedia Contents VII*, volume 5681, pages 631–642, San Jose, CA, January 16–20, 2005.

[95] J. Fridrich, T. Pevný, and J. Kodovský. Statistically undetectable JPEG steganography: Dead ends, challenges, and opportunities. In J. Dittmann and J. Fridrich, editors, *Proceedings of the 9th ACM Multimedia & Security Workshop*, pages 3–14, Dallas, TX, September 20–21, 2007.

[96] J. Fridrich and D. Soukal. Matrix embedding for large payloads. *IEEE Transactions on Information Forensics and Security*, 1(3):390–394, 2006.

[97] F. Galand and C. Fontaine. How Reed–Solomon codes can improve steganographic schemes. *EURASIP Journal on Information Security*, 2009. Article ID 274845, doi:10.1155/2009/274845.

[98] F. Galand and G. Kabatiansky. Information hiding by coverings. In *Proceedings IEEE, Information Theory Workshop, ITW 2003*, pages 151–154, Paris, March 31–April 4, 2003.

[99] S. I. Gel'fand and M. S. Pinsker. Coding for channel with random parameters. *Problems of Control and Information Theory*, 9(1):19–31, 1980.

[100] S. Gianvecchio and H. Wang. Detecting covert timing channels: An entropy-based approach. In P. Ning, S. De Capitani di Vimercati, and P. F. Syverson, editors, *Proceedings 14th ACM Conference on Computer and Communication Security (CCS)*, pages 307–316, Alexandria, VA, October 28–31, 2007.

[101] J. Giffin, R. Greenstadt, P. Litwack, and R. Tibbetts. Covert messaging through TCP timestamps. In R. Dingledine and P. Syverson, editors, *Proceedings Privacy Enhancing Technologies Workshop (PET)*, volume 2482 of Lecture Notes in Computer Science, pages 194–208, San Francisco, CA, April 14–15, 2002. Springer-Verlag, Berlin.

[102] M. Goljan, J. Fridrich, and T. Holotyak. New blind steganalysis and its implications. In E. J. Delp and P. W. Wong, editors, *Proceedings SPIE, Electronic Imaging, Security, Steganography, and Watermarking of Multimedia Contents VIII*, volume 6072, pages

1–13, San Jose, CA, January 16–19, 2006.

[103]   R. L. Graham and N. J. A. Sloane. On additive bases and harmonious graphs. *SIAM Journal on Algebraic and Discrete Methods*, 1(4):382–404, December 1980.

[104]   A. Gretton, K. M. Borgwardt, M. Rasch, B. Schölkopf, and A. J. Smola. A kernel method for the two-sample-problem. In B. Schölkopf, J. Platt, and T. Hoffman, editors, *Advances in Neural Information Processing Systems 19*, pages 513–520. MIT Press, Cambridge, MA, 2007.

[105]   H. Haanpää. Minimum sum and difference covers of abelian groups. *Journal of Integer Sequences*, 7(2), 2004. Article 04.2.6.

[106]   T. G. Handel and M. T. Stanford III. Hiding data in the OSI network model. In *Information Hiding, 1st International Workshop*, volume 1174 of Lecture Notes in Computer Science, pages 23–38, Cambridge, May 30–June 1, 1996. Springer-Verlag, Berlin.

[107]   J. J. Harmsen and W. A. Pearlman. Steganalysis of additive noise modelable information hiding. In E. J. Delp and P. W. Wong, editors, *Proceedings SPIE, Electronic Imaging, Security and Watermarking of Multimedia Contents V*, volume 5020, pages 131–142, Santa Clara, CA, January 21–24, 2003.

[108]   C. Heegard and A. A. El-Gamal. On the capacity of computer memory with defects. *IEEE Transactions on Information Theory*, 29(5):731–739, 1983.

[109]   Herodotus. *The Histories*. Penguin Books, London, 1996. Translated by Aubrey de Sélincourt.

[110]   S. Hetzl and P. Mutzel. A graph-theoretic approach to steganography. In J. Dittmann, S. Katzenbeisser, and A. Uhl, editors, *Communications and Multimedia Security, 9th IFIP TC-6 TC-11 International Conference, CMS 2005*, volume 3677 of Lecture Notes in Computer Science, pages 119–128, Salzburg, September 19–21, 2005.

[111]   F. S. Hill, Jr. *Computer Graphics Using Open GL*. Upper Saddle River, NJ: Prentice Hall, 2nd edition, 2000.

[112]   N. J. Hopper. On steganographic chosen covertext security. In *32nd Annual International Colloquium on Automata, Languages and Programming, (ICALP 2005)*, pages 311–321, Lisbon, July 11–15, 2005.

[113]   N. J. Hopper, J. Langford, and L. von Ahn. Provably secure steganography. In M. Yung, editor, *Advances in Cryptology, CRYPTO '02, 22nd Annual International Cryptology Conference*, volume 2442 of Lecture Notes in Computer Science, pages 77–92, Santa Barbara, CA, August 18–22, 2002. Springer-Verlag.

[114]   C. Hsu and C. Lin. A comparison of methods for multi-class support vector machines. Technical report, Department of Computer Science and Information Engineering, National Taiwan University, Taipei, 2001.

[115]   D. F. Hsu and X. Jia. Additive bases and extremal problems in groups, graphs and networks. *Utilitas Mathematica*, 66:61–91, 2004.

[116]   G. A. Francia III and T. S. Gomez. Steganography obliterator: An attack on the least significant bits. In *Proceedings 3rd Annual Conference on Information Security Curriculum Development (InfoSecCD '06)*, pages 85–91, Kennesaw State University, Kennesaw, GA, September 22–23, 2006.

[117]   D. Johnson. *White King and Red Queen: How the Cold War Was Fought on the Chessboard*. Houghton Mifflin Company, Boston and New York, 2008.

[118]   M. K. Johnson, S. Lyu, and H. Farid. Steganalysis of recorded speech. In E. J. Delp and P. W. Wong, editors, *Proceedings SPIE, Electronic Imaging, Security, Steganography, and Watermarking of Multimedia Contents VII*, volume 5681, pages 664–672, San Jose,

CA, January 16–20, 2005.

[119] N. F. Johnson and S. Jajodia. Exploring steganography: Seeing the unseen. *IEEE Computer*, 31:26–34, February 1998.

[120] N. F. Johnson and S. Jajodia. Steganalysis of images created using current steganography software. In D. Aucsmith, editor, *Information Hiding, 2nd International Workshop*, volume 1525 of Lecture Notes in Computer Science, pages 273–289, Portland, OR, April 14–17, 1998. Springer-Verlag, New York.

[121] N. F. Johnson and S. Jajodia. Steganalysis: The investigation of hidden information. In *Proceedings IEEE, Information Technology Conference*, Syracuse, NY, September 1–3, 1998.

[122] N. F. Johnson and P. Sallee. Detection of hidden information, covert channels and information flows. In John G. Voeller, editor, *Wiley Handbook of Science Technology for Homeland Security*. New York: Wiley & Sons, Inc, April 4, 2008.

[123] S. Katzenbeisser and F. A. P. Petitcolas, editors. *Information Hiding Techniques for Steganography and Digital Watermarking*. New York: Artech House, 2000.

[124] S. Katzenbeisser and F. A. P. Petitcolas. Defining security in steganographic systems. In E. J. Delp and P. W. Wong, editors, *Proceedings SPIE, Electronic Imaging, Security and Watermarking of Multimedia Contents IV*, volume 4675, pages 50–56, San Jose, CA, January 21–24, 2002.

[125] S. M. Kay. *Fundamentals of Statistical Signal Processing, Volume I: Estimation Theory*, volume II. Upper Saddle River, NJ: Prentice Hall, 1998.

[126] S. M. Kay. *Fundamentals of Statistical Signal Processing, Volume II: Detection Theory*, volume II. Upper Saddle River, NJ: Prentice Hall, 1998.

[127] A. D. Ker. Improved detection of LSB steganography in grayscale images. In J. Fridrich, editor, *Information Hiding, 6th International Workshop*, volume 3200 of Lecture Notes in Computer Science, pages 97–115, Toronto, May 23–25, 2004. Springer-Verlag, Berlin.

[128] A. D. Ker. A general framework for structural analysis of LSB replacement. In M. Barni, J. Herrera, S. Katzenbeisser, and F. Pérez-González, editors, *Information Hiding, 7th International Workshop*, volume 3727 of Lecture Notes in Computer Science, pages 296–311, Barcelona, June 6–8, 2005. Springer-Verlag, Berlin.

[129] A. D. Ker. Resampling and the detection of LSB matching in color bitmaps. In E. J. Delp and P. W. Wong, editors, *Proceedings SPIE, Electronic Imaging, Security, Steganography, and Watermarking of Multimedia Contents VII*, volume 5681, pages 1–15, San Jose, CA, January 16–20, 2005.

[130] A. D. Ker. Steganalysis of LSB matching in grayscale images. *IEEE Signal Processing Letters*, 12(6):441–444, June 2005.

[131] A. D. Ker. Fourth-order structural steganalysis and analysis of cover assumptions. In E. J. Delp and P. W. Wong, editors, *Proceedings SPIE, Electronic Imaging, Security, Steganography, and Watermarking of Multimedia Contents VIII*, volume 6072, pages 25–38, San Jose, CA, January 16–19, 2006.

[132] A. D. Ker. A capacity result for batch steganography. *IEEE Signal Processing Letters*, 14(8):525–528, 2007.

[133] A. D. Ker. A fusion of maximal likelihood and structural steganalysis. In T. Furon, F. Cayre, G. Doërr, and P. Bas, editors, *Information Hiding, 9th International Workshop*, volume 4567 of Lecture Notes in Computer Science, pages 204–219, Saint Malo, June 11–13, 2007. Springer-Verlag, Berlin.

[134] A. D. Ker. Optimally weighted least-squares steganalysis. In E. J. Delp and P. W. Wong, editors, *Proceedings SPIE, Electronic Imaging, Security, Steganography, and Watermarking of Multimedia Contents IX*, volume 6505, pages 6 1–6 16, San Jose, CA, January 29–February 1, 2007.

[135] A. D. Ker. Steganalysis of embedding in two least significant bits. *IEEE Transactions on Information Forensics and Security*, 2:46–54, 2007.

[136] A. D. Ker. The ultimate steganalysis benchmark? In J. Dittmann and J. Fridrich, editors, *Proceedings of the 9th ACM Multimedia & Security Workshop*, pages 141–148, Dallas, TX, September 20–21, 2007.

[137] A. D. Ker. Locating steganographic payload via WS residuals. In A. D. Ker, J. Dittmann, and J. Fridrich, editors, *Proceedings of the 10th ACM Multimedia & Security Workshop*, pages 27–32, Oxford, September 22–23, 2008.

[138] A. D. Ker and R. Böhme. Revisiting weighted stego-image steganalysis. In E. J. Delp and P. W. Wong, editors, *Proceedings SPIE, Electronic Imaging, Security, Forensics, Steganography, and Watermarking of Multimedia Contents X*, volume 6819, pages 5 1–5 17, San Jose, CA, January 27–31, 2008.

[139] A. D. Ker and I. Lubenko. Feature reduction and payload location with WAM steganalysis. In N. D. Memon, E. J. Delp, P. W. Wong, and J. Dittmann, editors, *Proceedings SPIE, Electronic Imaging, Security and Forensics of Multimedia XI*, volume 7254, pages 0A 1–0A 13, San Jose, CA, January 18–21, 2009.

[140] A. D. Ker, T. Pevný, J. Kodovský, and J. Fridrich. The Square Root Law of steganographic capacity. In A. D. Ker, J. Dittmann, and J. Fridrich, editors, *Proceedings of the 10th ACM Multimedia & Security Workshop*, pages 107–116, Oxford, September 22–23, 2008.

[141] Y. Kim, Z. Duric, and D. Richards. Modified matrix encoding technique for minimal distortion steganography. In J. L. Camenisch, C. S. Collberg, N. F. Johnson, and P. Sallee, editors, *Information Hiding, 8th International Workshop*, volume 4437 of Lecture Notes in Computer Science, pages 314–327, Alexandria, VA, July 10–12, 2006. Springer-Verlag, New York.

[142] G. Kipper. *Investigator's Guide to Steganography*. Boca Raton, FL: CRC Press, 2004.

[143] J. Kodovský and J. Fridrich. Influence of embedding strategies on security of steganographic methods in the JPEG domain. In E. J. Delp and P. W. Wong, editors, *Proceedings SPIE, Electronic Imaging, Security, Forensics, Steganography, and Watermarking of Multimedia Contents X*, volume 6819, pages 2 1–2 13, San Jose, CA, January 27–31, 2008.

[144] J. Kodovský and J. Fridrich. On completeness of feature spaces in blind steganalysis. In A. D. Ker, J. Dittmann, and J. Fridrich, editors, *Proceedings of the 10th ACM Multimedia & Security Workshop*, pages 123–132, Oxford, September 22–23, 2008.

[145] G. Kolata. A mystery unraveled, twice. *The New York Times*, pages F1–F6, April 14, 1998.

[146] G. Kolata. Veiled messages of terror may lurk in cyberspace. *The New York Times*, October 30, 2001.

[147] N. Komaki, N. Aoki, and T. Yamamoto. A packet loss concealment technique for VoIP using steganography. *IEICE Transactions on Fundamentals of Electronics, Communications, and Computer Sciences*, E86-A(8):2069–2072, 2003.

[148] O. Koval, S. Voloshynovskiy, T. Holotyak, and T. Pun. Information theoretic analysis of steganalysis in real images. In S. Voloshynovskiy, J. Dittmann, and J. Fridrich, editors,

Proceedings of the 8th ACM Multimedia & Security Workshop, pages 11–16, Geneva, September 26–27, 2006.

[149] C. Krätzer and J. Dittmann. Pros and cons of mel-cepstrum based audio steganalysis using SVM classification. In T. Furon, F. Cayre, G. Doërr, and P. Bas, editors, *Information Hiding, 9th International Workshop*, pages 359–377, Saint Malo, June 11–13, 2007.

[150] C. Krätzer, J. Dittmann, A. Lang, and T. Kühne. WLAN steganography: A first practical review. In S. Voloshynovskiy, J. Dittmann, and J. Fridrich, editors, *Proceedings of the 8th ACM Multimedia & Security Workshop*, pages 17–22, Geneva, September 26–27, 2006.

[151] M. Kutter and F. A. P. Petitcolas. A fair benchmark for image watermarking systems. In E. J. Delp and P. W. Wong, editors, *Proceedings SPIE, Electronic Imaging, Security and Watermarking of Multimedia Contents I*, volume 3657, pages 226–239, San Jose, CA, 1999.

[152] A. V. Kuznetsov and B. S. Tsybakov. Coding in a memory with defective cells. *Problems of Information Transmission*, 10:132–138, 1974.

[153] Tri Van Le. Efficient provably secure public key steganography. Technical report, Florida State University, 2003. Cryptography ePrint Archive, `http://eprint.iacr.org/2003/156`.

[154] Tri Van Le and K. Kurosawa. Efficient public key steganography secure against adaptively chosen stegotext attacks. Technical report, Florida State University, 2003. Cryptography ePrint Archive, `http://eprint.iacr.org/2003/244`.

[155] K. Lee, C. Jung, S. Lee, and J. Lim. New steganalysis methodology: LR cube analysis for the detection of LSB steganography. In M. Barni, J. Herrera, S. Katzenbeisser, and F. Pérez-González, editors, *Information Hiding, 7th International Workshop*, volume 3727 of Lecture Notes in Computer Science, pages 312–326, Barcelona, June 6–8, 2005. Springer-Verlag, Berlin.

[156] K. Lee and A. Westfeld. Generalized category attack – improving histogram-based attack on JPEG LSB embedding. In T. Furon, F. Cayre, G. Doërr, and P. Bas, editors, *Information Hiding, 9th International Workshop*, volume 4567 of Lecture Notes in Computer Science, pages 378–392, Saint Malo, June 11–13, 2007. Springer-Verlag, Berlin.

[157] K. Lee, A. Westfeld, and S. Lee. Category attack for LSB embedding of JPEG images. In Y.-Q. Shi, B. Jeon, Y.Q. Shi, and B. Jeon, editors, *Digital Watermarking, 5th International Workshop*, volume 4283 of Lecture Notes in Computer Science, pages 35–48, Jeju Island, November 8–10, 2006. Springer-Verlag, Berlin.

[158] X. Li, B. Yang, D. Cheng, and T. Zeng. A generalization of LSB matching. *IEEE Signal Processing Letters*, 16(2):69–72, February 2009.

[159] X. Li, T. Zeng, and B. Yang. Detecting LSB matching by applying calibration technique for difference image. In A. D. Ker, J. Dittmann, and J. Fridrich, editors, *Proceedings of the 10th ACM Multimedia & Security Workshop*, pages 133–138, Oxford, September 22–23, 2008.

[160] X. Li, T. Zeng, and B. Yang. Improvement of the embedding efficiency of LSB matching by sum and difference covering set. In *Proceedings IEEE, International Conference on Multimedia and Expo*, pages 209–212, Hannover, June 23–April 26, 2008.

[161] Tsung-Yuan Liu and Wen-Hsiang Tsai. A new steganographic method for data hiding in Microsoft Word documents by a change tracking technique. *IEEE Transactions on*

Information Forensics and Security, 2(1):24–30, March 2007.

[162] D. Llamas, C. Allison, and A. Miller. Covert channels in internet protocols: A survey. In M. Merabti and R. Pereira, editors, *Proceedings 6th Annual Postgraduate Symposium about the Convergence of Telecommunications, Networking and Broadcasting (PGNET)*, Liverpool, June 27–28, 2005.

[163] P. Lu, X. Luo, Q. Tang, and L. Shen. An improved sample pairs method for detection of LSB embedding. In J. Fridrich, editor, *Information Hiding, 6th International Workshop*, volume 3200 of Lecture Notes in Computer Science, pages 116–127, Toronto, May 23–25, 2004. Springer-Verlag, Berlin.

[164] M. Luby. LT codes. In *43rd Annual IEEE Symposium on Foundations of Computer Science, FOCS 2002*, pages 271–282, Vancouver, November 16–19, 2002.

[165] N. B. Lucena, G. Lewandowski, and S. J. Chapin. Covert channels in IPv6. In G. Danezis and D. Martin, editors, *Proceedings Privacy Enhancing Technologies Workshop (PET)*, volume 3856 of Lecture Notes in Computer Science, pages 147–166, Dubrovnik, May 30–June 1, 2006. Springer-Verlag, Berlin.

[166] A. Lysyanskaya and M. Meyerovich. Steganography with imperfect sampling. Technical report, Brown University, 2005. Cryptography ePrint Archive, http://eprint.iacr.org/2005/305.

[167] S. Lyu and H. Farid. Steganalysis using color wavelet statistics and one-class support vector machines. In E. J. Delp and P. W. Wong, editors, *Proceedings SPIE, Electronic Imaging, Security, Steganography, and Watermarking of Multimedia Contents VI*, volume 5306, pages 35–45, San Jose, CA, January 19–22, 2004.

[168] S. Lyu and H. Farid. Steganalysis using higher-order image statistics. *IEEE Transactions on Information Forensics and Security*, 1(1):111–119, 2006.

[169] W. Mazurczyk and K. Szczypiorski. Steganography of VoIP streams. In *Proceedings of the 3rd International Symposium on Information Security*, volume 5332 of Lecture Notes in Computer Science, pages 1001–1018, Monterrey, Mexico, November 10–11, 2008. Springer-Verlag, Berlin.

[170] A. D. McDonald and M. G. Kuhn. StegFS: A steganographic file system for Linux. In *Information Hiding, 3rd International Workshop*, volume 1768 of Lecture Notes in Computer Science, pages 454–468, Dresden, September 29–October 1, 1999. Springer-Verlag, Berlin.

[171] D. J. C. McKay. *Information Theory, Inference, and Learning Algorithms*. Cambridge: Cambridge University Press, 2003.

[172] S. Meignen and H. Meignen. On the modeling of DCT and subband image data for compression. *IEEE Transactions on Image Processing*, 4(2):186–193, February 1995.

[173] Y. Miche, B. Roue, A. Lendasse, and P. Bas. A feature selection methodology for steganalysis. In B. Günsel, A. K. Jain, A. M. Tekalp, and B. Sankur, editors, *Multimedia Content Representation, Classification and Security, International Workshop*, volume 4105 of Lecture Notes in Computer Science, pages 49–56, Istanbul, September 11–13, 2006. Springer-Verlag.

[174] J. Mielikainen. LSB matching revisited. *IEEE Signal Processing Letters*, 13(5):285–287, May 2006.

[175] M. K. Mihcak, I. Kozintsev, K. Ramchandran, and P. Moulin. Low-complexity image denoising based on statistical modeling of wavelet coefficients. *IEEE Signal Processing Letters*, 6(12):300–303, December 1999.

[176] D. S. Mitrinovic, J. E. Pecaric, and A. M. Fink. *Classical and New Inequalities in Analysis*. Kluwer Academic Publishers, Dordrecht, 1993.

[177] I. S. Moskowitz, R. E. Newman, D. P. Crepeau, and A. R. Miller. Covert channels and anonymizing networks. In S. Jajodia, P. Samarati, and P. F. Syverson, editors, *Proceedings Workshop on Privacy in the Electronic Society (WPES)*, pages 79–88, Washington, DC, October 30, 2003.

[178] P. Moulin, M. K. Mihcak, and G. I. Lin. An information-theoretic model for image watermarking and data hiding. In *Proceedings IEEE, International Conference on Image Processing, ICIP 2000*, volume 3, pages 667–670, Vancouver, September 10–13, 2000.

[179] P. Moulin and J. A. Sullivan. Information-theoretic analysis of information hiding. *IEEE Transactions on Information Theory*, 49(3):563–593, March 2003.

[180] P. Moulin and Y. Wang. New results on steganographic capacity. In *Proceedings of the Conference on Information Sciences and Systems, CISS*, Princeton, NJ, March 17–19, 2004.

[181] A. Munoz and J. M. Moguerza. Estimation of high-density regions using one-class neighbor machines. *IEEE Transactions on Pattern Analysis and Machine Intelligence*, 26(3):476–480, 2006.

[182] B. Murphy and C. Vogel. Statistically constrained shallow text marking: Techniques, evaluation paradigm, and results. In E. J. Delp and P. W. Wong, editors, *Proceedings SPIE, Electronic Imaging, Security, Steganography, and Watermarking of Multimedia Contents IX*, volume 6505, pages Z 1–Z 9, San Jose, CA, January 29–February 1, 2007.

[183] H. Noda, M. Niimi, and E. Kawaguchi. Application of QIM with dead zone for histogram preserving JPEG steganography. In *Proceedings IEEE, International Conference on Image Processing, ICIP 2005*, volume II, pages 1082–1085, Genova, September 11–14, 2005.

[184] H.-O. Peitgen, H. Jürgens, and D. Saupe. *Chaos and Fractals: New Frontiers of Science*. Berlin: Springer-Verlag, 1992.

[185] W. Pennebaker and J. Mitchell. *JPEG: Still Image Data Compression Standard*. Van Nostrand Reinhold, New York, 1993.

[186] F. Perez-Gonzalez and S. Voloshynovskiy, editors. *Fundamentals of Digital Image Watermarking*. New York: Wiley Blackwell, 2005.

[187] F. A. P. Petitcolas. MP3Stego software. 1998.

[188] K. Petrowski, M. Kharrazi, H. T. Sencar, and N. D. Memon. Psteg: Steganographic embedding through patching. In *Proceedings IEEE, International Conference on Acoustics, Speech, and Signal Processing*, pages 537–540, Philadelphia, PA, March 18–23, 2005.

[189] T. Pevný and J. Fridrich. Multiclass blind steganalysis for JPEG images. In E. J. Delp and P. W. Wong, editors, *Proceedings SPIE, Electronic Imaging, Security, Steganography, and Watermarking of Multimedia Contents VIII*, volume 6072, pages O 1–O 13, San Jose, CA, January 16–19, 2006.

[190] T. Pevný and J. Fridrich. Merging Markov and DCT features for multi-class JPEG steganalysis. In E. J. Delp and P. W. Wong, editors, *Proceedings SPIE, Electronic Imaging, Security, Steganography, and Watermarking of Multimedia Contents IX*, volume 6505, pages 3 1–3 14, San Jose, CA, January 29–February 1, 2007.

[191] T. Pevný and J. Fridrich. Benchmarking for steganography. In K. Solanki, K. Sullivan, and U. Madhow, editors, *Information Hiding, 10th International Workshop*, volume 5284 of Lecture Notes in Computer Science, pages 251–267, Santa Barbara, CA, June

19–21, 2008. Springer-Verlag, New York.

[192] T. Pevný and J. Fridrich. Detection of double-compression for applications in steganography. *IEEE Transactions on Information Forensics and Security*, 3(2):247–258, 2008.

[193] T. Pevný and J. Fridrich. Multiclass detector of current steganographic methods for JPEG format. *IEEE Transactions on Information Forensics and Security*, 3(4):635–650, December 2008.

[194] T. Pevný and J. Fridrich. Novelty detection in blind steganalysis. In A. D. Ker, J. Dittmann, and J. Fridrich, editors, *Proceedings of the 10th ACM Multimedia & Security Workshop*, pages 167–176, Oxford, September 22–23, 2008.

[195] T. Pevný, J. Fridrich, and A. D. Ker. From blind to quantitative steganalysis. In N. D. Memon, E. J. Delp, P. W. Wong, and J. Dittmann, editors, *Proceedings SPIE, Electronic Imaging, Security and Forensics of Multimedia XI*, volume 7254, pages 0C 1–0C 14, San Jose, CA, January 18–21, 2009.

[196] A. C. Popescu. *Statistical Tools for Digital Image Forensics*. PhD thesis, Department of Computer Science, Dartmouth College, 2005.

[197] S. Pradhan, J. Chou, and K. Ramchandran. Duality between source coding and channel coding and its extension to the side information case. *IEEE Transactions on Information Theory*, 49(5):1181–1203, 2003.

[198] N. Provos. Defending against statistical steganalysis. In *10th USENIX Security Symposium*, pages 323–335, Washington, DC, August 13–17, 2001.

[199] N. Provos and P. Honeyman. Detecting steganographic content on the internet. Technical report, 01–11, CITI, August 2001.

[200] R. Radhakrishnan, M. Kharrazi, and N. D. Memon. Data masking a new approach for data hiding? *Journal of VLSI Signal Processing Systems*, 41(3):293–303, November 2005.

[201] J. A. Reeds. Solved: The ciphers in Book III of Trithemius's *Steganographia*. *Cryptologia*, 22:291–319, October 1998.

[202] X.-M. Ru, H.-J. Zhang, and X. Huang. Steganalysis of audio: Attacking the Steghide. In *Proceedings of the International Conference on Machine Learning and Cybernetics*, volume 7, pages 3937–3942, Guangzhou, August 18–21, 2005.

[203] J. Rutenberg. A nation challenged: Videotape. *New York Times*, February 1, 2002.

[204] P. Sallee. Model-based steganography. In T. Kalker, I. J. Cox, and Y. Man Ro, editors, *Digital Watermarking, 2nd International Workshop*, volume 2939 of Lecture Notes in Computer Science, pages 154–167, Seoul, October 20–22, 2003. Springer-Verlag, New York.

[205] P. Sallee. Model-based methods for steganography and steganalysis. *International Journal of Image Graphics*, 5(1):167–190, 2005.

[206] K. Sayood. *Introduction to Data Compression (3rd edition)*. New York: Morgan Kaufmann, 2000.

[207] B. Schneier. *Applied Cryptography*. New York: John Wiley & Sons, 1996.

[208] D. Schönfeld and A. Winkler. Embedding with syndrome coding based on BCH codes. In S. Voloshynovskiy, J. Dittmann, and J. Fridrich, editors, *Proceedings of the 8th ACM Multimedia & Security Workshop*, pages 214–223, Geneva, September 26–27, 2006.

[209] T. Sharp. An implementation of key-based digital signal steganography. In I. S. Moskowitz, editor, *Information Hiding, 4th International Workshop*, volume 2137 of Lecture Notes in Computer Science, pages 13–26, Pittsburgh, PA, April 25–27, 2001. Springer-Verlag, New York.

[210] Y. Q. Shi, C. Chen, and W. Chen. A Markov process based approach to effective attacking JPEG steganography. In J. L. Camenisch, C. S. Collberg, N. F. Johnson, and P. Sallee, editors, *Information Hiding, 8th International Workshop*, volume 4437 of Lecture Notes in Computer Science, pages 249–264, Alexandria, VA, July 10–12, 2006. Springer-Verlag, New York.

[211] F. Y. Shih. *Digital Watermarking and Steganography: Fundamentals and Techniques.* Boca Raton, FL: CRC Press, 2007.

[212] B. Shimanovsky, J. Feng, and M. Potkonjak. Hiding data in DNA. In F. A. P. Petitcolas, editor, *Information Hiding, 5th International Workshop*, volume 2578 of Lecture Notes in Computer Science, pages 373–386, Noordwijkerhout, October 7–9, 2002. Springer-Verlag, Berlin.

[213] M. Sidorov. Hidden Markov models and steganalysis. In J. Dittmann and J. Fridrich, editors, *Proceedings of the 6th ACM Multimedia & Security Workshop*, pages 63–67, Magdeburg, September 20–21, 2004.

[214] G. J. Simmons. The prisoner's problem and the subliminal channel. In D. Chaum, editor, *Advances in Cryptology, CRYPTO '83*, pages 51–67, Santa Barbara, CA, August 22–24, 1983. New York: Plenum Press.

[215] A. J. Smola and B. Schölkopf. A tutorial on support vector regression. NeuroCOLT2 Technical Report NC2-TR-1998-030, 1998.

[216] T. Sohn, J. Seo, and J. Moon. A study on the covert channel detection of TCP/IP header using support vector machine. In S. Qing, D. Gollmann, and J. Zhou, editors, *Proceedings of the 5th International Conference on Information and Communications Security*, volume 2836 of Lecture Notes in Computer Science, pages 313–324, Huhehaote, October 10–13, 2003. Springer-Verlag, Berlin.

[217] K. Solanki, A. Sarkar, and B. S. Manjunath. YASS: Yet another steganographic scheme that resists blind steganalysis. In T. Furon, F. Cayre, G. Doërr, and P. Bas, editors, *Information Hiding, 9th International Workshop*, volume 4567 of Lecture Notes in Computer Science, pages 16–31, Saint Malo, June 11–13, 2007. Springer-Verlag, New York.

[218] K. Solanki, K. Sullivan, U. Madhow, B. S. Manjunath, and S. Chandrasekaran. Provably secure steganography: Achieving zero K–L divergence using statistical restoration. In *Proceedings IEEE, International Conference on Image Processing, ICIP 2006*, pages 125–128, Atlanta, GA, October 8–11, 2006.

[219] A. Somekh-Baruch and N. Merhav. On the capacity game of public watermarking systems. *IEEE Transactions on Information Theory*, 50(3):511–524, 2004.

[220] D. Soukal, J. Fridrich, and M. Goljan. Maximum likelihood estimation of secret message length embedded using $\pm k$ steganography in spatial domain. In E. J. Delp and P. W. Wong, editors, *Proceedings SPIE, Electronic Imaging, Security, Steganography, and Watermarking of Multimedia Contents VII*, volume 5681, pages 595–606, San Jose, CA, January 16–20, 2005.

[221] M. R. Spiegel. *Schaum's Outline of Theory and Problems of Statistics.* McGraw-Hill, New York, 3rd edition, 1961.

[222] M. Stamm and K. J. Ray Liu. Blind forensics of contrast enhancement in digital images. In *Proceedings IEEE, International Conference on Image Processing, ICIP 2008*, pages 3112–3115, San Diego, CA, October 12–15, 2008.

[223] I. Steinwart. On the influence of the kernel on the consistency of support vector machines. *Journal of Machine Learning Research*, 2:67–93, 2001. Available electronically at http://www.jmlr.org/papers/volume2/steinwart01a/steinwart01a.ps.gz.

[224] G. W. W. Stevens. *Microphotography – Photography and Photofabrication at Extreme Resolutions.* London, Chapman & Hall, 1968.

[225] K. Sullivan, U. Madhow, B. S. Manjunath, and S. Chandrasekaran. Steganalysis for Markov cover data with applications to images. *IEEE Transactions on Information Forensics and Security*, 1(2):275–287, June 2006.

[226] A. Tacticius. *How to Survive Under Siege/Aineas the Tactician.* Oxford: Clarendon Ancient History Series, 1990.

[227] C. M. Taskiran, U. Topkara, M. Topkara, and E. J. Delp. Attacks on lexical natural language steganography systems. In E. J. Delp and P. W. Wong, editors, *Proceedings SPIE, Electronic Imaging, Security, Steganography, and Watermarking of Multimedia Contents VIII*, volume 6072, pages 97–105, San Jose, CA, January 16–19, 2006.

[228] D. S. Taubman and M. W. Marcellin. *JPEG 2000 Image Compression Fundamentals, Standards, and Practices.* Kluwer Academic Publishers, Boston, MA, 2002.

[229] J. Taylor and A. Verbyla. Joint modeling of location and scale parameters of the t-distribution. *Statistical Modeling*, 4:91–112, 2004.

[230] J. Tobin and R. Dobard. *Hidden in Plain View: The Secret Story of Quilts and the Underground Railroad.* Doubleday, New York, 1999.

[231] B. S. Tsybakov. Defect and error correction. *Problemy Peredachi Informatsii*, 11:21–30, July–September 1975. Translated from Russian.

[232] R. Tzschoppe, R. Bäuml, J. B. Huber, and A. Kaup. Steganographic system based on higher-order statistics. In E. J. Delp and P. W. Wong, editors, *Proceedings SPIE, Electronic Imaging, Security and Watermarking of Multimedia Contents V*, volume 5020, pages 156–166, Santa Clara, CA, January 21–24, 2003.

[233] M. van Dijk and F. Willems. Embedding information in grayscale images. In *Proceedings of the 22nd Symposium on Information and Communication Theory*, pages 147–154, Enschede, May 15–16, 2001.

[234] L. von Ahn and N. Hopper. Public-key steganography. In C. Cachin and J. Camenisch, editors, *Advances in Cryptology – EUROCRYPT 2004, International Conference on the Theory and Applications of Cryptographic Techniques*, volume 3027 of Lecture Notes in Computer Science, pages 323–341, Interlaken, May 2–6, 2004. Springer-Verlag, Heidleberg.

[235] Y. Wang and P. Moulin. Steganalysis of block-structured stegotext. In E. J. Delp and P. W. Wong, editors, *Proceedings SPIE, Electronic Imaging, Security, Steganography, and Watermarking of Multimedia Contents VI*, volume 5306, pages 477–488, San Jose, CA, January 19–22, 2004.

[236] Y. Wang and P. Moulin. Statistical modelling and steganalysis of DFT-based image steganography. In E. J. Delp and P. W. Wong, editors, *Proceedings SPIE, Electronic Imaging, Security, Steganography, and Watermarking of Multimedia Contents VIII*, volume 6072, pages 2 1–2 11, San Jose, CA, January 16–19, 2006.

[237] Y. Wang and P. Moulin. Perfectly secure steganography: Capacity, error exponents, and code constructions. *IEEE Transactions on Information Theory, Special Issue on Security*, 55(6):2706–2722, June 2008.

[238] P. Wayner. Mimic functions. *CRYPTOLOGIA*, 16(3):193–214, July 1992.

[239] P. Wayner. *Disappearing Cryptography.* Morgan Kaufmann, San Francisco, CA, 2nd edition, 2002.

[240] J. Werner. *Optimization – Theory and Applications.* Braunschweig: Vieweg, 1984.

[241] A. Westfeld. High capacity despite better steganalysis (F5 – a steganographic algorithm). In I. S. Moskowitz, editor, *Information Hiding, 4th International Workshop*, volume 2137 of Lecture Notes in Computer Science, pages 289–302, Pittsburgh, PA, April 25–27, 2001. Springer-Verlag, New York.

[242] A. Westfeld. Detecting low embedding rates. In F. A. P. Petitcolas, editor, *Information Hiding, 5th International Workshop*, volume 2578 of Lecture Notes in Computer Science, pages 324–339, Noordwijkerhout, October 7–9, 2002. Springer-Verlag, Berlin.

[243] A. Westfeld. Space filling curves in steganalysis. In E. J. Delp and P. W. Wong, editors, *Proceedings SPIE, Electronic Imaging, Security, Steganography, and Watermarking of Multimedia Contents VII*, volume 5681, pages 28–37, San Jose, CA, January 16–20, 2005.

[244] A. Westfeld. Generic adoption of spatial steganalysis to transformed domain. In K. Solanki, K. Sullivan, and U. Madhow, editors, *Information Hiding, 10th International Workshop*, volume 5284 of Lecture Notes in Computer Science, pages 161–177, Santa Barbara, CA, June 19–21, 2007. Springer-Verlag, New York.

[245] A. Westfeld and R. Böhme. Exploiting preserved statistics for steganalysis. In J. Fridrich, editor, *Information Hiding, 6th International Workshop*, volume 3200 of Lecture Notes in Computer Science, pages 82–96, Toronto, May 23–25, 2004. Springer-Verlag, Berlin.

[246] A. Westfeld and A. Pfitzmann. Attacks on steganographic systems. In A. Pfitzmann, editor, *Information Hiding, 3rd International Workshop*, volume 1768 of Lecture Notes in Computer Science, pages 61–75, Dresden, September 29–October 1, 1999. Springer-Verlag, New York.

[247] E. H. Wilkins. *A History of Italian Literature*. Oxford University Press, London, 1954.

[248] F. J. M. Williams and N. J. Sloane. *The Theory of Error-Correcting Codes*. North-Holland, Amsterdam, 1977.

[249] P. W. Wong, H. Chen, and Z. Tang. On steganalysis of plus–minus one embedding in continuous-tone images. In E. J. Delp and P. W. Wong, editors, *Proceedings SPIE, Electronic Imaging, Security, Steganography, and Watermarking of Multimedia Contents VII*, volume 5681, pages 643–652, San Jose, CA, January 16–20, 2005.

[250] F. B. Wrixon. *Codes, Ciphers and Other Cryptic and Clandestine Communication*. New York: Black Dog & Leventhal Publishers, 1998.

[251] Z. Wu and W. Yang. G.711-based adaptive speech information hiding approach. In De-Shuang Huang, K. Li, and G. W. Irwin, editors, *Proceedings of the International Conference on Intelligent Computing*, volume 4113 of Lecture Notes in Computer Science, pages 1139–1144, Kunming, August 16–19, 2006. Springer-Verlag, Berlin.

[252] G. Xuan, Y. Q. Shi, J. Gao, D. Zou, C. Yang, Z. Z. P. Chai, C. Chen, and W. Chen. Steganalysis based on multiple features formed by statistical moments of wavelet characteristic functions. In M. Barni, J. Herrera, S. Katzenbeisser, and F. Pérez-González, editors, *Information Hiding, 7th International Workshop*, volume 3727 of Lecture Notes in Computer Science, pages 262–277, Barcelona, June 6–8, 2005. Springer-Verlag, Berlin.

[253] C. Yang, F. Liu, X. Luo, and B. Liu. Steganalysis frameworks of embedding in multiple least-significant bits. *IEEE Transactions on Information Forensics and Security*, 3:662–672, 2008.

[254] R. Zamir, S. Shamai, and U. Erez. Nested linear/lattice codes for structured multiterminal binning. *IEEE Transactions on Information Theory*, 48(6):1250–1276, 2002.

[255] T. Zhang and X. Ping. A fast and effective steganalytic technique against Jsteg-like algorithms. In *Proceedings of the ACM Symposium on Applied Computing*, pages 307–311, Melbourne, FL, March 9–12, 2003.

[256] T. Zhang and X. Ping. A new approach to reliable detection of LSB steganography in natural images. *Signal Processing*, 83(10):2085–2094, October 2003.

[257] W. Zhang, X. Zhang, and S. Wang. Maximizing steganographic embedding efficiency by combining Hamming codes and wet paper codes. In K. Solanki, K. Sullivan, and U. Madhow, editors, *Information Hiding, 10th International Workshop*, volume 5284 of Lecture Notes in Computer Science, pages 60–71, Santa Barbara, CA, June 19–21, 2008. Springer-Verlag, New York.

[258] X. Zhang, W. Zhang, and S. Wang. Efficient double-layered steganographic embedding. *Electronics Letters*, 43:482–483, April 2007.

Index

Printed in the United States
By Bookmasters